TECHNIQUES OF CHEMISTRY

ARNOLD WEISSBERGER, *Founding Editor*
WILLIAM SAUNDERS, *Editor*

VOLUME XIX

TECHNIQUES OF MELT CRYSTALLIZATION

TECHNIQUES OF CHEMISTRY

VOLUME XIX

TECHNIQUES OF MELT CRYSTALLIZATION

GILBERT J. SLOAN

E. I. du Pont de Nemours & Company, Inc.
Polymer Products Department
Wilmington, Delaware

ANDREW R. McGHIE

Laboratory for Research on the Structure of Matter
University of Pennsylvania
Philadelphia, Pennsylvania

A WILEY-INTERSCIENCE PUBLICATION

JOHN WILEY & SONS

New York · Chichester · Brisbane · Toronto · Singapore

Library of Congress Cataloging in Publication Data:

Techniques of chemistry. —   — New   York : Wi-
    ley,
       v.
    Irregular.
    Began in 1971.
    Description based on: Vol. 4, pt. 1.
    Photo-offset reprint separately cataloged and classified in LC.
    Editor: A. Weissberger.
    Merger of: Technique of inorganic chemistry; and: Technique of organic
chemistry.
    ISSN 0082-2531 = Techniques of chemistry.

    1. Chemistry—Manipulation—Collected works. I. Weissberger, Arnold,
1898–   . II. Title: Techniques of chemistry
    [DNLM: W1 TE197K]
QD61.T4   542—dc19   85-649508

ISBN 0-471-07875-1   AACR 2 MARC-S

Library of Congress   [8610]

Printed in the United States of America

10 9 8 7 6 5 4 3 2 1

In affectionate remembrance, this book is dedicated to
William G. Pfann. Bill Pfann's imprint is evident
throughout the work we discuss. He will be remembered
as an outstanding scientist, a prolific inventor,
and a trusted friend.

GILBERT J. SLOAN
ANDREW R. McGHIE

INTRODUCTION TO THE SERIES

Techniques of Chemistry is the successor to the Technique of Organic Chemistry Series and its companion—Technique of Inorganic Chemistry. Because many of the methods are employed in all branches of chemical science, the division into techniques for organic and inorganic chemistry has become increasingly artificial. Accordingly, the new series reflects the wider application of techniques, and the component volumes for the most part provide complete treatments of the methods covered. Volumes in which limited areas of application are discussed can be easily recognized by their titles.

Like its predecessors, the series is devoted to a comprehensive presentation of the respective techniques. The authors give the theoretical background for an understanding of the various methods and operations and describe the techniques and tools, their modifications, their merits and limitations, and their handling. It is hoped that the series will contribute to a better understanding and a more rational and effective application of the respective techniques.

Authors and editors hope that readers will find the volumes in this series useful and will communicate to them any criticisms and suggestions for improvements.

ARNOLD WEISSBERGER

Research Laboratories
Eastman Kodak Company
Rochester, New York

A NOTE FROM THE NEW EDITOR

Following in the footsteps of the man who originated this series and edited it for many years is no easy task. The aim remains the same: the comprehensive presentation of important techniques. At the same time, an effort will be made to keep volumes to a reasonable size by illustrating what can be done with a technique rather than cataloging all known applications. Readers can help with advice and comments. Suggestions of topics for new volumes will be particularly welcome.

WILLIAM SAUNDERS

Department of Chemistry
University of Rochester
Rochester, New York

PREFACE

Most chemists think of crystallization as something learned about in their science fair years, or at the latest, when they studied elementary organic chemistry. Crystallization from solvents is carried out casually, in much the same way as it was a century ago. In truth, the introduction of chromatographic methods has reduced—but not eliminated—the usefulness of fractional crystallization. In a sense this is unfortunate, because there are situations in which an "old fashioned" triangular fractional crystallization from an inexpensive solvent will lead to a desired separation very economically. Although elapsed time in crystallization is long, the total amount of work invested in a separation may be less than that required to develop a preparative chromatographic method. Nevertheless, few chemists today seem aware of the utility or practicality of traditional multiple-crystallization methods.

Crystallization of molten materials is also an ancient art, associated largely with metals. A number of chemists prepared single crystals from melts in the 1920s, and some excellent work was carried out in the same period on purification by crystallization of melts. The techniques that were developed, however, never came into wide use.

The situation changed in the 1950s, as large numbers of physicists and physical chemists became interested in the solid-state properties of organic materials. They soon realized that the properties under study were extraordinarily sensitive to impurities, and they needed effective methods of reaching very high purity. Fortunately, many of them, while studying organic solids, were intimidated by organic solvents. The bad odors, flammability, and toxicity associated with many solvents led them to look for ways other than chromatography or crystallization from solvents to get pure materials. As members of the solid-state community, they were well aware that silicon and germanium had been brought to very high purity by zone refining. It seemed that new methods based on crystallization of melts would be quick, safe, and effective.

Thus it was that the ultrapurification of organic (and some inorganic) chemicals was undertaken by people whose interests lay in characterization rather than in synthesis. The limitations of the method soon became evident (scale of operation, requirement of thermal stability), and these in fact guided the selection of materials for study. Nevertheless, a body of work on purification has emerged, largely unknown to organic and inorganic chemists, applicable to a broad range of materials and worthy of much wider use.

Our goal in *Techniques of Melt Crystallization* is to advance this cause. We describe a broad range of devices and methods, considering each in a somewhat "modular" approach. That is, we give designs of basic building blocks which can be assembled according to the needs of each experimenter. We have tried to provide illustrations that are clear enough to serve as shop drawings where custom fabrication is necessary. We have also tried to point the way to the use of commercially available devices as components of homemade assemblies for crystallization.

We are grateful to the many people whose ideas we have cited. We are grateful too, to all the members of our families, who have endured endless hours of discussion and debate about aspects of this book, and who have put up with a lot of domestic clutter as piles of references, reprints, notes, and illustrations accumulated.

If there are errors of fact or interpretation in our text, we will be pleased to hear of them.

GILBERT J. SLOAN
ANDREW R. McGHIE

Indian Field
Wilmington, DE
1986

CONTENTS

Chapter **I**

BASIC IDEAS IN CRYSTALLIZATION

The intent of this volume is to introduce the reader to the range of laboratory procedures involving crystallization of melts. We will develop the theoretical underpinnings only to the extent necessary for an understanding of the strengths, weaknesses, and ranges of applicability of the many techniques that have been described.

In both theoretical and experimental sections, the coverage is intended to be selective rather than exhaustive. In the experimental domain, we have provided descriptions of typical or preferred devices and techniques, rather than an encyclopedic report of everything that has been published. Further, we have tried to make the experimental descriptions sufficiently clear and detailed to allow the experimentalist to follow them, in most cases without recourse to the original literature.

Melts are crystallized for a great variety of reasons. Purification, while not the only reason, is certainly a major one. A number of cases are known in which properties of materials change drastically when some impurity is removed, and crystallization techniques are important in attaining the purity needed for measurement of intrinsic properties. Among impurity-conferred properties, one can mention a polymorphic transition in cesium nitrite. Various authors have observed such a transformation at temperatures between 353 and 393 K, but highly purified $CsNO_2$ did not show the transition at all (1). Another recent case of an impurity-conferred property involves the biological behavior of a coal tar heterocyclic. Commercially available acridine of $>98\%$ purity was found to be teratogenic to the cricket *Acheta domesticus*. Purification of the commercial product resulted in complete loss of teratogenicity, whereas the mixture of impurities recovered from the acridine showed a positive response (2).

In other work, impurities are of critical interest and crystallization offers many possibilities of concentrating them for easier identification and quantification. The growth of crystals also provides insights into the phase relationships of systems and hence information on thermodynamic quantities.

1   HISTORICAL BACKGROUND

It is instructive to consider the various possible consequences of solidifying a mixture of chemicals. If solidification takes place rapidly, then the local composition of the resulting solid will be very close to that of the original liquid. On the other hand, if solidification is slow, then the crystal architecture of the major component of the mixture may direct the chemical composition of the solid that forms. The crystal is, by definition, a regularly repeated array of a characteristic building block. The specific binding forces that lead to the regularity of the crystal lead to the exclusion of foreign molecules. Thus Figure 1.1 shows a mixture consisting of large and small spherical molecules. If the entire content of the test tube is melted (Figure 1.1a) and is crystallized directionally by lowering it slowly into a cooling bath, and if a crystal comprised of the large molecules forms first, the remaining melt will become richer in the smaller molecules (Figures 1.1b and c). The selectivity of crystallization depends upon the specific intermolecular forces acting among the constituent molecules. To a first approximation, these forces are determined by the relative sizes, shapes, and polarities of the constituent molecules. In the foregoing discussion, a distinction is made between rapid and slow crystallization. At this early point, it must be emphasized that these relative terms can have widely different quantitative meanings, depending upon the context. For example, metals may be solidified to good crystalline structures at linear rates of tens of centimeters per hour, while larger, less symmetrical organic molecules whose melts are of high viscosity, may not be solidified in an orderly manner at rates greater than millimeters or tenths of millimeters per hour. In the case of organic melts, attempts to promote rapid crystallization by increasing the thermal driving force, that is, the extent of cooling, may lead instead to the supercooling of the liquid with consequent formation of a glassy structure.

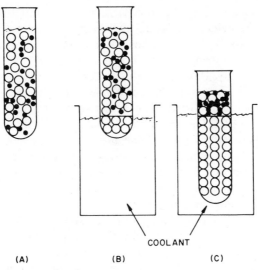

COOLANT

(A)   (B)   (C)

Fig. 1.1 Schematic representation of directional crystallization of a binary mixture. (*a*) Contents completely molten; (*b*) contents partially crystallized; (*c*) nearly complete crystallization.

Experimental examples of the power of directional crystallization are shown in Figure 1.2, a photograph of ingots containing azulene (ca. 5×10^{-4} mole fraction) in naphthalene and biphenyl, respectively. In the former case, the host and guest molecules are $C_{10}H_8$ isomers, but partial segregation occurs nonetheless, because of differences in shape and size (the volumes of naphthalene and azulene molecules are $0.1304 \, nm^3$ and $0.1364 \, nm^3$, respectively). In the latter case, the azulene is completely rejected from the growing crystal of biphenyl and appears at high concentration in the last portion to solidify.

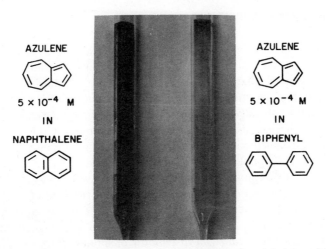

AZULENE

5×10^{-4} M

IN

NAPHTHALENE

AZULENE

5×10^{-4} M

IN

BIPHENYL

Fig. 1.2 Enrichment of azulene after directional crystallization of naphthalene and biphenyl.

Intuitively, one can accept that a large molecule will be rejected from a growing crystal of small molecules. A smaller solute may or may not be included in a growing crystal of a larger host. For example, in the naphthalene/2-chloronaphthalene system, the molecular volumes V_m are $0.130 \, \text{nm}^3$ and $0.145 \, \text{nm}^3$, respectively. Phase diagram studies (3, 4) show that 2-chloronaphthalene is rejected from naphthalene, while naphthalene is included in 2-chloronaphthalene. In a more extreme case, that of tetrazine ($V_m = 0.0664 \, \text{nm}^3$) and benzene ($V_m = 0.0845 \, \text{nm}^3$), no mutual solubility is observed.

In spite of the exceptions noted above, the generality is that differences in size, shape, and polarity do manifest themselves in rejection of foreign molecules from crystallizing melts. Even if a host lattice can accommodate impurity molecules, the solute concentration throughout the crystal is not likely to be the same as that in the initial melt, as seen in the azulene/naphthalene case in Figure 1.2. This property has long been used for purification, although the extent of use has been smaller than is warranted by the utility of the technique. In 1940 Schwab and Wichers reported the purification of benzoic acid by melting a charge in a glass tube which they lowered through a heating coil until solidification was complete (5). The last one-fourth to one-third of the resulting ingot was discarded. The melting and solidification were repeated and a product exceeding 99.999% purity was obtained. To achieve purification by this method, the impure fraction formed by crystallization of a portion of the melt must be physically removed from the purer solid. To attain still higher purity, the solid must be melted in its entirety and subjected to the solidification process again, with repeated physical removal of the impure portion of the charge. However effective, this method suffers from the disadvantages of:

- Low yield
- Slow speed
- Need to separate impurity after each solidification

In any case, relatively little attention was paid to this mode of purification for many years.

These disadvantages of purification by bulk crystallization of an entire charge are remedied by a remarkably simple strategem: that of melting only a small fraction of the total charge and moving the molten material directionally through the charge. The stunning simplicity of this concept makes it all the more remarkable that it was only discovered in relatively recent times and very slowly brought to widespread use. W. G. Pfann, the inventor of this zone-melting technique, has recalled an experiment which was in a sense premonitory, described in a 1928 publication by a Soviet physicist, Peter Kapitsa. In a situation that is fairly common in the sciences, Pfann was unaware of Kapitsa's work when it was carried out and reported, and only learned of it many years later. Kapitsa's experiment consisted of passing a single molten zone through an ingot with the intention of growing a single crystal. He missed entirely the potential of "the molten zone as a distributor of solutes" (6).

In the early 1930s, J. D. Bernal and his collaborators at Birkbeck College (University of London) used a single hot wire to pass a molten zone through a specimen of an organic compound on a microscope slide. Here too, the goal of the work was to grow single crystals for X-ray analysis, and to study phase transitions (7). Mackay has reviewed Bernal's experimental work and observed that Bernal was aware of the exclusion of impurities from a growing crystal (8, 9); the implicit conclusion is that Bernal's awareness of the ability of crystallization to effect purification constitutes discovery of zone melting. There is, however, no recorded indication that Bernal related crystal growth and purification. It seems clear that he, like Kapitsa, failed to grasp the potential of multiple passages of a molten zone for the attainment of high purity.

Pfann's earliest experiment in this area was similarly directed toward the growth of single crystals of uniform composition, and he employed a technique that later came to be known as zone leveling. These experiments date to 1939. It was some time after World War II that Pfann conceived the idea of using the molten zone for the removal of impurities. The essence of the zone-melting procedure can be grasped from Figure 1.3, showing a vertical cross section of an ingot in which a molten zone has been formed by an external heater at the top of the ingot. When the heater has moved down to the position shown, solid is melting at the advancing front and the melt is crystallizing at the receding interface. The events at the receding interface are in every way equivalent to those taking place at the solid/liquid interface described earlier, in the context of directional solidification. For a chemical system in which the minor component is rejected from the growing crystal, the movement of the molten zone has the following effect: solid of uniform starting composition is taken into the zone as it

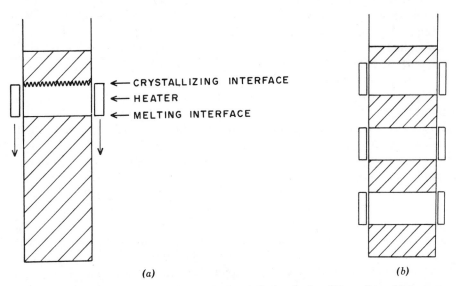

(a)   *(b)*

Fig. 1.3 Schematic representation of zone melting. (*a*) Single zone in midingot; (*b*) multiple zones.

advances, and simultaneously a solid forms at the receding edge of the zone, containing less of the impurity. Thus the zone acts as a "scavenger" of impurities in this kind of system and will produce a directional movement of impurity from one end of the ingot to the other.

The basic and decisive difference between this experimental operation and the directional crystallization described earlier is that a new zone may be introduced easily at the top of the ingot, and the crystallization experiment may be repeated as many times as desired. In fact, multiple zones may be passed through the ingot simultaneously (Figure 1.3b). Each solidifying interface acts as a purification stage independently of the others. The spacing at which these zones may be introduced is determined by substance-specific properties such as the thermal conductivity of the intervening solid, and by experimental parameters. Naturally, it is relative motion that is important, and either the ingot or the heaters may be moved. Thus zone melting relates to directional crystallization as fractional distillation relates to a simple, single-plate distillation. Alternatively, zone melting may be compared to partition chromatography, in the sense that the latter is an automatic iteration of a single extraction. Naturally, the number of single extractions required to equal the effectiveness of partition chromatography would be astronomical so much so that it is inconceivable to try to duplicate a chromatographic result by repeated single extractions. By the same token, zone melting in many cases produces effects that would not in any practical sense be duplicated by repeated single crystallizations.

We have used the term "zone melting" in this discussion without setting forth its position in the crystallization field. It should be made explicit that this term, sometimes used interchangeably with "zone refining," is a generic one, encompassing a rather wide range of procedures based physically upon the passage of a molten zone through a chemical charge. If the purpose of the passage of the molten zone is chemical purification, then the process is known correctly as zone refining. On the other hand, it is possible that a zone of melt may be moved sequentially in opposite directions through a charge of a chemical substance for the purpose of producing an impurity distribution which is as homogeneous as possible. In this case, the zone-melting process is referred to as zone leveling. Moreover, the molten zone may be passed through the charge for the purpose of converting a polycrystalline mass to a single crystal. The "charge" of material to be processed is often cast into an "ingot" and the two words are used more or less interchangeably.

The fundamental descriptive parameter of crystallization is the distribution coefficient, conventionally represented by the symbol k; it is the ratio of the solute concentration in a solid to the solute concentration in the melt from which the solid is formed. Thus $k = C_s/C_l$. The conventional phase diagram described in textbooks of physical chemistry is a graphic representation of the sequence of k's determined for a series of liquids of increasing solute content. Cooling a liquid of composition C_l until crystallization takes place results in the formation of crystals whose solute content C_s is different from that of the liquid (see Figure 3.1), and this is a quantitative description of the specific, crystallographically determined rejection of solute at a crystallizing interface, as depicted in Figure 1.1. We have shown here a particular phase diagram which is by no means universally descriptive, in that many different kinds of behavior can be and have been observed. These will be discussed in Chapter 2; however, a few additional details deserve discussion here. For example, some binary systems yield perfectly pure crystals of the major component upon solidification of the liquid. While it may be argued that the concept of "perfect purity" is more a philosophical than a scientific one, it is clear that absolute or perfect purity would require the growth of a crystal containing no defects. In the first place, of course, the grown sample must have a surface, and this departure from infinite extent comprises a basic and ineluctable deviation from perfection; it provides a possible residence for impurities even if the internal bulk of a sample is entirely pure. Furthermore most solids contain defects other than the macroscopic surface: there are one-, two-, and three-dimensional defects, all of which can act as mechanisms for the introduction of impurity into an otherwise perfectly pure lattice. In addition, there may be point defects, which likewise correspond to the presence of impurity.

2   PRINCIPLES OF SOLIDIFICATION

Of the three states of matter (gas, liquid, and solid), the solid state embodies the greatest stability and order. The ionic, atomic, or molecular building blocks of a solid are relatively immobile; they can move through only a small fraction of their lattice spacings, by vibration and rotation, and exchange positions only with difficulty. In liquids, on the other hand, their positions are random over times close to the reciprocal of the frequency of thermal vibration.

Solids may be distinguished from liquids on the basis of their response to stress: liquids change configuration under any stress, while solids are deformed elastically and shear strain is removed as stress is removed. (The class of materials known as plastic crystals constitutes an exception to this generalization.) A more functional distinction may be based on thermal considerations: heat must be supplied to convert a crystalline solid to a liquid at the same temperature. The amount of heat, of course, is the latent heat of fusion, which corresponds merely to the relocation of atomic or molecular units of the solid to states of higher potential energy, and does not include increase of energy owing to thermal vibration. Melting may be preceded by transitions to liquid-crystal or rotator phases.

Most elements occur only in the crystalline state, but a few appear in both amorphous and crystalline modifications (As, C, P, S, Sb, Se, and Tc). Many oxides tend to form amorphous structures only, and large numbers of organic compounds can be prepared in either crystalline or amorphous forms. Molecular size, structure, and symmetry, along with temperature, determine which form is preferred. Recent experiments have demonstrated that certain metallic alloys, normally crystalline, can be obtained in amorphous form by extremely rapid cooling of their melts.

Solids are formed from vapor, solution, or melt, through continuous or discontinuous transitions. The former process gives rise to amorphous bodies, which show only short-range order. The latter process gives crystalline bodies, which show long-range order. Liquid crystals comprise an intermediate state of matter; they show the flow properties of liquids, but are not isotropic. This behavior is the result of one- or two-dimensional order, while true crystals show three-dimensional order.

Methods of studying crystal structure and defects are beyond the scope of this book.

2.1 Nucleation

Crystals are formed by transition of ions, atoms, or molecules from an isotropic, unordered phase to an ordered phase. The parent phase can be gas, solution, melt, or even solid. In each case it is necessary that the phase equilibrium be disturbed by a supersaturation or supercooling which drives the transition process. The presence of such supercooling is not a sufficient condition for formation of solid in a liquid containing no crystalline solids; liquids can in fact exist for long times at temperatures well below their melting points. This well-documented fact opposes the thermodynamics of the situation, since at temperatures below the melting point, the free energy of the solid is less than that of the liquid and the liquid-to-solid transition is thus a spontaneous one.

To understand the metastability of supercooled melts, it is necessary to consider the genesis of a crystal. In fluid systems, there are constant statistical variations in local kinetic energy, density, or concentration, resulting in the assembly of numbers of elementary entities (ions, atoms, or molecules). The resulting cluster may form in the bulk fluid phase (homogeneous nucleation) or on an existing surface (heterogeneous nucleation). The number of such statistical clusters is very small compared with the numbers of atoms or molecules present; the critical characteristic that determines whether a cluster will attain detectable size is its curvature. This is so because equilibrium across a curved solid/liquid interface is different from that across a planar interface, since atoms in a curved solid have fewer nearest neighbors than those in a planar solid and can escape more readily. The result is that small clusters are less stable than macroscopic solid surfaces. Consequently, the rate of melting of a particle increases with decreasing size, and the temperature at which the melting rate equals the freezing rate is lower for small crystals than for large ones.

Ordered clusters are present in liquids under all conditions. If these are smaller than the critical size, they cannot grow. If they are larger than the critical size, nucleation takes place and observable crystals form. The formation of a region of crystalline solid within a fluid phase releases latent heat and consumes energy in the formation of a new interface. Thus the conventional Gibbs–Thomson formulation expresses the overall free-energy change ΔG resulting from formation of a solid sphere of radius r as the sum of a volume term and a surface term:

$$\Delta G = \tfrac{4}{3}\pi r^3 \,\Delta G_v + 4\pi r^2 \sigma \qquad (1.1)$$

where ΔG_v is the free-energy change that accompanies transition from liquid to solid, and σ is the specific interfacial free energy. The volume and surface terms are plotted in Figure 1.4 along with their sum. The sum is positive for very small r and negative for large r. The critical radius is defined as that for which either a decrease or increase in radius will result in a decrease in free energy. This is the point of maximum free energy, where $d(\Delta G)/dr = 0$:

$$-4\pi r^2\,\Delta G_v + 8\pi r\sigma = 0 \qquad (1.2)$$

Hence

$$r^* = \frac{2\sigma}{\Delta G_v} \qquad (1.3)$$

Assuming that the liquid and solid are incompressible,

$$\Delta G_v = \frac{\Delta H_f\,\Delta T}{T_f} \qquad (1.4)$$

Hence

$$r^* = \frac{2\sigma T_f}{\Delta H_f\,\Delta T} \qquad (1.5)$$

The radius of this spherical critical nucleus can be related to substance-specific properties and to the supercooling, ΔT (10). This inverse function of ΔT

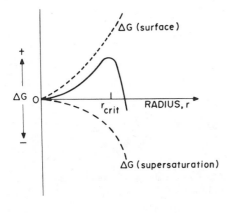

ΔG (surface)

ΔG (supersaturation)

Fig. 1.4 Free energy of formation of a spherical crystal, shown as the sum of surface and volume terms.

is plotted in Figure 1.5 along with a representation of the temperature dependence of *maximum* cluster radius r'. The intersection of these two curves defines the temperature at which a cluster attains critical radius and becomes a nucleus. If σ can be measured, r^* can be calculated. For homogeneous nucleation of copper, for example, ΔT is 236 K; this corresponds to $r^* \cong 10^{-7}$ cm, or a sphere containing about 360 atoms, a number that applies, approximately, to many metals.

At equilibrium, the number n_i^* of critical-size clusters containing i atoms is

$$n_i^* = n \exp\left(-\frac{\Delta G^*}{kT}\right) \tag{1.6}$$

where n is the number of atoms per unit volume in the system and ΔG^* is the excess free energy of the critical cluster. The rate of nucleation was first calculated by Volmer and Weber (11) for condensation of gases and then extended to the nucleation of melts by Volmer and others, especially Turnbull (12). The expression for nucleation rate is

$$\frac{dZ}{dt} = K \exp\left(-\frac{\Delta G^*}{kT}\right) \exp\left(\frac{\Delta G_A}{kT}\right) \tag{1.7}$$

where Z is the number of nuclei, ΔG_A is the activation energy for transfer of atoms from the melt to the crystal surface by diffusion, and

$$K = nn' \frac{kT}{h} \tag{1.8}$$

where n' is the number of atoms on the surface of the critical nucleus.

The nucleation energy E has been calculated to be

$$E = \frac{16\pi\sigma^2 v^2 T_f^2}{3\,\Delta H_f^2 (T_f - T)^2} \tag{1.9}$$

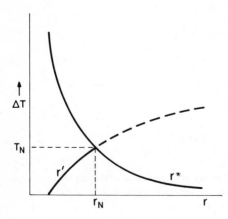

Fig. 1.5 Conditions defining the cluster of critical radius.

where σ is the surface energy per unit area, from which one finds that there is a minimum when $T/T_f = 0.33$. In this case, dZ/dt has a maximum. Experiment, however, shows that many organic substances show maximum nucleation rate when T/T_f is between 0.6 and 0.9, some two to three times the theoretical value. While experimental studies of homogeneous nucleation are beset by many problems, the expected maximum in nucleation rate with decreasing temperature is in fact observed. Usually a few Kelvins change in supercooling causes a few orders of magnitude change in (dZ/dt).

Glycerol, for example, has a maximum nucleation rate at $T = 0.73T_f$ (Figure 1.6a); at lower temperatures, nucleation is impeded by the reduced diffusitivity. As an experimental matter, it is important to note that nuclei cannot be counted as they form. In practice, a portion of melt is held at a given nucleation temperature for a fixed time interval and then heated to some higher temperature to allow the invisible nuclei to grow sufficiently to be counted.

The nuclei grow at a rate that increases with undercooling, to a maximum value (Figure 1.6b). At still greater undercoolings, growth rate drops. Growth-rate maxima are generally flatter than nucleation-rate maxima. Low undercoolings give slow growth because the heat of fusion liberated during crystallization is not removed rapidly enough in the small driving gradient. High undercoolings give slow growth because increased viscosity reduces matter transport to the solid surface (see Section 2, this chapter).

As mentioned earlier, the temperature of homogeneous nucleation is measurable in finely dispersed systems, in which the number of droplets is greater than the number of heterogeneous-nucleation sites. With this temperature and Eq.

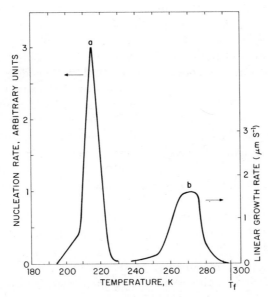

Fig. 1.6 Dependence of nucleation rate (a) and growth rate (b) on temperature, for glycerol.

1.7, it is possible to calculate the specific solid/liquid interfacial free energy, σ. For metals, σ is about half the latent heat of fusion ΔH_f, and for other materials is about one-third ΔH_f. From Equation 1.5 one finds that the reduced supercooling can be related approximately to the size of the atom or molecule:

$$\frac{a}{r^*} \cong \frac{\Delta T^*}{T_f} \qquad (1.10)$$

where $\Delta T^* = (T_f - T^*)/T_f$, T^* is the lowest temperature to which the liquid can be supercooled, and a is the atomic or molecular diameter.

Homogeneous melts can be nucleated at temperatures above T^* by cavitation. Microcavities, formed by ultrasonication, for example, experience large pressures as they collapse. The pressure pulses lower T_f and nucleation ensues.

Nucleation of a new solid phase can take place in a crystalline matrix as well as in liquid, solution, or vapor. Such nucleation is important in:

- Phase transformations
- Separation of the components of solid solutions with limited solubility
- Recrystallization of plastically deformed metals

As in other condensed-phase nucleations, a nucleation energy must be provided. Beyond this basic condition, however, many factors affect the process: lattice defects and local structural variations that generate stresses and deformations, are among these. The complexity of the problem has led to controversy over interpretation of experimental results and difficulty in reconciling theory with experiment. Two basically different assumptions have been made regarding nucleation within solids. One of these posits statistical nucleation analogous to the events in vapors and liquids (13), while the other assumes the presence of nuclei of the daughter phase within deformed regions of the parent phase (14).

Extensive studies of polymorphic transitions in layered molecular crystals provided no evidence of cooperative or displacive changes. Nor was any evidence found for "instantaneous" transitions that cannot be slowed by reduction of the driving force (15). Instead, it is argued that a phase transition is initiated at three-dimensional defects of some critical size, such that there is, on the one hand, sufficient space for local molecular rearrangement and, on the other hand, an activation energy for the transition that is lower than the sublimation energy. The daughter phase nucleates by an internal epitaxy across such a defect (15). Within this conceptual framework, two modes appear:

1. The daughter crystals grow in nonoriented fashion, so that their habit faces do not have rational indices in the parent lattice; this is the predominant mode.
2. Oriented growth of daughter crystals within the parent phase, made possible when both phases are layered and the structures of layers parallel to the cleavage plane are nearly the same in both phases.

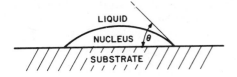

Fig. 1.7 Heterogeneous nucleation.

The second circumstance has been related to the presence of particular defects capable of surviving repeated cycling through a phase transition. The daughter phase appearing at a given location within a parent crystal was found to grow with the same orientation in each cycle (16).

Heterogeneous nucleation also takes place through configurational fluctuations in the liquid, but the fluctuations take place preferentially at an interface between the liquid and a solid. Heterogeneous nucleation generally takes place in bulk liquids at much smaller ΔT than is the case for small drops. This results from the fact that it is unlikely that an individual small drop will contain a nucleating center. Moreover, the bulk liquid is in contact with a container surface which comprises a major source of nucleation sites.

Independently of other considerations, critical nuclei are spherical. If a nucleus forms on a surface, however, its stability depends on its radius of curvature and on the stability of the line of contact with the substrate. If the substrate is rough, then the local angle of contact θ affects the nucleation properties. Effective nucleants have $\theta < 90°$; that is, θ, the solid/crystal interfacial tension, is less than the solid/liquid interfacial tension (Figure 1.7). The solid/solid interface between the nucleus and the substrate will have low energy if the two-dimensional lattices of the two materials in contact are similar in geometry and interatomic distance. Even if the match is good in only one dimension, some reduction of interface energy is expected, compared to that of the substrate/liquid interface.

2.2 Crystallization Rate and Interface Morphology

Crystals grow from melts at widely varying rates. Organic crystals typically grow at rates less than $3 \times 10^{-4}\,\mathrm{cm\,s^{-1}}$, ionic crystals at about $10^{-3}\,\mathrm{cm\,s^{-1}}$, and metals at rates up to $10^{-1}\,\mathrm{cm\,s^{-1}}$. The differences result from the differing activation energies required to move atoms, ions, or molecules from the melt to the crystal surface through the boundary layer around the crystal. The growth rates also depend on associated entropic changes. For small, symmetrical crystallizing units (atoms), the activation energy is low. For large, unsymmetrical units (organic molecules), it is high.

Calculation of crystallization rates has been approached from both thermodynamic and statistical viewpoints. In the former, the rate is calculated from the free energies of solid, melt, and an activated state in the melt. The resulting equation (17) predicts zero growth rate at $T = 0\,\mathrm{K}$ and when $T = T_f$, with a maximum at some intermediate temperature, $T_{\max}$ (T_f is the equilibrium crystallization temperature of the melt). If $\Delta G \gg R T_{\max}$, the maximum in the growth-rate curve is sharp; if ΔG_A is only slightly greater than RT_{max}, then there is a wide range in which growth rate is practically independent of T (ΔG_A is the difference between solid and liquid free energies at T_{max}). The domain of constant maximum growth rate is not uniquely a substance-specific property, but is dependent on the distribution of heat transfer at the interface.

The statistical approach to growth rate considers growth as the superposition of competing melting and crystallization rates. The probability of transition from melt to solid and vice versa can be given in terms of the vibrational energy of the atoms and the number of atoms having a particular energy. Both transitions are temperature dependent, but to differing degrees. Calculated curves for both rates are given for copper in Figure 1.8; again it is seen that the difference is zero at T_f and at very low temperatures.

The discussion thus far has referred to an isothermal situation in which solid, melt, and the boundary layer are at the same temperature. In an ongoing crystallization, heat of fusion is liberated at the interface, and it must be dissipated if crystallization is to proceed. At low growth rates, the heat of fusion is dissipated exclusively through the solid. The relationship between the latent heat and heat transport through the solid in this regime is given by Equation 1.11

$$V = \frac{K_s}{\Delta H_f \rho_s} \frac{dT}{dz} \tag{1.11}$$

where V = growth rate, K_s = thermal conductivity of the solid, ΔH_f = heat of fusion of the solid at its melting point, ρ_s = density of the solid, T = temperature, and z = axial coordinate parallel to the growth direction. It is assumed in deriving this equation that convection and radiation are negligible. Another underlying assumption is that heat is conducted axially from the

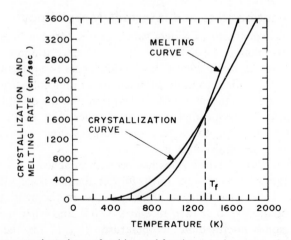

Fig. 1.8 Temperature dependence of melting and freezing rates in copper; after Chalmers (36).

interface through the crystal only (i.e., low growth rate); to the extent that this is true, crystallization rate is determined by the thermal conductivity of the solid. In fact, both axial conduction to the crystal and radial conduction to the wall of the crystallization apparatus must be considered. If the surface of the solid at the interface is convex, the container must be receiving heat from the surroundings: that is, the surroundings at the level of the interface are above the melting point of the solid. Conversely, if the surroundings are below the melting point of the solid, a concave solid interface results. The position of the planar isotherm (i.e., the level in the vertical growth apparatus at which a planar interface results) depends on the Biot number (ratio of the heat loss from the surface to heat conduction) of the sample. Naturally, the sensitivity of interface position to growth gradient also depends on the Biot number, increasing for:

- Small diameter
- High heat transfer from the container surface
- High thermal conductivity of the material

With increasing growth rate under constant ambient conditions, the interface becomes more concave because of:

- Decreasing interface temperature (owing to segregation and kinetic supercooling)
- Liberation of latent heat (18)
- Sensible heat carried by the container and its contents

The last-mentioned quantity (heat conduction by the container and its contents) varies during growth as the ingot descends into the cold zone. The changing distribution of temperature in the ingot has been analyzed as a function of its position, as well as its thermal and geometrical characteristics (19). By relating these to solidification rate, it is possible to choose parameters that make the translation rate equal to the crystallization rate. Table 1.1 shows the experimental conditions needed to achieve this equality for three ingot geometries. For high-conductivity materials, the ratio is close to unity only during growth of the central portion of the ingot. For low-conductivity materials, the ratio differs from unity only at the beginning and end of crystallization (i.e., where g, the fraction solidified, is close to zero or unity). The table shows that for a given material, the ratio is close to unity over a larger fraction of the length for a large length-to-diameter ratio than for a small one.

If it is necessary to maintain solidification rate equal to translation rate, the temperature of the heater can be programmed throughout the growth cycle (19). Since a convex solid favors growth of single crystals, it is desirable that the interface be situated in the heater.

The theoretical calculation of interface shape and position assumes some temperature distribution within the growing solid and nearby liquid. Experi-

Table 1.1. Relationship between Rate of Crystallization and Rate of Translation for Directional Crystallization of Materials of Differing Thermal Conductivity in Ingots of Differing Geometry (19)[a]

Material	Thermal Conductivity K	$L(2H/R)^{1/2}$	ψ	Values of g within which $V_i < 1.05V_t$
Cu	300	10	0.575	nil
		50	2.75	nil
		100	5.75	0.34–0.66
Al	100	10	1	nil
		50	5	0.38–0.62
		100	10	0.18–0.82
Sn	30	10	1.83	nil
		50	9.15	0.23–0.77
		100	18.3	0.09–0.91
Bi	10	10	3.16	nil
		50	15.8	0.11–0.89
		100	31.6	0.06–0.94

[a] K, $Jm^{-1}s^{-1}k^{-1}$. L is total ingot length, taken to be 5, 25, and 50 cm in the examples given above. H_h and H_c are the heat-transfer coefficients between the ingot and the heater and cooler, respectively; here $H_h = H_c = H = 10^2 Js^{-1}m^{-2}K^{-1}$. R is the radius of the ingot, in cm; in this case $R = 0.5$ cm. V_i is the rate of interface advance and V_t is rate of ingot translation. ψ is a geometric factor, equal to $L(2H/RK)^{1/2}$.

mental verification of these temperatures is not easily attained, because the measurement tends to disrupt the measured system. One attempt to overcome this problem (20) made use of a gradient method in which a crystal grows from a flat metal plate through a sequence of thermocouples mounted at known distances above the plate. Only for substances showing intrinsically small subcoolings and for low growth rates will the surface temperature equal the true melting temperature T_f. For other materials and/or for higher growth rates, a correction term is required. The form of the temperature-position relationship was calculated (solid line, Figure 1.9) and the temperatures were noted at thermocouples T_1 and T_2 at the instant when the interface was at X_3. Extrapolation of the thermocouple temperatures to X_3 (dotted line) gave an approximate interface temperature T_a, from which the true surface temperature could be calculated. For p-xylene growing from the plate at 283 K (3.26 K undercooling), the correction needed to convert T_a to the true surface temperature T_s ranged from about 0.05 K (near the start of growth) to about 0.01 K (after 1 cm of growth). At a higher undercooling (plate at 254.3 K), the corrections ranged from 0.6 to 0.07 K, respectively.

An indirect method has been described (21) which uses a very small (125-μm) quartz fiber terminated by a graphite bead. The bead was placed in the crystallizing melt and the fiber conducted the radiation emitted by the bead to a

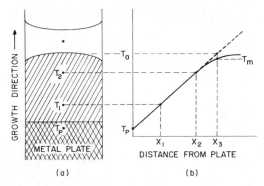

Fig. 1.9 (*a*) Schematic view of crystallization apparatus; *T*'s are temperatures measured with immersed thermocouples. (*b*) Plot of temperature against position in the solid; T_a is the extrapolated (approximate) temperature at the interface.

photodiode. It is claimed that this device introduced less distortion of the heat flow than a thermocouple and provided a ten-times-faster response, along with a noise level two times lower than that of a standard thermocouple probe.

Hot-stage, time-lapse micrography has been used in an elegant experimental study of the relationship between crystal growth rate and temperature in metasilicate glasses (22). Figure 1.10 shows a specimen of ellipsoidal shape near the junction of the measuring thermocouple. Several sources of error in the growth-rate measurements, including distortion of the growth front by con-

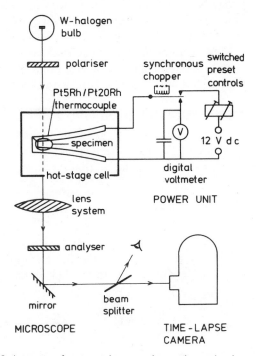

Fig. 1.10 Apparatus for measuring crystal growth rate in glassy materials.

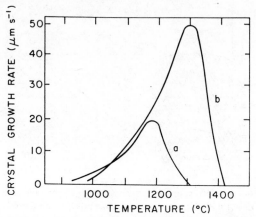

Fig. 1.11 Temperature dependence of crystallization rate in aluminosilicate glasses as a function of temperature: (a) 9.09 wt% Al_2O_3 and 45.45 wt% SiO_2; (b) 4.55 wt% Al_2O_3 and 50.00 wt% SiO_2. Both systems contain 36.36 wt% CaO and 9.09 wt% ZnO.

vection, and barrel distortion, had to be accounted for before reliable data could be obtained. Figure 1.11 shows representative curves for systems containing CaO, Al_2O_3, SiO_2, and ZnO.

It was pointed out earlier in this section that the maximum in the growth rate/supercooling curve may be sharp; this is especially true for materials with low entropy of fusion. The resulting difficulties in measurement of unconstrained crystallization rates may be responsible for the paucity and unreliability of the available experimental data. A recent approach makes use of two photocells and an oscilloscope to time the growth of a crystal, at known supercooling, through a known distance (23). Measurements on two substances with low entropy of fusion (cyclohexanol and succinonitrile), over a wide range of supercooling, gave true kinetic coefficients which were not in accord with theory (23).

2.3 Interface Morphology and Entropy of Fusion

The varying interface morphologies associated with different crystal structures have been related in turn to the thermodynamics of solidification. Jackson analyzed interface kinetics and found that the change in interfacial free energy σ is given by

$$\frac{\sigma}{NkT_f} = \alpha(1 - x)x + x \ln x + (1 - x) \ln(1 - x) \qquad (1.12)$$

where N is the number of surface sites, k is the Boltzmann constant, x is the fraction of surface sites that are occupied, and α is given by

$$\alpha = \left(\frac{\Delta H_f}{RT_f}\right)\psi \qquad (1.13)$$

where R is the gas constant and ψ is a crystallographic interface factor generally between 0.5 and 1.0 for low-index planes, and T_f is the equilibrium melting temperature. The physical significance of ψ is that it measures the fraction of total binding energy that binds a molecule to other molecules in the same layer, parallel to the crystallographic plane. It is larger for close-packed planes than for less closely packed ones.

Materials can be categorized with respect to smoothness of growth surface according to the value of the entropy factor, $\Delta H_f/RT_f$. For metals, acids and bases, silica, and a number of other inorganic glass-forming materials, and for certain categories of organic substances, the factor is less than 2. For a number of semimetals and semiconductors, it is between 2.2 and 3.2, while it is greater than 3.5 for most organic materials, silicates, borates, and other inorganic compounds (24).

For low-α crystal faces (of materials with low entropy of fusion, or high-index faces of other materials), even densely packed planes will be rough, and initiation of new growth layers will be easy. Hence growth on such faces will be rapid. Conversely, for materials with high α values, the closely packed planes are smooth, and initiation of new layers is difficult, while in these materials, less closely packed faces are expected to be rough. The importance of these thermodynamic considerations for crystal growth lies in their significance in establishing growth-rate anisotropy. By growth-rate anisotropy is meant the difference in rate of growth of different crystal faces at a given undercooling. For small anisotropy, growth in a relatively strong temperature gradient will proceed with an interface parallel to a growth isotherm and crystal advance will be approximately perpendicular to the isotherm. This leads to a planar interface, even on a microscopic scale.

For highly anisotropic systems (those with high entropy of fusion), the interface can be far from isothermal, and different parts of the interface can exist at substantially different temperatures, while advancing at the same interface velocity.

Most organic chemicals and many other substances of low crystal symmetry crystallize with high entropy of fusion. For materials with $\Delta H_f/RT_f$ greater than 4, the closest-packed faces should be smooth, while less densely packed ones should be rough; that is, both high-α and low-α faces will bound the crystal. In such cases, sizable growth-rate anisotropy is expected. An exception to this generalization exists in the category of rotator-phase organic crystals whose symmetry is relatively high and which crystallize with low entropy of fusion. These materials, which include substances such as carbon tetrabromide, cyclohexane, succinonitrile, triamantane, and pivalic acid show very nearly isotropic growth and are characterized by low entropy of fusion, generally less than 2 eu. Properties of these materials have been reviewed by Aston (25).

Theories of nucleation and crystal-growth kinetics have been reviewed by Jackson (26, 27), and the interested reader can use these reviews as guides to further study of these topics.

3   OVERVIEW OF CRYSTALLIZATION METHODS

Before going on to detailed description and discussion of the many ways in which liquids can be solidified usefully, it will be helpful to examine them briefly. Directional crystallization and the rudiments of zone melting have just been introduced, but will be described here somewhat further.

3.1   Directional Crystallization

Directional crystallization has been applied fairly extensively to the purification of both organic and inorganic chemicals, as well as metals (28). Most work has been carried out by lowering cylindrical glass containers from hotter to cooler environments. Attempts have been made to enhance the efficiency of the method by stirring the melt internally or by rotation of the container. A rather different approach to enhanced efficiency has been followed by Anderson, who used centrifugal force during crystallization (29). Other directional crystallization experiments have been designed primarily for the concentration and separation of trace impurities. Much work in directional crystallization has been devoted to studying the crystallization process itself, especially as it relates to characterizing phase diagrams, since it connects solidus and liquidus compositions. A more exotic, but nonetheless interesting application is to the measurement of thermal conductivities (30).

Directional crystallization has been carried out on milligram to multikilogram scale. While most directional crystallization has been carried out in the batch mode, a device has been described in which aluminum ingots were fed to a crystallizer from which a purified ingot emerged continuously (31).

3.2   Zone Melting

Most chemical zone melting is carried out vertically, in cylindrical glass tubes. However, sporadic reports of advantageous use of horizontal or nearly horizontal refiners have appeared (see Chapter 5). In general, molten zones are moved downward from a free surface into the ingot being processed. Containers of other shapes and materials have been used.

Zone melting has been carried out on milligram and even microgram charges (32). A number of attempts have been made to apply zone melting to multikilogram charges of chemicals (33). Continuous zone melting has been described (34, 35).

Vertical zone melting is usually carried out at 10^{-4} to 10^{-3} cm s^{-1}; speeds up to 3×10^{-2} cm s^{-1} have been used without loss of effectiveness in systems having effective mixing of the liquid zones (see Chapter 5).

3.3   Zone-Melting Chromatography

Thus far, we have referred to systems in which one component predominates. To cope with gross mixtures and with thermally unstable substances, a chromatographic mode of zone melting has been developed on the basis of a theoretical proposal by Pfann (see Chapter 7). The concept has been reduced to practice and consists essentially in forming a solvent column of a chemically inert host substance to which the mixture to be separated is applied in the form of a narrow zone. If the molten zone of mixed chemicals is moved through the host ingot, then the various components of the starting mixture will travel through the ingot at rates depending upon their respective distribution coefficients. This concept, while it has not been broadly used, has been applied especially to the separation of polymers according to molecular weight.

3.4   Traveling-Solvent Zone Melting

Yet another approach to the problem of zone melting a material that is not stable at its melting point has been described as "traveling-solvent zone melting." In this process the mixture to be separated is cast or compacted into ingot form and a relatively small amount of solvent is moved as a liquid zone through the solid charge. The effect is essentially that of fractional crystallization from solvent, but it is carried out in this configuration with considerable economy of time and material, as compared with conventional laboratory crystallization. Moreover, in this confined configuration, exclusion of potentially harmful atmospheric effects (such as oxidation) is easier than in conventional solvent crystallization (see Chapter 7).

3.5   Eutectic Zone Melting

Often the question arises whether materials unstable at their melting points may be purified by zone melting. Some possible approaches to this problem have been discussed above under zone-melting chromatography and the traveling-solvent method. Yet another approach is offered by eutectic zone melting. One considers a substance A, which decomposes at its melting point. It is usually possible to identify another substance B which will form a eutectic mixture with A. The eutectic mixture, by definition, has a lower melting point than either A or B and forms a solid of constant melting point whose composition is identical to that of its melt. Thus solidification of the mixture results in no change of composition, and one may consider the eutectic, for practical purposes, as the equivalent of a pure compound with regard to crystallization. Hence the solidification of a eutectic by a directional freezing or zone technique may offer the possibility of segregating minor components from the eutectic liquid, since these will be rejected as the eutectic solidifies. It is only necessary that component B, the eutectogen, be easily removed after the zone purification of the eutectic mixture. It is also necessary, of course, that the eutectogen not form solid solutions with the impurities. In addition, it is desirable that the eutectic mixture contain a substantial quantity of component A, whose purification is desired (see Chapter 7).

3.6   Temperature-Gradient Zone Melting

Temperature-gradient zone melting is another embodiment of the zone concept. This technique is based on the fabrication of a chemical system containing at least two components in which a compositional discontinuity exists and upon which a temperature gradient is imposed. The gradient is such that the lower-melting of the chemical constituents is melted and dissolves a quantity of the contiguous, higher-melting component. Diffusion of the dissolved, higher-melting major component across the width of the molten zone generates a transport of solute from one boundary of the zone to the other. The resulting change in composition causes movement of the molten zone through the charge, often without physical movement of the source of heat. That is to say, the gradient and the compositional changes brought about by diffusion collectively produce movement of the zone (see Chapter 5).

3.7 Centrifugal Solidification

As zone-melting techniques have evolved, specialized procedures have been introduced in response to particular requirements. In an attempt to improve the efficiency of impurity rejection at the solidifying interface, a number of workers have sought to use centrifugation as a means for enhancing matter transport (29) (see Chapter 5).

3.8 Column Crystallization

Column crystallization refers to a family of purification procedures in which a slurry of crystals and melt is subjected to countercurrent contact. In one system, for example, the charge of material to be purified is contained in an annular, cylindrical chamber which also houses a metal helix. Rotation of the helix drives the crystals in one direction, while melt moves countercurrently. Under an applied temperature gradient, a concentration gradient results. Thus the opposite ends of the column will contain material that is more (less) pure than the original. Continuous operation is relatively easily achieved by introducing feed at a central point and removing product and waste from the ends of the column (see Chapter 9).

3.9 Drum Crystallization

If a horizontally disposed cylinder is rotated while partially immersed in a melt, a portion of the melt can solidify on the (cooled) cylindrical surface. As the adhering solid film emerges from the melt, it can be removed mechanically or by melting. Under suitable conditions of solidification, the adhering solid film will be purer than the feed (see Chapter 9).

REFERENCES

1. S. C. Mraw and L. A. K. Stavely, *J. Chem. Thermodynam.*, **8**, 1001–1007 (1976).

2. B. T. Walton, *Science*, **212**, 51–53 (1981).

3. A. R. McGhie and G. J. Sloan, *J. Crystal Growth*, **32**, 60–67 (1976).

4. N. B. Chanh, Y. Bouillaud, and P. Lencrerot, *J. Chim. Phys. Physiochem. Biol.*, **67**(6), 1206–1212 (1970).

5. G. W. Schwab and E. Wichers, *J. Res. N.B.S.*, **32**, 253–259 (1944).

6. P. Kapitza, Proc. Roy. Soc., **119A**, 358–386 (1928).

7. J. D. Bernal and D. Crowfoot, *Faraday Soc. Trans.*, **29**, 1031–1049 (1933).

8. D. L. Mackay, *Trends Biochem. Sci.*, **4**(2), N33 (1979).

9. J. D. Bernal and W. A. Wooster, *Ann. Repts. Prog. Chem.*, **28**, 262–321 (1931).

10. B. Chalmers, *Principles of Solidification*, Wiley, New York, 1964, p. 67.

11. M. Volmer and A. Weber, *Z. physik. Chem.*, **119**, 277 (1926).

12. J. H. Holloman and D. Turnbull, *Progr. Met. Phys.*, **4**, 333 (1953).

13. J. E. Burke and D. Turnbull, *Prog. Met. Phys.*, **3**, 220 (1952).

14. W. G. Burgers, *Z. Elektrochem.*, **56**, 318 (1952).

15. Y. V. Mnyukh, N. A. Panilova, N. N. Petropavlov, and N. Suchvatova, *J. Phys. Chem. Solids*, **36**, 127 (1975).

16. Y. V. Mnyukh, *J. Crystal Growth*, **32**, 371–377 (1976).

17. D. Turnbull, *Thermodynamics in Physical Metallurgy*, American Society of Metals, Metals Park, Ohio, 1950, p. 282.

18. W. R. Wilcox and C. E. Chang, *J. Crystal Growth*, **21**, 135–140 (1974).

19. J. P. Riquet and F. Durand, *J. Crystal Growth*, **33**, 303–310 (1976).

20. J. A. De Leeuw den Bouter and P. M. Heertjes, *J. Crystal Growth*, **5**(1), 19–25 (1969).

21. D. Holmes, *Rev. Sci. Instr.*, **50**(5), 662–663 (1979).

22. A. Maries and P. S. Rogers, *J. Mater. Sci.*, **13**, 2119–2130 (1978).

23. G. A. Alfintsev, *J. Crystal Growth*, **52**, 76–81 (1981).

24. K. A. Jackson, D. R. Uhlman, and J. D. Hunt, *J. Crystal Growth*, **1**, 1–36 (1967).

25. J. G. Aston, "Plastic Crystals," in D. Fox, M. M. Labes, and A. Weissberger, Eds., *Physics and Chemistry of the Organic Solid State*, Vol. 1, Interscience, New York, 1963.

26. K. A. Jackson, *J. Educ. Modules Mater. Sci. Eng.*, **2**(3), 609–648 (1980).

27. K. A. Jackson, "A Tutorial Approach," in W. Bardsley, D. T. J. Hurle, and J. B. Mullin, Eds., *Crystal Growth*, North-Holland, Amsterdam, 1979.

28. V. N. Vigdorovich, A. E. Vol'pan, and G. M. Kurdyamov, *Directional Crystallization and Physicochemical Analysis*, Khimiya, Moscow, 1976.

29. E. L. Anderson, *Chem. Ind.*, 131–136 (1975).

30. J. C. Brice and P. A. C. Whiffin, *Solid State Electron.*, **7**, 183 (1964).

31. J. L. Dewey, U.S. Pat. 3,163,895, January 5, 1965.

32. H. Schildknecht, *Zone Melting*, Academic, New York, 1966, pp. 113–117.

33. J. C. Maire and M. A. Delmas, *Rec. Trav. Chim. Pays Bas*, **85**, 268–274 (1966).

34. A. R. McGhie, P. J. Rennolds, and G. J. Sloan, *Anal. Chem.*, **52**, 1738–1742 (1980).

35. G. H. Moates and J. K. Kennedy, "Continuous Zone Melting," in M. Zief and W. R. Wilcox, Eds., *Fractional Solidification*, Vol. 1, Dekker, New York, 1967.

36. B. Chalmers, *Principles of Solidification*, Wiley, New York, 1964, p. 37.

Chapter **II**

PHASE DIAGRAMS

1 Theory
    1.1 Unary Phase Diagrams
    1.2 Binary Phase Diagrams
         1.2.1 Components Immiscible in the Solid State
         1.2.2 Components Miscible in the Solid State
    1.3 Ternary and Higher Phase Diagrams
         1.3.1 Three Binary Eutectics Forming a Ternary Eutectic
         1.3.2 Three Binary Solid Solutions
         1.3.3 Two Binary Eutectics and a Eutectic System in Which a Congruently Melting Compound is Formed
         1.3.4 One System of Roozeboom Type I and Two Systems of Roozeboom Type V

2 Determination of Phase Diagrams
    2.1 Thermal Microscopy
         2.1.1 Preliminary Examination; Contact Preparation
         2.1.2 Establishment of Liquidus and Solidus Curves
    2.2 Thermal Analysis
         2.2.1 Differential Thermal Analysis
         2.2.2 Differential Scanning Calorimetry
         2.2.3 Ideal Cooling Curves
         2.2.4 Practical Cooling and Heating Curves
         2.2.5 Comparison of Heating- and Cooling-Curve Methods
         2.2.6 Thaw-Point Method
    2.3 Zone Melting
    2.4 Growth of Single Crystals
         2.4.1 Seeded Melt
         2.4.2 Directional Crystallization
    2.5 Miscellaneous Methods

References

In 1876, J. W. Gibbs derived a deceptively simple expression relating the number of components, the number of phases, and the number of independent intensive variables that must be specified to describe completely the state of a system in equilibrium; the expression is known as the Phase Rule. It is appropriate that a statement so fundamental and so broadly useful should have come from so polymathic a scientist. In fact, the historic course of the application of the Phase Rule reflects the major trends of science and technology in the last century. Until World War II, solid/liquid phase equilibra were of interest mainly to metallurgists and liquid/vapor equilibria were explored by chemical engineers. In the last 30 years, with the rise of solid-state physics and device-oriented electrical engineering, concern about phase equilibria has broadened to engage a much wider gamut of scientific and engineering specialties. The emergence of materials science as a discrete discipline is a manifestation of the growing need to understand chemical systems and their relationships to function. In turn, the materials scientist uses a knowledge of phase equilibria in synthesis, purification, and growth of crystals.

Phase diagrams, which are the graphical expression of the Phase Rule, have a central importance in the context of this book. If a phase diagram is known, then the separability of its components by a crystallization process can be predicted, assuming that kinetic and stability factors permit. More often, the phase diagram is known inadequately or not at all; in this case crystallization can provide insights into the nature of the phase diagram. Our treatment will stress this reciprocal relationship between phase diagrams and crystal-growth experiments. The body of knowledge afforded by crystallization studies related to the Phase Rule has scientific meaning which goes beyond questions of separation and purification. The results shed light on molecular sizes, configurations, and interactions and are important tools of thermodynamics.

The Phase Rule is stated as follows

$$P + F = C + 2$$

where P is the number of phases, C is the number of components, and F is the number of degrees of freedom. C refers to independently specified components; if a system contains three chemical species that are stoichiometrically related, then only two components need be specified to define the system. This restriction is discussed in detail by Campbell and Smith (1). The phases must be homogeneous and physically distinct. The number of degrees of freedom is the smallest number of independent, intensive variables needed to specify fully the state of the system. These may include pressure, temperature, concentration, electrical, magnetic, or gravitational fields, interphase energies, and so on. Normally only temperature, pressure, and concentration are considered to have an influence on phase relationships and the above equation is based on this assumption. If other variables become important, then the constant of this equation must be increased by one for each such variable. As the number of phases increases, the number of degrees of freedom decreases; that is, the condition of the system becomes more tightly defined. The derivation of the Phase Rule and discussion of its application are given in many textbooks of physical chemistry (1).

1   THEORY

1.1   Unary Phase Diagrams

A unary phase diagram shows the phase relationships of a single substance; since concentration is not a variable, the phase diagram consists of a temperature/pressure plot (Figure 2.1). The Phase Rule specifies that, since there is only one component, $P + F = 3$; hence if all of the sample is present in a single phase, the system is bivariant and can exist under a variety of pressure and temperature conditions in any of the solid, liquid, or vapor "fields" of Figure 2.1. Likewise, if two phases are present, the system is monovariant and only one variable can change along any interfield line segment. If one variable is chosen arbitrarily, the other is fixed. When three phases coexist, the system becomes invariant, as at a point of intersection of any two interfield line segments: B, C, and D. Under this circumstance, neither pressure nor temperature can be changed without the disappearance of a phase. Point C, at which solid, liquid, and vapor are in equilibrium, is known as the triple point.

Although the unary system contains only one component, the phase transitions that it embodies are the basis for separations in multicomponent systems. These are collected in Table 2.1. In multicomponent systems, partitioning of species generally accompanies phase transitions.

1.2   Binary Phase Diagrams

Purification normally concerns multicomponent systems. However, the behavior of such systems can best be described in terms of idealized binary systems. This approximation is valid, since in the absence of specific chemical interactions among solutes, each behaves independently of the others in its phase relationships with the major component. Because we are primarily concerned with crystallization of melts, we will not consider liquid/vapor

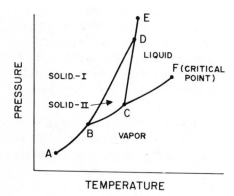

Fig. 2.1 Phase diagram of a one-component system in which there are two solid phases.

Table 2.1 Phase Transitions and Related Separation Processes

Transition	Line Segment in Fig. 2.1	Separation Process
Liquid → solid I	DE	Melt crystallization
Liquid → solid II	CD	Melt crystallization
Solid I → vapor	AB	Sublimation
Solid II → vapor	BC	Sublimation
Liquid → vapor	CF	Evaporation
Solid I → solid II	BD	Solid–solid transformation

equilibria, or solid/solid transitions except where the latter occur so close to the melting range as to influence solute segregation.

In a binary system, the Phase Rule indicates 3 degrees of freedom for a single phase, two for two phases, and one for three phases. The 3 degrees of freedom normally considered are pressure (p), temperature (t), and concentration (c). A complete phase diagram requires three-dimensional representation (Figure 2.2).

In general, solids are denser than the liquids from which they are formed. Thus dP/dT is positive and liquidus curves lie at higher temperatures with increasing pressure. Hence $T'_{M,B}$ is at a higher temperature than $T_{M,B}$, and boiling point $T'_{B,A}$ is likewise higher than $T_{B,A}$ (Figure 2.2). It is thus possible that a compound melting incongruently (see below) at ordinary pressures may melt

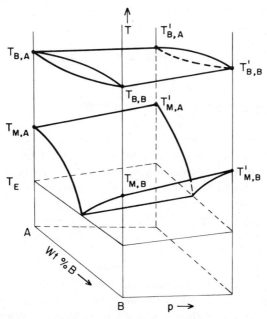

Fig. 2.2 Phase diagram of a binary system in which pressure and temperature are both variable. T_E is the eutectic temperature, $T_{M,A}$ and $T_{B,A}$ are the melting and boiling points, respectively, of component A and $T_{M,B}$, and $T_{B,B}$ are the corresponding values for component B.

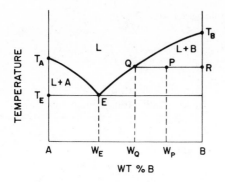

Fig. 2.3 Phase diagram of a binary system forming a simple eutectic.

stably at high pressure. In this case, a compound might be purified by crystallization under elevated pressure, while crystallization at ordinary pressure would cause decomposition.

Since almost all purification by melt crystallization is carried out at ambient pressure, the phase diagrams discussed below refer to constant, atmospheric pressure. In this case the Phase Rule reads

$$P + F = C + 1$$

since only temperature is variable. Figure 2.3 shows an isobaric section of a three-dimensional phase diagram, without the vapor/liquid equilibrium.

Given these assumptions, all binary systems may be described graphically in terms of a small number of characteristic relationships; these will be described in the following sections.

1.2.1 Components Immiscible in the Solid State

SIMPLE EUTECTIC

Consider mixtures of substances A and B, whose melting points are T_A and T_B, respectively. Partial solidification of liquid mixtures of these materials may give one or another of the pure crystalline substances, in contact with the molten mixture, at temperatures below T_A and T_B. Figure 2.3 shows graphically the relationship between composition and solidification temperature for all A/B mixtures. It is evident that there are two branches to the melting curve, intersecting at a point E which defines the lowest temperature at which a molten A/B mixture can exist. This point is called the eutectic (easily melted, from εὐ well and τηκεῖν to melt). Under our assumption of constant pressure, the eutectic is an invariant point; that is, so long as two phases are present, the temperature must remain constant. The melting behavior of the eutectic composition W_E mimics that of a pure compound, solidifying completely at T_E.

Above $T_A E T_B$, only liquid exists, and this curve is called the *liquidus*; it specifies the lowest temperature at which liquid of a given composition can exist as a single phase. Below $T_A E$ and above T_E, pure solid A coexists with liquid L of variable composition. Likewise, below T_BE and above T_E, pure solid B coexists with liquid L of variable composition. Thus the *liquidus* defines the temperature at which liquid of a given composition begins to solidify. Below T_E only solid exists, consisting of mixtures of crystals of pure A and pure B.

Temperature–composition diagrams give information regarding the relative amounts of solid and liquid phases in equilibrium. Consider a mixture of composition W_P, at some constant temperature between T_B and T_E, such as point P in Figure 2.3. At equilibrium, this system consists of g_B g of solid B in the presence of g_L g of liquid. The relative amounts of solid and liquid can be obtained by drawing a horizontal line through P, intersecting the liquidus at Q and the temperature axis at R; this horizontal is known as a "tie-line." The amount of A in the entire sample is given by

$$\frac{100 - W_P}{100} (g_L + g_B)\, \text{g}$$

while the amount of A in the liquid is given by

$$\frac{100 - W_Q}{100} (g_L)$$

Since A is present only in the liquid, these two quantities are equal. Hence

$$g_L(W_P - W_Q) = g_B(100 - W_P) \tag{2.1}$$

since $(W_P - W_Q) = PQ$ and $(100 - W_P) = PR$,

$$g_L \times P_Q = g_B \times PR \tag{2.2}$$

or

$$\frac{g_L}{g_B} = \frac{PR}{PQ} \tag{2.3}$$

Thus the weights of solid and liquid are inversely proportional to the distances between their respective compositions and the composition of the total system. If g_B g of solid and g_L g of liquid were placed at the ends of a lever, its fulcrum would lie at point P, the composition of the mixture. This geometric analogy has led to designation of Equation 2.3 as the *lever rule*. Note that the lever analogy applies only when composition is expressed as weight percent.

The actual shapes of curves T_AET_B are normally derived experimentally (see Section 2). If A and B form perfect solutions, the shape of the liquidus may be calculated from simple thermodynamics. (A perfect solution is one whose components obey the ideal-solution equation over the entire range of compositions.) If it is assumed that solid and liquid have equal heat capacities, then the temperature/composition curve is given by

$$\ln X = -\frac{\Delta H_{f,A}}{R(1/T - 1/T_A)} \qquad (2.4)$$

where X is the mole fraction of component A in the mixture whose liquidus temperature is T. $\Delta H_{f,A}$ is the molar latent heat of fusion of component A at its melting point T_A, and R is the gas constant.

Similarly the temperature/composition curve for melts in equilibrium with B is given by

$$\ln(1 - X) = -\frac{\Delta H_{f,B}}{R(1/T - 1/T_B)} \qquad (2.5)$$

It follows that if the heats of fusion and melting temperatures of both components are known, then the eutectic temperature and composition may be calculated. In practice however, the eutectic temperature is measured and the associated composition is calculated, using the known heat of fusion of one component.

Eutectics may crystallize in one of several morphologies. Principal among these are lamellar, rodlike, and discontinuous. In these the ordering is short range, with phase separations of the order of $1-20\,\mu m$ (2). This contrasts with the behavior of peritectic systems (see below).

Note that in the phase diagram under discussion (Figure 2.3), cooling a mixture whose composition lies to the left of the eutectic will give a solid consisting of pure A; similarly, liquid mixtures whose compositions lie to the right of the eutectic give pure B. Conventionally, mixtures are named according to whether their compositions lie to the left or right of the eutectic; the former are called hypoeutectic and the latter are called hypereutectic. Thus in terms of the distribution coefficient k introduced earlier (Chapter 1), these systems would have $k = 0$. In fact, equilibrium values of k are not normally measured. Instead, the effective distribution coefficient, k_{eff} is measured, which is greater than zero. In general it is not known whether the deviation from zero derives totally from experimental departure from ideality or whether it reflects a small but finite equilibrium solid solubility.

It has been argued that this diagram is an oversimplification in that it cannot depict a real chemical system. The essential point of the criticism is that total exclusion is not possible and that in every system, some of the minor component must be included as the major component solidifies. Thus for example, in the copper/germanium system the maximum solubility of copper in germanium has been found to be $< 10^{-7}$ atom fraction (3). For most purposes, exclusion of a foreign component to the extent of 1 part in 10^7 may be said to be total; however, there are systems in which contamination at this level may not only be detected, but may play a decisive experimental role.

Slight mutual solubility of two molecular species may easily be missed and total immiscibility may be attributed to a system erroneously. The polarization of absorption or emission spectra is a sensitive tool for detecting such slight solubility, especially in organic systems, since dissolved, oriented guest molecules in a host lattice will display characteristically polarized spectra (4).

EUTECTIC WITH POLYMORPHISM OF ONE COMPONENT

If one component of a binary system undergoes a polymorphic transition, the liquidus shows a characteristic change of slope which may be confused with an incongruently melting compound. In Figure 2.4 the dashed curves show extensions of the liquidus lines of two polymorphs. Cooling a liquid of composition C results in crystallization of pure B in form I, until the liquid attains the composition and temperature defined by point P. Here the total crystalline mass of B undergoes isothermal transition to form II, with evolution of the latent heat of transition. Subsequent crystallization upon further cooling yields crystals of B in form II until the eutectic E is attained. Depending on kinetic factors, form I of solid B may not transform at P, but may instead supercool along the dashed extension of $T_B P$.

The situation described here differs from a peritectic reaction (see Section 2.1) in that the latter necessarily involves a change in the composition of the solid phase during the isothermal transition. Moreover the composition of the solid formed during crystallization of a peritectic-forming system varies continuously with melt composition.

EUTECTIC WITH A CONGRUENTLY MELTING COMPOUND

Occasionally the two components, A and B, of a binary system interact to form a new compound of composition $A_x B_y$ where x and y are usually small integers. The resulting system may be considered as the combination of two simple-eutectic systems, with a maximum in freezing point. The freezing point of compound $A_x B_y$ may be higher or lower than those of A and B (see Figure 2.5).

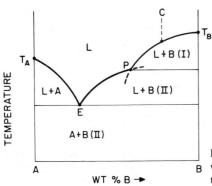

Fig. 2.4 Phase diagram of a eutectic system in which component B undergoes a polymorphic transition.

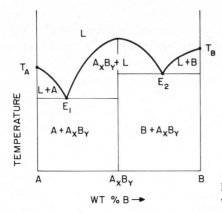

Fig. 2.5 Phase diagram of a eutectic system in which a congruently melting compound is formed.

EUTECTIC WITH AN INCONGRUENTLY MELTING COMPOUND

The pure components of a binary mixture may react to form a compound that is not thermally stable at its melting point. At some temperature lower than the hypothetical melting point, it dissociates to the pure components. In Figure 2.6 the dotted portion of the liquidus shows the theoretical maximum corresponding to the unstable compound. The dissociation is often referred to as a reaction and the temperature at which it occurs is designated T_R.

If solid compound A_xB_y is heated, a reaction ensues at T_R whereby liquid of composition R forms along with solid A:

$$A_wB_y \xrightarrow{\Delta} A(s) + R(l)$$

Further heating produces liquid whose composition follows the curve RT_A. A mixture whose composition lies between R and A_xB_y starts to melt at T_E, giving liquid of eutectic composition; the temperature of the system remains constant until all the eutectic has melted. The remaining solid is A_xB_y. Upon heating, the

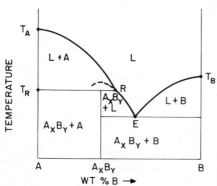

Fig. 2.6 Phase diagram of a eutectic system in which an incongruently melting compound is formed.

liquid composition changes, following liquidus ER, while the solid remains at composition A_xB_y. The fraction of liquid in the system is determined by the lever rule. When temperature T_R is reached, an isothermal transition occurs, yielding solid A and liquid R. Finally, further heating produces melting along RT_A until the liquid attains the same composition as the initial solid. Note that in this system a single liquid is in equilibrium with two solid phases at T_R. Further, at point R, two liquidus lines with slopes of the same sign intersect. These two conditions define what is known as a peritectic. Hence, this system is a special case of the peritectic reaction; see Section 2.1.

IMMISCIBLE LIQUIDS

It is common knowledge that liquids of radically differing chemical nature may be partially immiscible at certain temperatures. If the temperature range of immiscibility is completely above the liquidus, the phase separation has no effect on the crystallization behavior of the system. If, instead, the temperature range of immiscibility overlaps or intersects the liquidus of the T–C diagram, then a monotectic field is said to exist. In Figure 2.7, liquids whose compositions lie in region II yield a two-phase liquid system when cooled. Further cooling to temperature T_M brings about an equilibrium between solid A and the two liquids M and N. The reaction $L_M \rightarrow A + L$ takes place isothermally. During this process there are two liquid phases in equilibrium with one solid phase and the system is isobarically invariant. Still further cooling of liquid N yields more A and ultimately eutectic solid E.

Cooling a liquid whose composition lies in region I results in formation of solid A and a liquid progressively richer in B, until composition M is attained. At this temperature, as discussed above, $L_M \rightarrow A + L_N$, isothermally. Further cooling of L_N results in formation of additional pure A until the liquid reaches eutectic composition, whereupon the eutectic mixture of A and B crystallizes.

Chadwick has discussed the directional crystallization of a liquid in region I. If a tube of liquid is lowered into a cold zone, then solid A will form. When the liquid composition reaches M, liquid N will form. The nucleation of liquid N is governed by the relative interfacial tensions of the three phases: solid A, liquid

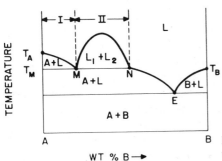

Fig. 2.7 Phase diagram of a binary system in which two immiscible liquids are formed.

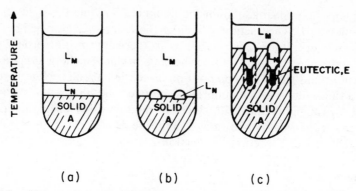

Fig. 2.8 Schematic representation of solidification of binary systems in which two immiscible liquids form: (a) L_N is denser than L_M and covers solid A; (b) L_N forms droplets on solid A; (c) droplets of L_N extend to form cylinders within solid A; the cylinders solidify in the temperature gradient.

M, and liquid N. Liquid N may:

1. Form discrete droplets within liquid M
2. Form sessile droplets upon the surface of solid A
3. Spread over the surface of solid A

If L_N is denser than L_M, then in cases 1 and 3, it will cover the surface of solid A and effectively isolate it from L_M. Thus continued growth of solid A is impossible, and intermittent freezing of A and L_M (above the $L_M L_N$ interface) takes place (Figure 2.8a). Case 2 yields a very different result. Sessile drops of L_N on A will remain liquid as the A interface advances around them (Figure 2.8b). These cylindrical inclusions of L_N, upon cooling, yield additional solid A (shown within the dashed boundary of the cylinder) and a core of liquid whose composition ultimately becomes that of eutectic E (Figure 2.8c).

1.2.2 Components Miscible in the Solid State

If the two components of a binary mixture are very similar in shape and size, they may form solid solutions. (In the German literature, this condition is known as "Mischkristallbildung" or "formation of mixed crystals" and the term "mixed crystal" has carried over into English.) As a general matter, solid solution may involve the inclusion of guest molecules either in interstitial voids of a host lattice, or by direct replacement of host molecules. In the case of binary organic systems, only the latter (substitutional) mode is known.

In metallic alloys, it has been found that there are few, if any, totally random solid solutions. While atoms of A and B may be homogeneously distributed on the macroscopic level, they are not likely to be homogeneously distributed atomically. Instead, there is a continuum of possibilities, from *short-range ordering*, in which unlike atoms tend to associate, through *clustering*, in which like atoms associate, to *long-range ordering* (compound formation). In metals and ionic compounds, it has long been known that solid solubility drops abruptly when the atomic radii of the partners in a binary mixture differ by more than 15%. If, on the other hand, the sizes are close, then a continuous series of solid solutions can form and the lattice constants of mixtures will vary linearly with composition. While these size criteria are important, relative electronegativities are also involved in determining the degree of solid solubility.

Similarity of size and shape are necessary for formation of solid solutions because only if these conditions are satisfied will the introduction of foreign molecules not lead to major changes in the number of contacts between adjacent molecules. Naturally, such increases in intermolecular contact will lead to a rise in the free energy of the solid-solution crystal as compared with the pure host crystal. In this connection, it should be noted that host molecules may be replaced more easily by smaller guest molecules than by larger ones (5, 6, 7, 8, 9).

In order for mutual solid solubility to be complete (i.e., continuous throughout the full range of composition), yet another condition must be satisfied: the structures of the components must have identical space group, the same number of molecules per unit cell, and similar molecular packing. The correctness of this assertion may be grasped intuitively from the fact that symmetry is a discontinuous property, and there cannot be a continuous change of structure between systems of differing symmetry. If two substances of differing symmetry mix to form solid solutions, then there must be a discontinuous change in crystal structure at some concentration. A more extensive discussion of the criteria for solid-solution formation has been given by Kitaigorodskii (5).

It is important to note that many published phase diagrams of organic systems violate Kitaigorodskii's rules and claim continuous solid solubility in systems in which it cannot exist, such as anthracene/carbazole (10) and anthracene/acenaphthene (11).

A number of apparent violations of this principle can be rationalized by consideration of the details of packing. Cases are known in which polar molecules form continuous solid solutions in nonpolar matrices and vice versa (10). Naphthalene/2-naphthol (12) and anthrone/anthraquinone (13) are examples of this behavior. The polar molecules (2-naphthol and anthrone) are statistically distributed to give an effectively centrosymmetrical packing. Intermediate cases are also known, in which mutual solid solubility is extensive, but incomplete. The systems *p*-dibromobenzene/*p*-nitrochlorobenzene and acridine/anthracene are of this type.

A study of the naphthalene/coumarin system showed that the former does not enter the lattice of the latter; this result is consistent with the noncentrosymmetric structure of coumarin. Additionally it is argued that the coumarin packing is stabilized by electrostatic dipole interactions, which prevent entry of the nonpolar naphthalene molecules (14).

The ultimate case of solid solubility is offered by crystals consisting of mixed isotopic species. In such mixtures there can be little question about crystallographic equivalency of the two chemical species comprising the mixture. Even here, however, it is possible to learn a great deal about the manner in which guest molecules are incorporated into host crystals. Imaginative application of optical and magnetic resonance spectroscopy has made it possible to identify isolated molecules, pairs, and higher clusters of guest in host. These assignments are based on the fact that energy migrates through molecular crystals; the mobile quanta ("excitons") act as sensitive probes of the identity of the species which ultimately trap or reemit the energy. Wolf and Port (15) have reviewed the results of such experiments. In naphthalene/perdeuteronaphthalene (N-h_8/d_8), only isolated guest molecules are observed at concentrations of 0.1% and lower. At higher concentrations (up to 20%), clusters of two, three, or more guest molecules predominate and grow in number as well as in magnitude, at the expense of "monomers." In fact, it is possible to distinguish host and guest and/or exciton states from each other over the entire range of concentrations.

The anthracene/perdeuteroanthracene system behaves differently, in that single molecules, pairs, and clusters partially lose their spectroscopic identity and are amalgamated in a common host–guest exciton band. A study of fluorescence spectra of this amalgamated exciton band over the whole concentration range has shown that no isolated clusters are observable.

The spectroscopic techniques applied to these systems may be more broadly applied to resolve questions about the mode of incorporation of guest molecules in host crystals. Doerner et al. (16) have carried out a study of the orientation of anthracene molecules in naphthalene and of tetracene molecules in anthracene single crystals. Their measurements were based on the fact that the angular dependence of the electron-spin-resonance spectra of molecules in triplet states yields information regarding the orientation of the principal molecular axes. Hence if the guest molecule whose orientation is in question can be uniquely excited to the triplet state, a study of the variation of the orientation of the crystal in the fixed magnetic field of the laboratory system can give unambiguous information regarding the orientation of guest molecules in the crystal. The crystals studied in this work were very dilute, about 10^{-5} moles of guest per mole of host. In both cases it was found that deviations from the substitutional position were less than 6°. These results are inconsistent with earlier measurements of Ostertag and Wolf (17), who found that anthracene molecules were misoriented by up to 30° from the naphthalene host molecules. The discrepancy may arise from the fact that in the earlier work, measurements were made only with respect to one axis, while in the later paper, rotation about all three crystal axes was described. Moreover it should be noted that the work reported by Ostertag and Wolf was carried out on crystals containing about 10^{-4} moles of guest per mole of host and it may be that the solubility limit was exceeded in these crystals and that the exsolved phase was misoriented. The critical point here is not so much the accuracy of a particular result, but that spectroscopic means, and in particular those relying on spin-resonance techniques, can provide detailed information on molecular orientation in dilute solid materials. Calculations of phase diagrams have been made difficult in the past by the mathematical problems encountered in estimating the free energies of equilibrium phases and by the need to know intermolecular potentials in order to carry out numerical calculations.

Kitaigorodskii and Yakushevich (18) have described an approach to the theoretical estimation of mutual solid solubility in binary systems. Their approach involved three assumptions:

1. Only nearest-neighbor interactions are considered.
2. At low concentrations, each impurity site is surrounded by matrix molecules only.
3. Electronic overlap is neglected and system energy is assumed equal to the sum of the binary interaction energies of the molecules.

Application of this method to the system biphenyl/1,1'-bipyridyl resulted in calculated solid-solubility limits ranging from zero at 0 K to about 10% biphenyl in 1,1'-bipyridyl and about 15% 1,1'-bipyridyl in biphenyl at 100 K. Experimentally, these values are not attained even at the melting temperatures of the respective solids; in fact, the experimental solid solubilities were 5 and 8% respectively. The discrepancy is attributed to the fact that vibrational contributions to the system's free energy were not taken into account in the model. Although the quantitative results were inaccurate, the model correctly predicted the asymmetry in solubility, which arises from a contribution to the entropy of the α-phase (1,1'-bipyridyl in biphenyl) by the statistical disorder of the guest molecules.

Systems showing total or partial miscibility in the solid state were classified in five categories by Roozeboom (19). His designations have been followed nearly universally, and we will do so here.

CONTINUOUS SERIES OF SOLID SOLUTIONS; ROOZEBOOM TYPE I

In this case, the pure compounds define the ends of the liquidus and solidus curves (Figure 2.9). The solidus is the curve that defines the highest temperature at which solid can exist with no liquid present. The pure components are connected by a single field of solid solution; cooling liquid C_l until solidification

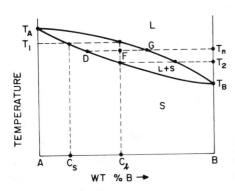

Fig. 2.9 Phase diagram of a binary system in which a continuous series of solid solutions is formed.

begins (at T_1) yields a solid whose composition is given by an isothermal (horizontal) tie line. Separation of solid C_s leaves a liquid enriched with respect to B, and further crystallization takes place at progressively lower temperatures until the last trace of liquid freezes at T_2, yielding a solid of composition C_l. Solid formed in this way will show a gradient of compositions from C_s to C_l.

If solidification occurs extremely slowly, the solid can attain compositional equilibrium by diffusion. In this case, the solid will show no concentration gradient. During crystallization, the relative amounts of solid and melt coexisting at any temperature are given by the lever rule;

$$\frac{\text{weight fraction solid}}{\text{weight fraction liquid}} = \frac{FG}{DF}$$

at some temperature T_n. At T_2 the entire solid has uniform composition C_l.

In practice, uniform concentration in the solid is rarely achieved, especially in organic crystals. In this case then, the solid will have an average concentration richer in A than is shown by the solidus curve. The lever rule, as applied to this more realistic situation, has been discussed by Rhines (20).

From the discussion above, it will be clear that solidification of a Roozeboom type I system can lead to meaningful separation. This is so because diffusivities in solids are typically about 10^5 times smaller than in liquids; hence solidification can be carried out slowly enough to approximate equilibrium in the liquid, but fast enough to avoid diffusive homogenization in the solid. Certain classes of crystals do not show this behavior; in plastic crystals and in liquid crystals, solid-state diffusivities are only 10^2 to 10^3 times less than liquid diffusivities (21, 22). In these materials, little or no concentration gradient will result from directional crystallization at very low speed.

If components A and B yield perfect solutions and $X_{A,s}$ and $X_{A,l}$ are the mole fractions of A in the solid and liquid phases, respectively, then for equilibrium at temperature T the chemical potentials (μ) of the components are equal in solid and liquid phases

$$\mu_{A,s} = \mu_{A,l} \tag{2.6}$$

and

$$\mu_{B,s} = \mu_{B,l} \tag{2.7}$$

Hence,

$$\mu_{A,s}^* + RT \ln X_{A,l} = \mu_{A,l}^* + RT \ln X_{A,l} \tag{2.8}$$

where μ_A^* is the molar free enthalpy of component A. Rearrangement gives

$$\ln \frac{X_{A,l}}{X_{A,s}} = \mu_{A,l}^* + \mu_{A,s}^* \tag{2.9}$$

Differentiation with respect to temperature and integration between limits gives

$$\ln \frac{X_l}{X_s} = -\frac{\Delta H_{f,A}}{R}\left(\frac{1}{T} - \frac{1}{T_A}\right) \tag{2.10}$$

Similarly

$$\ln \frac{1-X_l}{1-X_s} = -\frac{\Delta H_{f,B}}{R}\left(\frac{1}{T} - \frac{1}{T_B}\right) \tag{2.11}$$

If the heats of fusion $\Delta H_{f,A}$ and $\Delta H_{f,B}$ are known, then the solidus and liquidus curves can be constructed by solving Equations 2.10 and 2.11 simultaneously at selected intervals between T_A and T_B. Accurate values of ΔH_f are now readily available through differential scanning calorimetry. These data make it possible to calculate the theoretical shapes of solidus and liquidus for type I systems, assuming that perfect solutions are formed.

CONTINUOUS SERIES OF SOLID SOLUTIONS WITH MELTING-POINT MAXIMUM; ROOZEBOOM TYPE II

Type I solid solutions discussed above occur when the pure components are noninteracting (i.e., form ideal solutions). If an interaction exists, then in principle a maximum or minimum may ensue in the melting curve. The case of a melting-point maximum is designated Roozeboom type II, and only a single example of such behavior appears to have been reported. This is the case of d-carvoxime/l-carvoxime (23). So unusual is this situation that the accuracy of the original measurements has been challenged. However, the original work has been verified and explained by Oonk et al. (24). As may be seen from Figure 2.10, addition of either component to the other leads to an increase in melting point.

CONTINUOUS SERIES OF SOLID SOLUTIONS WITH MELTING-POINT MINIMUM; ROOZEBOOM TYPE III

If the two components of a binary system interact, in general the interaction is greater in the solid than in the liquid while the coordination number remains approximately unchanged. Hence the interaction parameter is greater for the

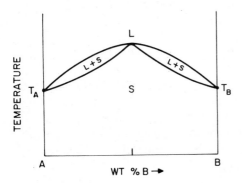

Fig. 2.10 Phase diagram of a binary system in which a continuous series of solid solutions shows a maximum melting point.

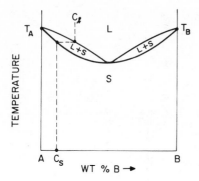

Fig. 2.11 Phase diagram of a binary system in which a continuous series of solid solutions shows a minimum melting point.

solid and both solidus and liquidus show minima at the point of contact, which is sometimes called the azeotropic point. This is an unfortunate use of the term because it refers properly only to boiling (Greek "ζειν," to seethe).

All compositions except for the pure components and the minimum-melting mixture show a range of melting or solidification. Solidification of composition C_l (Figure 2.11) yields solid of composition C_s and the composition of the remaining liquid moves toward the minimum. Thus this system mimics a solid-solution system with a eutectic (see below). Systems of this type are distinguished from eutectics by the existence of a *single* solid phase over all compositions while eutectic crystallization results in two solid phases. In the minimum-type solid solution, solidus and liquidus are continuous, in tangent contact at one point. Experimentally, the two cases can be differentiated by examining the solidification of the lowest-melting material in a contact preparation (see Section 2.1.1). A surer distinction can be made on the basis of X-ray diffraction studies of the respective solids (25). Thermodynamic relationships in binary systems with melting-point minima have been discussed by Kubaschewski and Heymer (26).

At this point is is expedient to digress into a discussion of the fate of a binary solid solution upon cooling. If the solid solution is perfect, then it is thermodynamically stable with respect to the individual components, and the solutions will persist at all temperatures. If the solid solution deviates from ideality, then upon cooling, the continuous solid solution may give way to a regime in which two solid solutions coexist, as illustrated in Figure 2.12. The process is called exsolution and the line separating single-solid-phase $S-I$ from the two-phase region is called the solvus.

PERITECTIC SYSTEM WITH PARTIAL MISCIBILITY IN THE SOLID STATE; ROOZEBOOM TYPE IV

If the solvus happens to intersect the solidus, then two situations may ensue. The first of these arises when the solvus intersects the solidus of a Roozeboom type I system. See Figure 2.13, in which the dashed curve represents the inaccessible part of the solvus. This geometrical circumstance implies that two solid phases are in equilibrium with a single liquid at some characteristic

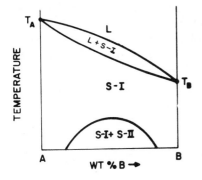

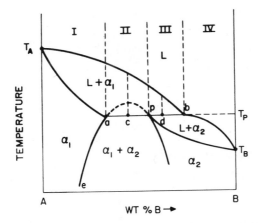

Fig. 2.12 Phase diagram of a binary system in which a single solid solution yields two solid solutions.

Fig. 2.13 Phase diagram of a peritectic system with partial miscibility in the solid state.

temperature, T_p. Above this temperature, the liquid is in equilibrium with solid α_1 and below it, the liquid is in equilibrium with solid α_2.

The solidification behavior of liquid L will depend upon the initial concentration. A liquid in concentration regime I, upon cooling, will yield solid α_1 only. The crystallization temperature will decrease, but the liquid will be exhausted above T_p. The single-phase solid, upon further cooling, may reach the solvus and undergo phase separation.

The details of the phase separation are often of little interest in studies of crystallization of melts. However, the phase separation is not without general interest, and methods of detecting it exist. X-ray techniques are quite generally applicable. For transparent materials, light scattering may be used to detect the appearance of a new phase. Kononenko and Heichler (27) used this method in the KCl/BaCl$_2$ system. They grew crystals of known concentration by the Kyropoulos method (see Section 4.2, Chapter 4); small parallelepipeds were cut from the crystals and heated in a furnace while the light scattering was monitored to obtain an approximate equilibrium temperature for the appearance of a new phase. For exact measurement of the equilibrium temperature, the samples were annealed in the furnace of the light-scattering photometer, then held for 1 h at progressively lower temperatures, in the neighborhood of the equilibrium temperature. Accuracy was increased by making both heating and cooling runs. Figure 2.14 shows a typical plot of scattered intensity against temperature. Under practical experimental conditions, directional crystallization yields an ingot in which the concentration of component B varies along solidus $T_A - a$ of Figure 2.13.

A liquid in regime II yields α_1 until T_p is reached, whereupon the reaction

$$L(b) + \alpha_1(a) \rightarrow \alpha_2(p)$$

takes place, in which $L(b)$ is liquid of composition b, $\alpha_1(a)$ is solid α_1 of composition a, and $\alpha_2(p)$ is solid α_2 of composition p. This process is known as a peritectic reaction and p is the peritectic point; the reaction continues at constant temperature until all the liquid has been consumed.

For a liquid of composition c, the amounts of the two solid phases present at T_p are given by the lever rule as

$$\frac{\alpha_1(a)}{\alpha_2(p)} = \frac{cp}{ac}$$

In regime III the behavior is similar to that of regime II; at T_p, however, the reaction

$$L(b) + \alpha_1(a) \rightarrow \alpha_2(p) + L(b)$$

takes place, in which the amount of $L(b)$ decreases as α_1 transforms to α_2. The relative amounts of liquid and solid α_2 for composition d are given by the lever rule as

$$\frac{L(b)}{\alpha_2(p)} = \frac{dp}{db}$$

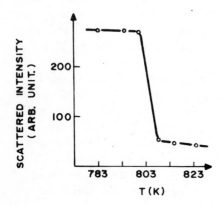

Fig. 2.14 Photometric detection of the formation of a new phase in the solid state.

In regime IV cooling of liquid yields α_2, of continuously changing composition.

With regard to directional crystallization under usual conditions (i.e., no diffusion in the solid and complete mixing in the liquid) the following comments are relevant: liquids in regions I, II, and III will also give ingots showing a gradient of rising concentration of B in A (i.e., α_1 followed by a region of crystallization of solid of peritectic composition). Compositionally, little is likely to change during further cooling of the ingot. That is to say, the $(T_A - a)$ gradient is "frozen in," and the possibility of phase separation upon later cooling depends on duration of annealing at a given temperature, on diffusivity, and nucleation kinetics of the second phase. Thus it cannot be stated *a priori* whether an ingot showing a concentration gradient along $(T_A - a)$, when cooled to a temperature below T_p, will undergo the thermodynamically indicated (local) phase separation to give an ensemble of crystals of compositions given by the α_1 and α_2 solvus lines.

Naturally, directional crystallization of liquid in regions I, II, and III after yielding ingots of compositions varying along $T_A - a$ will then give solid of peritectic composition in amount depending on the initial A/B ratio in the melt. In an ingot configuration, the early crystallized solid is physically inaccessible to the remaining peritectic liquid; hence no peritectic reaction can take place. As was stated above, the peritectic liquid yields peritectic solid isothermally at T_p. This solid may consist of lamellae of separately nucleated α_1 and α_2 crystals. Crystals of both phases coexist in microscopic regions of the ingot. The microstructure of peritectic systems is qualitatively different from that of eutectics in that the former show long-range distribution of the phases.

EUTECTIC SYSTEM WITH PARTIAL MISCIBILITY IN THE SOLID STATE;
ROOZEBOOM TYPE V

In the preceding section, the case is discussed in which the solvus intersects the solidus of a Roozeboom type I system. A similar situation results when the solvus intersects a Roozeboom type III system, resulting in the phase diagram shown in Figure 2.15, where again the inaccessible part of the solvus is dashed.

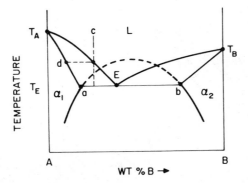

Fig. 2.15 Phase diagram of a binary eutectic system with partial miscibility in the solid state.

Systems such as these are similar to those showing formation of a simple eutectic, but in this case, liquids on either side of eutectic E yield solid solutions α_1 or α_2, rather than pure components A or B.

Cooling a liquid of composition c until it reaches the liquidus yields solid α_1, of composition d. Continued crystallization is accompanied by falling temperature, with continuing change in crystal composition along line da. The liquid composition likewise changes until it reaches point E, the eutectic. At E the eutectic reaction takes place:

$$L(E) \to \alpha_1(a) + \alpha_2(b)$$

Thus at the eutectic temperature, two solid solutions of compositions a and b are in equilibrium with liquid of composition E.

1.3   Ternary and Higher Phase Diagrams

The temperature/composition relationships of three-component systems may be represented by an equilateral prism in which the three sides of a triangular base represent compositions of binary mixtures and temperature is plotted on the altitude of the prism. Thus the ternary diagram is formed by the union of three binary diagrams for systems AB, AC, and BC (see Figure 2.16). It is customary to discuss the behavior of ternary systems in terms of triangular isothermal cross sections of the temperature/composition prism. These triangles, sometimes called Gibbs triangles, Figure 2.17, have three apexes corresponding to pure components; the lengths of the three sides are divided into 100 parts. Lines parallel to the three bases of the triangle, drawn through a point P, give the distances from point P to the sides. For example, horizontal lines (parallel to BC) designate constant percentages of A. Any mixture containing a given

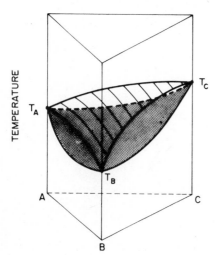

Fig. 2.16 Prism representation of a ternary system consisting of three binary solid solutions. The shaded surface represents the solidus.

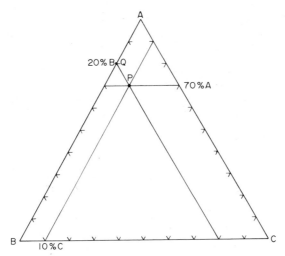

Fig. 2.17 Triangular representation of composition for a ternary system.

percentage of A will be plotted on the isocompositional line for that percentage. Hence a formulation containing all three components is represented by the intersection of three such isocompositional lines. For example, point P represents 70% A, 20% B, and 10% C. Similarly, point Q represents a binary mixture of 80% A and 20% B.

In the following sections we discuss four characteristic ternary relationships. Obviously many other combinations are possible, and some of these have been discussed in detail (20, 28). Our intention is merely to show that pure products can be obtained from such systems. The level of complexity of two-dimensional representations of ternary and higher-order systems rises rapidly. For systems consisting primarily of one component and small amounts of several impurities, it is possible to consider the crystallization behavior simply as that of an ensemble of noninteracting binary systems.

1.3.1   Three Binary Eutectics Forming a Ternary Eutectic

If one limits discussion to a ternary system comprising three binary eutectics and only one liquid phase, then a relatively straightforward crystallization sequence emerges (Figure 2.18). Cooling a liquid of composition X until its temperature intersects the curved liquidus surface results in an equilibration of a single three-component liquid with one solid phase consisting of component A. As solid A separates, the remaining liquid becomes depleted with respect to A and correspondingly enriched in B and C. Changing composition is represented by a line upon the curved surface, moving downward to a lower temperature. At D, this line intersects the line resulting from intersection of two curved liquidus surfaces, for example: $T_A E_1 E E_2$ and $T_C E_2 E E_3$ in Figure 2.18. This line of intersection, $E_2 E$, represents the equilibrium of the single liquid phase with two

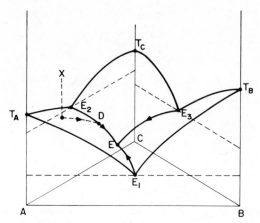

Fig. 2.18 Prism representation of a ternary system consisting of three binary eutectics.

pure solids, A and C in this case. Further cooling results in separation of $A-C$ eutectic until the intersection E of all three liquidus surfaces is reached, whereupon all the remaining liquid crystallizes as the ternary eutectic.

The events just described can be schematized in two dimensions by making an orthogonal projection of the liquidus surface onto the base of the temperature/composition prism. Because the interior of such a representation is of complex curvature, it is difficult to represent in two dimensions. Hence the custom has arisen of describing ternary systems through two-dimensional projections in which temperature is represented by curved isotherms. In this triangle E_1E, E_2E, and E_3E represent the curvilinear intersection of the three liquidus surfaces (Figure 2.19). Cooling a liquid of composition X until its temperature reaches the liquidus surface causes solidification of A, and the composition of the remaining melt follows XD (extension of line AX). When point D is reached, solid C begins to separate as well, and the composition of the melt follows line DE with continued decrease of temperature until the ternary eutectic E is reached. Thereafter, the system is invariant. Both temperature and composition remain unchanged until solidification is complete. In these two-

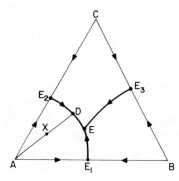

Fig. 2.19 Two-dimensional projection of the ternary system shown in Fig. 2.18. Arrows indicate direction of decreasing temperature.

dimensional isothermal representations, the direction of falling temperature is conventionally represented by arrows.

Directional crystallization of a ternary mixture of this type, with composition X, will produce an ingot showing three regions: (1) pure A, (2) a zone containing crystals of both A and C whose composition follows the line E_2E, and (3) the ternary eutectic.

1.3.2 Three Binary Solid Solutions

Another category of ternary systems is that in which all three binary solid pairs are completely mutually soluble. The three binary liquidus lines AB, AC, BC (Figure 2.20) are connected by a surface that represents the liquidus of all ternary mixtures. Similarly, the three binary solidus lines are connected to form another, lower surface. Cooling ternary liquid X will produce solid when its temperature reaches the liquidus surface at l_1; l_1 lies on an isothermal section *abcd* through the prism. This section intersects the solidus and liquidus surfaces along *ab* and *cd*, respectively. Tie lines in this plane, such as l_1s_1, indicate the concentrations of solid and liquid phases in equilibrium. As with binary solutions, the lever rule applies; a ternary liquid X, when cooled to X_1 will yield ternary solid solution α and ternary liquid, in the ratio $m_\alpha/m_l = X_1l_2/X_1s_2$.

The orientation of a tie line in an isothermal section cannot be calculated or predicted, but must be determined experimentally. In an experiment involving

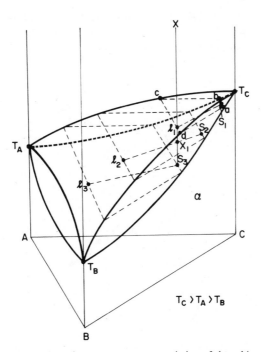

Fig. 2.20 Prism representation of a ternary system consisting of three binary solid solutions.

cooling of a ternary liquid, the changing state of affairs is shown by a sequence of isotherms in which the changing liquid/solid equilibrium is denoted by a series of tie lines (e.g., $l_1 s_1$, $l_2 s_2$, $l_3 s_3$). Equilibrium solidification of liquid X results in formation of solid at point l_1. The solid is of composition indicated by tie line $l_1 s_1$. Continued cooling causes further formation of solid with concomitant change of liquid composition shown at an intermediate stage of crystallization X_1. Liquid l_2 gives solid s_2. The last solid to form (s_3) has the same concentration as the original liquid and the last liquid to freeze has the composition l_3 (see Figure 2.20). The change in spatial orientation of tie lines $l_1 s_1$, $l_2 s_2$, and $l_3 s_3$ is the subject of Konovalov's rule, which states that the solid is always richer than the melt with which it is in equilibrium, in that component which raises the melting point when added to the system (29). Thus in the system under consideration, crystallization of s_1, which is enriched in component C, gives a liquid that is depleted with respect to C. Since $T_A > T_B$, Konovalov's rule tells us that s_2 is richer in A than s_1.

The para dihalogenated benzenes form ternary systems of this kind. For example, p-dichlorobenzene, p-dibromobenzene, and p-chlorobromobenzene have been reported to form a continuous ternary solid-solution system (30). Moreover, X-ray studies of the structure of p-chloroiodobenzene indicate that it is isomorphous with p-dichlorobenzene, p-dibromobenzene, and p-chlorobromobenzene, so that similar phase behavior could be expected in any group of three of these materials.

1.3.3 Two Binary Eutectics and a Eutectic System in Which a Congruently Melting Compound Is Formed

In a ternary system, a compound may be formed in one of the constituent binary systems, or it may arise only within the ternary field. Figure 2.21 shows

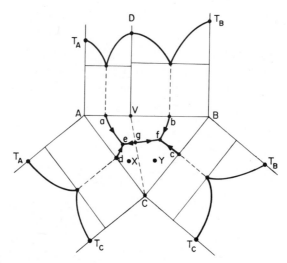

Fig. 2.21 Planar representation of a ternary system consisting of two binary eutectics and a eutectic system in which a congruently melting compound is formed.

the former case; this figure shows another mode of indicating the ternary interactions. The three prism faces have been folded back and the central triangle ABC contains a projection of the intersecting liquidus surfaces. The stoichiometry of the compound separates the ternary field into two areas, AVC and BVC. In the first of these, solids consist of the three crystalline species A, V, and C; in the second area, B, V, and C. In effect, the system consists of two ternaries and V takes part in both. There are four binary eutectics. The lines from these binary eutectics in the ternary field must be in the direction of diminishing temperature, since addition of a third component to any of the four binary eutectics causes a depression of freezing point. The binary eutectic lines converge at e, which is a ternary eutectic and hence a fixed point at which liquid is in equilibrium with all three solid phases, A, C, and V. Likewise, point f is a ternary eutectic of C, B, and V. Because the extension of the maximum D into the ternary field must also show a maximum, line ef must have a "saddle point" g as illustrated.

Directional crystallization of ternary liquid can give two compositional patterns. A liquid of composition X would first give pure C, then a regime of varying proportions of A and C, and finally a ternary mixture of crystals of A, C, and V. Liquid of composition Y would give pure C, variable $(B + C)$, and finally ternary $(B + C + V)$.

1.3.4   One System of Roozeboom Type I and Two Systems of Roozeboom Type V

Any liquid of composition within field abT_B will yield crystals of solid solution β (31), (Figure 2.22). Likewise any liquid of composition within the field abT_cT_A will yield crystals of solid solution α.

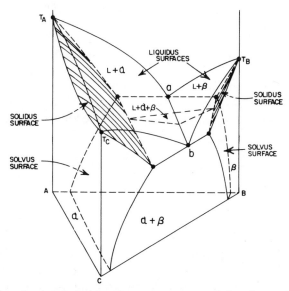

Fig. 2.22 Two-dimensional representation of a ternary system consisting of a complete series of solid solutions $(A-C)$ and two binary eutectic systems each having partial solid solubility $(A-B$ and $B-C)$.

2 DETERMINATION OF PHASE DIAGRAMS

The motivation for experimental studies of temperature/composition diagrams can vary. On the one hand, interest may center on the details of relationships between nearly pure materials and contaminants at low concentrations. This is the case in measurements of purity by thermal analysis. Producers of standard reference materials, fine chemicals, and pharmaceuticals are in this category. In pharmaceutical chemistry, interest is also centered on stabilization of desired polymorphs. Nearly pure substances are also of interest to spectroscopists attempting to relate the energetics of energy transfer to the geometric relationships of host and guest molecules in a crystal lattice. Likewise, those interested in solid-state properties generally find phase relationships in nearly pure materials to be important.

On the other hand, engineering design of crystallization systems requires knowledge of an entire phase diagram; eutectics, peritectics, and compound formation can exert decisive influence on the effectiveness of crystallization processes as industrial techniques. Current trends in industry favor increased use of melt crystallization for energy conservation. Consequently one can expect an increased need to explore the solid/liquid phase relationships of commercially important chemical systems.

This discussion is aimed at relating phase diagrams to purification processes. Hence attention will be focused primarily on establishment of solidus and liquidus curves. Crystallographic states of the solids as they are cooled are of less concern. Thus in general it does not matter that a solid formed from a melt may undergo a phase transformation at some lower temperature; however, if there is a phase transformation near the solidus, it may affect the course of purification in that a volume change or a heat effect associated with such a transformation could result in entrapment of impure melt at the interface.

One may wish to study the phase relationships of a particular form of a component which occurs in multiple polymorphic modifications. In such a case, the various polymorphs may be prepared by crystallization from a solvent or by sublimation.

We will describe four major techniques which may be used for establishing the solidus and liquidus curves of binary systems. These techniques are not all universally applicable, and some have only limited utility. They are:

1. Thermal microscopy
2. Thermal analysis
3. Zone melting
4. Single-crystal growth

The choice of method will depend on a number of factors. These include: availability of instrumentation, amount of material and time available, environmental sensitivity of the components, and temperature range.

Methods 1 and 2 may be carried out with milligram or even microgram quantities, while 3 and 4 normally require gram quantities; however, 1 and 2 require considerably more costly instrumentation than the other methods. It is fairly easy to carry out methods 3 and 4 in evacuated, sealed vessels. While the sample to be used in methods 1 and 2 may also be enclosed in a sealed container, the sample preparation must be carried out in a glovebox if it is necessary to exclude air completely from the sample. Naturally, method 1 could not be applied to a material or system that is photodegraded at or near its melting point.

We must point out that the four methods mentioned above are by no means a complete compendium of techniques applicable to the determination of phase diagrams. Any physical property that changes discontinuously during a liquid/solid phase transition can, in principle, be used to obtain liquidus and solidus curves. Line widths in nuclear magnetic resonance spectra (32, 33, 34), changes in electrical resistivity, mechanical properties, X-ray diffraction, dilatometry (20), and electrochemical methods (20) have all been used. These methods tend not to be generally applicable and will not be discussed in detail. In recent years, powerful methods have emerged for recognizing the appearance of new phases in very small samples. For example, electron diffraction carried out in an electron microscope can detect phase transitions in particles of nanometer size. These and other exotic methods have been reviewed by Rhines (35). Commoner methods are summarized in Table 2.2.

Maximum reliability in the determination of phase diagrams, both with respect to measurement of transition temperatures, and with respect to identifying the type of phase diagram, requires application of multiple techniques. The various methods of study have their advocates but no single method appears to be universally applicable to all problems of phase-diagram determination.

Table 2.2 Summary of Experimental Methods for Determination of Phase Diagrams

Technique	Range, %	Amount Required	Speed	Accuracy	Equipment Cost, $M, 1986
Thermal Microscopy	0–100	mg or less	High	High	5–7
DTA/DSC	0–100	Tens of mg	High	Moderate	20–30
Cooling/ heating curves	0–100	1–10 g	Low	Moderate	0.5–5
Zone refining	0– ~10	0.1–10 g	Low	Moderate	1–5
Temperature- gradient oven	0– ~10	1–10 g	Low	High	2–10
X-ray diffraction	0–100	mg	Low	Moderate	>10
Dilatometry	0–100	1–10 g	Low	Moderate	1–2

Even in cases in which more than one method of study is used, discrepancies and inaccuracies can arise. An important preliminary requirement is that the respective components used in the study must be pure at the outset. If small temperature effects and subtle phase changes are to be detected, then the properties of the pure components ought to be defined as reliably as possible. In modern terms, this means that the purity of each component should exceed 99.9% and the melting point should be determined with an accuracy of at least ± 0.1 K. There are numerous methods available for achieving and measuring these properties.

2.1 Thermal Microscopy

Microscopic observation of fusion and related phenomena, including polymorphism, offers numerous advantages:

1. Only small amounts of material are required.
2. Use of polarizing optics in transmission microscopy makes it easy to detect the appearance and disappearance of crystals on the basis of birefringence of the crystals. (This is true only of substances that form nonisotropic crystals. Substances that crystallize in the cubic system are isotropic and do not show characteristic birefringence.)
3. Microscope hot stages of low mass may be heated and cooled easily at variable rates.
4. With temperature-gradient microscope stages, crystallization kinetics and other crystal-growth properties may be investigated.
5. Sample purity and stability may be estimated easily.

We make no attempt here to describe the total range of usefulness of the microscope in the study of chemical systems. Scores of properties may be measured with the polarizing microscope, and these are described in textbooks of optical crystallography and chemical microscopy (36, 37, 38).

In spite of these merits, microscopic methods have been relatively little used by chemists. This is not to say that there is no background in the area. Numerous workers have made important contributions to the development of thermal microscopy (37) and recent advances in commercially available temperature-controlled stages make the technique more accessible than ever. Figure 2.23 shows a simple microscope stage with a thermometer to measure temperature; in this device, temperature is controlled by a variable voltage supply. Figure 2.24 shows a more sophisticated device in which the stage temperature may be programmed at preselected rates and which allows recording of critical temperatures on a digital display. In this unit, temperature is measured and controlled by a platinum resistance thermometer. While it is true that some instruction and practice in microscopy are desirable before attempting to use microfusion methods, it is also true that these methods are not among the most demanding.

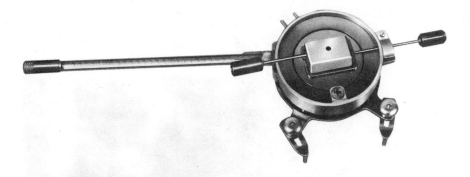

Fig. 2.23 Temperature-controlled microscope stage of the Thomas–Kofler design (photograph courtesy of Arthur H. Thomas Company, Philadelphia).

We will describe straightforward methods for establishing the phase relationships existing between components of binary mixtures, with especial reference to the Roozeboom classification of binary phase diagrams. The systems described in the following sections were selected to illustrate specific features that are commonly found in phase diagrams. Moreover, each system was selected to demonstrate wherever possible, a single characteristic. In practice, an individual phase diagram may contain several of these features.

It must also be pointed out that our attention is focused upon solid/liquid equilibria, since these are most important in the study of purification by melt crystallization. Solid/solid transitions are, of course, quite common and they may influence purification by crystallization. However, their influence will generally be minor, and we will not discuss them here.

2.1.1   Preliminary Examination; Contact Preparations

SIMPLE EUTECTIC

Much can be learned about phase relationships by merely allowing two substances to come into contact on a microscope slide. The first step is to prepare a single slide bearing both materials whose phase relationships are to be studied. This is done by placing a small amount of the higher-melting component (for example, benzoic acid) between a microscope slide and a cover slip. (We will adopt the convention of placing the higher-melting component on the left side of the figure.) Clearer boundaries are obtained when round cover slips are used rather than square ones. The preparation is heated until the benzoic acid melts and spreads to occupy about half the area of the cover slip. The melt is then allowed to cool and crystallize. The lower-melting component (naphthalene in this case) is then applied to the edge of the cover slip as shown in Figure 2.25a; then the preparation is heated until the naphthalene melts and is drawn by capillarity into the gap between the cover slip and the slide, and

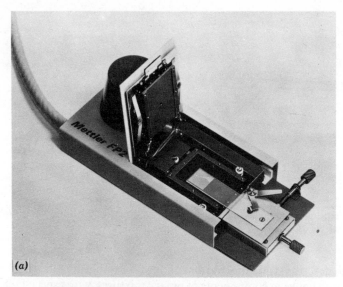

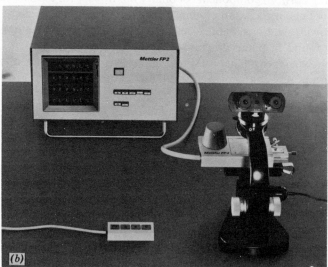

Fig. 2.24 Mettler microscope hot stage. (*a*) Detailed view of hot stage showing microscope slide and *X – Y* stage. (*b*) Overall view showing programming module and microscope assembly (photographs courtesy of Mettler Corporation).

contacts the benzoic acid (Figure 2.25*b*). These steps may be carried out on a small hot plate. The preparation is now heated in the microscope hot stage to melt a portion of the solid at the junction, and form a liquid zone of mixing between the two components (Fig. 2.25*c*). This liquid zone will contain all compositions from pure benzoic acid to pure naphthalene and will consequently contain any eutectic composition as well. If the preparation is cooled, crystals of

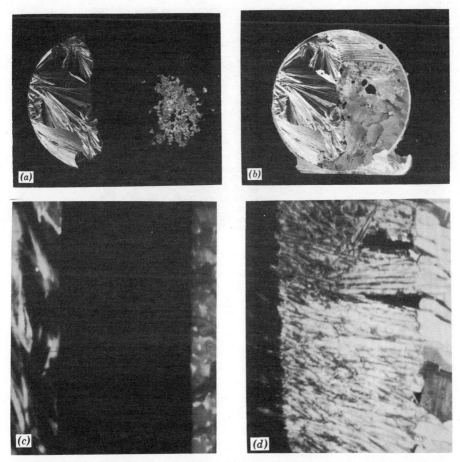

Fig. 2.25 Contact preparation. In (*a*) the higher-melting component has been melted on a microscope slide under a cover slip and the lower-melting component has been deposited next to the cover slip. In (*b*) the lower-melting component has been melted and drawn into contact with the higher-melting, without melting the latter. In (*c*) a liquid zone has been formed between the components. In (*d*) the liquid zone has solidified to eutectic crystals.

eutectic form from the central liquid (Figure 2.25*d*) and are recognizably different in morphology from the pure components.

The liquid zone of mixing in the contact preparation is a transient state of the system; if allowed to attain equilibrium, it would have uniform composition. Since mixing is limited in the thin liquid film, equilibrium is attained slowly and a concentration gradient may be maintained for some time. These facts can be used in characterizing an entire phase diagram, as seen in Figure 2.26, which shows the behavior of a simple eutectic system. A temperature gradient is applied perpendicular to the concentration gradient between pure components *A* and *B*. A solid/liquid interface is generated, whose position in the temperature

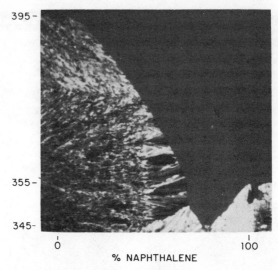

395 –

355 –

345 –

0   100

% NAPHTHALENE

Fig. 2.26 Photomicrograph of the naphthalene/benzoic acid system, representing the complete binary phase diagram obtained from a temperature-gradient microscope stage. Note the horizontal interfaces, corresponding to pure benzoic acid and naphthalene, at the 395 and 353 K isotherms, respectively.

gradient reflects the solidification temperatures of the various compositions. The lowest-melting solid shows the characteristic lamellar structure of the eutectic (39).

Temperatures observed during cooling may be substantially lower than equilibrium values because supercooling is needed to nucleate the crystalline phases. Consequently it may be preferable to cool the contact preparation until it is completely solid. Reheating reveals the eutectic melting point as the temperature of first appearance of liquid. This is most clearly seen when the observation is made between crossed polarizers, because the liquid appears as a region of minimum transmission in an otherwise bright, crystalline field.

CONGRUENTLY MELTING COMPOUND

A contact preparation of two substances which form a congruently melting compound (see Figure 2.5) can yield, in one sequence of measurements: the melting points of the individual components, the melting points of the two eutectics formed between the respective components and the congruently melting compound, and the melting point of the compound itself. It is instructive to consider a particular case of compound formation, namely that of stilbene, mp 397 K, with 2,4,7-trinitrofluorenone (TNF), mp 448 K. TNF is melted under a cover slip as described above, and then cooled. Stilbene may then be melted at the periphery of the cover glass and allowed to flow between the cover glass and the microscope slide until it contacts the solid TNF. The temperature of the preparation is then raised until the TNF also begins to melt. In this case, the 1:1 molecular addition compound, being higher melting than either component, begins to appear as a solid in the zone of mixing, even before cooling. Cooling the system from this condition results in a very informative sequence of events. Crystallization of the addition compound continues, and the excess component is rejected on the right and left sides of the preparation, respectively. Cooling the preparation to 373 K gives a totally solid field. Heating the preparation recapitulates the temperature/composition diagram, with the following events observed:

1. E_2 melts at 389 K (Figure 2.27a).
2. Stilbene melts at 397 K and addition compound appears as small laths in the melt on the right (Figure 2.27b).
3. E_1 begins to melt at 412.4 K (Figure 2.27c), and crystals of the addition compound have grown.
4. Addition compound melts at 422.5 K (Figure 2.27d).
5. TNF melts at 448 K.

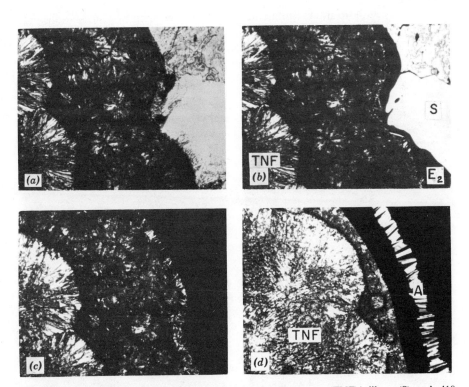

Fig. 2.27 Contact preparation of the system 2,4,7-trinitrofluorenone (TNF)/stilbene (S), mp's 418 and 397 K, respectively: (a) 383 K only solid is present; (b) 389 K E_2 melts; (c) 397 K stilbene melts; preparation heated close to mp of TNF, then cooled slightly, forming new crystals of A; (d) 417 K E_1 melts, leaving solid TNF and addition compound A.

A compound may melt congruently at a temperature lower than the melting point of either component. This behavior has been reported for the system rubidium nitrate/sodium nitrate (mp 583 K and 580 K, respectively) (40). Up to 444.4 K the entire system is solid; at this temperature a narrow band of melt begins to appear on the $NaNO_3$ side (Figure 2.28a). At a temperature only 1.1 K higher, this first band of melt has broadened and another has appeared on the $RbNO_3$ side (Figure 2.28b). The crystals in the center, which are said to have 2:1 stoichiometry, melt at 449 K (40).

INCONGRUENTLY MELTING COMPOUND

A somewhat different circumstance arises if the molecular addition compound that forms between the two reacting components melts incongruently. The biphenyl/TNF system is of this type. The contact preparation starts with the higher-melting compound, TNF, on the left and biphenyl on the right. After mixing along the line of contact at a temperature close to the melting point of TNF, the preparation is cooled. At the outset, the solid phase that forms upon cooling is pure TNF. At 405 K the incongruently melting addition compound is in equilibrium with the melt and it precipitates. Two crystalline phases exist contiguously with no intervening eutectic. Further cooling produces additional precipitation of the addition compound until, at length, the solidification temperature of pure biphenyl is attained, whereupon biphenyl begins to crystallize at the right extremity of the preparation, with eutectic remaining between the addition compound and the biphenyl. Continued cooling produces

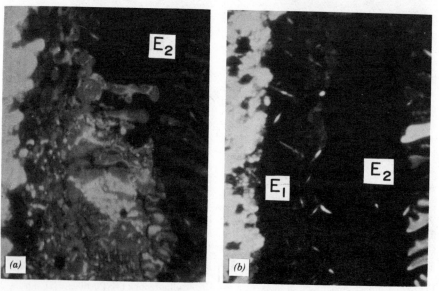

Fig. 2.28 Contact preparation of the system rubidium nitrate/sodium nitrate, mp's 583 and 580 K, respectively; a congruently melting compound is formed, with mp below that of both components: (a) 444.4 K E_2 melts; (b) 445.5 K E_1 melts.

final crystallization at 341 K, the crystallization temperature of the eutectic between biphenyl and the addition compound (Figure 2.29c). As in the earlier discussion, heating of this preparation results in the following events:

1. Eutectic melts at 341 K (Figure 2.29a).
2. Biphenyl melts at 342 K (Figure 2.29b), leaving a narrow barrier island of an addition compound. Heating above 405 K creates more of the addition compound, which crystallizes from the TNF surface (Figure 2.29c, at 393 K); no liquid exists between the two solid phases.
3. Addition compound melts at 405 K.
4. TNF melts at 448 K.

SOLID SOLUTIONS; ROOZEBOOM TYPE I

Consider the system stilbene/azobenzene, with mp 397 K and 341 K, respectively (see Figure 2.9). If stilbene is melted first and allowed to solidify, then the azobenzene may be melted in until it makes contact with the stilbene. The

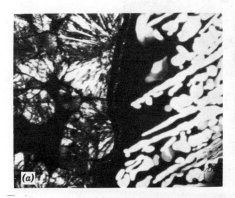
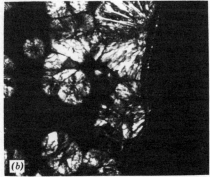

Fig. 2.29 Contact preparation of the system 2,4,7-trinitrofluorenone (TNF)/biphenyl (B), mp's 418 and 342 K, with an incongruently melting compound. (a) Contact zone at 341 K, then heated to 406 K; (b) cooled to 393 K; (c) addition compound crystallized on TNF at 405 K.

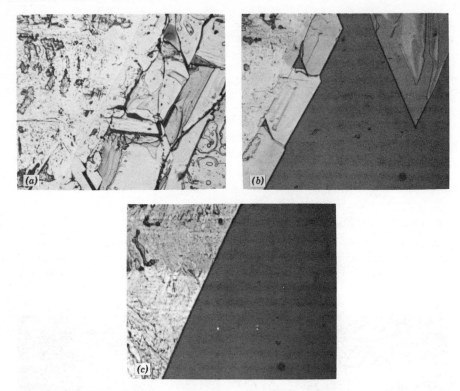

Fig. 2.30 Contact preparation of the system stilbene/azobenzene, mp's 397 and 344 K, respectively, of Roozeboom type I. (*a*) Cooled to 323 K; (*b*) heated to 342 K; (*c*) heated to 354 K.

preparation is cooled (Figure 2.30*a*) and then reheated. In this case the azobenzene is the lowest-melting material present and melts first (Figure 2.30*b*). The mixture, of increasing stilbene content, shows progressively higher-melting temperature and the liquid front advances steadily toward pure stilbene, which melts last (Figure 2.30*c*).

SOLID SOLUTIONS; ROOZEBOOM TYPE II

This is a very rare kind of system in which a solid solution with a maximum melting point is formed (Figure 2.10). For many years the only reported system showing this behavior was *d*-carvoxime/*l*-carvoxime, described by Adriani in 1900 (23). Figure 2.31 shows a contact preparation at 348 K, above the melting point of either component, with a band of solid persisting at the center. Cooling the preparation causes growth of solid having the same morphology *in both directions*, confirming that only a single solid-solution phase exists.

SOLID SOLUTIONS; ROOZEBOOM TYPE III

In this case, a minimum freezing temperature exists. This system differs from a simple eutectic, in which the liquid is in equilibrium with two solid phases,

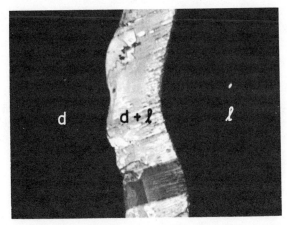

Fig. 2.31 Contact preparation of the system *d*-carboxime/*l*-carvoxime, of Roozeboom type II, showing maximum-melting central solid.

whereas in Roozeboom type III systems there is only one solid phase throughout the entire range of compositions. The two systems do, however, share the common feature that at the minimum-melting composition the liquid and solid phases have the same composition. Both systems may be said to mimic the behavior of a pure substance in that solidification takes place at constant temperature. In the contact preparation of a Roozeboom type III system, the higher-melting compound (for example, potassium nitrate, mp 607 K) is, as usual, melted in first, followed by the lower-melting (sodium nitrate, mp 580 K). The preparation is cooled and reheated slowly. Melting occurs first in the zone of mixing at the position corresponding to the lowest-melting mixture, 495 K (see Figure 2.32*a*). (41). Further heating causes the molten zone to broaden steadily in both directions until the two pure components melt. Upon cooling,

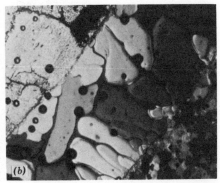

Fig. 2.32 Contact preparation of the system potassium nitrate/sodium nitrate, mp's 607 and 580 K, respectively, of Roozeboom type III. (*a*) Liquid appears in contact region at 496 K; (*b*) a single solid solution grows from both sides at 495 K.

on the other hand, solid solution of the same morphology will grow on both sides of the melt (Figure 2.32b). This behavior distinguishes type III from type V (see below and Figure 2.15) and from simple eutectics since in the latter cases, two different solid phases are formed at the edges of the central melt. Recent studies by calorimetry (42) and directional crystallization (43) of the $KNO_3/NaNO_3$ system gave results consistent with these conclusions.

Note that the closely related system $NaNO_3/RbNO_3$ has been studied with quite different results (see above). In this case the cations differ substantially in size and the phase diagram shows only slight mutual solid solubility, two eutectics and a $1:2$ molecular compound (40).

SOLID SOLUTIONS; ROOZEBOOM TYPE IV

Systems of this type are a variant of type I, in which a peritectic reaction takes place. The system stilbene/bibenzyl has been said to be of this type (44, 45). A contact preparation of the two substances (mp 397 K and 323.4 K, respectively), upon cooling, shows growth of crystals different from the high-temperature phase (Figure 2.33a). After further cooling, the bibenzyl-rich phase appears (Figure 2.33b).

SOLID SOLUTIONS; ROOZEBOOM TYPE V

Type V systems show eutectics, but the liquids on either side of the eutectic yield distinct solid solutions rather than pure compounds or the same solid solution upon solidification (see Figure 2.15). Figure 2.34 shows the contact zone of a preparation of anthracene (mp 491 K) and phenazine (mp 450 K), at 424.7 K, after it was heated to 438 K. Crystals of two solid solutions are seen growing into the eutectic; they differ from one another and from the original solids.

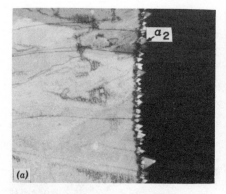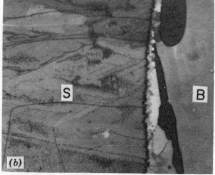

Fig. 2.33 Contact preparation of the system stilbene (S)/bibenzyl (B), mp's 397 and 325.4 K, respectively, of Roozeboom type IV. (a) Crystals of the low-temperature α_2 phase grow from the α_1 solid at left, at 326.4 K (see Fig. 2.13); (b) pure solid bibenzyl has formed at 324.2 K.

Fig. 2.34 Contact preparation of the system anthracene/phenazine, mp's 491 and 450 K, respectively, of Roozeboom type V, showing different solid solutions growing from the two components at 424.7 K.

IMMISCIBLE LIQUIDS

Some systems yield two mutually saturated liquids which are immiscible: resorcinol/triphenylmethane and phenol/water behave in this way (46). When heated, the mutual solubilities increase in these systems until the interphase meniscus disappears at the critical solution temperature (see Figure 2.7).

2.1.2 Establishment of Liquidus and Solidus Curves

The contact-preparation procedures just described give gross, qualitative features of phase relationships and are, in a sense, orienting in nature. In order to learn the shapes of liquidus and/or solidus curves, mixtures of known composition must be studied. In general, this will require the preparation of physical mixtures of the two components in various proportions.

The details of sample preparation may be critical and no single method is best for all compounds. Known amounts of components that are melt-stable may be melted together and agitated to assure homogeneity. The melt is then quenched rapidly by immersion in a cold bath.

Robinson and Scott (47) improved heat transfer in the quenching process by using a metal foil. A strip of aluminum foil (25 μm thick) was wrinkled by rolling it over glass paper; it was then degreased and etched with 1% nitric acid/1% hydrofluoric acid, and then rolled upon an aluminum wire of 0.3-cm diameter. The roll of foil was placed in a glass tube and the molten mixture was poured in to fill the open space within the foil; the tube was then quenched and disks were cut from the solidified composite. The appropriateness of using quenching to prepare samples for phase-diagram studies has been challenged (48, 49). When binary mixtures that form solid solutions were zone leveled to generate ingots of very uniform concentration, it was found that their melting behavior differed considerably from that of quenched samples (see Section 2.7.5, Chapter 5). A priori, there is no reason why the zone-leveled samples should behave differently from those taken from a carefully grown single crystal. Although the latter shows an overall concentration gradient, local sampling for removal of a few milligrams gives an essentially homogeneous preparation which is equivalent to a zone-leveled preparation.

In some cases, one component may be lost by sublimation or distillation, thus changing the composition of the mixture. This difficulty may be avoided by carrying out the mixing in small sealed tubes which are uniformly heated.

Melting may lead to thermal decomposition or to decomposition by interaction of the components or by reaction with atmospheric oxygen. In such cases, the components may be mixed loosely or ground gently together without melting. Yet another alternative is to dissolve both in a single, volatile solvent and recover the mixture by evaporation of the solvent. The melting behavior of mixtures prepared by grinding is identical to that of premelted samples with respect to the end of fusion (i.e., the determination of the liquidus point). Not surprisingly, at the start of melting such solid mixtures show different behavior from that given by premelted samples. In many cases, an initial sintering is observed, followed by resolidification and then a renewed onset of melting, which is most readily detected by disappearance of crystal edges. The early sintering corresponds to the melting point of the lower-melting component, or of the eutectic. The second onset of melting corresponds to a point on the solidus curve, as does the onset of melting of a premelted preparation (50).

Assuming that each solid mixture has the same uniform concentration as existed in the melt, then heating the solid in a microscope hot stage will yield a temperature at which melt first appears. This may be detected visually by the disappearance of birefringence or it may be detected more objectively by monitoring the optical transmission of the preparation with a photodetector. A plot of the transmission as a function of temperature in general is nearly horizontal until the liquidus temperature is reached, whereupon a very rapid decrease in optical transmission occurs as the birefringent crystals melt to form an isotropic liquid. (If the measurement is made between crossed polarizers, the liquid is opaque.) Because the heating causes disappearance of the birefringence originally observed in polarized light, the technique is sometimes called *thermal depolarization analysis.*

Vaughan provided an early description of simultaneous measurement of temperature and optical transmission in capillaries and in a microscope hot stage (51). A sequence of several compositions gives corresponding temperatures of the solidus curve. Note that the light transmission in Figure 2.35 is given in arbitrary units. The important point is that each curve shows the rapid change in slope at a characteristically higher temperature. This technique provides essentially the same information as the "thaw point" method, which is described in Section 2.2.6, but with much greater speed and reliability.

In the case of a simple eutectic system, all compositions will manifest first melting at the eutectic temperature.

In the case of a pure eutectic (Figure 2.3), a homogeneous mixture of A and B can never be obtained and microcrystals of A and B will always occur. If composition is off-eutectic, then excess of A or B will be present. On heating this solid, the regions of A and B crystals that are in contact will melt to form eutectic liquid. If the heating rate is very slow, eutectic will continue to be formed until the minor component is completely consumed, after which melt temperature will start to rise as the major component begins to melt. The melt composition will then follow the liquidus curve until the composition of melt in contact with pure major component equals that of the starting mixture, whereupon the solid major component is completely consumed. An alternative method of obtaining liquidus points is to note the temperature at which solid appears upon cooling melts of various known compositions. This procedure suffers, however, from the difficulty that supercooling may often occur. It is possible to overcome this difficulty by heating the solid until it is almost completely molten, then holding the small amount of remaining solid in contact with the melt until equilibration takes place. Cooling the preparation from this condition cannot give rise to supercooling and will result in overgrowth of the solid solution upon the remaining crystalline nucleus at the equilibrium temperature corresponding to the initial solid composition.

Thus a small number of mixtures in relatively small amount will provide

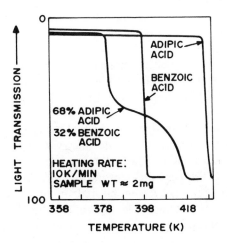

Fig. 2.35 Light transmission as a function of temperature, for adipic acid/benzoic acid mixtures.

information on eutectic composition and on temperature/composition relationships for both the solidus and liquidus.

It is necessary to point out that observations made during the cooling of molten preparations on the microscope hot stage must be interpreted with care because deviation from equilibrium behavior may exist and may produce misleading effects. Thus, for example, cooling the eutectic mixture of a Roozeboom type V system may not produce the characteristic, lamellar, fine-grained eutectic dispersion, but instead may result in the formation of super-cooled liquid of eutectic composition, which upon nucleation will yield coarse crystals of the terminal solid solutions. Likewise, in a system showing peritectic reaction, cooling a liquid whose composition is that of the incongruently melting compound may not produce solidification as the temperature crosses the liquidus line, but instead, the liquid may supercool until the unstable compound nucleates.

There is no absolute assurance that these disturbing effects can be avoided, but a great increase in reliability can be attained by carrying out both heating and cooling at very low rates and by approaching equilibrium from both directions.

Yet another word of caution is required regarding determination of phase diagrams by general microscopic methods, for systems involving polymorphic forms. One polymorph of compound A may give a continuous series of solid solutions with another compound B, while a second polymorph of A may give a eutectic system with B. Thus polymorphic modifications may appear to show the behavior of two totally unrelated materials and care is required to establish that, in fact, the two polymorphs are chemically identical. McCrone (52) refers to the system 1,3,5-trinitrobenzene/picric acid. The former exists in three modifications and the latter in two. The two materials, depending on the crystal forms present, may exhibit several diverse phase diagrams.

2.2 Thermal Analysis

Historically the term "thermal analysis" has referred to the measurement of time/temperature plots during freezing or melting of chemical substances. Such experiments were carried out on quantities ranging from a few grams to some tens of grams, and were of long (hours) duration.

These methods are still practiced, but the advent of commercial instrumentation for the rapid measurement of changes of enthalpy of small chemical samples has opened wide new horizons for the facile determination of phase relationships.

There are two basic instrumental modes for carrying out thermal analysis: differential thermal analysis (DTA) and differential scanning calorimetry (DSC). In DTA a sample that is to undergo a phase transition is heated or cooled along with a thermally inert reference material. Temperature sensors measure the temperature of each. The difference in temperature, ΔT, is monitored, amplified, and recorded during a programmed heating or cooling cycle. Figure 2.36 shows a block diagram of a differential thermal analysis system.

BLOCK CELL CROSS SECTION [TA]

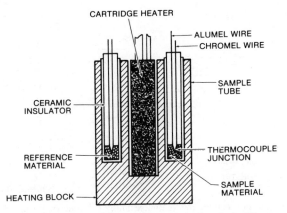

CARTRIDGE HEATER

ALUMEL WIRE
CHROMEL WIRE

SAMPLE TUBE

CERAMIC INSULATOR

THERMOCOUPLE JUNCTION

REFERENCE MATERIAL

SAMPLE MATERIAL

HEATING BLOCK

Fig. 2.36 Cross section of a cell block for differential thermal analysis. (Courtesy E. I. du Pont de Nemours & Company, Inc.)

This procedure results in much higher sensitivity than classical thermal analysis. It is also possible to construct a cell and to calibrate ΔT of the sensors in such a way that ΔQ, the enthalpy change associated with a transition, can be measured. This enables the system to be used as a calorimeter, allowing absolute determination of such quantities as latent heats of fusion, sublimation, vaporization, and other first-order transitions. In addition, the dynamic specific heat can be obtained as a function of temperature by comparing an unknown with a standard reference material (such as sapphire). For phase-diagram determinations therefore, DTA is of considerable utility as a rapid method of obtaining accurate information.

In DSC both sample and reference are maintained at the same temperature during programmed heating or cooling. If the sample undergoes a transition having a latent heat, then the heat flux applied to the sample must be modulated to compensate for the heat of transition. The amount of heat required to maintain isothermal conditions is recorded as a function of time or temperature. A differential scanning calorimeter cell is shown in Figure 2.37. DSC is supplanting DTA because of the greater ease of extracting quantitative information regarding changes in enthalpy from the experimental plot.

Both instrumental modes are able to provide much useful information regarding phase diagrams on the basis of analysis of samples weighing only a few milligrams. Endothermic and exothermic phenomena are clearly visualized in the instrumental plots. Commercial instrumentation for this kind of work is available from numerous manufacturers (see Appendix II); the temperature range of approximately 120 to 2300 K may be covered.

DSC CELL CROSS-SECTION $\boxed{\text{DSC}}$

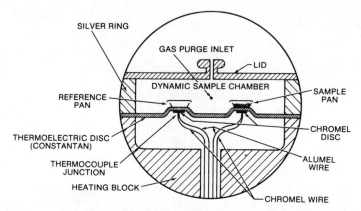

Fig. 2.37 Cross section of a cell for differential scanning calorimetry. (Courtesy E. I. du Pont de Nemours & Company, Inc.)

We will not go into a discussion of experimental details of instrumentation. Instead the interested reader is referred to the literature provided by the manufacturers of commercial instruments and to numerous textbooks available in this area (53).

2.2.1 Differential Thermal Analysis (DTA)

Although DTA has been widely used in determining phase diagrams, relatively few publications have dealt with the details of the relationship between DTA curves and phase diagrams. Moreover the experimenter must be aware that sensitivity and accuracy of DTA curves may be influenced by many factors, including sample size and sample geometry, heating rate, thermal conductivity of sample container, and method of measurement of sample temperature (54).

When the two components of a binary system are mixed in varying ratios for later thermal characterization, it must be borne in mind that the thermal results to be obtained may be influenced by the extent of "demixing" produced during the solidification and subsequent storage of the samples.

Etter et al. (54) have described a hypothetical phase diagram (Figure 2.38) showing features common to many phase diagrams, along with the DTA curves that would result from selected compositions of this hypothetical system. Heating a sample having composition A shows only the fusion of solid solution α. The onset of melting (first deviation from baseline) is taken as the solidus temperature of the sample and the return to the baseline corresponds approximately to the liquidus.

Composition B, having the stoichiometry of an incongruently melting compound, gives a sharp endotherm at the temperature of the peritectic halt, T_3.

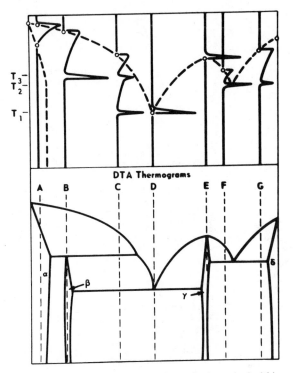

Fig. 2.38 DTA traces observed for various compositions of a hypothetical binary phase diagram (54).

Further heating causes melting, at an increasing rate, until the liquidus is reached.

Composition C is located on the eutectic side of the incongruently melting compound. The first endotherm corresponds to melting at the eutectic temperature T_1. On further heating, the remaining β-phase solid melts at progressively higher temperatures until the peritectic halt is reached, whereupon the remaining phase decomposes into α phase and liquid. Continued heating produces further melting of the α phase until the liquidus is reached.

Composition D, the β–γ eutectic, shows a single endotherm at T_1. The magnitude of the eutectic endotherm is greatest for this composition.

Composition E, corresponding to a congruently melting compound, likewise gives a single endotherm at its melting point.

F and G both show eutectic endotherms and continued melting of phases γ and δ, respectively, until the liquidus is reached.

DTA has been applied to determination of the naphthalene/benzoic acid phase diagram. Eight scans sufficed to give both branches of the liquidus and the eutectic composition and temperature, as shown in Figure 2.39a (55). Figure 2.39b shows the DTA curves for two hypoeutectic compositions, containing 10

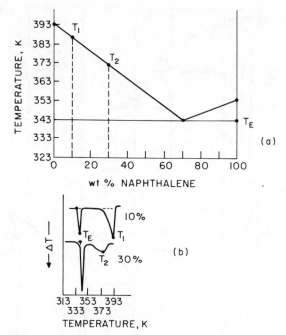

Fig. 2.39 Melting behavior of various compositions of the benzoic acid/naphthalene system. (*a*) Benzoic acid/naphthalene phase diagram; (*b*) thermograms of mixtures containing 10 and 30% naphthalene, respectively.

and 30% naphthalene, respectively. Both show eutectic melting at 343 K; the 30% mixture shows a larger eutectic endotherm, since it is closer to the eutectic composition. In this work, the peak of the first endotherm was taken to be the eutectic temperature and the peak of the second endotherm was taken to be the liquidus temperature for each composition. This assignment is not consistent with the practice of later workers (56) who assumed that the liquidus temperature was that at which the endotherm returned to the baseline.

The thermal history of samples subjected to DTA may bear importantly on the results. Both peak shapes and peak temperatures of DTA curves may be altered by thermal treatments. Consequently, the feature of the curve which is related to a characteristic temperature of the phase diagram (e.g., eutectic temperature or liquidus temperature) should be specified in reports of DTA studies.

This problem has been attacked in a very direct way by Van Tets and Wiedemann (57), who carried out simultaneous thermomicroscopic and DTA investigations using a Mettler FP-2 hot stage with an object carrier provided with a thin-film (gold–nickel) thermocouple.

The correlation between thermal effect and extent of melting may be made still firmer by monitoring the optical transmission of the crystalline preparation during melting. If the initial observation is made with crossed polarizers in the

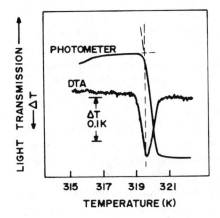

Fig. 2.40 Comparison of optical and thermal observation of the monoclinic-to-octahedral phase transition in tetrabromomethane.

microscope, the birefringent crystals will raise the optical transmission to a maximum. As soon as any (isotropic) melt begins to form, the optical density increases. The transmission is then recorded as a function of temperature; a separate optical record of the melting process is thus available for comparison with the DTA curve.

Vaughan (51) has published results of such measurements, using the monoclinic-to-octahedral transformation of tetrabromomethane as a test transition. Figure 2.40 shows that the extrapolated onset of the photometric curve corresponds to the peak temperature of the DTA curve. The delay in the onset of the photometric curve compared to the DTA curve may result from a thick sample, in which some solid remained in the light path after partial transformation.

2.2.2 Differential Scanning Calorimetry (DSC)

It is pointed out in Section 2.1 that the shapes of the two branches of the liquidus curve for a sample eutectic system can be calculated by Eqs. 2.4 and 2.5, the so-called Le Chatelier–Schroder–van Laar (CSL) equations. From these, Smith (58) derived an equation relating DSC signal (dq/dT) to mole fraction X of component B, for a simple eutectic (see Figure 2.3). For starting composition X_B (e.g., W_Q of Figure 2.3)

$$\frac{dq}{dT} = \frac{\Delta H_{f,B}(1 - X_B)X}{RT^2(1 - X)^2} \qquad (X_B > X_E) \tag{2.12}$$

where $\Delta H_{f,B}$ is the heat of fusion of component B and

$$X = \exp\left[\frac{\Delta H_B}{RT_B}\left(1 - \frac{T_B}{T}\right)\right] \qquad (X > X_E) \tag{2.13}$$

and T_B is the melting point of component B. (Note that Figure 2.3 shows composition as wt%, but the CSL equation is derived in terms of mole fraction.) The DSC curve calculated by means of Equations 2.12 and 2.13 agrees well with the experimental curve for a liquid-crystal mixture showing a simple-eutectic behavior. Figure 2.41 shows that the peak of the eutectic endotherm may be taken as the eutectic temperature and the return to baseline of the posteutectic melt may be taken as the liquidus temperature (T_l).

Robinson and Scott (56) have applied DSC extensively to the determination of binary organic phase diagrams. Figure 2.42 illustrates the method as applied to the determination of the anthracene/carbazole phase diagram. They distinguish between two concentration regimes. In the low-concentration regime, the melting curve corresponds to that obtained for a pure material; the leading edge has constant slope, with a rapid return to baseline after complete melting (Figure 2.42a). In the high-concentration regime, the leading edge of the curve deviates markedly from that of the pure compound (Figure 2.42b). In Figure 2.42a, the solidus is taken to be the intersection, S, of the baseline with the extrapolated leading edge of the melting curve and the liquidus, L, is taken to be the intersection of the baseline with a line, parallel to the leading edge, through the point at which the melting curve begins to deviate from the upward extension of the trailing edge. In Figure 2.42b the solidus S and liquidus L are taken as the points of departure from and return to the baseline, respectively.

Perhaps the most painstaking approach to ascertaining characteristic temperatures from thermograms is that described by Radomska and Radomski (59). Thermograms were measured for each of a series of compositions at heating rates of 4, 2, 1, and 0.5 K min^{-1} and for each sample the temperatures of onset

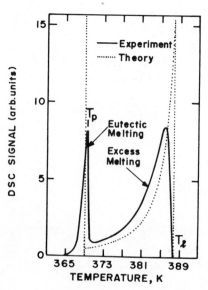

Fig. 2.41 Experimental and calculated thermograms for a binary eutectic liquid-crystal system.

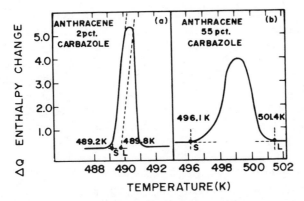

Fig. 2.42 DSC curves of anthracene/carbazole mixtures. (*a*) In the low-concentration region, the solidus is the intersection of the extrapolated leading edge and baseline; the liquidus is the intersection of the baseline with a line through the top of the trailing edge, parallel to the leading edge. (*b*) In the high-concentration region, the solidus is the point of departure from the baseline, and the liquidus is the point of return to the baseline.

and termination of melting were plotted against the square root of scan rate and extrapolated to zero heating rate.

Kolkert (60) has analyzed the temperature change taking place during the melting of solid solutions, in terms of sample history. If a solid sample is prepared by slow cooling of liquid and no supercooling takes place, the first solid to form will have the concentration given by the isothermal tie line through the liquidus. Because diffusion in the solid is negligible, the remaining liquid is not in equilibrium with the entire solid, but only with the most recently frozen solid. As a result, the last liquid freezes at a temperature below that which would have prevailed under true equilibrium conditions; thus the final solidification takes place *below* the solidus temperature of the initial mixture. When the resulting solid is heated, melting takes place below the true solidus temperature of the average sample composition and an artificially broad DSC curve results.

This difficulty in direct measurement of solidus temperatures may be overcome by a thermodynamic calculation (61). The method is based on the equal G curve (EGC), that is, the curve giving the temperatures, T_{EGC}, at which the liquid and solid fractions of binary mixtures have equal Gibbs energies. The respective Gibbs energies are formulated in terms of known heats and entropies (ΔH^* and ΔS^*) of melting for the pure components:

$$T_{EGC}(X) = \frac{(1 - X)\,\Delta H_1^* + X\,\Delta H_2^* + \Delta H^E(X)}{(1 - X)\,\Delta S_1^* + X\,\Delta S_2^* + \Delta S^E(X)}$$

where X is the mole fraction of the second component and the superscript E refers to excess thermodynamic functions. Because it is not possible to derive uniquely two functions (ΔH^E and ΔS^E) of X from a single other function (T_{EGC}) of X, an iterative method was developed, using intermediate phase-diagram calculations. The iteration is continued until an EGC is found that gives optimum agreement between the experimental and the calculated liquidus.

If a zone-leveled sample is melted (see Section 1.3, Chapter 5), the first liquid to form has the same composition as the starting solid, again because of the slowness of diffusion in the solid. Now melting takes place at some constant temperature, and the resulting liquid is supercooled. As long as the supercooling persists, the melting is isothermal, because the melting solid retains its constant composition. Ultimately the supercooled liquid will crystallize, causing an increase in temperature and final solidification at the true liquidus temperature of the original, homogeneous solid.

Van Genderen et al. (48) have confirmed that the DSC trace of a solid-solution sample depends markedly on the manner of preparation. When zone-leveled samples of p-dichlorobenzene/p-dibromobenzene were characterized by DSC, the curves resembled those of pure compounds in shape, although they were somewhat broader. The sharp thermograms result from the fact that rapid heating (i.e., rapid with respect to solid-state diffusion) of a microscopically homogeneous solid can only give liquid of the same composition as the solid. It has been pointed out that the solid and liquid mixtures, having the same composition at a given temperature, have equal Gibbs energies. Hence the temperature of such an equilibrium is that of the EGC (61). Quenched samples, in contrast, gave grossly broader DSC traces. The "melting point" of each of the former samples, which defines the solidus for a given concentration, was taken to be the intersection of the extrapolated leading edge with the baseline, in accord with earlier practice (56).

In a DSC study of another system, naphthalene/camphor, in which only narrow regions of solid solution exist, all mixtures in the region 12–94% naphthalene showed primary eutectic melting. Figure 2.43 shows a typical melting curve for a hypo- or hypereutectic mixture; in this case solidus and liquidus are taken as points of departure and return to the baseline, as above.

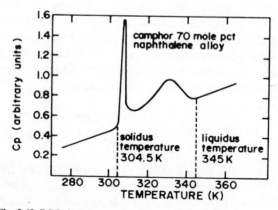

Fig. 2.43 DSC thermogram for the camphor/naphthalene system.

Precise determination of the onset and termination of melting is a general problem. A variety of techniques can be used in conjunction with DSC to aid the interpretation of DSC curves. Several of these methods can be used independently and have been discussed.

Great care must be used in the interpretation of DSC traces, since non-equilibrium artifacts can occur. Brennan et al., in a study of the benzene/hexafluorobenzene phase diagram, observed two eutectic endotherms in mixtures that should have given only one. Cycling the temperature through the melt/solid transition changed the relative amounts of the two eutectics, but both persisted. Evidently, rapid freezing of supercooled melt did not yield the true equilibrium mixture of solid plus eutectic. Samples of about equimolar composition gave two eutectic peaks as well as the major posteutectic melt (62).

Yet another problem can arise in systems with large deviations from ideality. The system carbon disulfide/hexafluorobenzene manifests its large positive deviation from ideality in a very unusual phase diagram (Figure 2.44). One result is the very small slope of the liquidus over a range of about 0.7 in mole fraction; another is a eutectic composition containing > 0.99 mole fraction CS_2, whose freezing temperature is nearly indistinguishable from that of pure CS_2. In common with other systems containing C_6F_6, this one evidences formation of a molecular compound; it has the composition $2C_6F_6 \cdot CS_2$ and is responsible for the peritectic at 266.1 K (63).

It is generally reported that calorimetric data obtained by DSC have a precision of 1–5%, while conventional calorimetry using large samples is frequently claimed to offer precision of a few tenths percent. Brennan and Gray, however, have studied the influences of calorimetric and noncalorimetric factors upon the precision of DSC measurements. (Noncalorimetric factors include the

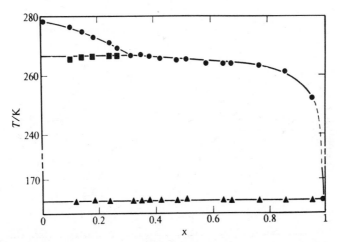

Fig. 2.44 Phase diagram of the system hexafluorobenzene/carbon disulfide: ●●● freezing curve, ■■■ peritectic, ▲▲▲ eutectic.

weight of the sample and measurement of the area under calorimetric curves.) When substantial care was applied to weighing, recording, and data analysis, it was found that calorimetric values subject to random errors of only a few tenths percent could be obtained by DSC. Apparently, the factors limiting the precision of calorimetric measurements by DSC are the precision and accuracy of the recorder used to obtain the data. If a digital recorder of high sensitivity is coupled directly to the calorimeter output, then substantial gains in precision and accuracy are available with DSC instrumentation.

2.2.3 Ideal Cooling Curves

PURE COMPOUND

Gibbs' Phase Rule predicts and experiment confirms that the time/temperature relationships observable in solidifying melts may be described qualitatively by only a few curves. Thus experimental measurements of temperature during solidification of binary mixtures can yield much information about the phase relationships that exist between the components of the mixture. It should be noted that phase relationships are generally determined at atmospheric pressure; hence the number of degrees of freedom is given by the phase rule as

$$F = C - P + 1$$

where C and P are the numbers of components and phases, respectively.

If a sample of a pure material is melted and stirred as it is allowed to cool, its temperature may be monitored as a function of time. This kind of experiment is often carried out in a bath whose temperature is reduced at a constant rate, to give an approximately linear cooling curve for the melt. The cooling is characterized by three distinct regimes, shown as curve I in Figure 2.45. While

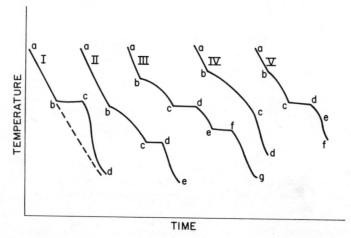

Fig. 2.45 Idealized cooling curves for various features of phase diagrams.

only liquid (one phase) is present, $F = 1$ and the melt may cool as indicated by segment *ab*. After the appearance of solid, the two-phase, one-component system becomes invariant; its temperature remains constant as the heat of fusion is evolved, until solidification is complete (segment *bc*). After complete solidification, the temperature may again drop at a rate determined by the specific heat of the solid and by experimental conditions. Recall that the bath in which the measurement is being made is cooling linearly; at the end of solidification (point *c*), the temperature difference between sample and bath is maximum. Consequently, more rapid cooling of the solid is observed immediately after this point. Since solids generally have lower specific heats than their melts, they may be expected to cool more rapidly under equilibrium conditions.

SIMPLE EUTECTIC

If a second component is added to a pure compound, the time/temperature behavior will be different. Let us assume that the two components, designated as *A* and *B*, are completely immiscible in the solid state and form a simple eutectic system (Figure 2.3). A liquid mixture whose composition lies between pure *A* and eutectic *E* gives cooling curve II shown in Figure 2.45. The onset of crystallization *b* has taken place at a lower temperature than in the case of pure *A*. Here the two-phase system contains two components and retains 1 degree of freedom. Hence the temperature of the system continues to decrease during crystallization. Other factors remaining constant, the rate of cooling may be expected to increase as the fraction of solid increases, because of the solid's lower specific heat. During solidification along segment *bc*, pure *A* is crystallizing at progressively lower temperature and the melt becomes richer in *B* until composition *E* is reached. At this point, eutectic solid is formed whose composition is the same as that of the liquid. The system is again invariant, and its temperature remains constant during eutectic solidification, as indicated by segment *cd*.

Another melt composition of this system, closer to the eutectic, will show qualitatively the same cooling curve as the first mixture. However point *b*, corresponding to the onset of crystallization, will occur at a lower temperature. The eutectic "halt" will occur at the same temperature as for the first composition but it will be of greater duration, owing to the greater amount of eutectic that can form. For a mixture of composition *E*, the cooling curve will be qualitatively similar to that shown by the pure components (curve I), except that the crystallization halt will take place at T_E.

Tammann proposed long ago that the duration of the eutectic halt is approximately proportional to the fraction of eutectic in a given mixture (64). Consequently, a plot of the duration of eutectic halt against composition shows a maximum at the composition of the eutectic itself (see Figure 2.46). This technique is rather cumbersome, but it does provide useful information using only simple apparatus. Naturally, if instrumentation is available for the measurement of DTA or DSC curves, it is possible to make a more reliable plot representing the actual amount of eutectic present in mixtures and thus to

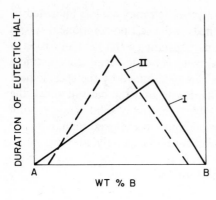

Fig. 2.46 Tammann plot for a simple eutectic system (I), and for a system of Roozeboom type V (II).

ascertain the composition of the eutectic. The intercepts on the composition axis indicate the range of eutectic formation. For a simple eutectic showing no solid solution, there is a eutectic halt of some duration at all concentrations (curve I). If there is solid solubility (Roozeboom type V), then the eutectic halt does not span the entire concentration range (curve II) (see sections on solid solutions below).

CONGRUENTLY MELTING COMPOUND

Consider now a binary system whose components are completely miscible in the liquid, completely immiscible in the solid, and which form a congruently melting compound, as in Figure 2.5. Five compositions of this system will give cooling curves (Figure 2.45) of type I: A, B, E_1, E_2, and A_xB_y. All others will give cooling curves of type II.

INCONGRUENTLY MELTING COMPOUND

The situation in the case of a system with an incongruently melting compound (Figure 2.6) is somewhat more complex. Compositions A, E, and B will show cooling curves of type I. Compositions between B and R and between A_xB_y and A will show cooling curves of type III. Segment bc corresponds to primary crystallization of component A and continuous change of melt composition with cooling, until point R is reached. The system remains invariant as compound A_xB_y is crystallized (segment cd), then cools again (segment de) until the eutectic temperature and composition are reached, whereupon the eutectic halt ef is observed.

In actual practice, inflections such as that shown at R in Figure 2.6 are frequently not sharply defined. The observation of two halts occurring only in the cooling curves of mixtures in a central composition range (R to A_xB_y) is an unambiguous indication of the existence of an incongruently melting compound. It should be noted as well that a similar "double" halt in the cooling curve would result from a polymorphic transition of component A followed by eutectic crystallization. This circumstance is easily distinguished from the case of an incongruently melting compound, since in the former case, the eutectic halt would be observed at T_E for all compositions from A to B.

Tammann's method is applicable to systems showing formation of a compound between the components. In this case, too, a plot is prepared of the duration of the eutectic halt (or the area of the eutectic endotherm, for DSC) as a function of composition (Figure 2.47). Zero duration is observed for both pure components and for any stable compounds. If an unstable compound exists, then the composition corresponding to the "hidden" maximum yields a maximum peritectic halt and zero eutectic halt.

CONTINUOUS SOLID SOLUTIONS; ROOZEBOOM TYPES I, II, III

In systems showing continuous solid solubility, the pure components and maximum- or minimum-melting mixtures all show cooling curves of type I (Figure 2.45). All other compositions of such systems will show cooling curves of type IV. These show an inflection at b corresponding to the onset of crystallization and another inflection at c, corresponding to the end of solidification. In principle, point c of curve IV corresponds to the solidus point of the starting composition. However, this is rarely the case in practice, because the composition of the solid is rarely uniform.

INTERRUPTED SOLID SOLUTIONS; ROOZEBOOM TYPES IV AND V

In solid solutions of Roozeboom type IV (see Figure 2.13), compositions in the solid-solution regions T_Bb and PaT_A show cooling curves of type IV. Compositions within the range bP show a different behavior, corresponding to curve V of Figure 2.45. Primary crystallization occurs at point b of curve IV, with formation of solid solution α_1. Concurrently the melt becomes richer in B, until its composition reaches point b (Figure 2.13). At this composition, the system becomes invariant while solid solution α_1 reacts with melt b to form solid solution α_2. The temperature of the system remains constant, as shown by the halt cd in the cooling curve, until all available α_1 has been transformed. After this the remaining liquid may cool, as solid solution α_2 is formed, and the

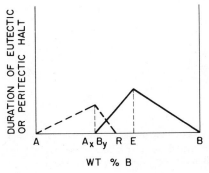

Fig. 2.47 Tammann plot for a binary system with an incongruently melting compound. The solid curve shows the eutectic halt and the dashed curve shows the peritectic halt. The abscissa designations correspond to those of Fig. 2.6.

temperature follows *de* in cooling curve V. After complete solidification, the temperature of the system drops along segment *ef* of curve V (Figure 2.45).

Solid solutions of Roozeboom type V (Figure 2.15) have three compositions that give cooling curves of type I in Figure 2.45: the two pure components and the eutectic. Other compositions give cooling curves of type II.

Application of the Tammann method to systems of this type gives a Λ-shaped figure whose peak corresponds to the composition of the pure eutectic and whose intercepts correspond to the extremities of eutectic composition. Thus in the case of no solid solution, the two branches of the curve will intersect with the composition axis at values corresponding to the pure components. Conversely, if substantial regions of solid solubility exist, the two branches will intersect the composition axis at values corresponding to the solid-solution limits.

2.2.4   Practical Cooling and Heating Curves

In our discussion up to this point, we have assumed several system characteristics which may not be realized in actual chemical systems. These implicit assumptions include:

- No supercooling of the melt
- Complete compositional equilibration between solid and melt
- Absence of polymorphic transitions in the solid
- Temperature equilibrium between system and sensor

In fact, many chemicals, especially organics, can be supercooled far below their equilibrium solidification temperatures. When crystallization does occur, the latent heat of fusion is evolved rapidly, and the temperature rises. Because of the low heats of fusion and the low thermal conductivities of organic compounds and because of heat losses from the system, the maximum temperature may not reach the equilibrium value. This leads to rounding of the transition points and this course of events is shown in Figure 2.48 as a solid line, with the theoretical curve shown as a dashed line.

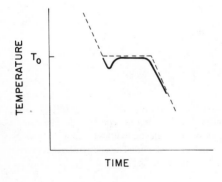

Fig. 2.48 Comparison of theoretical(----) and experimental (——) cooling curves for a pure substance.

If supercooling is experienced or anticipated, its effects may be minimized by seeding the melt at a temperature slightly below that at which crystallization is expected. Supercooling can lead to yet another phenomenon which may obscure the phase relationships of binary pairs, namely the crystallization of a polymorph that is stable only below the crystallization temperature of the highest-melting solid. McCrone has discussed this phenomenon and suggested means for deliberately nucleating low-temperature crystalline modifications (37).

In our discussion of ideal systems exhibiting solid solubility, it is implied that a liquidus and a solidus point can be established from each cooling curve. The slowness of solid-state diffusion prevents complete equilibration; hence the last material to solidify will not have the composition of the original melt and its temperature will be lower than that of the solidus corresponding to the original melt. As the deviation from compositional equilibration becomes greater, the point corresponding to the end of solidification becomes less clear. This fact means that it is often impossible to determine solidus points of solid solutions from their cooling curves.

A related difficulty can arise in simple eutectic systems because of incomplete mixing in the liquid. This leads to entrapment of melt in the solid, which in the ideal case would be a pure, single component. The trapped impurity results in behavior similar to that of a solid solution. This manifests itself experimentally in the absence of a eutectic halt. Thus curve I of Figure 2.45 would be distorted to show only a change of slope at point c, rather than a horizontal segment.

In principle it might appear that the cooling curve of Figure 2.45 could be observed in reverse; one might start with a solid sample and observe its temperature during heating, until the sample is completely molten. In practice this method is not feasible for bulk samples because it is extremely difficult to maintain compositional equilibrium throughout the sample, since microscopic homogeneity depends on the way in which the solid sample was frozen. In turn, the temperature at which melting begins depends on whether or not the sample contains a segregated, impurity-rich fraction, because mass transfer in solids cannot be promoted by external means such as stirring and because diffusion in solids is extremely slow. It must be concluded that mass transfer within solids is negligible under the experimental conditions usually used in the determination of heating curves (65). Thus only the surface layer of the solid is in thermodynamic equilibrium with the melt and the bulk of the solid is not in equilibrium during the determination of the heating curve.

In spite of these difficulties, the heating-curve method is usable with very small samples. Naturally small samples offer the additional advantages of speed and economy of material. Moreover the heating process shows no phenomenon analogous to supercooling. In fact, Smit (65) has argued that for purposes of purity determination at least, the melting mode is preferable to the freezing mode.

On the basis of these concepts, Smit has proposed the use of temperature/heat content curves for purity determination, using a thin film of solid on the bulb of a thermometer (66). Application of this approach to phase-diagram studies is obvious. Both heating and cooling curves may be obtained with small samples, without resorting to Smit's thin-film method, by using minor modifications of the traditional capillary melting-point method. If a sample is placed in a capillary and a small thermocouple is centered in it, the capillary may be heated rapidly in a microoven of low mass. The thermocouple output may be recorded, and the eutectic temperature and solidus points at least, may be determined in a few hours using only tens of milligrams of material. The onset and end of melting corresponding to points on the solidus and liquidus, respectively, may be determined by extrapolation. Since closure of the capillary/thermocouple assembly is difficult, the method is in fact limited to materials that melt without reaction with atmospheric gases.

2.2.5 Comparison of Heating- and Cooling-Curve Methods

The classic cooling-curve methodology is simple and reasonably accurate, and may still be used if adequate quantities of material are available. The instrumental requirements are minimal. A heater surrounding a test tube containing a thermometer and stirrer are all that is required to work with samples of 3–5 g (50). Mixtures of known composition are melted completely and stirred rapidly as cooling begins, to minimize supercooling. The onset of crystallization is always clearly evidenced as a sharp inflection in the time/temperature plot. This temperature corresponds to one point on the liquidus curve. The time/temperature relationships during the later stages of crystallization may not correspond to equilibrium because of low rates of crystallization. This fact has led Van Wijk and Smith (65) to suggest that a melting curve is preferable to a cooling curve. Moreover the end of crystallization is much harder to discern than is the onset. In macrooperation this is the result of the growth of a thermally insulating layer upon the temperature sensor (typically a mercury-in-glass thermometer). In any case, no clear significance can be assigned to the temperature corresponding to the end of crystallization, except for eutectics, because in solid-solution systems the final temperature of crystallization corresponds to a point on the solidus only if compositional equilibrium is maintained throughout the crystallization process. In practice, this is not usually the case owing to the low diffusivity in most solids. Smit has pointed out that this problem is not encountered in the determination of cooling curves starting with completely molten samples.

There are arguments on both sides of the heating curve/cooling curve question. Neither method is without difficulties, both theoretical and experimental. Which method is selected will depend on the properties of the particular system under study. Rates of crystallization and diffusivities must be considered in making a selection.

The problems of bulk diffusion in solid layers have also been considered by Smit in the context of cryometric determination of purity. Smit's analysis gave an optimum layer thickness of $0.2r$ for use with a mercury-in-glass thermometer whose bulb has radius r (66). The details of sample preparation may be critical, and it may be that no one method is best for all compounds.

A comparison of cooling curves, heating curves, and microscopic observation is useful. As mentioned earlier, the end of the cooling procedure is not a valid method of locating points on the solidus. This determination is better made by observing the onset of melting, either by the "thaw" method (see below) or by direct microscopic observation.

It should be noted that with eutectic systems it is easy to miss the eutectic melting in mixtures containing only small amounts of the second component, that is, in nearly pure materials. Naturally, this is because the amount of eutectic melting is small, which may lead the experimenter to conclude erroneously that a solid solution is present.

Aleksandrov has argued that heating curves are preferable to cooling curves because they yield a closer approximation to equilibrium and that the melting-point method requires a smaller amount of material and makes use of simpler apparatus. All of these assertions have been challenged by Anikin (67).

In judging the relative merits of these two methods, it must be borne in mind that essentially two separate ranges of impurity content have to be dealt with. The most sensitive methods deal with 10^{-2} to $10^{-3}\%$ impurity, while most "industrial" methods deal with 10^{-1} to $10^{-2}\%$ impurity.

2.2.6 Thaw-Point Method

Rheinboldt has extended the heating-curve method and developed from it a technique he named the "thaw-point method" because the early stages of melting are reminiscent of the thawing of snow (68, 69). This technique has an antecedent in the "sinter-point curves" of Stock (70). Stock applied the method only to the case of formation of a congruently melting compound P_4S_7 from P_4S_3 and P_2S_5, while Rheinboldt applied it broadly to the general problem of determining phase diagrams.

The thaw point may be determined, as described in the sections on heating curves, by extrapolating the linear segments of a time/temperature plot or by direct visual observation. The thaw-point method gives the start and end of melting; in addition it yields eutectic and/or transformation horizontals. The end of melting gives points on the liquidus curve.

In principle, binary mixtures that form simple eutectic systems, show "thawing" at the eutectic temperature, for all compositions. In practice, in mixtures containing only a small amount of eutectic (nearly pure components), it is hard to see the appearance of eutectic melting, and the observed "thaw points" are displaced to higher temperatures. The failure to see eutectic melting in the nearly pure components leads to an erroneous impression that solid solutions exist.

A binary system that contains a congruently melting compound affords the thaw-point diagram shown in Figure 2.49. The curve indicates that a compound exists, and the maximum reveals its composition.

If the system contains an incongruently melting compound, the thaw-point diagram again shows clearly the stoichiometry of the unstable compound. This

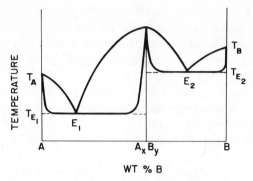

Fig. 2.49 Phase diagram determined by the thaw-point method. The solid solidus lines near the pure compounds deviate from the true equilibrium lines (shown dashed).

is the case because the upper transformation horizontal intersects with the curve rising from the eutectic horizontal at the composition of the compound.

Solid-solution systems show thaw points corresponding to the solidus, and melting points (end of melting) corresponding to the liquidus. Thaw-point determinations may be made as follows.

Mixtures of known composition may be prepared by any of the methods discussed earlier. Portions of powdered mixtures are tamped into thin-wall capillaries to a depth of about 3 mm. These are heated in a suitable air or liquid bath, close to the bulb of the measuring thermometer. The early stages of melting are best observed in oblique, lateral illumination, with magnification. Upon slow heating ($5 \times 10^{-3} \, \text{K s}^{-1}$), the mass becomes moist, and this effect is seen most clearly by reflection from the capillary wall; this temperature is the "thaw point." Continued heating leads to rapid sintering, and then to a turbid liquid. The final disappearance of the last crystals is best followed with transmitted light. The mixture in the capillary may be stirred by rotating within it a fine glass fiber or platinum wire. The temperature at which the last crystal disappears is said to be the melting point of the mixture. It corresponds to a point on the liquidus curve.

Colored substances show the thaw point most clearly as a sudden deepening of color. If there is only a slight difference between the refractive index of the melt and that of the solid, it is advantageous to add a small amount of an inert, inorganic pigment.

2.3 Zone Melting

Zone melting can produce accumulation of impurity at the end of an ingot. Since liquid of eutectic composition yields solid of the same composition, continued zoning of a system in which the liquid has attained the eutectic composition produces no further change in concentration. Thus the composition of the eutectic mixture may be determined from an analysis of samples taken   from   several   positions   in   a   zone-melted   ingot.   The acenaphthene/anthracene (11) and azulene/naphthalene (71) cases are instructive in this connection.

In principle, the eutectic composition may be determined from geometrical considerations only, without direct chemical analysis. If the eutectic is distinguishable from the pure components by color or morphology, then the eutectic concentration is related to the fractional length of the ingot showing uniform (eutectic) composition as follows

$$C_E = \frac{C_0 L}{l_E} \tag{2.14}$$

where l_E and L are the lengths of the eutectic and total ingot, respectively, C_0 is the initial, uniform starting concentration and C_E is the eutectic composition (Figure 2.50). It should also be noted that for systems in which the distribution coefficient is close to unity, conventional methods of studying phase diagrams fail because the effects of the impurity on measured properties are small. Thus directional crystallization may produce a negligibly small concentration gradient and the freezing point depression of the major component may be very small. The iterative effect of zone melting may be helpful in such cases. After a number of passes, it is usually fairly easy to decide whether the impurity concentration in the head of the ingot is enriched or depleted with respect to the starting concentration. This knowledge of whether k is greater than or less than unity gives a critical insight into the nature of the phase diagram: solute transport in the direction of zoning means $k < 1$ and solute transport in the opposite direction means $k > 1$. In the case of acenaphthene/anthracene mentioned earlier, the erroneous assignment (11) of this system to the class of continuous solid solutions (Roozeboom I) would not have been made had the authors been aware that k_{eff} for anthracene is <1 in acenaphthene at low concentration.

It is also possible to detect peritectic reactions by analysis of zone-melted ingots, in that systems showing such behavior exhibit a discontinuity in the concentration profile. This fact has been used by Joncich and Bailey (72) in their

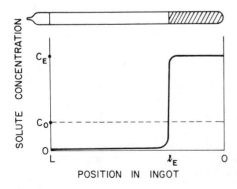

Fig. 2.50 Concentration distribution in an ingot of a eutectic system, after zone refining.

study of the anthracene/phenanthrene system, to identify a peritectic containing about 25% anthracene at 413 K. The existence of the peritectic was clearly observable in the concentration profile of a zone-melted ingot as a discontinuity between 40 and 55% anthracene (Figure 2.51). Figure 2.55 shows how zone melting (and directional crystallization) can be used to discern limiting concentrations in two types of phase diagrams.

Chistozvonova et al. (73) studied the acetonitrile/benzene system in much the same way. Zone-melting experiments confirmed the temperature and composition of the eutectic (8.5 mole% benzene at 223.7 K) as measured by thermal analysis. The starting charges for the zone-melting experiments were selected to lie on either side of the eutectic; the final concentration was the same in both cases. It was found that the system shows no solid solubility. Nevertheless, liquid mixtures containing about 50% benzene gave solid consisting of pure benzene only at the start of crystallization, which occurred at 246.7 K. A discontinuous jump was observed in the solid composition, to about 35%. This composition corresponds to that of the incongruently melting compound.

Even differences as subtle as isotopic variants may be detected by this means. Benzoic acid/benzoic acid-D and H_2O/D_2O have been investigated in this way (74). Isotopic enrichment has been effected by zone melting of zinc and cadmium; the light and heavy isotopes showed k less than and greater than unity, respectively (75, 76).

Eutectics containing only a small amount ($<5\%$) of one component are common in organic systems and the eutectic temperature is only slightly less than the melting point of the pure component (especially if solid solutions are formed). Such eutectics are easily overlooked in thermal analysis but are readily detected in zone-melted ingots. For example, the anthracene/carbazole system

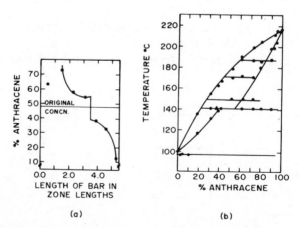

(a)   (b)

Fig. 2.51 Application of zone melting to detection of a peritectic reaction in the anthracene/phenanthrene system; (a) shows a discontinuity in the anthracene concentration near the center of the ingot and (b) shows the resulting phase diagram.

(56) shows a eutectic at 4.5% carbazole, whose freezing point is less than 1K below that of pure anthracene.

Zone melting an ingot of a eutectic-forming system gives an ingot segment of uniform (eutectic) composition. Segments of uniform concentration also result when ingots of other systems, showing maximum or minimum freezing points, are processed. (The constant compositions that can be obtained include: solid solution—maximum, solid solution—minimum, and congruently melting compound.)

A study of the bibenzyl/diphenylacetylene system showed that it consists of a continuous series of solid solutions with a minimum melting point (Roozeboom type III). The temperature and composition of the minimum were defined accurately by multipass zone melting of mixtures containing slightly more and slightly less than the estimated concentration (about 25 wt%) at the minimum. Figure 2.52 shows the concentration profile in ingots with starting concentrations of 25 and 30 wt% diphenylacetylene, respectively. The profiles converge at a concentration of 26.4 wt% and a temperature of 316.4 K (77).

Visual observation of zone-melted ingots can often provide a basis for distinguishing between a eutectic system and a system of solid solutions with a minimum melting point. The former will generally be polycrystalline while the latter may solidify as an aggregate of large crystals.

Except for water, the systems discussed in this section are organic. Considerable work has been carried out on application of the same ideas to alloys and inorganics. In an early application of the method, Yue and Clark used zone melting to confirm the concentration of the eutectic composition of the Mg/Al system. With ingots 1.6 cm in diameter and 23 cm long, with 6.4-cm zones, it was necessary to carry out two passes at $0.28 \, \mathrm{cm \, h^{-1}}$ to reach the eutectic concentration over the equilibrium length of the ingot. The Mg/Al/Zn system was studied by zoning a few arbitrarily selected compositions, which showed

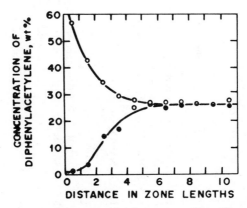

Fig. 2.52 Concentration profiles of ingots resulting from zone melting of hypo- and hypereutectic mixtures (open and closed circles, respectively).

Table 2.3 Zone Melting to Limiting Compositions (81)

System	Type	C_0	Concentration after Zoning[a]		Literature Value of Limiting Concentration
			At $g = 0$	At $g = 1$	
NaNO$_3$/NaCl	Eutectic	1.0 wt% NaCl	0.02	4.83	4.80 wt% NaCl
		10.0 wt% NaCl	23.5	4.86	
KNO$_3$/Ba(NO$_3$)$_2$	Eutectic	9.8 eq.% Ba(NO$_3$)$_2$	0.98	23.84	22.5 eq.% (Ba(NO$_3$)$_2$
		32.0 eq.% Ba(NO$_3$)$_2$	39.7	22.88	
KNO$_3$/NaNO$_3$	Minimum-melting solid solubility (Roozeboom V)	42.0 mole% NaNO$_3$	38	47.7	48–50
		59.0 mole% NaNO$_3$	62	49.8	
KCl/K$_2$Cr$_2$O$_7$	Eutectic, with limited solid solubility (Roozeboom V)	16.5 wt% KCl	21.4	8.45	8.77
		3.9 wt% KCl	2.6	8.39	

[a]Units are the same as for C_0.

segregation toward a common composition. Additional mixtures were zoned until a composition was reached in which zoning generated no segregation (78).

Agoshkov zone melted a number of binary and ternary inorganic systems and showed that limiting concentrations (eutectics and solid-solution minima) could be reached from differing starting concentrations (see Table 2.3). Three differing compositions of the system $CdCl_2/NaCl/KCl$ were zone refined and all converged to a common terminal ternary eutectic, which melted at 629 K in all three cases (79).

2.4 Growth of Single Crystals

In the discussion of the principles of solidification in Chapter 1, we observe that slow crystallization of a binary mixture may result in selective inclusion or rejection of the minor component. It is also observed that the distribution coefficient k is given by C_s/C_l, where C_s is the concentration of the minor component in the solid and C_l is the uniform concentration of the same component in the liquid, at the liquid/solid interface. Under equilibrium conditions, $k = k_0$ and represents the ratio of solidus composition to liquidus composition at constant temperature. Hence slow solidification of a binary melt will yield a crystal of initial composition $C_s = k_0 C_l$. Crystal growth from a series of binary melts containing increasing solute concentrations will yield a family of tie lines which define the solidus and liquidus curves.

There are two methods of obtaining distribution coefficients by melt crystallization. In the first of these, only a small fraction of a molten mixture is solidified on a seed. The grown crystal has essentially constant composition because only a small fraction of the nutrient melt is consumed in the growth of the crystal. In the second method, commonly known as the Bridgman–Stockbarger method (Section 4.1, Chapter 4), all of the molten starting material is converted to a crystal whose composition must necessarily change during the solidification.

2.4.1 Seeded Melt

A mixture of the two components is prepared in a cell which is immersed in a temperature-controlled bath, with provision for visual monitoring of the experiment. Crystal nuclei are allowed to form on a metal needle that is inserted into the molten mixture. At temperatures above the liquidus, these added nuclei disappear. If the inoculation procedure is repeated at progressively lower temperatures, a critical temperature is reached at which the inoculated seeds will survive. This is a measure of the liquidus temperature for the initial composition. Slight cooling (0.1 to 0.2 K) produces slow growth of the nuclei and crystals of millimeter dimensions are formed in 1.8 to 3.6×10^4 s. Analysis of the chemical composition of the grown crystals gives the solidus point for the original mixture.

In this method, only a small fraction of the starting melt is crystallized, so that the composition of the melt remains approximately constant during crystal growth. Kitaigorodskii has used this method extensively to determine organic phase diagrams and points out that, while it is laborious, it offers a high degree of accuracy (± 0.05 K, $\pm 0.05\%$ in concentration) (80).

Entrapment of melt in the growing crystal is a potential source of error in this technique and growth conditions must be chosen to minimize this effect. These include: minimum temperature fluctuation, efficient stirring in the melt, and slow growth. This effect is believed to account for the erroneous phase diagram determined by method A for the 2-chloronaphthalene/naphthalene system (85), which recently has been redetermined by two different methods (30, 8).

2.4.2 Directional Crystallization

It was pointed out earlier that the directional crystallization of a binary mixture is accompanied by a continuing change in composition as solidification proceeds. The change in concentration reflects the phase relationships existing among the components of the mixture. Conversely, unknown phase relationships may be deduced from an experimental study of solute distribution during directional crystallization. A crystal is grown and samples are taken sequentially from a number of positions along its length and later analyzed. A plot is made of relative concentration as a function of position in the ingot. Both the intercept and the slope of the log/log plot are measures of k, the distribution coefficient of the solute in the major component. This measurement gives the terminal points of a tie line; it relates solidus and liquidus compositions. Hence directional crystallization affords clear and reliable insights into phase relationships. Growth of single crystals by this method is discussed in greater detail in Section 4, Chapter 4.

An exception arises when peritectics exist. In such cases, the liquid solution crystallized above the peritectic point can "zone," that is, yield a concentration gradient in the solid during directional crystallization, which in turn prevents the peritectic reaction from going to completion, since only the solid at the solid/liquid interface can take part in the peritectic reaction. Solid remote from the interface retains the composition that existed at the time of crystallization and can equilibrate only by diffusion.

Aloi and Kirgintsev (40) used directional crystallization to study phase relationships in the controversial system $NaNO_3/RbNO_3$ (see Section 1.2.2). Samples were crystallized in a horizontal apparatus at 5.6×10^{-5} cm s^{-1}. The resulting ingots were sectioned and analyzed by flame photometry and by a radiotracer method. Rb in $NaNO_3$ gave k values of about 0.025 over the concentration range 10^{-3} to 25 wt%. Na in $RbNO_3$ gave k values of about 0.04 at low concentrations (2–7 wt%) and substantially lower values (to 0.005) at higher concentrations (to 20 wt%). It is clear that mutual solid solubility in this system is minimal. The critical central region of the diagram was explored by crystallizing two samples containing 42 wt% and 85 wt% of $RbNO_3$, respectively. One of these gave an ingot which reached a constant (eutectic) composition of 56.8 wt% $RbNO_3$ and the other a constant (eutectic) composition of 71 wt% $RbNO_3$. This evidence then, points to the existence of a congruently melting compound of composition $NaNO_3:2RbNO_3$.

Systems derived from lithium chloride and other chlorides (magnesium, calcium, copper, strontium, barium, and manganese) have been studied in this way. From experimentally measured k values and the known liquidus curve, the solidus was derived for $LiCl/MgCl_2$. The measured k values revealed solid solutions in the LiCl systems with $CuCl_2$, $SrCl_2$, and $BaCl_2$ (81).

Directional crystallization experiments do not, however, specify the temperature at which the tie line compositions are in mutual equilibrium. To determine these temperatures, separate experiments must be carried out, as described earlier in the discussion of thermal analysis. It is possible to combine the composition and temperature measurement in a single experimental sequence. To this end, a specially designed gradient oven is required. Limited regions of a phase diagram may be explored by growing single crystals in such an oven. The essential feature of the modified oven is its ability to accommodate several tubes simultaneously in a stable growth gradient (8).

Several Pyrex tubes 0.6 cm in outer diameter and 5 cm long (shown schematically in Figure 2.53 and marked 1, 2, 3) are charged with carefully weighed amounts of the components and sealed off under 5×10^4 Pa of helium. Each tube has a "tail" of solid Pyrex rod fitting snugly in stainless-steel holder 4, along with one tube containing the pure major component and another containing a pure reference material whose melting point is slightly lower than the solidification temperature of the lowest-melting binary mixture.

The assembly is lowered through a Bridgman oven made of two heaters separated by a transparent baffle which consists of three concentric Pyrex rings (7, 8, 9) held between grooved support plates (10, 11) of Transite. The temperatures in the upper and lower sections of the oven are controlled to ± 0.02 K by separate controllers actuated by thermocouples (12, 13). An illumination port and a viewing port are provide provided in the outer jacket (14) of the oven so that the growing crystals may be illuminated by lamp (I) and observed over a length of about 1 cm through cathetometer (c). The sample holder is rotated slowly to ensure that each crystal experiences the same thermal environment and to bring each tube into view between the illumination and viewing ports.

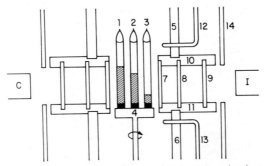

Fig. 2.53 Cross section of a differential Bridgman oven for determination of phase diagrams. (Number designations are explained in the text.)

As the crystals are lowered through the oven, the interfaces in the reference tubes remain at constant height. Positions of the interfaces in binary mixtures become lower as solidification proceeds. The interface height may be measured periodically with the cathetometer and the corresponding interface temperatures are obtained by interpolation between the solidification temperatures of the pure reference materials. The position of the interface in each mixture is noted with respect to the bottom of the ingot as well as with respect to the reference interfaces.

Crystals are grown in temperature gradients of $5-10 \, \text{K cm}^{-1}$, and interface position may be measured to about $\pm 0.01 \, \text{cm}$. The derived temperatures consequently are precise to about $\pm 0.1 \, \text{K}$.

This system is not suitable for growth under high-temperature-gradient conditions because the rotating sample holder creates turbulence in the air space between the high- and low-temperature regions. An improved version has been constructed by Sloan, in which four crystals may be grown simultaneously in an oven consisting of two temperature-controlled metal blocks separated by a transparent spacer. Four symmetrically disposed ports are provided as shown in cross section in Figure 2.54, through which closely fitting crystal tubes may be lowered. The crystals are lowered simultaneously by attachment to a screw-driven stage. The stage is equipped with a reversible, variable-speed motor

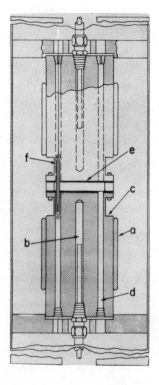

Fig. 2.54 Cross section of a variable-gradient oven for simultaneous growth of four crystals.

which drives four independent spindles connected to the crystal tubes. Using this apparatus, gradients of $<0.5\,\mathrm{K\,cm^{-1}}$ to $>100\,\mathrm{K\,cm^{-1}}$ have been achieved.

Figure 2.55 shows how concentration profiles after directional crystallization and single-pass zone refining can define the eutectic concentrations in a phase diagram of Roozeboom type V (82). Figure 2.56 shows similar diagrams for a phase diagram of Roozeboom type IV. Figure 2.57 shows limiting distributions for eight classes of phase diagrams after:

- Directional crystallization (column b)
- Zone refining (column c)
- Zone leveling (column d) (82)

Note that each mode of crystallization gives a break in the concentration profile after attainment of the limiting solubility.

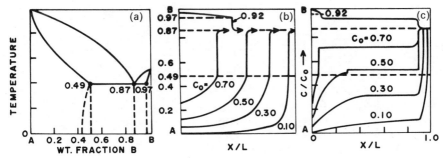

Fig. 2.55 Use of directional crystallization and zone refining to establish limiting solubility in a phase diagram of Roozeboom type V. The parameter shown for each curve is the starting concentration C_0 of solute B, expressed as weight fraction (after Vigdorovich). Note that for each value of C_0, the concentration changes discontinuously at a different position in the ingot, but at the same value of concentration. Diagrams (b) and (c) refer to directional crystallization and single-pass zone refining, respectively. The relative concentration scale is the same for both diagrams.

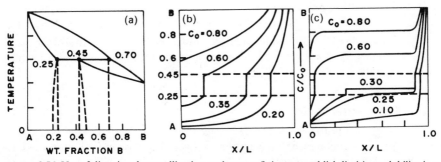

Figure 2.56 Use of directional crystallization and zone refining to establish limiting solubility in a phase diagram of Roozeboom type IV. As in Fig. 2.55, diagrams (b) and (c) refer to directional crystallization and single-pass zone refining, respectively.

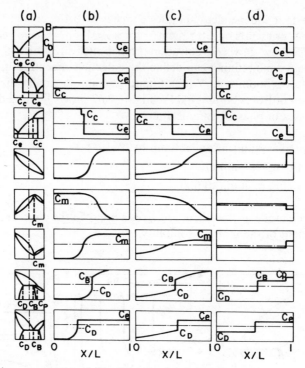

Fig. 2.57 Limiting concentration profiles for various phase diagrams after (b) directional crystallization, (c) zone refining, and (d) zone leveling. The broken lines in both the phase diagrams and the concentration profiles indicate the initial, uniform concentration.

The liquidus curve of a binary system is readily established on the basis of melting-point measurements carried out on microscopic samples. Schildknecht has shown that single-pass zone refining can be used to derive points of the solidus from this known liquidus. If a small amount (3–5 mg) of a mixture of known composition C (see Figure 2.9) is charged into a capillary and a single molten zone is passed through the ingot, the first solid to form can be recovered and its final melting point determined. From this value and the known liquidus curve, its composition C_s is established, and can then be related to that of the initial sample taken from zone melting, along isotherm T_1. In this way, a complete phase diagram can be built up from a small number of microsamples of known composition.

2.5 Miscellaneous Methods

No other technique is as generally applicable as thermal analysis, but a variety of physicochemical measurements have been proposed and used to monitor the progress of melting.

For substances showing nonzero nuclear spin, that is, exhibiting nuclear magnetic resonance (NMR) absorption, the amplitude of the NMR signal will be proportional to the fraction of a sample that is liquid. This method has been applied by Burnett and Muller (83). Herington (30) used a related method based on the large motional line narrowing which accompanies melting. Generally the line widths decreased by three orders of magnitude upon melting. The fraction melted is proportional to the area under the narrow line. A liquid fraction of less than 0.01 can be observed readily, so that this method provides a sensitve technique for the determination of solid/liquid phase diagrams. It has been applied to the systems phenol/m-cresol, phenol/o-cresol, and 1-methylnaphthalene/2-methylnaphthalene (32, 34).

Many substances fluoresce in the crystalline state but not in the liquid state. This represents yet another sensitive technique for determining the onset of melting, with even greater sensitivity for determining the end of melting, as fluorescence emission can be detected at extremely low levels. This technique can be coupled to DSC using a DSC cell with a transparent window and suitable measuring optics.

The onset of crystallization manifests itself optically as well as thermally. If a melt is cooled and continuously inoculated with seeds, the seeds will melt until the equilibrium temperature is reached. At and below the equilibrium temperature, the seeds are viable and can grow. If a light beam is passed through the melt, it will experience little or no scattering until viable nuclei persist. Hence, the onset of scattering can be used as a sensitive means of detecting crystallization. Somewhat the same methodology has been used by Kononenko (27) to detect exsolution in a solid.

REFERENCES

1. A. N. Campbell and N. O. Smith, *Alexander Findlay, The Phase Rule and Its Applications*, Dover, 1951.

2. G. A. Chadwick, "Constitutional Supercooling and Microsegregation," in M. Zief and W. R. Wilcox, Eds., *Fractional Solidification*, Dekker, New York, 1976.

3. C. D. Thurmond and J. D. Struthers, *J. Phys. Chem.*, **57**, 831–835 (1953).

4. M. Grossmann, S. G. Elkomoss, and J. Ringeissen, Eds., *Molecular Spectroscopy of Dense Phases*, Elsevier, Amsterdam, 1975.

5. A. I. Kitaigorodskii, *Molecular Crystals and Molecules*, Academic, New York, 1973, pp. 94–105.

6. A. I. Kitaigorodskii and R. M. Myasnikova, *Dokl. Phys. Chem.*, **129**, 963–966 (1959).

7. A. I. Kitaigorodskii and R. M. Myasnikova, *Kristallografiya*, **5**(4), 638–642 (1960).

8. A. R. McGhie and G. J. Sloan, *J. Cryst. Growth*, **32**, 60–67 (1976).

9. N. B. Chanh, Y. Haget, F. Hannoteaux, and N. Chezeau, *J. Chim. Phys. Physicochim. Biol.*, **72**(5), 670–674 (1975).

10. V. M. Kravchenko *Ukrainskii Khim. Zhur.*, **19**, 599–609 (1953).

11. R. P. Rastogi and K. T. Rama Varma, *J. Chem. Soc.*, 2097–2101 (1956).

12. F. Baumgarth, N. B. Chanh, R. Cay, J. Lascombe, and N. LeCale, *J. Chim. Phys. Physicochim. Biol.*, **66**, 862–868 (1969).

13. J. W. Harris, *Nature*, **206**, 1038 (1965).

14. A. I. Kitaigorodskii, R. M. Myasnikova, I. E. Kozlova, and A. I. Prtsin, *Sov. Phys. Cryst.*, **21**, 511–514 (1976).

15. H. C. Wolf and H. Port, "Excitons in Aromatic Crystals," in M. Grossmann, S. G. Elkomoss, and J. Ringeissen, Dds., *Molecular Spectroscopy of Dense Phases*, Elsevier, Amsterdam, 197.

16. H. Dorner, R. Hundhausen, and D. Schmid, *Chem. Phys. Letters*, **53**, 101–104 (1978).

17. R. Ostertag and H. C. Wolf, *Chem. Phys. Letters*, **22**, 65–67 (1973).

18. A. I. Kitaigorodskii and L. V. Yakushevich, *Sov. Phys. Solid State*, **13**, 1998–2002 (1972).

19. H. W. B. Roozeboom, *Z. Phys. Chem.*, **10**, 145–164 (1892).

20. F. N. Rhines, *Phase Diagrams in Metallurgy*, McGraw-Hill, New York, 1956, p. 25.

21. J. N. Sherwood, *Surface Defect Prop. Solids*, **2**, 250–268 (1973).

22. J. N. Sherwood, *Proc. Brit. Ceram. Soc.*, **9**, 233–242 (1967).

23. J. H. Adriani, *Z. Phys. Chem.*, **33**, 453–476 (1900).

24. H. A. J. Oonk, K. H. Tjoa, F. E. Brants, and J. Kroon, *Thermochim. Acta*, **19**, 161–171 (1977).

25. Y. Haget, N. B. Chanh, A. Corson, and A. Meresse, *Mol. Cryst. Liq. Cryst.*, **31**, 93–104 (1975).

26. O. Kubaschewski and G. Heymer, *Acta Met.*, **8**, 416–423 (1960).

27. V. G. Kononenko and W. Heichler, *Kristall Technik*, **12**(10), 1031–1036 (1977).

28. A. Prince, *Alloy Phase Equilibria*, Elsevier, Amsterdam, 1966.

29. Ibid, p. 129.

30. N. B. Chanh, Y. Bouillaud, and P. Lencrerot, *J. Chim. Phys. Physiochem. Biol.*, **67**(6), 1206–1212 (1970).

31. F. N. Rhines, *Phase Diagrams in Metallurgy*, McGraw Hill, New York, 1956, p. 128.

32. E. F. G. Herington and I. J. Lawrenson, *J. Appl. Chem.*, **19**, 341–344 (1969).

33. E. F. G. Herington and I. J. Lawrenson, *Nature*, **219**, 928 (1968).

34. E. F. G. Herington and I. J. Lawrenson, *J. Appl. Chem.*, **19**, 337–341 (1969).

35. F. N. Rhines, "An Overview of the Determination of Phase Diagrams," in *Applications of PhaseDiagrams in Metallurgy and Ceramics*, Proceedings of a Workshop at National Bureau of Standards, Gaithersburg, MD, Jan. 1977, NBS-SP496.

36. A. N. Winchell, *Optical Properties of Organic Compounds*, Academic, New York, 1954.

37. W. C. McCrone, *Fusion Methods in Chemical Microscopy*, Interscience, New York, 1957.

38. N. H. Hartshorne and A. Stuart, *Crystals and the Polarizing Microscope*, 4th ed., Arnold, London, 1970.

39. K. A. Jackson and J. D. Hunt, *Acta. Metal.*, **13**, 1212–1215 (1965).

40. A. S. Aloi and A. N. Kirgintsev, *Akad. Nauk SSSR Sib. Otd. Ser. Khim. Izv.*, **1**, 59–61 (1977).

41 A. Kofler, *Monatsh. Chem.*, **86**, 643–652 (1955).

42. C. M. Kramer and C. J. Wilson, *Thermochim. Acta*, **42**, 253–264 (1980).

43. A. N. Kirgintsev and V. E. Kosyakov, *Izv. Akad. Nauk SSSR. Ser. Khim.*, 2208–2213 (1968).

44. H. Nojima, *Bull. Chem. Soc. Japan*, **51**, 2513–2517 (1978).

45. N. Ia. Kolosov, *Kristallografiya* **3**, 700–705 (1958).

46. A. N. Campbell and A. J. R. Campbell, *J. Am. Chem. Soc.*, **59**, 2481–2488 (1937).

47. P. M. Robinson, H. G. Rossell, and H. G. Scott, *Mol. Cryst. Liquid Cryst.*, **10**, 61–74 (1970).

48. A. Nguyen Van Mau, M. Averous, and G. Bougnot, *Mat. Res. Bull.*, **8**, 717–720 (1973).

49. W. J. Kolkert, *J. Crystal Growth*, **30**, 213–219 (1975).

50. A. G. Grimm, M. Gunther, and A. Tittus, *Z. Physikal. Chem.*, **14**, 169–218 (1931).

51. H. P. Vaughan, *Thermochim. Acta*, **1**, 111–126 (1970).

52. W. C. McCrone, *Fusion Methods in Chemical Microscopy*, Interscience, New York, 1957, p. 154.

53. W. W. Wendlandt, *Thermal Methods of Analysis*, Wiley, New York, 1974.

54. D. E. Etter, P. A. Tucker, and L. J. Wittenberg, "Application of Differential Thermal Analysis to the Study of Phase Equilibria in Metal Systems," in R. F. Schwenker, Jr. and P. D. Garn, Eds., *Thermal Analysis*, Vol. 2, Academic, New York, 1969, pp. 829–850.

55. M. J. Visser and W. H. Wallace, *Du Pont Thermogram*, **3**(2), 9–11 (1966).

56. P. M. Robinson and H. G. Scott, *Mol. Cryst. Liquid Cryst.*, **5**, 387–404 (1969).

57. A. Van Tets and H. G.Wiedemann, "Simultaneous Thermomicroscopic and Differential Thermal Investigations of Melting and Freezing," in R. G. Schwenker, Jr. and P. D. Garn, Eds., *Thermal Analysis*, Vol. 1, Academic, New York, 1969, pp. 121–135.

58. G. W. Smith, *Mol. Cryst. Liq. Cryst.*, **42**, 307–318 (1977).

59. M. Radomska and R. Radomski, *Thermochim. Acta*, **40**, 415–425 (1980).

60. W. J. Kolkert, *J. Crystal Growth*, **30**, 225–232 (1975).

61. J. A. Bouwstra, N. Brouwer, A. G. G. Van Genderen, and H. A. J. Oonk, *Thermochim. Acta*, **38**, 97–107 (1980).

62. J. S. Brennan, N. M. D. Brown, and F. L. Swinton, *J. Chem. Soc. Faraday Trans. 1*, **70**, 1965–1970 (1974).

63. R. G. Hill, E. O'Kane, and F. L. Swinton, *J. Chem. Thermo.*, **10**, 1205–1207 (1978).

64. G. Tammann, *Z. Anorg. Chem.*, **37**, 303–313 (1903).

65. H. F. Van Wijk and W. M. Smit, *Anal. Chim. Acta*, **23**, 545–551 (1960).

66. W. M. Smit, *Rec. Trav. Chim.*, **75**, 1309–1320 (1956).

67. A. G. Anikin, *J. Anal. Chem.*, **31**, 999–1000 (1976).

68. H. Rheinboldt, K. Hennig, and M. J. Kircheisen, *J. Prakt. Chem.*, **111**(2), 242–272 (1925).

69. H. Rheinboldt, "Thermal Analysis, Determination of Organic Molecular Compounds," in E. Müller, Ed., *Die Methoden der Organischen Chemie*, 4th ed., Vol. 2, Part 1, Thieme, Stuttgart, 1953, pp. 827–865.

70. A. Stock, *Ber.*, **42**, 2059–2061 (1909).

71. K. Rokos, P. M. Robinson, and H. G. Scott, *Mol. Cryst. Liquid Cryst.*, **24**, 331–336 (1973).

72. M. J. Joncich and D. R. Bailey, *Anal. Chem.*, **32**, 1578–1581 (1960).

73. O. S. Chistozvonova, G. M. Dugacheva, and A. G. Anikin, *Zh. Fiz. Khim.*, **41**, 42–47 (1967).

74. G. J. Sloan, *J. Am. Chem. Soc.*, **85**, 3899–3900 (1963).

75. O. A. Troitskii, *Zh. Tekh. Fiz. pis'ma*, **2**(24), 1126–1129 (1976).

76. O. A. Troitskii and V. P. Shcheredin, *Zh. Tekh. Fiz.*, **47**(2), 469 (1977).

77. H. Nojima and S. Akehi, *Bull. Chem. Soc. Japan*, **53**, 2067–2073 (1980).

78. N. Yonehara, M. Kamada, and K. Fukunaga, *Bunseki Kagaku*, **30**, 619 (1981).

79. V. M. Agoshkov, *Doklady Akad. Nauk SSSR*, **152**(1), 96–99 (1963).

80. A. I. Kitaigorodskii and V. M. Kozhin, *Kristallografiya*, **4**, 209–213 (1959).

81. I. F. Kravchuk, *Zh. Neorg. Khim.*, **24**(10), 2764–2769 (1979).

82. V. N. Vigdorovich, *Akad. Nauk. SSSR Inst. Met.*, 24–31 (1972).

83. L. J. Burnett and B. H. Muller, *Nature*, **219**, 59–60 (1968).

Chapter III

DIRECTIONAL CRYSTALLIZATION

1   SOLUTE REDISTRIBUTION

The goal of the study of solute redistribution is to understand observed solute distributions in terms of basic thermodynamic relationships and to predict distributions in terms of measured system properties. The discussion in Chapter 1 shows qualitatively that a solid formed at a solid/liquid interface can have a composition different from that of the liquid; this results in enrichment or depletion of solute in the interfacial liquid region. The concentration gradient thus formed in the liquid is reduced by diffusive flow of solute into the bulk liquid. The goal of theory is to understand and predict the balance between the two processes: solute enrichment or depletion resulting from crystal growth, and liquid homogenization resulting from diffusion. This understanding is required in order to be able to predict the overall solute distribution resulting from fractional solidification.

Several factors must be considered:

1. The degree of mixing in the liquid
2. The shape of the interface
3. Substance-specific properties such as diffusivity, thermal conductivity, viscosity, and latent heat of fusion

1.1   Degree of Mixing in the Liquid

In considering what the concentration profile of a solute at a solid/liquid interface will look like, three conditions must be examined:

- Completely mixed liquid
- Quiescent liquid
- Partially mixed liquid

In Chapter 1, the concept of a distribution coefficient k is introduced. For the simplest of phase diagrams (and one that is prevalent in organic systems), the impurity is totally insoluble in the solid host (see Fig. 2.1); here $k_0 = 0$ and $k_{eff} \simeq 0$. The equilibrium distribution coefficient k_0 has the value C_s/C_l derived from the phase diagram and it is a characteristic of the binary system. Hence C_s is the concentration of impurity in the solid at the interface and C_l is the (uniform) concentration in the liquid at equilibrium; k_{eff} is defined as the distribution coefficient that is measured after solidification under a given set of experimental conditions. As a measured quantity, its value can vary and is influenced by the experimental conditions of solidification; it is defined as $k_{eff} = C_s/C_0$ where C_0 is the average solute concentration in the melt, excluding the interface transient (see below).

Consideration of the phase diagrams described in Chapter 2 leads quickly to an awareness that k_0 and k_{eff} can be greater than or less than unity, depending on the slopes of the liquidus and solidus. In any case, k_{eff} always lies between k_0 and unity (see below).

More generally, the phase diagram may show formation of a solid solution between the two components, and the slopes of liquidus and solidus may be negative or positive. That is, the mixtures may show lower or higher freezing points than the major component.

For a phase diagram of type I in the Roozeboom classification (see Figure 2.9), the solidus and liquidus may be represented by straight lines over a small concentration range. For negative slope, k_0 is less than 1; that is, solidification results in formation of a solid containing less of the solute than the original liquid. Cooling a liquid of composition C_l denoted by the vertical line in Fig. 3.1a results in formation of a solid of composition C_s. For a phase diagram of positive slope (Figure 3.1b), cooling liquid of composition C_l results in formation of solid of composition C_s, which is richer in solute. It should be noted that several names are applied to "k" as defined here. It has been variously called segregation coefficient, partition coefficient, and distribution coefficient. The last of these appears to be most widely used and is preferred.

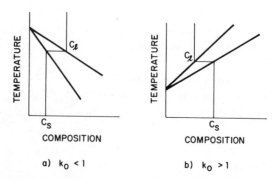

a) $k_0 < 1$ b) $k_0 > 1$

Fig. 3.1 Schematic representation of the termini of a phase diagram of Roozeboom type I, showing distribution coefficients less than and greater than unity. (a) $k_0 < 1$; (b) $k_0 > 1$.

If complete equilibrium is maintained in both solid and liquid phases, then solute concentration in the solid changes toward homogeneity, by diffusion during solidification. After partial crystallization of a portion of liquid, the remaining liquid can be separated from the solid and it will be richer in solute for $k_0 < 1$ (and, of course, poorer in solute for $k_0 > 1$). In the phase diagram shown in Figure 3.2, the liquid C_l first yields solid of composition C_s at temperature T_1. As solidification proceeds, the melt becomes richer in solute; at a later stage of crystallization such as T_2, the system consists of liquid of composition M and solid of composition N in the ratio given by the lever rule as NL/ML. The equilibrium solid has uniform composition N. Separation of the phases at this instant would provide solid that is purer than the original liquid. If crystallization is continued until the entire charge solidifies, then its concentration is given by Q, and is the same as that of the starting liquid. Hence true equilibrium solidification gives a solid whose composition is uniform and equal to that of the starting liquid.

Experimentally this situation is rarely, if ever, achieved, nor is it often desired. Most crystallizations are carried out for the purpose of achieving a concentration difference between solid and liquid. In fact, it is possible to do so because of the large difference between diffusive mixing in the solid and convective mixing in the liquid. Even in the hypothetical case in which solid diffusivity equals that of the liquid, the solid diffusivity has only a small effect on the grown-in solute distribution for $k_0 < 1$. For $k_0 > 1$ on the other hand, there is a substantial effect. Normally the composition of the liquid is at least approximately uniform, while the concentration profile of the solute in the solid is "frozen in."

The concentration in an ingot undergoing *equilibrium* solidification, after a fraction g has solidified, is given by (2)

$$C = \frac{k_0 C_0}{1 + g(k_0 - 1)} \tag{3.1}$$

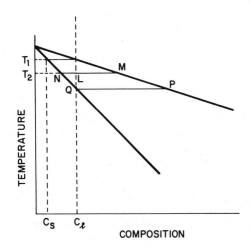

Fig. **3.2** Equilibrium solute distribution during solidification.

In this equation C, C_0, and k_0 are expressed in consistent units such as atom, mole, or weight fraction. These units are independent of density and are especially suitable for use with organic chemicals, which typically show large changes in density upon solidification.

1.1.1   Completely Mixed Liquid

The case considered in the last section was an extreme, not generally achieved in practice. Considering now the other extreme, we assume that there is no diffusion in the solid and that mixing in the liquid is perfect. If a cylindrical charge of liquid is solidified directionally in this regime, the solid will show a concentration profile defined by the slope of the liquidus line of the phase diagram. Thus, in Figure 3.3a, liquid of composition C_0 has begun to solidify. The bar below the graph shows the extent of solidification in the ingot. In this regime, the concentration profile in the solid is accompanied by a concentration discontinuity at the interface. The ratio of solid-to-liquid concentration after solidification of fraction x is $C_s(x)/C_l(x)$, and is the distribution coefficient k_0 defined by the phase diagram. Figure 3.3b shows solute distribution after further solidification.

Of the two assumptions on which this analysis is based (namely no diffusion in the solid and perfect mixing in the liquid), the former is more nearly satisfied in practice. Mixing in the liquid may provide homogeneity in the bulk, but there is always a layer of liquid in contact with the solid interface in which transport is diffusive and limited. Hence the solid/liquid discontinuity shown in Figure 3.3a is not achieved in practice.

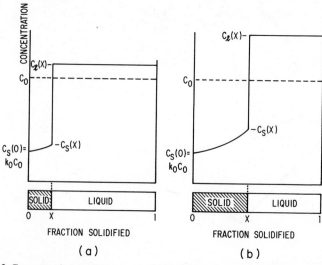

Fig. 3.3 Concentration profile near a solid/liquid interface, at two stages of solidification.

Under the assumptions made here, the concentration profile generated by solidification is given by

$$C_s = k_0 C_0 (1 - g)^{(k_0 - 1)} \qquad (3.2)$$

where C_s is the solute concentration in the solid at the interface after fraction g of the original sample has solidified. The profile is shown in Figure 3.4 as the curve designated "Equation 2" (see below).

Real chemical systems, of course, rarely show constant k_0; that is, the slopes of solidus and liquidus vary with concentration, and so must k_0. Nevertheless, k_0 is a useful quantity in describing the inherent separability of binary systems, since it defines the "best" separation that can be achieved in a single crystallization. If k_0 differs sufficiently from unity, then practical purification and/or solute enrichment may be achieved in such a single-stage crystallization (see Chapter 4). Even if k_0 is close to unity, repeated partial crystallization, with rejection of the molten portion of the charge, can lead to substantial purification. However, the operation can become tedious and time consuming, and zone refining might be preferred. (See Chapter 5).

Equation 3.1 cannot be valid for all concentrations. In the first place, k_0 cannot be constant over the entire range of composition, although it may be assumed to be constant over small concentration ranges. Second, the equation does not apply at eutectic or peritectic points, since no solute distribution takes place upon solidification of these compositions.

Attempts to deal with the real world of varying k have been of two kinds: phenomenological and mechanistic. Several of the former approaches will be described. These arbitrarily assume some nonlinear temperature/composition relationship in the phase diagram and calculate an appropriate k or set of k's. The mechanistic approaches attempt to explain varying k by considering kinetic as well as diffusive processes at the solid/liquid interface.

Li (3) has treated solute distribution arising from directional crystallization, without assuming constancy of k_0. Instead, he described solidus and

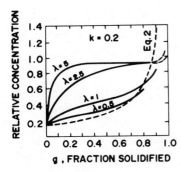

Fig. 3.4 Concentration profiles after directional crystallization with varying degrees of mixing in the melt, for $k = 0.2$; λ is discussed in ref. 9.

liquidus by empirical equations that are quadratic in temperature:

$$C_s = a_s T + b_s T^2 \tag{3.3}$$

$$C_l = a_l T + b_l T^2 \tag{3.4}$$

where a_s, b_s, a_l, and b_l are curve-fitting constants for solidus and liquidus, respectively. From these, an equation was derived relating the instantaneous temperature T to g, the fraction solidified.

$$(1 - g) = \left(\frac{T}{T_0}\right)^x \left[\frac{a_l - a_s + (b_l - b_s)T}{a_l - a_s + (b_l - b_s)T_0}\right]^y \tag{3.5}$$

where T_0 is the temperature at the start of crystallization (i.e., when $g = 0$) and

$$x = \frac{a_l}{a_s - a_l} \quad \text{and} \quad y = \frac{a_l b_l + a_l b_s - 2b_l a_s}{(a_s - a_l)(b_l - b_s)}$$

Li applied Equations 3.3, 3.4, and 3.5 to the solidification of a germanium/antimony alloy containing 10% antimony. He used published data on the quadratic fit to solidus and liquidus curves to calculate the fraction of an arbitrary sample remaining molten, $(1 - g)$, at a series of temperatures. Naturally, the values of $(1 - g)$ calculated from the empirically fitted curves give the "exact" answer. These data were compared with values of $(1 - g)$ calculated respectively by the Runge–Kutta and finite-difference methods, using the lever rule and the classical directional crystallization equation (Equation 3.2), assuming constant k. The fractional differences from the exact values of g were -4.4×10^{-6}, 1.9×10^{-3}, 2.2×10^{-2}, and 3.9×10^{-2}, respectively [4].

Other treatments of the problem of varying k_0 have been provided by Wilcox [5]. One of these involves integration of the material-balance relationship

$$d \ln k = \left(1 - k + \frac{d \ln X_s}{d \ln(1 - g)}\right) d \ln(1 - g) \tag{3.6}$$

by numerical methods. If solute concentration in the ingot is known as a function of fraction solidified, g, and the value of k is known at $g = 0$, then k_i, the value upon solidification of the ith interval of the ingot, is given by

$$\ln k_i = \ln k_{i-1} + \left\{1 - k_{i-1} + \frac{\ln(C_{s,i}/C_{s,i-1})}{\ln[(1 - g_i)/(1 - g_{i-1})]}\right\} \ln \frac{1 - g_i}{1 - g_{i-1}} \tag{3.7}$$

The second method involves direct integration of the concentration profile. C_s and g can be measured in the solidified ingot. The instantaneous liquid concentration at any value of g is given by

$$C_l = \frac{\int_g^1 C_s \, dg}{1-g} = \frac{C_0 - \int_0^g C_s \, dg}{1-g} \tag{3.8}$$

where C_0 is the initial concentration in the liquid and $k \doteq C_s/C_l$. Figure 3.5 shows the areas represented by the integrals.

In a general theory of solute redistribution, Matz (6) introduced the idea of a separation factor, ϕ, for a binary system undergoing solidification; ϕ involves the relative concentrations of both components in both phases

$$\phi = \frac{C_{A,l}/C_{B,l}}{C_{A,s}/C_{B,s}} \tag{3.9}$$

where $C_{A,l}$ and $C_{A,s}$ are the concentrations of the lower-melting component in the liquid and solid phases, respectively and $C_{B,l}$ and $C_{B,s}$ are the corresponding values for the higher-melting component. This definition derives from distillation theory, where a similar relationship between liquid and vapor concentrations is called relative volatility. By analogy, ϕ is called relative meltability. If $X_{A,l}$ and $X_{A,s}$ are the mole fractions of A in liquid and solid, respectively, then

$$\phi = \frac{X_{A,l}(1 - X_{A,s})}{X_{A,s}(1 - X_{A,l})} \tag{3.10}$$

and the solid/liquid equilibrium is given by

$$X_{A,l} = \frac{\phi X_{A,s}}{1 + (\phi - 1)X_{A,s}} \tag{3.11}$$

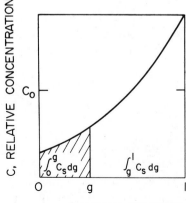

Fig. 3.5 Evaluation of k by the integral-impurity method.

A material balance in a partially solidified charge containing N moles of solute in the melt shows that

$$X_{A,l} = \frac{N v_l}{(v_0 - v)} \tag{3.12}$$

where v_l is the molar volume of the remaining liquid, v_0 is the original volume of liquid, and v is the volume solidified. If differential volume dv solidifies, containing concentration $X_{A,s}$ of solute, then

$$dN = -\frac{X_{A,s} dv}{v_s} \tag{3.13}$$

where v_s is the molar volume of the solid. Differentiation of Eq. 3.12 gives

$$dN = [(v_0 - v) \, dX_{A,l} - X_{A,l} dv]\left(\frac{1}{v_l}\right) \tag{3.14}$$

From Equation 3.11, after separation of variables v and $X_{A,s}$, and equating v_s with v_l, one obtains

$$\frac{dX_{A,s}}{[1 + (\phi - 1)X_{A,s}]X_{A,s}(1 - X_{A,s})} = \frac{\phi - 1}{\phi} \cdot \frac{dv}{v_0 - v} \tag{3.15}$$

Equation 3.15 may be integrated by parts at constant ϕ (i.e., when ϕ is independent of $X_{A,s}$) to give

$$\frac{[1 + (\phi - 1)X_{A,s}]^{\phi-1}(1 - X_{A,s})/(1 - C_0)}{(\phi X_{A,s}/C_0)^{\phi}} = \left(\frac{v_0 - v}{v_0}\right)^{\phi - 1} \tag{3.16}$$

Equation 3.16 may be evaluated numerically for a sequence of assumed values of $X_{A,s}$, with ϕ as parameter. Figure 3.6 shows curves obtained for $1.2 < \phi < 1000$ and $C_0 = 0.1$. It should be noted that when $[(\phi - 1)X_{A,s}] \ll 1$ and $C_0 \ll 1$, then Eq. 3.14 gives

$$X_{A,s} = \frac{C_0}{\phi(1 - v/v_0)^{(1/\phi) - 1}} \tag{3.17}$$

By setting $1/\phi = k$, Equation 3.17 gives Eq. 3.2, since $v_s = v_l$ and $X_{A,s}$ is the solute concentration in the solid, expressed as mole fraction.

In Matz's treatment, no assumption is made regarding constancy of ϕ. In fact, if k is constant and equal to $X_{A,s}/X_{A,l}$, then differentiation of Equation 3.10 with respect to $X_{A,s}$ shows that $d\phi/dX_{A,s} > 0$ for $k < 1$, and correspondingly, for

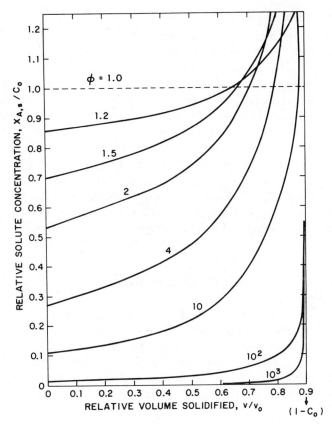

Fig. 3.6 Solute concentration profiles as a function of fraction solidified, for various separation factors and an initial concentration $C_0 = 0.1$ (after Matz).

$k > 1$, $d\phi/d(1 - X_{A,s}) > 0$. If in Eq. 3.10, $X_{A,l}$ is replaced by $X_{A,s}/k$, then

$$\phi = \frac{1 - X_{A,s}}{k - X_{A,s}} \tag{3.18}$$

Equation 3.18 may be approximated by a geometric series in $X_{A,s}/k$ and for dilute solutions ($X_{A,l} \ll 1$), higher-order terms may be ignored, yielding

$$\phi = \frac{1}{k}\left[1 + \left(\frac{1}{k} - 1\right)X_{A,s} + \frac{1}{k}\left(\frac{1}{k} - 1\right)X_{A,s}^2\right] \tag{3.19}$$

If $\phi_0 = 1/k$, then

$$\phi = \phi_0[1 + (\phi_0 - 1)X_{A,s} + \phi_0(\phi_0 - 1)X_{A,s}^2] \tag{3.20}$$

1.1.2 Quiescent Liquid

If the melt is totally quiescent, then solute transport at the interface is completely controlled by diffusion and the bulk of the melt retains its original composition, C_0. Upon solidification, the solute concentration in the transient layer $C_l(x)$ increases until its value is C_s/k_0, that is, $k_{eff} = 1$. This extreme situation is not often realized in practice, nor is it desired. Generally k_{eff} lies between k_0 and unity. However, if solidification were carried out at a finite speed under conditions of no mixing, only an initial transient region would show purification. The bulk of the ingot would have the concentration of the original melt and material balance requires a terminal concentration gradient corresponding to deposition of the solute removed in the initial transient. The transient is typically of small dimension, for example, 0.01–0.1 cm. The concentration profile resulting from directional crystallization in this regime is approximated in Figure 3.4 by the curve for $\lambda = 5$ (see Section 1.2.3).

1.1.3 Partially Mixed Liquid

If mixing is incomplete at the interface, then some of the impurity rejected from recently formed solid will remain close to the interface, generating an impurity-rich boundary layer in the liquid. The resulting concentration gradient in the liquid is naturally opposed by diffusion, which is, by definition, a process leading to equalization of concentration within a single phase.

Tiller et al. (7) analyzed the dynamics of solute distribution in this regime and derived the following relationship between position x in the liquid (measured from the solid/liquid interface) and solute concentration $C_l(x)$ at that point (see Figure 3.7):

$$C_l(x) = C_l(0) \exp\left(-\frac{V_x}{D}\right) + C_0 \tag{3.21}$$

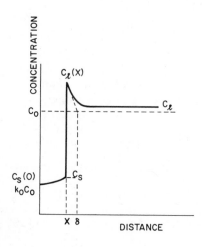

Fig. 3.7 Enhanced solute concentration in the liquid near solid/liquid interface.

where $C_l(0)$ is the instantaneous solute concentration in the liquid at the interface, V and D are growth rate and diffusion coefficient, respectively, and C_0 is the solute concentration in the initial melt. Equation 3.21 describes an exponential decay of solute concentration in the liquid adjacent to the interface; its form is determined by the ratio (V/D). Since D is a constant, growth rate V becomes the controlling factor determining the ultimate effect of diffusive mixing upon the distribution.

Under steady-state conditions, $C_s = C_0$ and $C_l(0) = (C_0/k_0) - C_0$, which upon substitution into Equation 3.21 gives

$$C_l(x) = C_0\left[1 + \frac{1 - k_0}{k_0} \exp\left(-\frac{Vx}{D}\right)\right] \qquad (3.22)$$

Assuming that the rise of C_s is of exponential form, and applying boundary conditions, Tiller et al. derived an equation for the solute concentration in the solid, C_s, at any point x in the crystal (measured from the start of crystallization):

$$C_s = C_0\left\{(1 - k_0)\left[1 - \exp\left(-k_0\frac{Vx}{D}\right)\right] + k_0\right\} \qquad (3.23)$$

This equation describes an asymptotic rise of solute concentration from k_0C_0 at $x = 0$, toward C_0.

The general correctness of the description embodied in Figure 3.7 and Equation 3.21 has been confirmed by experiment with crystals grown from dilute solutions of radioactive zinc (^{65}Zn) in tin (8). The growth ampoules were quenched after about half the charge had crystallized, then the distribution of solute was measured by autoradiography and by sectioning the monocrystalline and polycrystalline solid on either side of the interface. Figure 3.8 shows one of the measured profiles. The measured rise at the interface is not discontinuous; this may be a result of segregation during the quench, or of the fact that the interface was not exactly perpendicular to the growth axis.

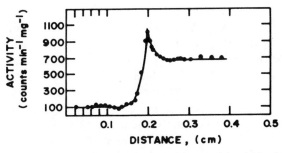

Fig. 3.8 Concentration profile in a quenched specimen of tin containing 0.05 wt% ^{65}Zn, after growth at 2.44×10^{-4} cm s^{-1}.

Kulik and Zil'berman (9) have given a more general form of Equation 3.23 in which the exponential term contains a dimensionless parameter λ, which characterizes the extent of diffusive mixing in the liquid. It is defined by

$$\lambda^2 = \frac{VL}{4D} \qquad (3.24)$$

where L is the total length of the charge.

The general distribution equation is plotted in Fig. 3.4 with λ as the parameter. The dashed curve for very small λ is the one given by Eq. 3.2 for complete mixing in the liquid. Intermediate values of λ give curves with an asymptotic rise toward C_0.

Under conditions of incomplete mixing, the concentration of impurity in the advancing crystal is not determined by the impurity concentration in the bulk liquid but rather in the "diffusion layer" adjacent to the solid. Burton, Prim, and Slichter (BPS) (10) have analyzed impurity distribution at the interface of a rotating crystal being withdrawn from a melt. It has been found experimentally that the BPS treatment applies equally well to crystals grown by the Bridgman–Stockbarger technique (see Section 4, Chapter 4). The BPS treatment was based on the assumption that fluid flow is laminar, that is, that there is no turbulence in the liquid. They assumed also that flow velocity normal to the immersed, planar face of the crystal is independent of distance from the rotation axis. It is important to note that the BPS treatment is one-dimensional; it is thereby limited to the case of a radially uniform solute concentration in the crystal. We shall soon see that this condition is not always met.

In the regime posited by BPS, solute partitioning at the interface, as mentioned above, is determined by the solute concentration in the transient layer at the interface rather than by concentration in the bulk. The interface diagrams shown in Figure 3.3 must be modified to reflect these conditions, as in Figure 3.7. The BPS equation relating k_{eff} to the conditions of crystallization is

$$k_{eff} = \frac{k_0}{k_0 + (1 - k_0)e^{-\Delta}} \qquad (3.25)$$

where $\Delta = V\delta/D$, V is the growth rate, and δ is defined as the thickness of the layer in which solute transport is diffusion controlled (see Figure 3.7). D is the diffusion coefficient of the solute, which is assumed to be independent of concentration. This equation has been widely used in the analysis and interpretation of solidification processes, since the quantity (δ/D) characterizes the nature of an experimental crystallization process. While diffusion coefficients are not generally known accurately, values of (δ/D) may be determined by measuring k_{eff} for crystals grown at different rates under the same mixing conditions. Such experiments are analyzed by plotting $\ln(1/k_{eff} - 1)$ against V, giving a line whose slope is $(-\delta/D)$ (11)

$$\ln\left(\frac{1}{k_{eff}} - 1\right) = \ln\left(\frac{1}{k_0} - 1\right) - V\delta/D \qquad (3.26)$$

This procedure is discussed more fully in Chapter 9.

The BPS treatment has been applied fruitfully to many kinds of materials. For example, Zharinov et al. (12) made a careful study of the directional crystallization of aqueous sodium chloride solutions at various speeds and concentrations, with vigorous mixing of the liquid. Their results are shown in Figure 3.9 where each curve refers to crystallizations of a single concentration at different rates. Table 3.1 shows characteristics of each curve. For a concen-

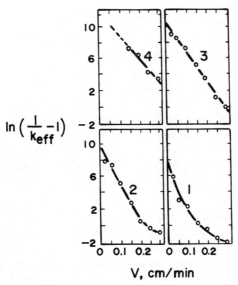

Fig. 3.9 BPS plots for directional crystallization of stirred NaCl solutions (12); 1(0.38%), 2(0.12%), 3(0.05%), 4(0.0025%).

Table 3.1. Solidification of Aqueous Sodium Chloride Solutions

C_0, wt%	$\delta/D \times 10^{-3}$, s cm^{-1}	k_0 (extrapolated)
0.38	—	10^{-4}
0.12	27	6×10^{-5}
0.05	21	4×10^{-5}
0.0025	19.4	0.5×10^{-5}

tration of 0.12% at $V > 2.5 \times 10^{-3}$ cm s^{-1}, and for a concentration of 0.38% at virtually all the rates investigated, δ/D varied markedly with concentration.

Wilson (13) has proposed a modification of the BPS treatment, substituting a quantity $\bar{\delta}$ for the δ term of BPS; the former is defined in terms of the physical properties of the liquid, and is not necessarily the same as the boundary-layer thickness. At low growth rates, the two definitions are equivalent.

The BPS theory predicts that k_{eff} will lie between k_0 and 1, assuming constant growth conditions. In practice, growth conditions often fluctuate periodically, causing periodic oscillations in solute distribution. The differential equation describing solute distribution in the diffusive boundary layer was solved, with consideration of time dependence and solid-state diffusion (14). Under conditions of periodic oscillation, a surprising result ensued: k_{eff}, related to the mean solid-state concentration, can be higher or lower than the steady-state value, or even lower than k_0 (for $k_0 < 1$).

1.2 Constitutional Supercooling

Figure 3.10b repeats the concentration profile of Figure 3.7 along with the relevant portion of the equilibrium phase diagram (Figure 3.10a) and the impressed temperature gradient relative to the equilibrium crystallization temperatures in the interface region (Figure 3.10c).

Examination of this figure leads to a number of insights into the details of interface morphology. It is evident immediately that every point on the concentration transient in the interface liquid corresponds to a particular point on the liquidus of the phase diagram. In other words, the liquid of highest solute concentration, immediately adjacent to the solid (C_1 in Figure 3.10b) has the lowest equilibrium crystallization temperature (T_1 in Figure 3.10a). Liquid of progressively lower solute concentration, farther from the solid interface, has higher equilibrium crystallization temperature; for example, concentrations C_2 and C_3 correspond to equilibrium crystallization temperatures T_2 and T_3. In Figure 3.10c, curve A shows the equilibrium crystallization temperature as a function of position in the interfacial region; curve B shows an actual (experimentally imposed) temperature distribution in the same region. It follows that some of the liquid near the interface is at a temperature below the equilibrium crystallization temperature (shaded region); hence it is supercooled. This condition exists in spite of the presence of the adjacent solid; it differs from supercooling that results from the absence of a nucleus, and is called "constitutional supercooling" to direct attention to the fact that the supercooling derives from the constitution or composition of the liquid (15). The existence of this region of constitutional supercooling can give rise to three effects (16):

- Cellular structure
- Cellular-dendritic structure
- Nucleation ahead of the solid/liquid interface

These will be treated briefly in turn.

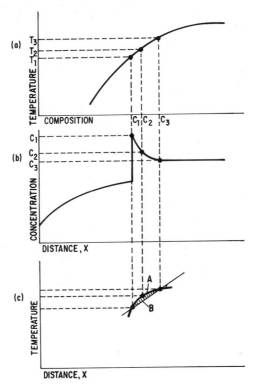

Fig. 3.10 Origin of constitutional supercooling in the liquid near a crystallizing interface. (*a*) Equilibrium phase diagram; (*b*) concentration profile in ingot; (*c*) imposed temperature gradient.

1.2.1 Cellular Structure

Under conditions of constitutional supercooling, a planar interface is unstable, and local shape perturbations occur, which lead to growth of projections from the solid/liquid interface. If the supercooled layer is thin, then these projections can grow into the liquid only to the thickness of the constitutionally supercooled layer. The projections contain less solute than the liquid, and the liquid surrounding them is correspondingly enriched, as shown by the arrows in Figure 3.11. The equilibrium crystallization temperature of the solute-rich liquid between the projections is depressed and a stable, nonplanar interface results. The projections tend to assume uniform spacing and each becomes a discrete "cell" separated from other cells by liquid regions of high solute content.

A quantitative experimental study of interface morphology has been carried out and the effect of convection was evaluated (17). At the onset of instability, approximately sinusoidal undulations appeared, with amplitudes that increased exponentially with time and underwent harmonic distortion as instability developed. Depressions and protrusions were characterized by compositional

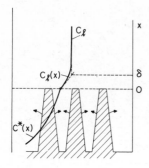

Fig. 3.11 Rejection of impurity leading to cellular structure at a solidifying interface. $C_l(x)$ is the solute concentration profile in the liquid region near the cells and $C^*(x)$ is the corresponding profile in the intercellular liquid.

maxima and minima, respectively, in accord with steady-state equilibrium behavior. The evolving heat of fusion and the difference in thermal conductivity of solid and liquid were shown to promote morphological stability under conditions of incipient constitutional supercooling. A related theoretical study confirmed the sinusoidal nature of the instability (18). Segregation was found to depend on the parameter $(V\lambda/D)$ where V is rate of solidification, λ is the wavelength of the sinusoidal perturbation, and D is the diffusivity of the solute in the liquid. Both very large and very small values of the parameter lead to a small degree of segregation which is proportional to the sinusoidal amplitude. For intermediate values of the parameter, depending on k, segregation is greater and no longer proportional to sinusoidal amplitude.

The form of the cellular structure depends on the crystallographic requirements of the growth process and on the crystal structure of the growing solid. Thus for example, a tin/lead alloy shows a fairly regular hexagonal solute substructure (Figure 3.12).

Hexagonal cells were likewise observed in transverse sections of aluminum/copper ingots which were quenched after directional solidification at varied growth rates, in various temperature gradients. Longitudinal sections showed solute trails extending up to 2 mm behind the quenched interface; these trails were regularly spaced at about 0.1 mm, and showed no thickening with increasing distance from the growth front (25). On the contrary, the cell boundaries became progressively thinner, starting at about mid-depth of the trails. In sum, the experiments supported the conclusion that changes in the solute distribution pattern behind the interface result from solute diffusion from enriched intercellular liquid toward the surrounding solid. Moreover, the surface is nonuniformly depressed along cell boundaries, with the greatest depression in solute-rich nodes (19). Figure 3.13 shows the rectangular solute substructure found in a crystal of germanium grown in a [100] direction (20). Anthracene, a monoclinic organic crystal, shows a quite different decanted surface (Figure 3.14) when grown under conditions that give constitutional supercooling.

Extension of the mathematical treatment of solute distribution from the case of a planar interface to the more realistic case of a cellular or dendritic surface

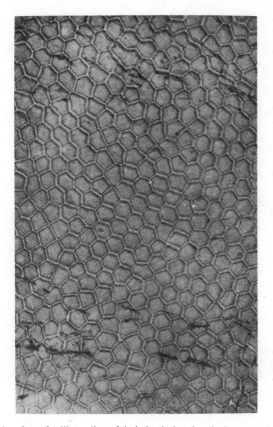

Fig. 3.12 Decanted surface of a dilute alloy of tin in lead, showing the hexagonal solute substructure.

poses a severe challenge. A steady-state solution was derived, taking into consideration that impurity transport is convective in the melt, but diffusive in the cellular interface region. Good agreement between theory and experiment was obtained for both a simple eutectic system (naphthalene/benzoic acid) and a solid-solution system (naphthalene/β-naphthol) (21, 22).

1.2.2 Cellular-Dendritic Structure

It has been observed that dendritic growth can occur in a direction of rising temperature (23, 24). Consider crystallization taking place radially in a spherical or cylindrical vessel. The medium surrounding the vessel provides the driving force for crystallization and is, accordingly, at a lower temperature than the crystallizing system. After a layer of solid has formed, the heat of fusion liberated as crystallization proceeds inwardly, must be dissipated in the interior liquid or conducted through the solid "skin." Under such circumstances, the temperature gradient becomes small and the region of constitutional supercooling becomes correspondingly larger. The cells that characterize crystallization under

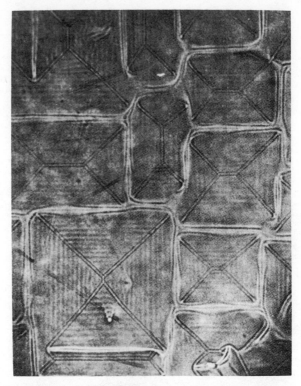

Fig. 3.13 Rectangular solute substructure in a gallium-doped germanium crystal (19).

moderate constitutional supercooling give way to structures called "cellular dendrites" to denote physical difference between this morphology and that of "free" dendritic growth, which results from thermal supercooling.

It is apposite to consider quantitatively the conditions under which a planar interface will become cellular and under which a cellular interface, in turn, will become cellular dendritic. The factors governing the transition are of two kinds: experimental and substance specific. Of the former, the growth rate V and imposed temperature gradient G are critical. The latter category consists of the binary diffusion coefficient D, the equilibrium distribution coefficient k_0, and the slope of the liquidus, m. The transition between noncellular and cellular structure occurs when the impressed temperature gradient at the solid/liquid interface is tangent to the existing equilibrium freezing curve at the interface, and is represented by the equation

$$\frac{G}{V} = \frac{mC(1 - k_0)}{k_0 D} \tag{3.27}$$

If G/V is less than the right side of the equation, then constitutional supercooling is present. Thus constitutional supercooling is brought about by a high growth

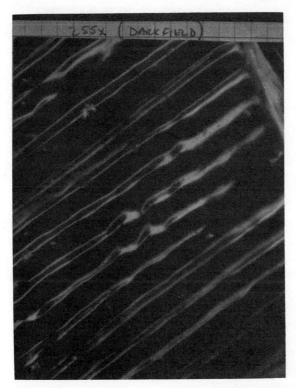

Fig. 3.14 Decanted surface of an anthracene crystal grown in a 5 K cm^{-1} gradient. The striations are about 3.7×10^{-3} cm wide.

rate, poor stirring, and a shallow temperature gradient at the interface. These are experimental quantities subject to control. An additional factor that predisposes toward constitutional supercooling is a large value of $|1 - k_0|$, which is an inherent property of each binary system. Avoidance of constitutional supercooling requires slow growth, efficient stirring, and a steep temperature gradient.

Many detailed studies have been made on the effects of varying the operative factors in this mode of growth: G, V, and C_0. Sharp and Hellawell (25) crystallized aluminum/copper alloys at steady state, then increased the growth rate either instantaneously or through a period of gradual acceleration (see also ref. 19). The ingots were quenched and sectioned longitudinally to follow the morphological changes and impurity distributions at the solid/liquid interface. Figure 3.15a shows the surface as it appeared when quenching took place 30 s after instantaneous acceleration from $1.67 \, \mu\text{m s}^{-1}$ to $16.7 \, \mu\text{m s}^{-1}$. Figure 3.15b shows the corresponding copper profile as measured by electron microprobe analysis. The broken line shows the profile before acceleration; the shaded areas show the solute redistribution following acceleration. In like manner, Figure 3.16a shows the surface as it appeared when quenching took place 60 s after the instantaneous acceleration and Figure 3.16b shows the relevant profile; the

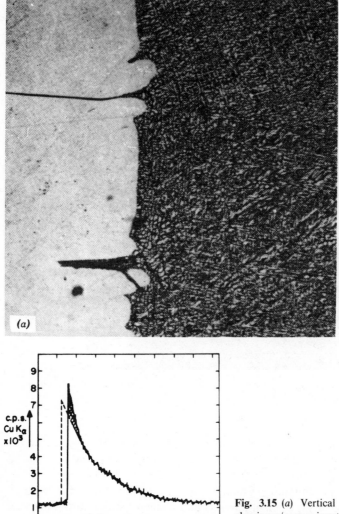

Fig. 3.15 (*a*) Vertical cross section of an aluminum/copper ingot, quenched after rapid increase in growth rate (180 ×) (25). (*b*) Profile of the copper concentration in the ingot of Fig. (*a*), measured by electron microprobe.

lower solid curve represents the copper concentration within the cells, while the upper solid curve is the mean of inter- and intracellular solid concentrations. In all, the experimental results show that the system achieves steady state by adjustment of tip-curvature in such a way as to minimize constitutional supercooling. The cell (or dendrite) spacing changes rapidly under transient conditions, but under steady-state conditions at a given temperature gradient, the spacing is insensitive to the product VC_0.

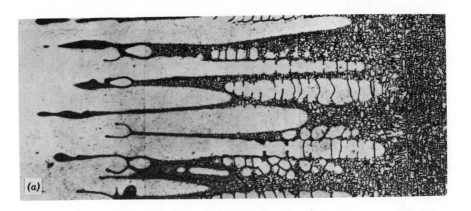

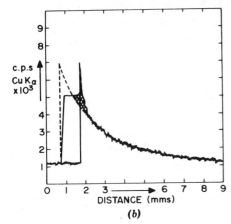

Fig. 3.16 (*a*) Vertical cross section of an aluminum/copper ingot, quenched 60s after rapid increase in growth rate. (*b*) Profile of the copper concentration in the ingot of (*a*), measured by electron microprobe.

The constitutional supercooling conditions cannot exist at the onset of crystallization and "incubation" is required for formation of an enriched boundary layer. An interesting direct observation of this phenomenon was reported by Bardsley et al. (26), who found that the formation of a cellular interface was accompanied by a marked change in the appearance of the external cylindrical surface of a crystal being pulled by a Czochralski method (see Section 4.2, Chapter 4). They grew gallium-doped germanium crystals and observed that surface corrugations appeared suddenly during growth. Withdrawal of the growing crystal after this onset revealed a cellular interface; rapid withdrawal (διακεκωμενη κρνσταλωσισ) before the onset of corrugation, gave a smooth surface. The cellular structure was thus correlated with the onset of constitutional supercooling.

Organic compounds lend themselves particularly well to observation of constitutional supercooling. This is true because organic crystals are often transparent, so that impurities can be located visually. Moreover many analytical techniques are available for the facile detection of concentration gradients in organic crystals (11, 27, 28, 29, 30).

1.2.3   Nucleation Ahead of the Interface

If constitutional supercooling is sufficiently large in the region near the interface, then heterogeneous nucleation of crystals may take place within the supercooled liquid. Once formed, these crystals grow rapidly and all of the supercooled liquid in the interface region solidifies. The liberation of the latent heat of fusion raises the temperature of the interface material to the equilibrium value and the resulting solid shows the concentration gradient formerly present in the liquid ($C_1C_2C_3$ of Figure 3.10b).

The next solid to form in this system will derive from liquid of bulk composition C_3. Hence it will contain substantially less impurity than the solid formed from the liquid between C_1 and C_3. This sequence of impurity accretion at the interface, nucleation of the resulting constitutionally supercooled layer, and later crystallization of purer solid is repeated cyclically, leading to a "banded" impurity distribution. Such banding has been observed in many systems (11, 12, 31).

1.3   Radial Solute Distribution

It was observed earlier that the BPS theory was derived on a one-dimensional basis, with the implication that solute concentration in the crystal is radially uniform. A number of works have appeared, both experimental and theoretical, showing that the assumption is not universally valid and that large radial inhomogeneities can arise from crystal growth under commonly used conditions (32, 33).

In the studies of Bardsley et al. (19) on gallium-doped germanium, it was found that crystals rotated at about $0.2\,\mathrm{rev\,s^{-1}}$ showed poor radial uniformity. This work is discussed in the previous section in connection with constitutional supercooling. The gallium concentration was highest in the center of the growing crystal, and the elevated concentration was associated with the fact that the cell structure forms first at the center. This, in turn, may result from the fact that mixing is poorest at the center of a crucible with a crystal rotating in only one direction.

A different pattern of radial inhomogeneity has been observed in chromium-doped ruby crystals grown by the Verneuil method. The chromium, isomorphically substituted in the host crystal, was neutron activated, making it possible to obtain contact macroautoradiograms. When the crystals were grown with a convex isotherm, chromium concentration in a transverse cross section increased gradually from the center and then more rapidly near the edge; the maximum concentration at the edge was about 2.7 times higher than in the center. In crystals grown in a convex–concave isotherm, chromium concentration rose toward the center of the radius and then fell markedly near the edge, mirroring the shape of the isotherm. In some crystals the exterior was covered with Cr (34).

Wilcox has attributed the relationship between interface curvature and segregation to four factors (ref. 35, Chapter 3):

1. The first has to do with the time sequence of solidification in cross sections of an ingot perpendicular to the growth axis. This is illustrated in Figure 3.17a where P_1 represents a point at the apex of a convex crystal surface. The uniform concentration over the interface at the time t_1 is given by

$$C_1 = kC_0(1 - g_1)^{k-1} \qquad (3.28)$$

where g_1 is the fraction of the total charge solidified at time t_1. Consider growth until a later time t_2 such that bb' contains apex P_1. The concentration along the interface is now given by

$$C_2 = kC_0(1 - g_2)^{k-1} \qquad (3.29)$$

The curvature of the interface may be measured in terms of the incremental quantity of solid $\Delta g = (g_2 - g_1)$. The concentration gradient along section bP_1b' is defined by the ratio C_2/C_1:

$$\frac{C_2}{C_1} = \left(\frac{1 - \Delta g}{1 - g}\right)^{k-1} \qquad (3.30)$$

For a concave interface (Fig. 3.17b), the analysis refers to section aP_2a'. Mil'vidski (36) has plotted this function for convex and concave interfaces for various radii of curvature with $k = 0.5$ as shown in Figures 3.18a and 3.18b, respectively. The numbers above the curves are the respective values of the parameter Δg, which was defined above as a measure of interface curvature. It should be stressed that this treatment assumes uniform solute concentration over the curved interface. The predicted radial inhomogeneity is a result of geometrical considerations only and turns out to be small.

If a solute is being segregated during solidification, then its concentration in the melt must increase with time. Consequently, the impurity concentration at point P_2 in Figure 3.17 will be higher than at point P_1. Thus the curvature of the

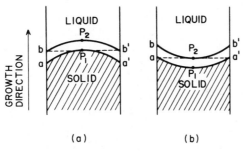

(a)   (b)

Fig. 3.17 Effect of interface curvature on radial solute distribution. (a) Convex solid leads to enhanced surface concentration; (b) concave solid leads to reduced surface concentration.

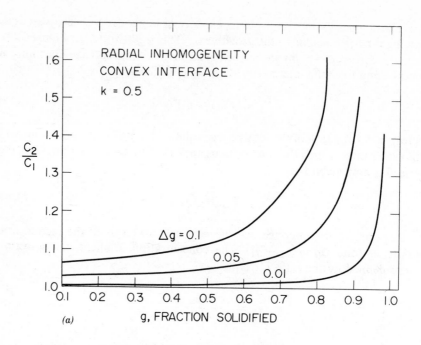

(a)

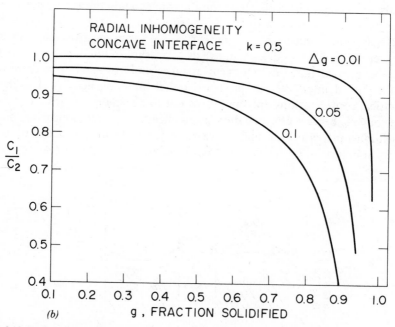

(b)

Fig. 3.18 Radial inhomogeneity brought about by segregation of an impurity with $k_{\text{eff}} = 0.5$, during directional crystallization with various degrees of interface curvature (36). (*a*) Convex solid; (*b*) concave solid.

122

interface gives rise to a radial concentration gradient, with higher concentration near the periphery of the cylindrical crystal for a convex interface. Similarly, a concave interface shows higher concentration at the center than at the periphery.

This effect must increase with increasing curvature of the interface. It has been observed for crystallization of sodium nitrate containing 3–4 wt% of $^{89}Sr(NO_3)_2$. While equilibrium crystallization (planar interface) gave a homogeneous radial distribution, crystallization at $7.8 \times 10^{-4} cm s^{-1}$ gave a concave solid/liquid interface which in turn led to a radial inhomogeneity. Radiographic darkening in the center of the image of a cross section was six to seven times more intense than at the periphery (37).

2. The second of the four effects is related to variation in growth rate with distance from the crystal axis. In Figure 3.19, elemental volume A grows through a distance R_1 The lower growth rate of elemental volume B, normal to the growth surface, corresponds to more complete segregation of solute, that is, to an effective distribution coefficient farther from unity than is the case along the crystal axis. This effect naturally leads to greater inclusion of solute in the growing crystal with increasing distance from the crystal axis.

The overall solute distribution in the analysis of Kulik and Zil'berman (9) contains components deriving from a planar interface, from radial temperature gradient, and from interface curvature. Their results included a concentration term proportional to $r^2/R'V$ (where r is the radial coordinate in the cylindrical crystal of radius R, and R' is the radius of curvature of the interface) even with a perfectly mixed melt. For a convex/solid interface, the concentration at the periphery is higher than in the center. Experimental conditions of heat flow determine the position of the interface in the growth apparatus and consequently its radius of curvature. As a convex interface rises (for example, in the Bridgman mode described in Section 4.5), the radius of curvature decreases, resulting in a more pronounced radial inhomogeneity (see Figure 3.20) (9).

Jaccard (32) gave an analysis very similar to that of Kulik and Zil'berman, but carried it further to the definition of a quantity describing the radial

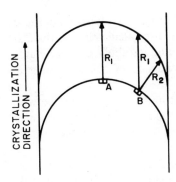

Fig. 3.19 Radial inhomogeneity resulting from varying growth rate normal to a convex solid/liquid interface.

dependence of the relative concentration in the solid:

$$S'_0(r) = \frac{aV \exp(aVr^2)}{\exp(aV) - 1} \tag{3.31}$$

Here "a" is a measure of the relative interface curvature: as before, V is growth rate and r is the radial coordinate of the cylindrical crystal. The expression was calculated for a parabolic interface, in the case of a system whose effective distribution coefficient is zero or nearly so. Convex surfaces are characterized by $aV > 0$ and concave ones by $aV < 0$. The two exponential factors in Equation 3.29 introduce a very strong dependence on radius. Realistic values of growth parameters can lead to an aV equal to 10, giving values of S'_0 of 4.5×10^{-4} at the crystal axis and 10 at the crystal periphery. Hence concentration can differ by several orders of magnitude between axis and surface of a cylindrical crystal.

Blicks et al. (38) in fact observed radial concentration differences greater than an order of magnitude in ice crystals doped with HF and NH_4F. Similar effects were observed in an organic system in which k was very small.

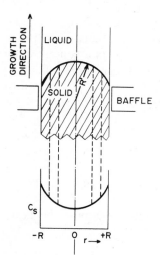

Fig. 3.20 Radial variation of solute concentration at a curved interface.

Figure 3.21a shows the radial-dependence factor $S'_0(r)$ as a function of radius for various values of the parameter aV(32). Jaccard also calculated quantity $S'_0(r)$ for an interface that has a flat center of radius r_1 and parabolic curvature at its periphery. Figure 3.21b shows the radial dependence of $S'_0(r)$ for such an interface, with $aV = 5$ and varying values of r_1.

3. A third effect of interface curvature arises from trapping of melt between the interface and the container in a region of poor mixing. This would appear to be a more substantial problem in the case of a convex solid than in the case of a concave solid, as shown in Figure 3.22.

4. The fourth interface-shape effect has to do with the appearance of facets on the freezing surface. In Czochralski growth (see Section 4.2, Chapter 4), if the

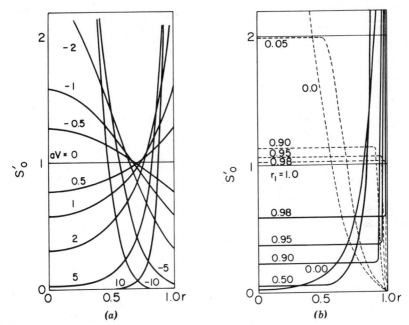

Fig. 3.21 Predicted radial variation of solute concentration, as a function of curvature parameter aV. (a) Parabolic interface; (b) interface with flat center of radius r_1 and parabolic curvature ($aV = 5$) at the periphery.

crystal surface is convex in the melt, and if a central facet exists, then the buildup of solute concentration will be greatest in the region below the central facet because of the relatively poor mixing there. Hence constitutional supercooling is greatest there; the new layers that form from the supercooled melt grow rapidly in the radial direction, and k_{eff} approaches unity because of impurity entrapment at the facet edges. This effect has been observed and confirmed in compound semiconductors, wherein k_{eff} can differ by nearly an order of magnitude between on-facet and off-facet segregation. In some cases k_{eff} for on-facet growth is greater than unity, while k_{eff} for off-facet growth is less than

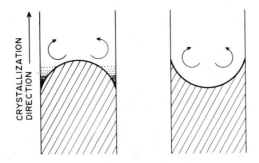

Fig. 3.22 A convex solid surface traps stagnant liquid at the periphery.

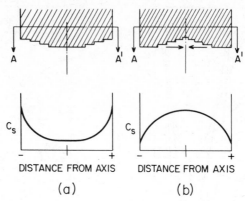

Fig. 3.23 Effect of faceting on radial solute distribution. (*a*) Central facet on convex solid leads to solute depletion at center of ingot; (*b*) absence of central facet on concave solid leads to solute accumulation at center of ingot.

unity. It is clear that both results cannot derive from equilibrium solidification of the same binary system. That is, to say, a single phase diagram cannot give rise to $k > 1$ and $k < 1$ for the same composition; it has been concluded that preferential adsorption on the faceted faces is responsible for the difference. Figure 3.23*a* shows that growth on the central facet is axial, while growth off the facet is radial. If the interface is concave (Figure 3.23*b*), there is no central facet; each horizontal surface of the interface experiences nucleation and growth by layer spreading toward the growth axis. The result is a buildup of impurity in the center of the ingot. These concepts, collectively, explain the phenomenon observed by Bardsley et al. in Czochralski solidification of gallium-doped germanium, in which the ingot contained elevated solute concentration in its center.

Lyubalin et al. (39) have carried out extensive studies on the orientation dependence of k_{eff} along 12 crystallographic directions in antimony-doped germanium and found an overall variation from k_{eff} (110) = 5.48×10^{-3} to k_{eff} (111) = 8.89×10^{-3}, which correlated with supercooling at the respective growth surfaces.

1.4 Particle Pushing

The discussion thus far might have conveyed the idea that solute redistribution occurs through a few, well understood processes. In fact, real experimental studies must take into account a variety of phenomena that can have a profound influence upon solute redistribution during crystallization. Microscopic studies of zone-melting phenomena have been carried out by Wilcox (40), who observed (a) liquid and gaseous inclusions, (b) impure liquid drawn by capillary forces into cracks in the solid, (c) incorporation or rejection of insoluble particles at the interface, and (d) random agitation of the melt by vapor-bubble movement. Most, if not all, binary systems undergoing solidification contain insoluble particles or gas bubbles. Although these inclusions may not be solutes, they are treated in this chapter because they can exert a strong influence on the nature of the solidification process. Insoluble particles may be rejected or captured by the freezing interface, and they may act as nucleation centers. Bubbles of gaseous contaminants can cause agitation of the melt near the interface and can give rise to selective growth of volatile components. One of the earliest accounts of such "particle pushing" during crystal growth is due to Stöber (41). Copper and glass spheres dropped on a crystal of sodium nitrate, which was growing from the bottom of a crucible, were pushed up by the rising crystal surface. The same effect was noted with bubbles of air and water that were present in the starting material.

Buckley (42) has provided a fairly extensive discussion of the appearance of inclusions in various natural and synthetic crystals. In some cases at least, the inclusions started as liquid trapped during growth, which, at the temperature of crystallization, just filled the cavity within the crystal. On cooling, solute precipitates from the included solution and if the solubility is high, the liquid volume in the inclusion may diminish considerably, leaving a void. Buckley discussed two separate mechanisms which might govern the inclusion or rejection of solid particles. One of these is a thermodynamic argument based on the interfacial tensions existing among the phases. If the sum of the interfacial tensions (solid I/solution) + (solid II/solution) is greater than the interfacial tension (solid I/solid II), then the total free energy of the system is decreased by the inclusion. The other mechanism is essentially chemical in nature, and argues that microscopic particles in a liquid phase may have molecules adsorbed on their surfaces, oriented in such a way that they resemble the solid being formed, and are hence more likely to be included in the solid.

Uhlman, Chalmers, and Jackson (43) carried out an extensive systematic study on the interaction between particles and the solid/liquid interface. They made microscopic observations of the behavior of insoluble particles in thin films of liquid undergoing directional crystallization. They studied particles of solid materials (graphite, magnesium oxide, silt, silicon, tin, diamond, nickel, zinc, iron oxide, and silver oxide) in several organic matrices and in water. The organic materials grow from the melt exclusively with smooth interfaces; water is expected to have a smooth interface only for growth normal to the basal plane, and a rough interface for other growth directions. They used a simple microscope-stage heater, which made it possible to observe the interface in motion at various rates, at magnifications up to about $500 \times$. Characteristic critical velocities were observed for each type of particle. Below this velocity, particles were rejected by the interface and above it, they were entrapped in the solid. The distinction between rejection and capture applied to both colloidal and larger-than-colloidal particles. For particles smaller than 15-μm diameter, however, the critical velocity was independent of particle size, while for larger particles, critical velocity was approximately inversely proportional to particle size and somewhat dependent on particle shape as well.

It is tempting to relate the critical velocity for particle capture to the critical velocity for interface breakdown from planar to cellular structure, but the critical velocity for particles in the ice/water system exceeded the critical velocity for interface breakdown at a given temperature gradient. In the transition region in which interface velocity is greater than the critical velocity for interface breakdown but less than that for particle capture, the interface continued to push particles between the cell boundaries while trapping them at the boundaries. The resulting solid contained lines of particles separated by the cell dimensions (typically about 50 μm).

For a given matrix material, various particles showed different critical velocities. Likewise, for a given particle, different matrices showed different velocities. In general, the critical velocity was inversely related to the viscosity of the matrix.

The interpretation of the results was based on the axiom that particle pushing requires a force opposing incorporation of the particle into the solid and a supply of fresh material to the interface zone immediately behind the center of the particle. Electrical measurements indicated that electrostatic interaction is not the limiting process that prevents incorporation of particles. Rather, surface energy effects provide the repulsive interaction as described earlier by Buckley (see above). If, on the other hand, liquid cannot diffuse sufficiently rapidly to the growing solid behind the particle, then the particle becomes incorporated in the solid. A theoretical treatment based on conservation of mass in the contact region at the interface showed that at low velocities, the interface remains fairly flat; at higher velocities, it curves in accordance with the shape of the particle. The separation between the particle and the interface decreases with increasing speed, and the chemical potential of the system increases accordingly. The critical velocity, then, corresponds to the point at which further decrease of the separation between particle and interface results in a net lowering of the chemical potential of the system. This value of the separation was calculated to be about 10^{-7} cm.

Omenyi and Neumann (44) carried out a careful study of the behavior of small particles at solidification interfaces of biphenyl and naphthalene. They approached the problem from a thermodynamic viewpoint, measuring the contact angles and interfacial tensions of the particle material with the host compounds, and calculating from these the net free energy of engulfment. If this quantity is negative, particles will be engulfed, as suggested earlier by Buckley and by Uhlman et al. Because the relevant contact angles and interfacial tensions are those at the melting temperature of the host, it was necessary to measure their temperature dependence and extrapolate to the melting point. Hydrophobic particles (of Teflon polytetrafluoroethylene resin or of silicone-treated glass) were engulfed, even at very slow rates of solidification. Hydrophilic particles (of nylon or acetal resins, or of untreated glass) on the other hand, were rejected by the advancing crystallization front at low and moderate growth rates. In both cases, the behavior agreed quantitatively with predictions based on thermodynamic theory. Nevertheless, the critical pushing velocity for a given particle material depended on the identity of the host, indicating that host/guest mechanical relations need to be considered along with the interface thermodynamics.

The results of Omenyi and Neumann were reexamined by adding spherical particles of various kinds and sizes to naphthalene undergoing zone melting in a rotating horizontal tube. The rate of zone transport was increased periodically in 0.1-cm increments and the onset of trapping was noted by direct microscopic observation (45). After solidification the ingot was sectioned, the naphthalene was sublimed, and the mass of beads trapped in each section was measured. Small numbers of beads were trapped below V_c and large numbers were pushed even above V_c. Hence trapping probability is finite at all rates, and V_c is to be looked on as a rate at which trapping probability rises sharply without reaching unity.

In horizontal rotating tubes, hydrophobic particles were not trapped even at the highest speeds used, 6.9×10^{-4} cm s^{-1} (2.5 cm h^{-1}). In vertical, nonrotating tubes, on the other hand, such particles were trapped at speeds below 5.6×10^{-5} cm s^{-1} (0.2 cm h^{-1}). These results disagree with the earlier measurements of Omenyi and Neumann. The discrepancy is thought to be caused by impurities in the naphthalene used in the later study.

Wilcox (46) reported detailed observations of particle rejection and particle capture during solidification of thin films of organic liquids. He found that bubbles and insoluble particles both serve to make visible any erratic fluid behavior at the interface. Bubbles were seen to oscillate or move rapidly, sometimes in the axial direction and sometimes normal to the axis. Because gases are much less soluble in solids than in the associated melt, gas bubbles tend to accumulate in the melt during solidification. Such bubbles are susceptible to violent movement within the melt; melt continguous to the bubble may then move around the bubble. The driving force for such mobility may be evaporation of volatile components of the melt from one side of the bubble to another in the interfacial temperature gradient. The resulting variation of surface tension, from one side of the bubble to another, in turn produces the rotational force. The motion of bubbles can have a dramatic effect on solute distribution at an interface. First, mass transfer may be reduced simply by reduction in the available area, and heat transport may be reduced as well. The reduced heat transport can, in turn, cause surface roughening by changing local growth rates.

In an extension of this study, Wilcox (47) observed that the critical velocity for particle trapping at a convex interface is lower near the wall than at the center. This effect is doubtless related to the enhanced inclusion of melt-soluble impurities at the exterior of crystals growing with convex interfaces. Critical trapping velocity for carbon particles in naphthalene was found to vary by about 20% near variously oriented grains at the interface.

The practical exigencies of growing a magnetoelectric composite material led Van den Boomgaard (48) to a monumental study of the ways in which growth rate and gas pressure influence the properties of the resulting crystal. In contrast to the observations with solid particles, gas bubbles were formed in the liquid at the solidification front below, rather than above, a critical growth rate that depended on the partial pressure of oxygen present in the system. When the same material was grown in the Bridgman mode, gas bubbles continued to be formed at higher growth rates than in the EFG mode (edge-defined film-fed growth).

2   EXPERIMENTAL PROCEDURES

The family of techniques known generically as directional crystallization has been in use for decades as a laboratory procedure; as a common technological operation, it has probably been in use for centuries. Many separations that are laborious or even impossible by other means are fairly readily effected by this technique. It remains a curiosity that so old and so useful a family of procedures remains so little used. It is our intention in this chapter to review the historical record and to describe the experimental possibility in sufficient detail to enable the experimenter to assemble easily the necessary equipment and carry out appropriate experiments.

In considering what techniques to use for a particular separation problem, a number of questions arise almost universally when melt crystallization is considered. These are tabulated below as a means of indicating the scope of the content to follow:

- How can one tell if the method is applicable?
- How rapidly can it be carried out?
- What quantity can be handled?
- How can one tell if purification has been effected?
- Can a grossly impure material be handled?
- How can one tell if impurities are being created in the process?
- What categories of materials can be handled?

Only single-stage solidification will be considered under the heading of "directional crystallization." The term directional crystallization is meant to indicate the exclusion of slurry crystallization and solution-based techniques. Directional crystallization may be carried out in three ways:

1. *Cylindrical/axial.* A cylinder of liquid, at a temperature slightly above the crystallization temperature of the contents, is moved through a temperature gradient into a cold zone in such a way that the contents of the tube crystallize (Figure 3.24a). Because the solidifying interface is perpendicular to the axis of the container, this process has also been called "normal freezing."

Another mode of generating a solid cylinder from a melt is to "pull" a crystal on a cooled rod from a vessel containing the melt. This tactic, long used for growth of single crystals, is known as the Czochralski method (Section 4.2, Chapter 4).

2. *Cylindrical/radial.* This procedure may be carried out in one of two modes, namely, radially inward and radially outward (Figure 3.24b). In the

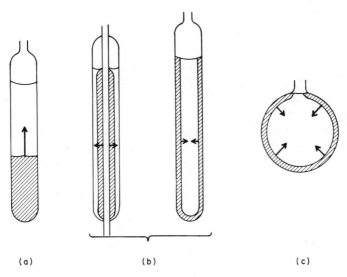

Fig. 3.24 Modes of directional crystallization. (*a*) Cylindrical/axial; (*b*) cylindrical/radial; (*c*) spherical/radial.

former, a cylindrical sample container is immersed in a thermostat bath whose temperature is slightly lower than the freezing temperature of the contents of the tube. This geometry produces gradual inward solidification from the cylindrical wall at a rate that diminishes with time, as a result of the thermal impedance of the solid that is forming. In the latter, a cylindrical sample container is provided with a hollow axial tube through which coolant may be circulated in such a way that outward solidification will proceed on the inner tube, toward the wall of the cylindrical container.

    3. *Spherical/radial.* In this configuration, the sample is contained in a sphere, which may be a round-bottom flask; it is immersed in a cooling bath and a heat source is placed at its center. Radial solidification takes place inwardly, leaving a small fraction molten in the vicinity of the central heater (Figure 3.24c).

    By far the preponderance of experimental work and theoretical analysis has been devoted to the cylindrical/axial method.

    No single apparatus has emerged as best for carrying out directional crystallization. In fact, it is not possible to buy a complete assembly of the few components needed to conduct such experiments. For this reason, we describe the components separately and only later comment on combinations that have been used and described. Since directional crystallization requires an advancing solid/liquid interface, the experimental requirements are essentially three:

1. A heat source to melt the sample
2. A heat sink to cause solidification
3. A means of producing relative motion between the solid/liquid interface and the heating/cooling mechanism

2.1 Heating

2.1.1 Electrical Resistance Heating

A cylindrical sample tube may be raised or lowered from the furnace, as described by Jansson and Lunden (49) for determination of the distribution coefficient of $^6LiNO_3$ in $^7LiNO_3$ (mp 528 K). The furnaces or ovens used for growing single crystals by the Bridgman method (Section 4.1, Chapter 4) are suitable for use in directional crystallization. Such ovens, generally designed to operate between about 325 and 625 K, are often made by wrapping two cylindrical tubes with resistance wire and assembling them in such a way as to give a variable gradient in the space between them. A device of this kind was described by Sherwood and Thomson (50); it offers the particular feature of transparency, allowing the solidification process to be observed (Figure 3.25).

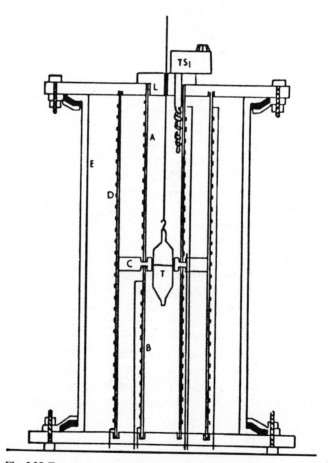

Fig. 3.25 Transparent two-zone furnace for directional crystallization.

Heating coils A and B (2.2 m and 3.6 m, respectively) of Nichrome ribbon (0.005 cm × 0.064 cm) were wound on Pyrex tubes of 5-cm diameter. They gave 500 and 800 W at 250 V. The ribbons were wound nonuniformly to provide the most uniform temperature possible. The two heater sleeves were separated by a baffle C of insulating board. The outer diameter of C made a close fit inside Pyrex tube D, upon which a 600-W heater was uniformly wound. All three heaters were seated in grooves machined in top and botton insulator plates and were enclosed in a Pyrex pipe E, of 30-cm diameter. The central opening was closed by lid L in which thermostat TS_1 was mounted. Heaters A and B were energized by an adjustable voltage supply and D was controlled by the thermostat. It should be noted that in this design the controlled heater is physically remote from the controlling sensor, a situation not conducive to close control with rapid response and minimal overshoot.

It is possible to achieve the transparency of glass construction without the nonuniformity introduced by helically wrapped heating elements. Transparent coatings of tin oxide on the outer surfaces of Pyrex cylinders afford high wattage density and good uniformity (11). For temperatures up to 535 K, convenient heaters are commercially available in the form of metal grids printed on plastic films or rubber sheets. If a plastic film of "Kapton" is used as the substrate, the resulting heater is essentially transparent; furthermore, the intrinsic color of "Kapton" provides a filtering effect, excluding light of wavelengths below about 550 nm (Minco Products, Inc., 7300 Commerce Lane, Minneapolis, MN 55432).

Yet another approach to construction of transparent furnaces, applicable up to relatively high temperatures, makes use of the infrared-reflective property of thin gold films (51). One such furnace is available commercially (Trans-Temp Co., 155 Sixth Street, Chelsea, MA 02150) and consists of a quartz tube wrapped with a platinum resistance element, enclosed in a Pyrex jacket bearing a gold deposit 20 nm thick. Above 970 K, the jacket is transparent and the furnace is illuminated by its own radiation. At lower temperatures, it may be illuminated by a light pipe or an external source. Such furnaces are available with inner diameters of 3.7 and 5.7 cm, and lengths of 30 and 50 cm; they are suitable for long-term use at 1270 K and offer uniformity within ± 2 K over 85% of their length. Suitably adjusted, they provide gradients up to 48 K cm^{-1}.

Figure 3.26 shows a two-zone furnace based on these principles; it offers still higher gradients, which can be varied by moving the relective mirrors, by wrapping insulation around the gap, and by varying the power to each coil.

Although these ovens are generally transparent, central baffles may obscure the region of greatest interest, namely the solid/liquid interface. Several approaches have been used to overcome this difficulty. For example, the central baffle may be made of a solid glass plate separating two independently controlled metal heater blocks. The periphery and central bore of the glass baffle plate must be polished to assure transparency. For operation at low temperature, a plastic spacer may be used. Alternatively, a transparent central baffle may be made of several concentric rings of plastic, glass, or quartz (see Section 2.4.2, Chapter 2). An opaque central baffle disk may be provided with

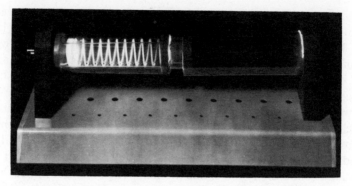

Fig. 3.26 Two-zone Transtemp furnace using an infrared-reflective jacket.

diametrically opposed viewing ports. This approach introduces a nonuniformity in the heat field, which may be overcome by rotating the sample tube.

Metal blocks may be heated by clamp-on heaters, which are commercially available from many manufacturers in a wide range of sizes and wattages (Watlow Electric Manufacturing Co., 12001 Lackland Road, St. Louis, MO 63141). Modular heating units containing wound-wire elements in refractory housings are available in a wide variety of sizes, shapes, and wattages. These can be used to make furnaces of square, rectangular, or circular cross section (Thermcraft, Inc., Winston-Salem, NC 27105).

A simple tubular heater suitable for directional crystallization can be made by wrapping a commercial heater tape or cord (Glas-Col Apparatus Co., 711 Hulman St., Terre Haute, IN 47802) on a metal tube. The manufacturer's recommendations should be followed carefully to avoid electrical contact between the heater element and the metal tube. The assembly can be enclosed in ordinary pipe insulation, which is available in glass-fiber (upper limit about 530 K) or hydrous calcium silicate construction (upper limit about 1120 K). Each of these is available with a variety of coverings including paper, aluminum, and stainless steel (Johns-Manville, P.O. Box 5108, Denver, CO 80217). The thermal conductivity of the glass-fiber type varies from 0.033 to 0.144 W m^{-1} K^{-1} at room temperature to 0.05 at 425 K, while that of the hydrous calcium siliate type varies from about 0.05 at room temperature to 0.09 at 650 K. Other materials can be used for higher temperatures. These include alumina (1675 K) and zirconia (2475 K) and are available as cylinders and boards (Zircar Products, Inc., 110 North Main Street, Florida, NY 10921).

In constructing heaters based on glass tubing, it is useful to know which glass size will fit most closely in the available insulation; Table 3.2 shows the metric diameters of pipes for which sleeve insulation is commercially available. For example, glass tubing of 2.5-cm diameter will fit readily in pipe insulation designed for $\frac{3}{4}$-in. pipe. The insulation is easily worked with simple tools, so that viewing windows can be made and diameters can be changed.

In addition, it is useful to know what the surface temperature of the insulation will be for various internal working temperatures, in order to ensure safety from

Table 3.2. Diameters of Pipes for Which Insulation is Available

Nominal Pipe Size, in.	Actual od, cm
$\frac{1}{2}$	2.13
$\frac{3}{4}$	2.67
1	3.34
$1-\frac{1}{4}$	4.22
$1-\frac{1}{2}$	4.83
2	6.03
$2-\frac{1}{2}$	7.30
3	8.89
$3-\frac{1}{2}$	10.16
4	11.43
5	14.13
6	16.83

burns. Figure 3.27 shows these temperatures for insulation made of hydrous calcium silicate. The upper family of lines is for 2.5 cm of insulation on tubing of 1-in., 2-in., and 4-in. nominal diameter; the lower family of lines is for 5 cm of insulation on the same sizes of tubing.

Riccius et al. (52) have described a furnace in which temperature gradients up to 200 K cm^{-1} were attained. It consisted of a stainless-steel cylinder inside a

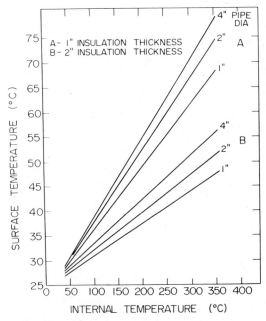

Fig. 3.27 Surface temperature plotted against internal temperature for pipe insulation of various diameters in two thicknesses.

resistance-wound Alundum tube. The stainless-steel cylinder was mounted on a ceramic ring which, in turn, rested on a water-cooled copper tube. The lower portion of the stainless-steel cylinder, closest to the ceramic ring, was provided with an auxiliary heater winding inside the Alundum tube. The combined assembly was well insulated.

An internal heater has been used to provide a planar interface (53). After establishment of an interface by external heating, a flat spiral of platinum wire was positioned horizontally within the melt, just above the interface, with the outer diameter of the spiral fitting closely to the container wall. This internal heater was kept at fixed elevation while the sample tube was lowered.

2.1.2 Vapor-Bath Heating

Boiling liquids have long been used as thermostats. A vertical arrangement of two such baths provides a simple and inexpensive method of forming a stable gradient for crystallization. Figure 3.28 shows such a two-vapor thermostat (54).

The all-glass device consists of an inner tube A, about 4.5 cm in diameter by 50 cm long, with standard-taper joints at both ends. Outer tube B (about 6.5 cm in diameter) is sealed to both joints and centrally divided by sealed-in septum C. Vapors from boiling liquids in flasks D pass through the jackets to condensers (not shown) in standard-taper joints E. Trap F allows condensate to return by overflow. To assure maximum temperature stability, the central tube and the vapor lines should be well insulated.

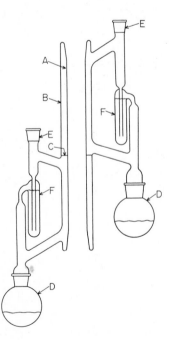

Fig. 3.28 Boiling liquids used to maintain constant temperature in a two-chamber vessel.

This device is a versatile one, allowing establishment of widely differing gradients. With the higher-boiling liquid refluxing in the upper jacket, it is useful for directional crystallization, especially for growth of single crystals by the Bridgman method (see Section 4.1, Chapter 4). With the lower-boiling liquid refluxing in the upper jacket, it provides an "inverted" gradient suitable for growth of crystals from the vapor.

It is important that the liquids be selected with careful consideration of thermal stability, both to assure long-term constancy of temperature and to avoid potential hazards. For example, peroxide-forming liquids should be avoided.

To be sure, the boiling temperature of each liquid is sensitive to variations in atmospheric pressure, but both liquids will respond simultaneously and gradual changes are not likely to cause rapid variation of the interface position. In growing single crystals by this method, even modest changes may be detrimental, but for purification, they are not likely to be a major problem.

A useful compilation of solvent properties is available in microfiche and lists a large number of materials, many of which are available commercially, in four separate sequences. The lists are arranged in order of increasing density, refractive index, boiling point, and dielectric constant, respectively (55). Another source of information about solvents useful in directional crystallization is the book by Riddick and Bunger (56).

2.1.3   Constant-Temperature Liquid Baths

Especially for materials melting near or below ambient temperature, crystallization gradients can be established by circulating liquids at controlled temperature, or simply by lowering the crystallization vessel into a low-temperature bath.

Hayakawa et al. (57) circulated water from two sources (Figure 3.29) through jackets A and B, separated by rubber sheet C, through which a close-fitting tube

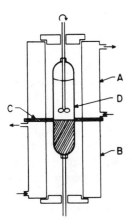

Fig. 3.29 Liquids circulated from two constant-temperature sources through chambers separated by a rubber septum.

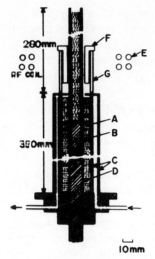

Fig. 3.30 Directional crystallization at high temperature, with liquid-metal coolant.

D was lowered. This method provides good heat transfer through the cylindrical sample tube, and a very sharp gradient at the interface.

Figure 3.30 shows a device for the directional crystallization of metals (58) in which a liquid-metal coolant A comes into direct contact with alumina sample tube B. The coolant is retained by an O-ring seal in the base plate. The coolant is held at constant temperature by water circulating from an external thermostat through the double-walled, stainless-steel jacket C. Support D is of the same diameter as B, so that lowering the sample produces no displacement of liquid A; D is connected to a lowering mechanism. Melting is brought about by induction heating; coil E energizes graphite susceptor F, within support G (which also serves as a heat shield). Gradients as high as 400 K cm^{-1} were attained.

2.1.4 Two Immiscible Liquids

Horton has described an apparatus for establishing a sharp temperature gradient at the interface between two immiscible liquids. Essentially the apparatus consists of a vertical column containing two immiscible liquids which are heated by independent resistance elements. A sample tube may be lowered through the interface, and solidification results if the lower liquid is below the freezing point of the sample. If the diameter of the sample tube is considerably smaller than that of the heater column, then the change in height of the interface resulting from liquid displacement may be ignored. If on the other hand, the sample tube is nearly as large as the heater column, a means must be provided to maintain constant interface level in the outer heated column. Horton (59) and Horton and Glasgow (60) have described one such method, involving an elaborate pumping system to recirculate the displaced liquid. A simpler method for maintaining a nearly constant level is shown in Figure 3.31. Here the sample tube descends in a close-fitting, two-liquid bath, the lower level of which

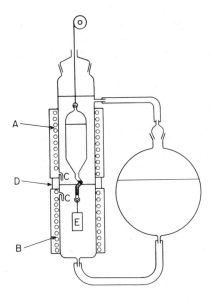

Fig. 3.31 Two immiscible liquids, separately heated, provide a sharp interface gradient.

communicates with a large reservoir of the same liquid. In this way, only the narrow column must be thermostatically controlled, as the reservoir acts to minimize displacive motion of the interface. Separate heater windings A and B are powered by controllers actuated by thermocouples placed in wells C. The central region, between the heater insulations, may be provided with a viewing baffle D. The crystal tube must be equipped with a weight E to assure smooth descent. A related device has been described in which an outer annular vessel contains two immiscible liquids that are independently heated and stirred, while an inner tube, containing stagnant layers of the same liquid, is used as the growth vessel within which a crystal tube is lowered.

A variant of this two-liquid system has been described by Guillaud et al. (61), who used mercury as a lower liquid and glycerol or a silicone oil as the upper liquid. The mercury was heated by a resistance heater wound on a copper sleeve. The upper layer was surrounded by a disk of asbestos whose thickness equalled the height of the layer. A second resistance heater was placed above the disk. With a glycerol layer 1.5 cm deep, a gradient of about 70 K cm^{-1} was achieved with upper (air) temperature at 473 K, and the mercury layer at about 350 K.

2.2   Cooling

2.2.1   Consumable Coolants

Ice, solid carbon dioxide, and liquid nitrogen may be used to cool liquid baths into which crystallization vessels are lowered. Direct introduction of coolant can cause disruptive changes in the level of the solidification interface. The problem may be solved by circulating a liquid or gaseous heat-exchange medium through a coolant reservoir and then through a heat exchanger immersed in the crystallizer. Constant-level devices are available for use with liquid nitrogen (JC Sales Limited, 1500 Rose Avenue, Pleasanton, CA 94566). Constant liquid level may be achieved by using a central standpipe in a coolant slurry (see Section 2.5.2, Figure 3.39).

2.2.2 Other Modes of Cooling

Refrigerated baths are available from many vendors, in a wide range of sizes and cooling capacities. Of course, domestic refrigerators, appropriately modified for safety, can be used as cold chambers for directional crystallization (see Appendix II). Refrigerated probes are available, which may be inserted in any suitable vessel, for operation at temperatures as low as about 175 K (see Section 2.2.1, Chapter 5, and Figure 5.23).

Thermoelectric cooling is discussed in Section 2.2.4, Chapter 5.

2.3 Transport

2.3.1 Small Motors

Most laboratory-scale directional crystallization operations are carried out on a fairly small scale, typically less than 0.3 kg. For this scale, synchronous ("clock") motors or instrument motors are adequate. The sample tube is attached to a spindle by a filament and the tube is lowered by slow rotation of the spindle; the lowering rate is determined by spindle diameter. Programmed lowering rates may be achieved by using a tapered spindle. Alternatively, the lowering motor may be controlled by a cycle timer, which provides a variable duty cycle. If the total cycle time is sufficiently short relative to the growth rate, the discontinuity in growth rate is negligible.

Variable solidification speed may also be provided by direct-current motors or by stepping motors. For larger containers, more substantial motors are needed, and these, too, are available in constant- and variable-speed versions (Hurst Manufacturing Corp., Princeton, IN 47670). Somewhat less convenient is the use of a counterweight and pulley assembly whereby a small motor may still be used to move a substantial weight.

It should be recalled that in an early and very successful application of directional crystallization, the sample tube was supported on a platform floating in a tank of water; the water was drained slowly from the bottom of the tank to provide the required slow lowering rate (63).

2.3.2 Leadscrew or Rack-and-Pinion

Leadscrew or rack-and-pinion drives may be used to provide positive transport, especially for larger crystallization vessels, or to move a heating or cooling mechanism. Linear slides, advanced by screw mechanisms, are commercially available (Velmex, Inc., P.O. Box 38, E, Bloomfield, NY 14443). Such screw-driven slides may be used to propel a heater or cooler mounted on a track. Because crystallization is carried out slowly and because the motion need not be as smooth as required for single-crystal growth, sturdy but inexpensive devices may be used. Thus, for example, drawer guides function quite well (Chassis-Trak, Dept. EM, P.O. 19188, Indianapolis, IN 46219).

Leadscrew mechanisms designed as microsyringe pump drives are also suitable for many crystallization applications. For example, The Houston Atlas Model 1001 Syringe Drive (Houston Atlas, Inc., 9441 Bay Thorne Drive, Houston, TX 77041) provides a wide range of forward speeds with a rapid reverse speed, and a maximum stroke of 9.5 cm. Complete crystal-pulling mechanisms, providing both transport and rotation, are available commercially (see Appendix II).

2.3.3   Moving Gradient

A solidification interface can be caused to move through a sample without moving either the sample or a heater/cooler. If a gradient is once established in a heater/cooler system, the level corresponding to the freezing isotherm of the sample can be moved simply by reducing the heat input. In a variant of this method, a large number of separate heating coils may be deenergized sequentially to produce a "stepped" solidification.

Another method of moving an interface through a stationary charge has been demonstrated by Karl (64) who used a "reflector furnace." in which a gold-coated semitransparent heat reflector is raised around the copper portion of a uniformly wound cylindrical heater. The reflected heat causes melting in the upper section. A related method has been described and designated "thermic screen translation" (TST) technique. In this method a cylindrical screen is interposed between the sample crucible and an external heater. The screen is formed by wrapping three layers of molybdenum foil around a graphite tube; the layers are separated from each other by a few turns of molybdenum wire. The screen method was applied to crystallization in the range 1000–2000 K (65).

2.4   Containers and Filling Procedures

2.4.1   Cylindrical Tubes

Materials that are stable in air at their melting points may be poured into test tubes; those stable at room temperature but not at their melting points may be placed in tubes that may be sealed after displacement of the air.

For greater assurance of the absence of air or other contaminants, and to contain the large volume of a fluffy solid, the vessel shown in Figure 3.32 may be used. The round flask contains the fluffy solid; after evacuation and admission of an inert gas, the solid is melted and allowed to flow into the appended tube. Melting may be effected by careful, direct application of a flame. A safer procedure is to use an electrically heated "air gun"; this may provide an air flow at temperatures up to about 825 K. An electrical heating tape may also be wound around the entire tube; it may be desirable to cover the tube with an aluminum-foil wrapping before applying the heater tape, to improve the

Fig. 3.32 Flask for charging a directional crystallization/zone-melting tube.

uniformity of heating. Simple, lightweight ovens can be built of insulating board coated with metal foil to assure high infrared reflectivity. A 250-W lamp can give temperatures above 575 K in such an enclosure with dimensions of 30 cm × 30 cm × 60 cm (66).

After resolidification, the tube is again evacuated; solid adhering to the uppermost constriction is sublimed or distilled away by careful heating. The tube is sealed by flame fusion at the constriction, either in vacuum or after readmission of inert gas. A glass loop may be attached after seal-off, or a wire hook may be cemented in place. After directional crystallization of most of the sample, remaining melt is decanted into the bulb by inverting the tube; the bulb may be removed by flame fusion at the constriction. Of course, additional bulbs may be provided for repeated crystallization (60). With this tube, directional crystallization can be used as a preliminary purification before zone melting the final ingot.

2.4.2   Spherical Containers

Spherical containers offer two advantages over tubes: larger quantities can be handled and the ratio of crystallization area to sample volume is greater. Schwab and Wichers (63) used a long-neck, 5-L flask embedded in an insulating powder, with a sleeve heater surrounding the base of the neck to minimize heat loss through the neck. (The use of this apparatus is described in Section 1.3, Chapter 4.)

2.4.3 Other Containers

Metal containers, with small height-to-radius ratio, have been used for directional crystallization with rotation. Plastic containers are discussed in Section 2.4, Chapter 5.

Very small amounts of liquid may be solidified directionally as thin films on microscope slides. For the most part, this procedure has been used more for the microscopic study of the solidification process than for preparative purposes. Nevertheless, the same techniques are applicable (67, 68, 46). It is now possible to buy glass capillary tubes with rectangular cross sections having path lengths of 50, 100, 200, 300, and 400 μm (Vitrodynamics, Rockaway, NY).

2.5 Mixing

The effectiveness of impurity rejection in directional crystallization depends strongly on the thickness of the stagnant melt at the interface; in turn, the thickness of the stagnant layer depends on the degree of mixing in the bulk melt. For this reason, no experimental aspect of directional crystallization has been more thoroughly studied than the problem of mixing the melt. Various methods of homogenizing the melt are discussed below.

2.5.1 Container Rotation

Given the convenience of carrying out the crystallization in a closed cylindrical tube, it is natural to consider first the rotation of a tube about its long axis. In a study of distribution coefficients of impurities in anthracene, Sloan used a stirrer motor (5 rev s^{-1}) mounted on a carriage driven by a leadscrew placed below the crystallization oven (11). Kirgintsev and Aladko used rotation and counter-rotation through 1.25 rev at 4 rev s^{-1} (69) in a study of the partition of Sr during directional crystallization of NaNO$_3$. Turner et al. have described a circuit that provides reversing rotation with controlled acceleration (70). Of particular interest in this context is the recent availability of variable-speed permanent-magnet motors in the "pancake" configuration. These lightweight motors are available in various sizes; their flat armatures offer low mass and high torque. The low inertia of the armature yields fast response (PMI Motors, 5 Aerial Way, Syosset, NY 11791); these motors operate at no-load speeds up to about 85 rev s^{-1}.

Combined rotation/transport mechanisms are available commercially (Crystar, P.O. Box 262, Cupertino, CA 95024; Centorr Associates, Route 28, Suncook, NH 03275; Metals Research, Melbourn, Royston, Herts. SG 86EJ). These mechanisms are designed primarily for single-crystal growth and do not offer the high rotation speeds necessary for effective mixing.

Anderson has described (71) a rotating-container method in which the height of an aluminum container is small relative to its radius. Crystallization takes place in an annular charge in a direction parallel to the axis of rotation (see Figure 3.33). Rejection of impurity and liberation of latent heat of fusion at the

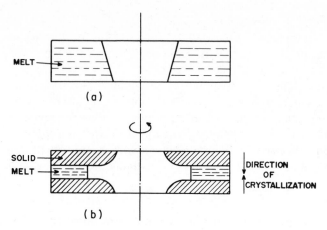

Fig. 3.33 Directional crystallization in a flat, rotating container.

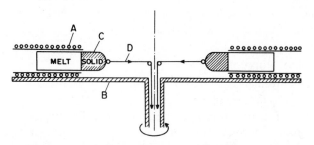

Fig. 3.34 Directional crystallization in tubes undergoing centrifugation.

solidifying interface generate a density difference between bulk and interface liquid. In the gravitational field, the effect of this small density difference is multiplied by the acceleration in the radial sense, and strong mixing currents ensue. Tangential shear between the rotating solid and the inertially restrained melt reinforces the mixing.

Afanasiadi et al. (72) have described another centrifugal mixing method in which crystallization takes place in tubes that rotate in a plane perpendicular to the advancing crystal front; the assembly is shown schematically in Figure 3.34. A is affixed to plate B, which rotates about the vertical axis at speeds of 12–35 rev s^{-1}. Tubes C (1.4 cm diameter × 8 cm long) are pulled out of the heaters toward the central axis by a cable drive D, which runs through the hollow rotation shaft. Crystallization was effected at 1.3×10^{-3} cm s^{-1}.

2.5.2   Internal Stirrers

Ideally, the most intense stirring should occur at the solidifying interface. If crystallization is taking place in a descending tube, the solid/liquid interface will remain at an essentially constant elevation, determined by the heater/cooler temperatures, freezing point of the sample, rate of descent, and degree of mixing of the melt. Hence an agitator must be sealed at its point of entry into the crystallization vessel, yet allow the vessel to descend freely.

Gouw (73) used an impeller driven by a variable-speed motor; the stirrer shaft was sealed into the sample-tube closure, through which a flow of nitrogen was introduced. In reported runs, the impeller was operated at 35 rev s^{-1}. Hayakawa et al. (57) used a similar arrangement, with stirring speeds to 25 rev s^{-1} (see Figure 3.29).

Powers and his co-workers (74) used a stainless-steel rod with transverse pins B as a stirrer (Figure 3.35). The stirrer motor D could be blocked at fixed level by stop G or could follow the descending tube A into the cold bath, guided by rods F.

Schildknecht, Rauch, and Schlegelmilch (75) described a similar and more elegant approach to this mode of stirring; it is shown in Figure 3.36. The apparatus consists of three main parts: (1) the cooling bath, (2) the lowering device, and (3) the stirrer. The cooling bath does not require a circulating refrigerant; a constant solid/liquid level is maintained with a fixed charge of coolant, which may be liquid nitrogen, ice, or a mixture of a solvent with solid CO_2. The components are identified in the figure captions.

The solid/liquid interface is defined approximately by the top of the copper tube. The stirrer is mounted so that its blades end 0.1–0.3 cm above interface.

Liquids in quantities up to 5×10^3 cm^3 have been directionally crystallized with mixing by a vibrating perforated plate A (Figure 3.37) driven by a Vibro-Mixer B (Chemap AG, CH-8708, Mannedorf, Switzerland) with its shaft sealed by septum C. The kettle was suspended from ring assembly D, whose three tie-

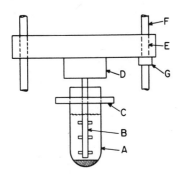

A   TEST TUBE
B   STIRRER
C   SUPPORT RACK
D   MOTOR
E   MOTOR MOUNT
F   GUIDE RODS
G   ADJUSTABLE STOP

Fig. 3.35 Internal stainless-steel stirrer in directional crystallization.

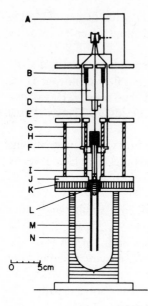

Fig. 3.36 Low-temperature directional crystallization with internal stirring.

A clock motor
B stirrer-motor clamp
C variable-speed motor
D lowering suspension
E stirrer shaft
F support plate with a bore for the sample tube
G guide rods
H stabilizer rods
I sample tube
J plastic plates that support the superstructure
K insulation
L central seal
M copper tube
N Dewar vessel

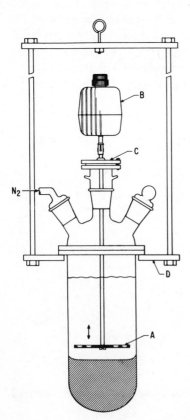

Fig. 3.37 Vibro-Mixer used in large-scale laboratory directional crystallization.

rods surround the Vibro-Mixer. Use of a vibrator in crystallization has been described by Borschevskii and Tretyakov (76).

Pfann and Carides (77) have described a method for scraping the surface of a growing, solid ingot. The intent of their experiments was twofold: (1) to produce fine-grained equiaxed microstructures in eutectic or other multiphase solids and (2) to improve the distribution coefficients of impurities in single-phase solids. The scrapers consisted of a variety of tools including a rotary file, a steel-wire cup brush, and a Kovar wedge; these were rotated at speeds of 0.2 to 15 rev s^{-1} under vertical loads of 0.06 to 1.2 kg. Growth rates of 5.5×10^{-4} to 60×10^{-4} cm s^{-1} were used. For single-phase tin or lead, scraping produced no major reduction in grain size or shape, while with eutectic and off-eutectic alloys the scraping produced the anticipated fine-grained, equiaxed structure with no trace of lamellar structure.

Schwab and Wichers (63) used a Pyrex diffuser tip through which a stream of nitrogen was introduced into the melt undergoing directional crystallization. This method of stirring offers the added advantage of continuously maintaining an inert atmosphere above the melt.

Pumping of a melt through an external circuit can produce effects similar to those of stirring. A liquid stream may be pumped from the melt and returned in the form of a jet, impinging directly on the solid/liquid interface (78).

Directional crystallization has been carried out in the radial mode on a rotating cylinder. The cylinder A (Figure 3.38) rotates within a vessel B, stirring the charge of melt, C. Coolant is introduced axially into the cylinder to cause growth of a crystalline layer D. When the layer has reached the desired thickness, the temperature of the cylinder is raised to expand and weaken the attached layer. An increase in rotation rate then causes the layer to break away and fall to the collection funnel near the bottom. A vibrator E compacts the solid deposit, which is then melted by an external heater, and collected through V_1.

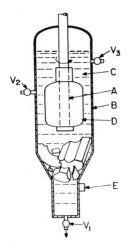

Fig. 3.38 Directional crystallization on an immersed rotating cylinder.

Liquid feed is admitted through V to restore the volume of C, and impure melt overflows through V_3. During melting of the solid product, a portion of the newly molten material rises and countercurrently washes descending solid resulting from the next freezing cycle (79).

2.5.3   Magnetic Levitation

A simple way to maintain stirring at a fixed level in a descending tube is to levitate the stirrer magnetically; the stirrer may be sealed within its container, thereby obviating sliding/rotating seals (80). Figure 3.39 shows such a stirring assembly. Glass tube A (2.8 cm od by 40 cm long), with a tooled opening for a stopple closure, contains a magnet B (covered with Teflon) to which a ribbon of twisted, stainless-steel gauze is attached. The stirrer assembly is inserted before closure of the bottom of A. A platinum wire (0.1 cm diameter) is sealed through the bottom of A to avoid supercooling the liquid. Tube A is lowered by motor C through a diametrically magnetized annular magnet D (Permag Corp., Jamaica, NY; No. CG 706) to which a brass spur gear E is attached. Gear E is driven at 1.3 rev s^{-1} by motor F through an idler gear G. Inner magnet B floats in the rotating field of D at a level such that the bottom of the stainless-steel stirrer is about 0.3 cm above the solid/liquid interface. Thus tube A slides down, past rotating magnet B, which remains levitated at constant height; A is hung from a wire cradle bearing a transverse member H, which slides along a vertical guide to prevent rotation of the tube.

A Dewar container I is used to hold the coolant, which is cracked ice for compounds melting above 280 K or a slurry of solid carbon dioxide in trichloroethylene for compounds melting above 213 K. To assure smooth

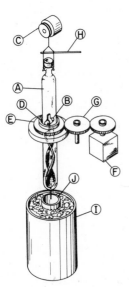

Fig. 3.39 Directional crystallization with a magnetically levitated internal stirrer.

descent and constant coolant level, tube A is not lowered directly into the slurry, but into a centrally positioned metal tube J, filled with water or trichloroethylene.

2.5.4   Ultrasonication

Static magnetostriction in metals can produce vibrational amplitudes of about 10 μm. The amplitude can be increased by use of a tuned velocity transformer consisting of a solid, tapered metal horn. Such a horn can produce strong local turbulence at a solid/liquid interface. In a study of crystallization of acetamide (81), it was found that the effective distribution coefficient of impurities went through a minimum value with increasing amplitude of the ultrasonic signal.

2.5.5   Periodic Regrowth (Meltback)

Jackson and Miller have described a technique for directional crystallization with meltback. Growth was carried out in a thin film on a microscope stage. Application of a jet of cold nitrogen caused the crystal front to advance rapidly; stopping the nitrogen flow caused rapid remelting. In growth of salol, a cycle time of about 15 s was used. The net growth rate, controlled by transport of the microscope stage, was 1.39×10^{-4} cm s^{-1}. Improved crystal morphology resulted from the meltback (82). In a cellular interface, meltback produces liquid of lower impurity content than the interfacial liquid layer, a situation giving rise to concentration-driven mixing.

2.6   Ancillary Techniques

2.6.1   Thermal Stability of Chemical Substances

It is desirable that a material be thermally stable at its melting point to purify it by a melt-crystallization technique. It is not necessary, however, that the material be completely stable because the effects of decomposition at the melting point can be circumvented in at least two ways. One of these is based on the fact that it may be possible to remove an impurity more rapidly than it is generated, so that the net effect of crystallization is enhancement of purity. Another approach requires the formation of a eutectic system which may be crystallized from the melt at a temperature substantially below the decomposition point of the desired component. This and other added-component techniques are discussed in Chapter 8.

As a general matter, however, it is desirable to test the thermal stability of materials that are candidates for melt crystallization to assure success of the experiment and safety of the experimenter. Naturally, in all experiments making use of melts, it is desirable to limit the overheating of the melt, since decomposition rate can be a strong function of temperature. A simple method requiring little time or material is as follows:

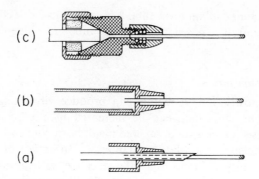

(c)

(b)

(a)

Fig. 3.40 Methods of evacuating capillaries for stability testing.

A conventional melting-point capillary is charged with a column of the substance under investigation, to a height of several mm. The capillary is then evacuated and refilled to a pressure of about one-half atmosphere of inert gas. Figure 3.40 shows two means of evacuating melting-point capillaries. In Figure 3.40a, a large-bore hypodermic needle has been passed through a rubber cap, and the loaded capillary has been inserted in the needle. The needle is carefully withdrawn, leaving the capillary gripped by the cap. In Figure 3.40b the cap has been attached to an evacuable glass or metal tube. A more convenient method for connecting the capillary to a vacuum system is shown in Figure 3.40c; one end of the fitting (Omnifit, Inc., P.O. Box 268, 217 Washington Avenue, Cedarhurst, NY 11516) accepts tubing up to 1.1 cm od, and the other, containing three O-rings, accepts tubing of 0.05–0.4 cm od. The capillary is then sealed by flame fusion and the resulting sealed ampoule is heated in a conventional melting-point apparatus for observation of the melting point and postmelt behavior. Attention must be directed toward discoloration and gas evolution from the melt. The sample may then be held for several hours at a temperature a few degrees higher than the melting point in order to confirm long-term stability. Finally, the molten sample should be resolidified and the melting point redetermined in order to learn whether thermally generated impurities have depressed the melting point. Naturally a given sample or sequence of samples may be held at progressively higher temperatures above the melting point in order to learn the temperature range of stability above the melting point.

2.6.2 Prepurification

A wide variety of techniques may be used to purify a material before directional crystallization. Crystallization from a solvent, distillation, treatment with selective adsorbents, and solvent extraction should all be considered. There are numerous books dealing with various purification techniques (56, 83, 84, 85).

On occasion, chemicals decompose when an attempt is made to melt them. Careful study may show that they are not in fact unstable, but may react with adsorbed or entrapped gases, or that they contain impurities that catalyze decomposition. Simple sublimation may free them from such impurities; ideally, the sublimate should be transferred to the crystallization vessel without exposure to air.

The device shown in Figure 3.41 allows this. Sublimand C is charged into the vessel through O-ring joint A and is covered with glass wool or other filtration medium B, which may serve the purely mechanical function of removing particulate contaminants. It may, on the other hand, be chosen to adsorb selectively an impurity of volatility comparable to that of the main component. This method is an extension of the "adsorption–sublimation" technique described long ago by Koffler (86). Crystallization tube E is attached at O-ring joint F. The system is evacuated through O-ring joint G. Coolant is circulated through tubing D, of semicircular cross section, and the sublimand is heated by a cylindrical heater (not shown). When sublimation is complete, tubing D is removed and the sublimate is melted into tube E. The operation is completed as described in Section 2.4.1.

Yet another method of charging freshly sublimed material into a crystallization tube is shown in Figure 3.42. The impure sublimand A is charged through tube B, which may be flame sealed. The entire assembly is wrapped with a heating tape and evacuated through joint C, causing condensation on water-cooled flask D. Upon completion of sublimation, D is drained, the system is filled with an inert gas and then heated until the sublimate melts and flows into crystallization vessel E. The remainder of the procedure for removal of vessel E is as described in Section 2.4.1.

It is possible to combine the loading of a directional crystallization vessel with a preliminary treatment with a selective adsorbent by passing the molten compound through an adsorbent bed (Figure 3.43). The material to be purified

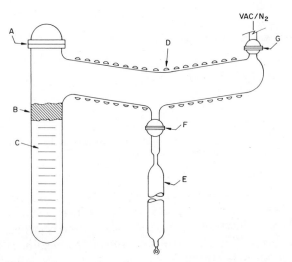

Fig. 3.41 "Sausage" tube for charging a directional-crystallization tube with sublimed material.

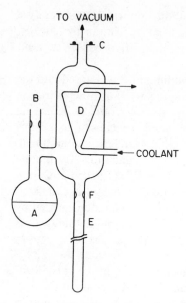

Fig. 3.42 An alternate method of charging a tube with sublimed material.

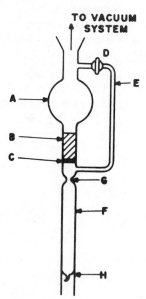

Fig. 3.43 Adsorptive pretreatment in charging a directional-crystallization tube.

is added to flask A, and rests upon adsorbent bed B, which is supported by frit C. The system is evacuated with stopcock D open, so that zone-refining tube F is evacuated via line E. Inert gas is admitted and the solid is melted; reevacuation of the system causes the melt to flow through the adsorbent into tube F. Pressure may be applied to the melt, with stopcock D closed. After solidification of the ingot, the tube F is sealed off at constriction G. Zoning is carried out with breakseal H at the top. Alternatively, the material to be purified may be sublimed through an adsorbent bed. For example, in Figure 3.44, sublimand A is evaporated by external heater F through adsorbent B and filter-medium C. The vapors are condensed on D, with additional heat provided by mantle E. Upon completion of sublimation in vacuum, inert gas is admitted through the upper stopcock and a crystallization vessel is attached in place of the sublimation tube.

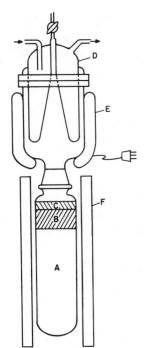

Fig. 3.44 Sublimation through an adsorptive layer.

2.6.3   Chemical Treatment

Impurities that are not removed by physical methods (including crystallization) can be removed during directional crystallization by subjecting the sample to a selected chemical reagent or physical effect. Crystallization is ineffective if an impurity is so similar to its host that $k_{eff} \simeq 1$. In such cases, chemical treatment before or during crystallization may be useful in improving the separability of the system. The cases described here are merely illustrative, and are by no means exhaustive. The use of these methods presupposes some knowledge of the nature and the amount of impurities present. If such information is available, it is often possible to select a reaction that will facilitate separation or make separation unnecessary by converting an impurity to the major component. An example of this category is aromatization of an aliphatic impurity in an aromatic host (e.g., dehydrogenation of cyclohexadiene to benzene). Another example is hydrogenation of the stilbene impurity contained in bibenzyl, to convert it to bibenzyl.

The polarity of an impurity may be changed by oxidation or reduction. Anthracene is hard to remove from phenanthrene by crystallization, but it is easy to oxidize anthracene selectively to anthraquinone, which is readily rejected during crystallization of phenanthrene. In the same sense, fluorene may be oxidized to fluorenone to improve its separability from phenanthrene or dibenzofuran.

An impurity may be converted to a larger, more readily segregated species. Halogen can be added to an olefinic impurity in a saturated host to give a larger, polar molecule which is easily segregated. Another approach is to use Diels–Alder addition to form a larger molecule (e.g., maleic anhydride reacts selectively with anthracene in phenanthrene to give an adduct that is separable). Yet another tactic applicable to the same system is photochemical dimerization at a wavelength specific for the reaction of the anthracene.

An impurity can be induced to react with its host. For example, if an impurity has a lower ionization potential than its host, treatment of the mixture with an alkali metal will convert the impurity to a radical ion which will react with host molecules to form a larger, separable product. If the impurity has lower thermal stability than the host, it may be "destroyed" simply by heating the mixture.

A complex may be formed with an impurity: if one component of a mixture reacts preferentially with a donor (D) or acceptor (A), to yield a D/A complex, the product may be more easily removed by crystallization (11).

On occasion, commercially available chemicals contain impurities that are not removable by crystallization or any other technique. In such cases it is sometimes possible to obtain the desired material by purifying a chemical precursor and ultimately generating the material of interest by a subsequent chemical step. This approach was used to obtain fluorene, which is usually accompanied by hard-to-remove dibenzofuran. To avoid this problem, fluorenone, prepared by cyclization of 2-phenylbenzoic acid, was purified by crystallization and chromatography or by zone refining. The purified ketone was reduced to fluorene, which was then further purified (87).

2.6.4 Product Recovery

Crystallized ingots that are liquid at room temperature can be melted, in portions, and decanted or siphoned from the container. Solids may be melted by appropriate heaters, and similarly decanted or poured. Alternatively, the sample container may be broken into segments whose contents may be sublimed or dissolved. Especially when dealing with larger samples, it is not practical to break the sample container; instead, the contents may be sublimed in portions. For this, the modified McCarter sublimer (88) shown in Figure 3.44 can be used. The crystallization tube is connected to the lower opening of the sublimer via the standard-taper joint shown or an o-ring-sealed connector such as an Ultra-Torr union (Cajon Co. 32550 Old South Miles Rodd, Sonon, Ohio 44139). After the system is evacuated, portions of the ingot are sublimed in sequence, starting at the top. The amount of ingot sublimed is determined by the length of heater; a single heating tape may be used, covering progressively larger fractions of the ingot.

2.6.5 Separation of Impurities from Enriched Fraction

Many techniques may be adapted to the separation of enriched minor components. Crystallization from solvents, solvent extraction, and sublimation have been useful. Chromatographic methods, given their power and versatility, are preferred. In favorable cases the separation and identification can be combined by applying gas chromatography/mass spectroscopy (GC/MS) directly to the enriched fraction. If it is desired or necessary that μg-to-mg amounts of an enriched impurity be separated, it is possible to trap individual components exiting from a gas chromatograph (89).

The "casual" user of GC, who is not an expert in separation science, may be intimidated by the plethora of GC media that are now available commercially. It should be comforting then that it now emerges that nearly all GC analyses can be performed on one or another of the six polymers listed in Table 3.3 (90). All six can be used over a wide temperature range and collectively cover the entire field of chemical interaction. A separation that is not effected by any of these will seldom be improved by use of still other stationary phases; instead, the temperature, molecular weight of the polymer, the solid support, or some other column condition should be changed to seek improved separation. Naturally,

Table 3.3. Primary List of Preferred Stationary Phases

(1) Dimethylsilicone (a) (e.g., OV-101, SP-2100, SE-30, SF-96)
(2) 50% phenylmethylsilicone (a) (e.g., OV-17, SP-2250)
(3) Polyethylene glycol (b, c), MW > 4000 (e.g., Carbowax)
(4) Diethyleneglycol succinate (b, d)
(5) 3-cyanopropyl silicone (b, e) (Silar 10C, Apolar 10C, SP-2430)
(6) Trifluoropropyl methyl silicone (OV-210, SP-2401)
    (a) (1) and (2) are chromatographically very similar. Inadequate separation on one will seldom be much improved by trying the other.
    (b) (3), (4), and (5) are chromatographically similar, but less so than (1) and (2).
    (c) Epoxy bridging of polyethylene glycols to form higher polymers such as Carbowax 20M does not affect the retention properties, nor does termination of the polymer with terephthalic acid lead to improved stability.
    (d) Stabilization by attachment to dimethyl silicone changes the retention properties slightly.
    (e) There are a number of cyanoalkyl silicones currently available, varying widely in their retention properties. The 3-cyanopropyl silicone was chosen because it is the most polar silicone having a definite composition.

the list of primary preferred phases may be expanded in the future, or some of its members may be displaced as new phases are described. The primary list does not include phases appropriate to very high or very low temperature, nor to some highly corrosive chemicals. For these and other special purposes, a secondary list has been prepared (90).

2.6.6   Hard-to-crystallize Melts

Tipson (91) has reviewed strategies for inducing nucleation in melts that do not crystallize readily. The first suggestion is that impurities that may retard crystallization should be removed. This may be done by differential reaction, distillation, adsorption, or derivative formation; naturally, it is necessary to remove all traces of solvent from materials that are to be crystallized. Another procedure, little used but possibly of quite general validity, is the formation of a precipitate in a solution that is to be decolorized. The nascent precipitate, as it is forming, adsorbs colored materials. A recent theoretical study has analyzed the problem of inducing crystallization in glass-forming systems (92).

It is helpful to cool the melt strongly and then to allow the melt to warm slowly to some temperature substantially below the melting point and hold it there. The melt should then be warmed slowly to a second, higher temperature and held there. The intent of this procedure is to cause the material to spend some time at an unknown temperature which is optimal for the formation of nuclei and then at some higher temperature favorable for growth (see Section 2.1, Chapter 1). Throughout these procedures, it is necessary to exclude moisture rigorously. Alternatively, it is possible to subject the melt to a temperature gradient such that part of it is at a temperature ideal for nucleation and another part at a temperature suitable for growth.

It has also been found helpful to extract the material to be crystallized with a solvent, which does not dissolve the crystallizand and which may leach out impurities. During this procedure of contacting the crystallizand with "poor" solvent, it is possible to agitate the mixture and/or to add crystalline material such as kieselguhr, silica gel, or charcoal, and to rub the melt in the presence of these. In addition, as a last resort, it is possible to expose a thin film of the crystallizand to random dust particles. The addition of isomorphic compounds of crystalline homologs, derivatives, isomers, and/or compounds that yield molecular compounds with the crystallizand can also be helpful. Scratching the surface of the container may induce crystallization, either by injection of particles or by electrostatic interaction. Other techniques that have been used to induce crystallization include: continued agitation, bubbling of gas through the crystallizand, ultrasonication, irradiation with beta rays, X-rays, and application of electric fields.

Another procedure that is useful in "difficult" crystallization is based on isothermal distillation. A nearly saturated solution of the solute is prepared in a good solvent. The open container of solution is placed in a vessel containing a portion of another solvent in which the solute is less soluble. It is desirable that the poor solvent should be somewhat more volatile than the good solvent. The outer vessel is closed and allowed to stand at constant temperature. The outer (poor) solvent condenses in the solution and creates a concentration gradient which moves slowly downward into the solution. When the solubility in the mixed solvent is exceeded, nucleation can take place. Newly formed crystals fall to the bottom of the inner, open container, where the surrounding solvent is unsaturated. Partial dissolution takes place until the solvent gradient reaches the saturation value, whereupon growth resumes. The slow passage through saturation can bring about crystallization in systems with strong tendency to form supercooled liquid.

REFERENCES

1. W. A. Tiller and R. F. Sekerka, *J. Appl. Phys.*, **35**, 2726–29 (1964).

2. W. G. Pfann, *Zone Melting*, 2nd ed., Wiley, New York, 1966, pp. 10–12.

3. C. H. Li, *Brit. J. Appl. Phys.*, **18**, 359–360 (1967).

4. C. H. Li, *Phys. Stat. Sol.*, **15**, 3–56 (1956).

5. W. R. Wilcox, *Mat. Res. Bull.*, **13**(4), 287–291 (1978).

6. G. Matz, *Chemie-Ing. Technik*, **36**, 381–394 (1964).

7. W. A. Tiller, K. A. Jackson, J. W. Rutter, and B. Chalmers, *Acta Met.*, **1**, 428–37 (1953).

8. M. Krumnacker and W. Lange, *Kristall Technik*, **4**(2), 207–220 (1969).

9. I. O. Kulik and G. E. Zilberman, "Impurity Segregation During Growth of a Crystal from a Melt," in A. V. Shubnikov and N. N. Sheftal, Eds., *Growth of Crystals*, Vol. 3, Consultants Bureau, 1963.

10. J. A. Burton, R. C. Prim, and W. P. Slichter, *J. Chem. Phys.*, **21**, 1987–1996 (1953).

11. G. J. Sloan, *Mol. Cryst.*, **1**, 161–194, (1966).

12. V. I. Zharinov, E. E. Konovalov, Sh. I. Peizulaev, and T. V. Kayurova, *Russ. J. Phys. Chem.*, **48**(4), 572–574 (1974).

13. L. O. Wilson, *J. Crystal Growth*, **44**, 247–250 (1978).

14. A. M. J. G. Van Run, *J. Crystal Growth*, **47**, 680–692 (1979).

15. J. W. Rutter and B. Chalmers, *Can. J. Phys.*, **31**, 15–39 (1953).

16. B. Chalmers, J. Metals, *Trans. AIME*, **200**, 519–532 (1954).

17. D. T. J. Hurle, E. Jakeman, and A. A. Wheeler, *J. Crystal Growth*, **58**, 163–179 (1982).

18. S. R. Coriell, R. F.Boisvert, R. G. Rehm, and R. F. Sekerka, *J. Crystal Growth*, **54**(2), 167–176 (1981).

19. C. E. Schvezov and D. Fainstein-Pedraza, *J. Crystal Growth*, **34**, 55–60 (1976).

20. W. Bardsley, J. S. Boulton, and D. T. J. Hurle, *Solid State Electronics*, **5**, 395–403 (1962).

21. N. V. Lapin, V. A. Malyusov, and N. M. Zhavoronkov, *Proc. USSR Acad. Sci.*, **256**(3), 9–12 (1981).

22. N. V. Lapin and V. A. Malyusov, Teor. Osn. Khim. Tekhnol., **15**(6), 836–42 (1981), through CA **96**:13761f (1982).

23. E. L. Holmes, J. W. Rutter, and W. C. Winegard, *Can. J. Phys.*, **35**, 1223–1227 (1957).

24. T. S. Plaskett and W. C. Winegard, *Trans. AIME*, **51**, 222 (1950).

25. R. M. Sharp and A. Hellawell, *J. Crystal Growth*, **6**, 334–340 (1970).

26. W. Bardsley, J. M. Callan, H. A. Chedzey, and D. T. J. Hurle, *Solid State Electronics*, **3**, 142–154 (1961).

27. A. C. M. Irons and D. J. Morantz, *Transitions Non-Radiat. Mol., Réunion Soc. Chim. Phys.*, 20th, 158–163 (1969).

28. D. J. Morantz and K. Mathur, *Nature*, **226**, 638–639 (1970).

29. R. A. M. Scott, D. J. Morantz, and K. K. Tay, *Solid State Comm.*, **9**, 1559–1561 (1971).

30. M. E. Glicksman, R. J. Schaefer, and J. A. Blodgett, *J. Crystal Growth*, **13–14**, 68–72 (1972).

31. A. N. Kirgintsev, B. M. Shavinskii, and E. D. Abramovich, *Zhur. Fiz. Khim.*, **42**, 1092 (1968).

32. C. Jaccard, *Phys. Kondens. Materie*, **4**, 349–354 (1966).

33. A. R. McGhie and G. J. Sloan, *J. Crystal Growth*, **32**, 60–67 (1976).

34. I. N. Pribyl'skii, E. A. Zhadanov, and Ye. M. Lobanov, *Dokl. Akad. Nauk BSSR*, **21**(10), 896–899 (1977).

35. W. R. Wilcox, "MasTransfer in Fractional Solidification" in W. R. Wilcox and M. Zief, Eds., *Fractional Solidification*, Dekker, New York, 1967.

36. M. G. Mil'vidskii, *Sov. Phys. Crystallog.*, **6**, 647–648 (1961).

37. L. I. Isaenko and A. N. Kirgintsev, *Akad. Nauk SSSR (Sib.)*, 148–149 (1977).

38. H. Blicks, H. H. Egger, and N. Riehl, *Phys. Kondens. Mater.*, **2**, 419–422 (1964).

39. M. D. Lyubalin, V. N. Tretjakov, and V. A. Mokievskii, *Kristall Technik*, **13**(10), 1181–1186 (1978).

40. W. R. Wilcox, *U.S. Clearinghouse Fed. Sci. Tech. Inform.*, AD 1968, AD-685162, 27 pp.

41. F. Stöber, *Chem. Erde*, **6**, 357–367 (1931).

42. H. E. Buckley, *Crystal Growth*, Wiley, New York, 1951.

43. D. R. Uhlmann, B. Chalmers, and K. A. Jackson, *J. Appl. Phys.*, **35**(10), 2986–2993 (1964).

44. S. N. Omenyi and A. W. Neumann, *J. Appl. Phys.*, **47**, 3956–3962 (1976).

45. R. B. Fedich and W. R. Wilcox, *Sep. Sci. Technol.*, **15**(1), 31–38 (1980).

46. W. R. Wilcox, *Separation Science*, **4**(2), 95–109 (1969).

47. V. H. S. Kuo and W. R. Wilcox, *Ind. Eng. Chem.*, *Process Des. Develop.*, **12**(3), 376–379 (1973).

48. J. Van Den Boomgaard, *Philips J. Res.*, **33**, 149–185 (1978).

49. B. Jansson and A. Lunden, *Z. Naturforsch.*, **25a**, 697–699 (1970).

50. J. N. Sherwood and S. J. Thomson, *J. Sci. Instr.*, **37**, 242–245 (1960).

51. M. Volmer, *Ann. Phys.*, **42**, 485 (1913).

52. H. D. Riccius, W. E. E. Berger, and C. J. Van der Hoeven, *Phys. Stat. Sol. (a)*, **1**, K63–K64 (1970).

53. S. D. Gromakov, V. N. Kurinnaya, Z. M. Latypov, and M. A. Chvala, *Russ. J. Inorg. Chem.*, **9**(5), 1305–1306 (1964).

54. B. J. McArdle, J. N. Sherwood, and A. C. Damask, *J. Crystal Growth*, **22**, 193–200 (1974).

55. R. L. Schneider, *Eastman Organic Chemicals Bulletin*, **47**(1), (1975).

56. J. A. Riddick and W. B. Bunger, *Techniques of Organic Chemistry–Organic Solvents*, Vol. 3, 3rd ed., Wiley, New York, 1970.

57. T. Hayakawa, M. Matsuoka, and K. Satake, *J. Chem. Eng. Japan*, **6**(4), 332–337 (1973).

58. C. J. May, *J. Phys. E, Sci. Instr.*, **8**, 354–355 (1975).

59. U.S. Patent 2,754,180, July 10, 1956.

60. A. T. Horton and A. R. Glasgow, *J. Res. N.B.S.* **69C**(3), 195–198 (1965).

61. G. Guillaud, M. LeHelley, and G. Mesnard, *Bull. Soc. Fr. Mineral Crystallogr.*, **93**, 131–132 (1970).

62. R. Khanna and J. M. Low, "Laboratory Heat Transfer," in E. S. Perry and A. Weissberger, Eds., *Techniques of Chemistry*, Vol. 13, Wiley, New York, 1979.

63. F. W. Schwab and E. Wichers, *J. Res. N.B.S.*, **32**, 253–259 (1944).

64. N. Karl, "High Purity Organic Molecular Crystals," in H. C. Freyhardt, Ed., *Crystals, Growth, Properties, Applications*, Vol. 4, Springer Verlag, New York, 1980.

65. H. LeGal and Y. Grange, *J. Crystal Growth*, **49**, 449–457 (1979).

66. L. E. Prescott, *Rev. Sci. Instr.*, **33**, 485–486 (1962).

67. A. J. Lovinger, J. O. Chua, and C. C. Gryte, *J. Phys. E, Sci. Instr.*, **9**, 927–928 (1976).

68. C. E. Miller, *J. Crystal Growth*, **42**, 357–363 (1977).

69. A. N. Kirgintsev, *Deposited Publication*, VINITI, 6136–73, 17 pp. (1973).

70. C. E. Turner, A. W. Morris, and D. Elwell, *J. Crystal Growth*, **35**, 234–235 (1976).

71. E. L. Anderson, *Chem. Ind.*, 131–136 (1975).

72. L. I. Afanasiadi, A. B. Blank, V. Ya. Vakulenko, and N. I. Shebtsov, *Industrial Laboratory*, **40**, 1791–**1792 (1974).**

73. T. H. Gouw, *Separation Science*, **2**(4), 431–437 (1967).

74. P. W. Liscom, C. B. Weinberger, and J. E. Powers, *A.I.Ch.E.–I. Chem. E. Symposium Series No. 1*, 90–104 (1965).

75. H. Schildknecht, G. Rauch, and F. Schlegelmilch, *Chemiker Ztg.*, **83**, 549–552 (1959).

76. A. S. Borschevskii and I. Tretyakov, *Sov. Phys. Solid State*, **1**, 1360–1362 (1959) (English transl.).

77. W. G. Pfann and J. N. Carides, *Mat. Res. Bull.*, **5**, 31–36 (1970).

78. W. G. Pfann, *Zone Melting*, 2nd ed., Wiley, New York, 1966, p. 106.

79. C. C. Adams, U.S. Patent 3,966,445, June 29, 1976.

80. G. J. Sloan, *Anal. Chem.*, **38**, 1805–1806 (1966).

81. K. B. Yurkevich, E. N. Ozerenskaya, and S. N. Afanas'ev, *Sb. Mosk. Inst. Stali Splavov*, **77**, 194–197 (1974).

82. K. A. Jackson and C. E. Miller, *J. Crystal Growth*, **42**, 364–369 (1977).

83. D. D. Perrin, W. L. F. Armarego, and D. R. Perrin, *Purification of Laboratory Chemicals*, Pergamon, Oxford, 1966.

84. E. S. Perry and A. Weissberger, *Technique of Chemistry, Vol. 12, Separation and Purification*, 3rd ed., Wiley-Interscience, New York, 1978.

85. B. L. Karger, L. R. Snyder, and C. Horvath, *Introduction to Separation Science*, Wiley-Interscience, New York, 1973.

86. W. Koffler, *Monatsh. Chem.*, **80**, 694–701 (1949).

87. D. L. Horrocks and W. G. Brown, *Chem. Phys. Letters*, **5**(2), 117–119 (1970).

88. R. J. McCarter, *Rev. Sci. Instr.*, **33**, 388–389 (1962).

89. L. S. Ettre and W. H. McFadden, Eds., *Ancillary Techniques of Gas Chromatography*, Wiley-Interscience, New York, 1969.

90. S. Wold and J. Yancey, *J. Chromat. Sci.*, **13**, 115–117 (1975).

91. R. S. Tipson, "Crystallization and Recrystallization," in A. Weissberger, Ed., *Techniques of Organic Chemistry*, Vol. 3, Part 1, 2nd ed., Wiley Interscience, New York, 1956.

92. I. Gutzow, *Contemp. Phys.*, **21**(2), 121–137 (1980).

Chapter **IV**

APPLICATIONS OF DIRECTIONAL CRYSTALLIZATION

We have already alluded to the use of directional crystallization in purification. Less obvious uses have emerged in connection with enhanced analytical sensitivity, measurement of thermodynamic properties including phase relationships, study of crystal-growth mechanisms, and preparation of single crystals. In the following sections, we review these applications through selected examples. The intent is not to cover the field exhaustively, but rather to show the range of possibilities.

1   PURIFICATION

Many chemicals form simple eutectics with all or most of the impurities that accompany them; these may be purified by a single, carefully executed directional crystallization. The interest of the experimenter may focus primarily on the pure material for use as a reagent, as an analytical standard or for a physical measurement. On the other hand, there may be greater interest in the impurities rejected during crystallization, since qualitative detection and quantization are easier in the concentrate than in the original mixture.

In this review of the use of directional crystallization, emphasis is on materials of chemical interest. The extensive work on crystallization of metals and metalloids is highly specialized in technique and is treated extensively in the literature of metallurgy and solid-state science; accordingly, metals will be discussed here only in connection with techniques of particular interest.

1.1   Elements

ALUMINUM

Dewey has described a method for purification of aluminum by continuous casting (1). Although the apparatus would not be considered for use in the chemical laboratory, the idea of continuous directional crystallization from a crucible merits description.

Molten metal was fed through heated inlet A (Figure 4.1) into molding chamber B, whose outlet was equipped with an annular cooling jet C. Stirrer D was positioned vertically so that fluted impeller E agitated the liquid at the interface; it caused the melt to move downwardly along the wall, then radially inward toward the axis and upwardly away from the interface. The impure melt overflowed through outlet F. The casting chamber was heated at 1144 ± 0.25 K by a furnace (not shown), which provided bulk melt and overflow temperatures of 945 and 1001 K, respectively, during steady-state operation.

The product ingot G (of 20-cm diameter) was lowered from the molding chamber by a hydraulic mechanism (not shown) at rates between 7×10^{-4} and 7×10^{-3} cm s^{-1}. By varying the feed and growth rates, the apportionment between product and overflow was adjusted so that starting materials of varying purity could be processed. Table 4.1 shows the impurity concentrations achieved in several runs under different conditions. The axial and radial homogeneities of the ingots were evaluated by multiple spectrographic analyses. In one run, total

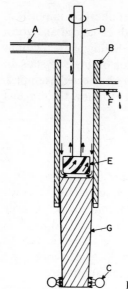

Fig. 4.1 Device for continuous casting of an aluminum ingot.

impurity concentration along the product ingot averaged 172 ppm. The standard deviation was 35 ppm. The average total impurity on a given cross section was 157 ppm, with a standard deviation of 18 ppm.

The work described here has been extended to ingots of 50-cm diameter (2). The product ingots in the three runs described in Table 4.1 comprise only 13–36% of the aluminum fed to the apparatus.

ANTIMONY

Lead, copper, and iron were segregated in the last-frozen end of an antimony ingot (3).

ARSENIC

Kirgintsev et al. (4) directionally crystallized arsenic in a gradient heater under high pressure in a quartz ampoule at 6.2 MPa. This was cooled from 1123 to 1023 K at 4×10^{-3} K s^{-1}. A float consisting of a disk of sodium chloride was used to indicate completeness of melting.

BISMUTH

Neutron-activation analysis has been used to study the distribution of impurities in directionally crystallized bismuth. Easily oxidized elements were present at higher concentrations than relatively inert elements (5). Radiotracers were used to measure k_0 for a number of solutes in Bi: Cu(0.0052), Ge(0.11), Pb(0.015), Po(0.15), Se(0.030), Sn(0.08), and Te(0.27), As(0.72), Ni(3.6×10^{-4}), Ag(2.9×10^{-4}), Au(2.7×10^{-5}), Ir($\leqslant 10^{-4}$), Hg($< 10^{-3}$), and Pt(2×10^{-4}) (6, 7).

Table 4.1. Impurity Concentrations in Directionally Crystallized Aluminum

Feed rate, kg s⁻¹		5.06×10^{-3}			2.5×10^{-3}			31.6×10^{-3}	
Purity, %		99.9			96			99.9	
Overflow rate, kg s⁻¹		4.42×10^{-3}			1.61×10^{-3}			25.9×10^{-3}	
Growth rate, cm s⁻¹		7.06×10^{-4}			1.0×10^{-3}			$5.56\text{--}6.94 \times 10^{-3}$	

Concentration, ppm

Impurity	Feed	Product	Overflow	Feed	Product	Overflow	Feed	Product	Overflow
Si	348	46	395	16,500	7,400	21,800	400	91	410
Fe	420	25	469	16,200	5,500	19,900	350	116	390
Cu	18	3	19	5,700	2,400	6,900	15	10	18
Mg	9	3	7	—	—	—	—	—	—
Zn	68	37	71	230	200	360	120	40	140
Ga	71	32	86	—	—	—	90	38	100

IODINE

Kirgintsev et al. purified crude iodine by first filtering and then crystallizing it. Filtration reduced the nonvolatile residue and the chlorine content, giving a product of about 99.6 wt% purity from starting material of 96.9 wt% purity. The prepurified iodine was sealed in a glass ampoule (25 cm × 1.4 cm diameter) at 13.3 Pa; the charge weighed 80 g and filled the ampoule to a depth of 17 cm. A heater/cooler assembly was moved horizontally over the ingot at 1.1×10^{-4} cm s^{-1}; the heater was maintained at 428 K and the ampoule was cooled to 283 K. Not surprisingly, directional crystallization was more effective when the ampoule was rotated about its axis at 4.8 rev s^{-1}, than in a stationary ampoule. In the former case, the yield of analytical-grade product (99.8 wt% purity) was 77%. A substantial fraction of the impurity was rejected completely and deposited on the wall of the ampoule above the ingot (8).

KRYPTON

Pruett studied electron drift velocity in solid and liquid krypton and found that the number of carriers $N_0(E)$ was sensitive to impurity concentration. A given sample of gas, crystallized slowly, gave $N_0(E)$ up to 30% larger than that obtained after rapid growth. The difference was attributed to segregation of impurity by directional crystallization (9).

SULFUR

The separation of ^{32}P from neutron-irradiated sulfur has been studied; directional crystallization was found to be more effective than zone refining for concentrating the microcomponents (10, 11).

TELLURIUM

Kirgintsev and Lastushkin purified Te by directional crystallization in a 4.5-cm tube rotating at 4.0×10^{-2} rev s^{-1}. The product was 4N6 pure (12, 13).

Directional crystallization has also been applied in a quantitative study of impurity enrichment in Te. Te was directionally crystallized with added ^{203}Hg; k_0 was found to be 0.11 (14).

URANIUM

Poeydomenge et al. (15) melted a 0.6-kg ingot of electrolytic U in a UO$_2$ boat inside an alumina tube. The molten U was held at 1523 K for 3.6×10^4 s, then was solidified at 1.39×10^{-4} cm s^{-1} by transport of the furnace. Impurity migration was studied by metallography and measurement made of electrical resistivity and microhardness.

1.2 Inorganic Compounds

ALKALI−METAL SALTS

Lithium nitrate has been directionally crystallized in a study of the enrichment of ^{6}Li in ^{7}Li. It had been found earlier that multipass zone refining was troubled by the large change in volume accompanying melting. Hence the distribution coefficient was measured by directional crystallization, in spite of its nearness to unity. Crystallization was carried out by raising the sample container from a furnace and a slight increase in ^6Li concentration was observed in the first-frozen solid. The concentration profiles after crystallization at 1.19×10^{-4} and at 0.528×10^{-4} cm s^{-1} gave k_{eff}'s of 1.0025 and 1.0030, respectively. In turn, these values were used to calculate δ/D and k_0 of the BPS equation (see Section 1.2.3, Chapter 3), which were 2.8×10^3 s cm^{-1} and 1.0035, respectively. The value of D has not been measured at the melting point, but extrapolation of low-temperature data gave $D = 1.36 \times 10^{-5}$ cm^2 s^{-1}, from which δ is calculated to be 0.038 cm (16).

In an early study of directional crystallization of sodium nitrate, distribution coefficients were measured for 15 impurities (17), and it was concluded that repeated directional crystallization is an effective method for purifying this compound. The dependence of the k's upon crystallization conditions was explored for Li, K, Rb, Cs, Ca, Pb, and Cu nitrates. Some of these systems were later studied in greater detail (18). The shape of the solid/liquid interface, which depends on growth rate, was found to determine the value of the distribution coefficient. Equilibrium distribution coefficients were measured at growth rates $< 1.94 \times 10^{-3}$ cm s^{-1}. The system $NaNO_3/RbNO_3$ was studied by determining distribution coefficients in directional crystallization experiments. The low mutual solid solubility was said to prove the impossibility of forming a continuous series of solid solutions with minimum melting point (19) in this system.

Kirgintsev has studied the radial distribution of $^{89}Sr(NO_3)_2$ in $NaNO_3$, directionally crystallized at 7.78×10^{-4} cm s^{-1}. A plot of rate of crystallization against initial melt concentration indicated that there are regions of equilibrium and nonequilibrium distribution, separated by a hyperbola. The $Sr(NO_3)_2/NaNO_3$ system was further examined (20) to learn the effects of container diameter (d), orientation (ϕ), rotation frequency (ω), and oven temperature (T) on the distribution coefficient k_{eff}. It was found that k_{eff} increased toward unity as tube size was decreased below 2 cm; k decreased to a constant value as ω was increased to 2.33 rev s^{-1}. Without rotation, interface shape was determined by gravitational effects for orientations above and below horizontal ($\phi < 0°$ and $\phi > 0°$, respectively). With rotation, the best interface was obtained at $\phi = 0°$. As T increased, k_{eff} increased.

The optimum conditions established in the work just described were used in the study of directional crystallization of $NaNO_3$ and $NaCl$ containing, in each case, a major and a minor impurity. In the former case, both $Sr(NO_3)_2$ and $CsNO_3$ were used as major and minor impurities; in the latter case, $SrCl_2$ and $CsCl$ were major and minor impurity, respectively. In the equilibrium–crystallization regime, the distributions of major and minor impurities were mutually independent, while in nonequilibrium crystallization, k_{eff} of the major impurity increased toward unity and that of the minor impurity decreased as crystallization proceeded.

Sodium nitrate has also been purified by directional crystallization in a centrifugal field (21); the removal of oppper and magnesium ions was studied. At a crystal growth rate of 1.33×10^{-3} cm s^{-1} (4.8 cm h^{-1}) k_{eff} of copper was 0.1. A corresponding experiment involving vertical directional crystallization without centrifuging, employing growth rates of 1.1×10^{-4} to 1.3×10^{-3} cm s^{-1} (0.4 to 4.8 cm h^{-1}) gave distribution coefficients ranging from 0.2 to 0.5. For magnesium, vertical directional crystallization without centrifugation, at 4.4×10^{-4} cm s^{-1} (1.6 cm h^{-1}), gave a k_{eff} of 0.5. With centrifugation, at the same crystallization rate, k_{eff} was 0.1. A partial phase diagram was constructed for the system cesium iodide/thallium iodide, on the basis of directional crystallization of mixed melts. Equilibrium distribution coefficients were measured for TlI in CsI, using a wide range of starting compositions, and the variation with concentration is shown in Figure 4.2 (22).

SALT HYDRATES

(See Section 1.3.1, Chapter 8) Süe et al. (23) recognized that ice-salt eutectics are amenable to purification by melt techniques at temperatures below the melting points of the salts themselves. Thus potassium chloride (mp 1049 K) was dissolved in water (24.1 g KCl/100 g H$_2$O) and aliquots of solution were doped with radioactive contaminants (0.1 wt% referred to dry KCl) before directional crystallization at 1.56×10^{-4} cm s^{-1}. Distribution coefficients measured for Na, Cs, Ca, and Sr chlorides were 0.32, 0.18, 0.44, and 0.36, respectively, while those of potassium sulfate and phosphate were 0.34 and 0.43, respectively. It is interesting that the measured distribution curves corresponded to complete mixing in the liquid, although no attempt was made to stir the remaining liquid during solidification (see Section 1.2, Chapter 3). In fact, measurement of the variation of k_{eff} with crystallization rate showed a rapid rise from 0.1 to 0.4 between 0.56×10^{-4} and 2.2×10^{-4} cm s^{-1}, followed by a slower rise toward unity at higher speed.

Blank et al. also studied aqueous solutions of KCl and KBr of cryotectic composition and measured distribution coefficients of nitrates and sulfates with and without stirring. Stirring was found to decrease the thickness of the diffusion layer by an order of magnitude (24).

Similar experiments have been carried out with sodium and cesium iodide dihydrates (25, 26). Kingintsev et al. extended this work to Sr(NO$_3$)$_2$,

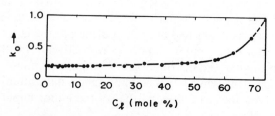

Fig. 4.2 Equilibrium distribution coefficient of TlI in CsI, as a function of liquid composition, from directional crystallization experiments.

$(NH_4)_2SO_4$, NH_4F, $NaCl$, and $CsCl$ and compared the experimental results with theoretical predictions based on the hydrodynamics of the crystallization interface (27).

1.3 Organic Compounds

Schildknecht has observed that purification by crystallization is an area in which industrial operations have led laboratory practice (28). Petroleum oils were dewaxed by chilling as early as the second decade of this century. Naphthalene and anthracene have been frozen out of the so-called "anthracene oil" of coal tar for about the same length of time. The experiments of Gaubert (29) on rejection of dyes by freezing organics were among the first laboratory applications of directional crystallization. He found that when dyes (e.g., rosaniline, rhodamine, etc.) were added to various organics at high concentrations, pure, transparent crystals grew at small subcooling. When the preparations were strongly cooled, an apparently homogeneous mass resulted.

The systematic use of directional crystallization for purification of organics may be said to date from the classic work of Schwab and Wichers, which appeared in 1940 (30). They were led to "fractional freezing" by their awareness of the usefulness of freezing range as a criterion of purity. Three cylindrical/axial crystallizations of benzoic acid with stirring, gave product of $>5N$ purity (i.e., $>99.999\%$). The experiments were carried out on 50-g samples in 2-cm tubes, then on a somewhat larger scale, using 0.5-kg samples in 4.6-cm tubes (31). In these experiments the remaining liquid was agitated by a gas stream and acid of 3N1 purity was brought to 4N7 purity by three solidifications, after each of which one-fourth of the charge was discarded. Ultimately, much larger charges were crystallized in the spherical/radial mode, with mechanical stirring, in an apparatus which has already been described (see Section 2.4.2, Chapter 3). The sole heat input was provided by the small heater at the base of the neck. It partially compensated the heat lost through the insulation and controlled the rate of solidification, which corresponded to a linear rate of 0.40×10^{-4} cm s^{-1}. After about two-thirds of the charge had solidified, the stirrer was replaced by a centrally positioned 45-W lamp, which melted about 100 g of surface solid. This "washing" was repeated three times, to give a 60% yield of product of 4N purity. Repetition of the operation with the combined product of two runs gave a final product of 4N8 purity.

Schwab and Wichers defined the efficiency of purification as the percentage of total impurity present that is removed in a single operation. For their cylindrical/axial and spherical/radial experiments, they calculated efficiencies of 68 and 85%, respectively. The difference derived mainly from the stirring and slower rate of freezing in the second mode. More interesting, however, is a comparison of these efficiencies with those resulting from careful crystallization from solvents, by the same workers. The figures were 14 and 29% for crystallization from water and benzene, respectively. Hence 25 recrystallizations from water would have been required to match the increase in purity afforded by two freezings; for benzene, 11 recrystallizations would have been needed.

The crystallization of naphthalene/biphenyl melts on the outer surface of a water-cooled tube was studied as a step toward a commercial crystallizer (32) (see Chapter 9). A vertical tube of 2.5-cm diameter and 2.1-m length was surrounded with a concentric, electrically heated jacket which was provided with viewing ports for observation of nucleation and growth. Cooling water was passed through the inner tube at a flow rate that was high enough to maintain constant temperature over the length of the tube; the temperature of the coolant was reduced linearly. The test mixture was sprayed onto the top of the tube at a measured rate until a solid layer formed. The thickness of the layer was monitored by measuring the increasing diameter of the cylinder. Samples were removed from various positions along the tube for analysis and measurement of k_{eff}.

The test mixtures contained 1, 5, and 10 wt% biphenyl. The Reynolds numbers of the flowing films were 200, 650, and 1050, corresponding to a range of hydrodynamic properties. The melts were overheated by 3 or 8 K above the crystallization temperature, giving growth rates of 2.8×10^{-4} to 8.3×10^{-4} cm s^{-1} (1 to 3 cm h^{-1}) or 6.9×10^{-4} to 1.4×10^{-4} cm s^{-1} (2.5 to 5 cm h^{-1}).

Visual observation of the growing crystal layer showed that over the upper portion of the tube, the naphthalene grew as a compact, glass-clear layer, with some cracks caused by thermal stress. Figure 4.3 shows the change of k_{eff} along the length of the tube for runs with the three starting concentrations shown; the region of nearly constant k_{eff} corresponds to the clear, solid deposit. The transition in crystal morphology corresponds to the onset of constitutional supercooling. An equation was derived for calculation of this critical crystalli-

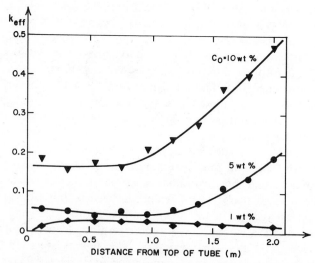

Fig. 4.3 Variation of k_{eff} during outward radial directional crystallization in a vertical tube, as a function of position, with starting concentration as parameter.

zation length, from feed point to the appearance of constitutional supercooling.

Naida et al. compared distillation and directional crystallization for purification of 85% formic acid (33). Vacuum distillation gave 99% formic acid; directional crystallization at 3.1×10^{-4} cm s^{-1} gave 95% formic acid ($k_{\text{eff}} = 0.32$). It is probable that k_0 for this system is zero or close to zero; hence more efficient directional crystallization should give nearly pure product.

Schildknecht installed the apparatus described in Section 2.5.2, Chapter 3 in a drybox in order to use it for purification of low-melting organic solvents, with liquid air as coolant. Solvents of low polarity (carbon tetrachloride, methylene chloride, and chloroform) were readily solidified at 1.67×10^{-3} cm s^{-1} with stirring at 23.3 to 50 rev s^{-1}. More polar liquids (methanol and ethanol) were less suitable and ethylene was totally unsuitable for purification because of the viscosity of their melts (34).

1.4 Polymers

The application of directional crystallization to purification of polymeric materials requires some adjustment of outlook and expectations. First, it must be recognized that even "pure" polymers are heterogeneous in the sense that they contain more-or-less broad distributions of molecular weight. These mutually soluble molecular-weight fractions are separable and hence may be thought of as mutal contaminants. Of course, conventional impurities also may be present in polymers: unreacted monomer, unpolymerizable components of the monomer(s), catalyst, and, if a commercial material is under consideration, additives such as colorants, antioxidants, plasticizers, and so on. Second, the size and length of polymer molecules lead to sluggish crystallization kinetics, with molecular entanglement at the interface a near certainty. Given these considerations, it is not surprising that relatively little use has been made of directional crystallization for purification of polymers.

Electron microscopy of melt-grown polytetrafluoroethylene crystals revealed extended-chain crystals of differing thickness in the molecular-chain direction. This led to the speculation that "the molecules may be to some extent locally segregated into regions of roughly uniform molecular length" (35). Slowly crystallized polyethylene showed similar segregation by molecular weight (36, 36, 38).

Moyer and Ochs (39) provided conclusive evidence of segregation of both added impurities and low-molecular-weight polymer during crystallization. In one experiment, they used tritiated dilauryl thiodipropionate as an additive in polypropylene. (Untagged dilauryl thiodipropionate is commonly used as an antioxidant in polypropylene.) In the case of polyethylene, a fraction of low molecular weight was tritiated to give a sample whose number-average molecular weight was 320. Finally, isotactic polystyrene was doped with tritiated, atactic polystyrene.

Samples of the mixtures, in some cases after deposition from solution, were melted between microscope slides and cooled slowly. After removal of the upper slide, autoradiographic stripping film was applied to the polymer and allowed to

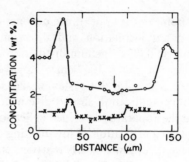

Fig. 4.4 Radial segregation of a nickel-containing additive in a spherulitic crystal of isotactic polypropylene, partially crystallized at 403 K and quenched. ooo $C_0 = 4$ wt%, xxx $C_0 = 1$ wt%. Arrows indicate centers of spherulites.

stand until suitably exposed ($3-20 \times 10^5$ s). The developed emulsions were dried and photographed. Autoradiograms of the polypropylene spherulites showed that the tagged dilauryl thiodipropionate was concentrated in relatively large deposits, leaving much of the spherulite essentially free of additive. Similarly autoradiograms of spherulites grown from polyethylene containing 1% tritiated low-molecular-weight polymer showed exclusion of the smaller molecules from the central areas. The atactic polystyrene was not only rejected from the central area, but primary crystallization was found to proceed in six geometrically uniform directions (39).

Calvert and Ryan (40) studied impurity distribution in isotactic polypropylene. To facilitate measurement of the impurity profiles, impurities were selected which were UV absorbing, fluorescent or metal containing; they were incorporated in the polymer by joint dissolution in dichloromethane. Samples were molded at 493 K, then melted at 513 K and crystallized isothermally. After partial spherulitic crystallization at temperatures around 400 K, the samples were quenched and analyzed. Figure 4.4 shows the profiles of a nickel-containing additive, measured by energy-dispersive X-ray spectrometry across a spherulite diameter. In each case, the spherulite itself was depleted, with an apparent central "dip" in concentration and the area in contact with the perimeter of the spherulites was strongly enriched in impurity.

The observed distributions were compared with those calculated on the assumption that the impurity's distribution coefficient is given by the fractional content of amorphous material at the periphery of the spherulite. Under this assumption, with growth rate and spherulite radius known, it was possible to use the diffusion coefficient as a fitting parameter in mating calculated and observed profiles. The computed profiles matched the experimental ones very well, except that the former showed no central dip. The disparity was attributed to variation of crystallinity within the spherulites, not considered in the computed model. The diffusion coefficients that gave good fits ranged from 10^{-8} to 10^{-7} cm^2 s^{-1} and decreased with increasing molecular weight of the additive.

2 IMPURITY ENRICHMENT; ENHANCEMENT OF ANALYTICAL SENSITIVITY

As more effective purification techniques come into use, the difficulty of measuring the concentration of residual impurity becomes greater. The residual impurities in highly purified materials are generally "homologues" of the major component; their determination is made difficult by the similarity of their chemical properties to those of the main component. Directional crystallization offers a method of concentrating the microcomponents for analytical purposes (41); it can be carried out with little effort and is noncontaminating because no reagent need be added. It also has the advantage of economy and makes little or no contribution to background impurity level.

Preconcentration of impurities is especially important in the analysis of complex mixtures. Once an impurity is identified, if its distribution coefficient in the host is known, its concentration can be calculated at levels below analytical detectability. Moreover, initial preconcentration can aid in the qualitative analysis of complex mixtures.

Blank (41) has considered the conditions under which it may be assumed that all of the impurity is concentrated into a small fraction of the original sample. Starting with the directional-crystallization equation (Eq. 3.2, Chapter 3), for nonequilibrium conditions (i.e., $k_{eff} = \text{const} \neq k_0$),

$$C_s = k_{eff} C_0 (1 - g)^{k_{eff} - 1} = k_{eff} C_l \qquad (4.1)$$

it is possible to define an extraction coefficient, K_{ext} as follows:

$$K_{ext} = \left(\frac{C_l}{C_0}\right) \cdot (1 - g) = (1 - g)^{k_{eff}} = \left(\frac{m}{M}\right) K_{eff} \qquad (4.2)$$

where

m = the weight of the concentrate
M = the total weight of the sample
C_l = the impurity concentration in the impure end
C_0 = the original impurity concentration

M/m is defined as K_{enr}, the enrichment coefficient. For all of the impurity to be in the concentrate (i.e., $K_{ext} \geqslant 0.9$) after a tenfold concentration ($M/m = 10$), the distribution coefficient k_{eff} must be $\leqslant 0.04$, and be even lower for higher K_{enr}.

Kirgintsev (42) has tabulated values of m/M required to assure essentially complete concentration of impurity ($K_{ext} > 0.95$) for various values of k_{eff}. Figure 4.5 shows plots of similar data. For example, in order to have 95% of the impurity present in the concentrate from a system whose k_{eff} is 0.02, it is necessary to take nearly $\frac{1}{10}$ of the original charge as concentrate. It is clear that

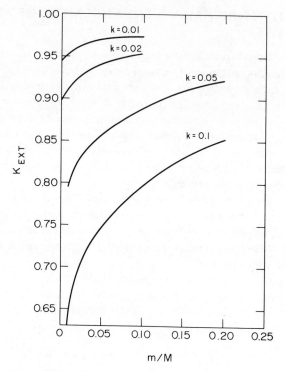

Fig. 4.5 Extraction coefficient, K_{ext}, as a function of enrichment with k_{eff} as parameter.

systems with higher k_{eff} are not amenable to this treatment. Hence it is desirable to have a more general relationship between the analyte concentration measured in the enriched concentrate and that in the original sample.

If k_{eff} is constant, the impurity concentration in the impure end of a directionally crystallized ingot is given by

$$C_I = C_0\left(\frac{M}{m}\right)^{1-k_{eff}}$$

(4.3)

Hence

$$C_0 = \frac{C_I}{K_{ext}K_{enr}}$$

(4.4)

K_{ext} can be obtained from measurements on synthetic mixtures or it can be calculated from a measured value of k_{eff} using Eq. 4.2. Thus Eq. 4.3 makes it possible to calculate C_0 from C_I after producing a concentrate containing any arbitrary, known fraction of the original impurity. Figure 4.6 shows plots of C_I/C_0, the concentration enhancement for various values of m/M, with k_{eff} as the parameter.

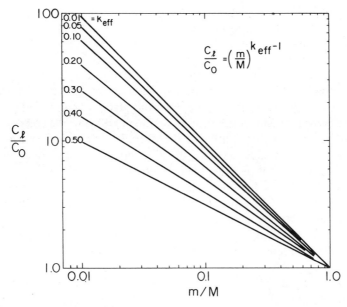

Fig. 4.6 Concentration enhancement, C_l/C_0, as a function of enrichment, with k_{eff} as parameter.

As a practical matter, it should be pointed out that the radial distribution of impurity must be considered in evaluating concentration data from directional crystallization. In a simple-eutectic system, a substantial fraction of the impurity may appear on the surface of the entire ingot, rather than in the terminal eutectic concentrate (43).

Blank (44) has calculated theoretically the increase in sensitivity which can be obtained by preconcentration. Under optimum conditions, a twentyfold reduction in the limit of detection requires an enrichment coefficient $K_{enr} = M/m \sim 100$. Thus it is necessary to achieve high and accurate values of K_{enr}. This can be done by using a crystallization tube in which the concentrate is deposited in a section of reduced diameter. Calculations for some alkali halides showed that the optimum diameter of the narrow tube is about 0.25 ± 0.05 cm.; A 2.5-cm length of this tubing contains about 0.5 g. In order to achieve the desired hundredfold enrichment, the main ingot must weigh about 50 g, and is conveniently grown in a tube of 2.5-cm diameter.

2.1 Inorganic Salts

In practice, charges of sodium iodide (30–50 g) were gradually heated under vacuum in tared ampoules until the salt was molten; an inert gas was admitted, and crystallization was carried out by raising the ampoule from an oven at 4.4×10^{-4} cm s^{-1}. When the crystallization front reached the narrow tip of the ampoule, it was cooled and weighed to give M, the original weight of the sample. The tip was broken off and its contents were dissolved in water; the weight m of the concentrate was found by difference. Flame photometry was used to determine Li, K, Rb, and Cs, and the original concentrations were calculated from Eq. 4.1. For determination of about $10^{-4}\%$ of these impurities, after an enrichment of 100, the maximum relative error was about 15% at 0.95 confidence level (44).

The increase in sensitivity attainable from a preconcentration method depends on the ratio of the errors of the direct and preconcentration methods, respectively. A consideration of the various factors contributing to the cumulative error of the preconcentration method (44) led to the conclusion that preconcentration contributes little to the overall error and the ratio is near unity. Hence it was possible to reduce the determinable minima of impurities in sodium iodide by a factor of 20 for Li and K (to $1.0 \times 10^{-5}\%$), and by 50 for Rb (to $2.0 \times 10^{-6}\%$) and by 100 for Cs (to $2.0 \times 10^{-6}\%$). Chloride impurity in NaI has also been segregated by directional crystallization, and k_{eff} was found to be 0.43 ± 0.02 for starting concentrations between 10^{-4} and 2×10^{-2} wt%, with growth rates of 8.3×10^{-5} cm s^{-1} to 1.1×10^{-3} cm s^{-1} (0.3 to 4 cm h^{-1}). Sensitivity of turbidimetric determination of Cl$^-$ was increased to 2×10^{-5} wt% (44).

This preconcentration method has been improved by the use of centrifugal directional crystallization (see Section 2.5.1, Chapter 3), which reduces the time required for crystallization from 1.2×10^4 s (at 4.4×10^{-4} cm s^{-1}) to about 3.6×10^3 s (at 1.3×10^5 cm s^{-1}) (45). The improved method was applied to spectrographic determination of divalent impurities in sodium nitrate. It was possible to quantify Mg, Cu, Pb, Mn, Co, and Ni down to $2-8 \times 10^{-6}\%$. The relative error from three parallel measurements was less than 20–25% after treatment of the results at a confidence level of 0.95. The analysis of $NaNO_3$ for traces of $Sr(NO_3)_2$ was facilitated by upward directional crystallization at 5.56×10^{-4} cm s^{-1} in a rotating container with periodic reversal of rotation direction (46).

With potassium bromide, directional crystallization afforded a decrease in the minimum determinable concentration of lithium (by a factor of 10^2, to $\leqslant 2 \times 10^{-6}$ wt%) and of rubidium (by a factor of 5 to $\leqslant 6 \times 10^{-5}$ wt%). The distribution coefficients for these two solutes were found to be 0.07 ± 0.02 and 0.72 ± 0.01, respectively, in concentrations up to about 10^{-3} wt% (47).

Blank et al. (48) have described the use of directional crystallization to concentrate chloride impurities in ammonium alum. Because experimental problems limit the turbidimetric analysis of chloride to small samples, the lowest measurable concentration is about $10^{-3}\%$. Since the salt hydrate melts at 367 K without decomposition, preconcentration by directional crystallization may be considered. Samples were crystallized in two ways: (1) vertically, from top to bottom, and (2) with centrifugation. The variation in the distribution coefficient with rate of crystallization is shown for the two cases in Fig. 4.7. It is clear that the second mode produced greater depletion in the bulk of the ingot; at $V = 1.4 \times 10^{-4}$ cm s^{-1}, "extraction" could be virtually complete, although the operation would require about 3.6×10^4 s. Accordingly, a somewhat faster rate

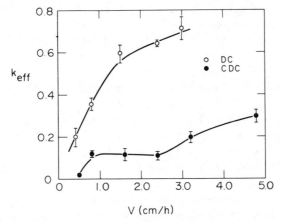

Fig. 4.7 Distribution coefficient as a function of crystallization rate. DC is directional crystallization and CDC is centrifugal directional crystallization.

was used, and it was still possible to lower the limit of measurement by a factor of about 60, to $2 \times 10^{-5}\%$.

The determination was carried out on 20 g of molten salt in an ampoule with a conical tip. A small seed crystal affixed to a platinum-wire holder was immersed in the top of the melt. After centrifugal directional crystallization, the ampoule was broken and the concentrate (0.2 g) was separated, then dissolved in dilute HNO_3 and diethylene glycol. Silver nitrate solution was added and the absorbance was measured in a filter photometer at 360 nm in a 5-cm cell. The chloride content was read from a calibration curve. Statistical analysis of the results showed negligible systematic analytical error and random error less than 30 relative %.

Traces of Co, Cr, Cu, Ga, Mn, Mg, Mo, Ni, Pb, Ti, and V were determined in ammonium alum after preconcentration of a 15-g sample to 0.3–0.5 g. The detection limits were about 10^{-6}–$10^{-8}\%$ (49).

A similar method was used in crystallizing the hydrate of cesium iodide in order to concentrate lithium and rubidium (50). Of course, the hydrate (freezing point-269 K) can be crystallized at a much lower temperature than the pure salt, and this factor is an advantageous convenience. Beyond this, it should be noted that the segregation of rubidium from molten cesium iodide is not very efficient since $k \simeq 0.7$. Crystallization of the hydrate, on the other hand, results in strong rejection of both lithium and rubidium, with k_{eff} varying from 0.06 to 0.14, depending on the temperature of the crystallization bath. It was found possible to lower the determinable limit for Li and Rb by a factor of 20–25, to about 10^{-5} wt%. The crystallization experiments were simple. A 6.9-g portion of the analyte was dissolved in 18.1 mL of water and the salt hydrate was crystallized at 2×10^{-4} cm s^{-1} by immersion in a bath at 260 K. When about $\frac{1}{20}$ of the charge remained liquid, it was decanted, diluted with a like volume of hot water, and then diluted to give a 2% (wt/vol) solution which was analyzed by flame photometry. Chlorides were concentrated in like manner by directional crystallization of the eutectic salt solution of CsI, and the detection limit was found to be $3 \times 10^{-6}\%$. Carbonates in CsI were determined with a detection limit of $3 \times 10^{-5}\%$, after preconcentration (51).

The hydrate of cadmium nitrate was directionally crystallized to concentrate metallic impurities (52). The detection limits of the following elements were reduced twenty- to fiftyfold: Bi, Cr, Co, Ga, In, Mg, Mn, Mo, Ni, Pb, Ti, V, and Zn. Ni for example, was detected at $2 \times 10^{-5}\%$ without concentration and at $3 \times 10^{-7}\%$ with concentration. The quantitatively determinable amounts were also reduced: Ni could be determined at $6 \times 10^{-7}\%$ with a relative standard deviation of 0.15–0.20.

Related work on the use of directional crystallization to increase analytical sensitivity has been described by Kirgintsev et al. (53). Their attention was focused on the way in which crystallization rate changes as a function of the fraction of the ingot that has solidified. In order to keep the experimental growth rate consistent with the equilibrium crystallization rate, a circuit was designed to program the oven temperature on the basis of a signal from an optical sensor. The horizontal optical path traversed the oven through periodic viewing ports; it traversed the nearly horizontal sample ampoule above the axis of the latter and so was able to detect the interface between the growing solid and the remaining melt, through the inert atmosphere above the melt. (The solid fills the cross section of the tube and rotates with it while the liquid fills only part of the cross section.) To aid in avoiding supercooling, the starting end of the rotating ampoule was fitted with a propeller. Charges of 250 g of cadmium were solidified until about 2 g of melt remained. The melt was decanted and heated in the rotating tube at 723 K for 1.1×10^4 s, to distill the more volatile components, including Cd. In this way, the sensitivity of the spectrographic determination of Bi, Cu, Ir, Ni, Pb, and Li was increased from $(0.05–1.0) \times 10^{-4}\%$ to $(1.5–3) \times 10^{-8}\%$.

2.2 Aqueous Systems

Water offers particularly favorable possibilities for enrichment of impurities by directional crystallization. Liquid water, as is known, is a very potent solvent for a wide variety of substances. Solid water, with its close-packed, hydrogen-bonded lattice, excludes nearly all solutes. These properties have been used to good effect.

Schildknecht and Mannl found that solidification of an aqueous solution of the enzyme catalase gave essentially pure ice and a concentrated solution (54). Schildknecht et al. (55) studied directional crystallization as a means of concentrating various solutes in water. In particular, they studied the effects of several operating variables, such as starting concentration, stirrer speed, and rate of crystallization (lowering rate).

Solutions of potassium permanganate in 1.2-cm tubes were frozen at 1.67×10^{-3} cm s^{-1} with stirring at 18.3 rev s^{-1}. The results were said to show increasingly effective enrichment with decreasing concentration. However, the data show clearly that if the same volume fraction of concentrate had been taken in each case (i.e., equal K_{enr}, see above), then the enrichment factors would have been independent of concentration. Crystallization of potassium permanganate solutions at various speeds (55) was said to show enhanced enrichment at higher stirrer speeds. Again, if the same volume fraction of concentrate had been taken, then the enrichment factors would have been independent of stirrer speed in the range $23.3-50$ rev s^{-1}. In acetic acid solutions of 9.1, 4.5, and 0.7%, there was in fact a significant inverse dependence of enrichment factor upon starting concentration. Solidification of 0.7% acetic acid at 5.6×10^{-4} cm s^{-1} gave no enrichment without stirring; when the solution was stirred at 36.7 rev s^{-1}, a marked enrichment was observed (see Figure 4.8).

Shapiro has carried out directional crystallization of aqueous solutions in volumes from 0.1 to 300 L (56). Small volumes were solidified in flasks attached to a wrist-action shaker mounted in such a way that several samples were simultaneously solidified in a freezer at 243 K. After twentyfold concentration, more than 99% of the solute was recovered in the remaining liquid.

Larger amounts of water were frozen in a cylindrical metal container whose

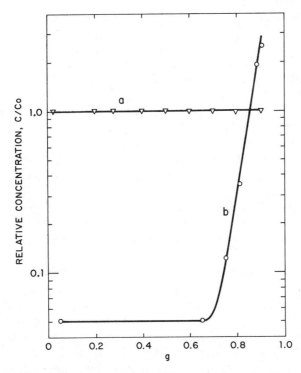

Fig. 4.8 Directional crystallization of dilute acetic acid: (*a*) without stirring; (*b*) with stirring. g is the fraction crystallized.

top was just below the freezing level. The solution was stirred during solidification and the stirrer was raised periodically as solid grew inwardly from the sides and bottom of the container. This procedure gave a shallow conical core in which solutes were concentrated. In this way, a total of 48 L was concentrated to 1 L in 1.8×10^5 s. If contamination from the metal container is troublesome, the cylinder can be lined with a plastic bag.

In a later publication (57), Shapiro described improvements in his apparatus for concentrating impurities from large volumes of water. A brine bath in a 55-gal drum was cooled to about 267 K by a mechanical refrigeration unit. Samples of natural water (up to 200 L) were solidified batchwise in a 5-gal container of stainless steel to isolate humic acids which were initially present at 1–20 mg L^{-1}.

Volatile solutes can also be recovered by freezing of aqueous solutions: iodine was recovered in 94% yield from water containing 12.7 ppm, after about fourfold concentration. Likewise, five-fold concentration of 15 L of water containing acetone, isopropanol, and sodium chloride (50, 100, and 10 ppm, respectively) gave recoveries of 102, 89.7, and 89.4% (58).

Kobayashi and Lee (59) used the earlier Shapiro procedure to concentrate ionic materials from water. Rhodamine B (about 100 μg) was dissolved in 5 L of water; concentration to 0.5–1 L gave recoveries of 82–94% in the remaining liquid. With sodium chloride as the solute, amounts of 2.6 to 11.0 mg were recovered quantitatively from 2.2 L of solution by freeze-concentration to 0.19 to 0.35 L.

Freezing of 5-L samples of distilled water until 0.05 L of liquid remained, followed by analysis of the concentrate, showed that the chloride content of the original sample was 3.6 ppb. In like manner it was established that redistilled water contained 1.6 ppb chloride (60).

Wilson et al. carried out freeze concentration of an aqueous glucose/folic acid solution (61). The concentrations were 10 μg mL^{-1} and 5×10^{-3} μg mL^{-1}, respectively. Twelve-liter batches were concentrated 20-fold and 26.7-fold, respectively. The distribution of the recovered solutes is shown in Table 4.2. The ice residue contained only 2.91% of the folic acid activity; some activity might have been lost through oxidative inactivation. Following this successful concentration, Wilson et al. applied the same method to concentration of a yeastlike fungus, *Candida albicans*. The reduction in volume was limited to fivefold because other components of the solution caused frothing and entrapment in the ice. Nevertheless a 60% recovery was attained.

Baker (62) has applied directional crystallization to the preconcentration of contaminants in water before analysis by gas chromatography. He used a particularly simple, yet effective apparatus. The dilute mixture in a round-bottom flask was attached to a rotary evaporator. The rotating flask was partially evacuated, then immersed in a low-temperature bath. Very low temperatures (e.g., 195 K) were unsuitable because they resulted in too-rapid freezing. In fact, an ice-salt bath at 267 K was found most suitable. Synthetic mixtures were prepared containing phenol, *m*-cresol, and 2,4-dichlorophenol.

Table 4.2. Recovery of Glucose and Folic Acid from Water by Directional Crystallization (61)

Sample–Volume Reduction	Volume, mL	Glucose Recovered			Folic Acid Recovered		
		Concentration, $\mu g\,mL^{-1}$	Total amount, mg	%	Concentration, $\mu g\,mL^{-1}$	Total amount, μg	%
Initial	12,000	10.0	120	100	5×10^{-3}	60	100
20-fold	600	200.1	120.06	100.05	0.0925	55.5	92.5
26.7-fold	450	260	117	97.5	0.1215	54.0	90.1

Portions of solution (0.2 L) were frozen in a 1-L round-bottom flask until 3.5 mL of liquid remained. Chromatographic analysis showed that the recoveries of the three organics were 63, 72, and 40%, respectively.

Janson, Ersson, and Porath (63) have described an apparatus providing for the directional crystallization of about 22 L of solution in 12 vessels (8-cm diameter × 40 cm long). The ensemble of vessels was lowered into a large cooling bath containing 130 L of ethanol at controlled temperatures down to 198 K. Each of the cylindrical vessels was equipped with a stainless-steel stirrer operating according to the centrifugal-pump principle. The stirrers were mounted at fixed elevation, while the cylindrical crystallization vessels were lowered. The distance between the stirrer and the solid/liquid interface was varied between 0.5 and 2.5 cm without substantial effect on the effectiveness of separation. Since water was the solvent, solidification produced an increase in volume of about 8%, and the excess liquid was allowed to overflow from the crystallization vessel. In addition to stainless-steel crystallization vessels, others made of transparent acrylic plastic were used. These were provided with removable stainless-steel bottoms in order to reduce the risk of supercooling and to promote prompt crystallization at the bottom of the vessel.

After completion of crystallization, unfrozen liquid was siphoned off and the outside of the vessel was then warmed with running water so that the solid ingot could be slid out of the vessel. It was then possible to slice the cylindrical ingot into portions with a band saw and to transfer the slices to numbered cups for later analysis. It was found that the overall effectiveness of the concentration could be increased if continuous dialysis was performed on the unfrozen liquid fraction. The unfrozen liquid was removed from the crystallizing vessel with a peristaltic pump, passed through the dialysis vessel, and then through an ice bath to precool it before it was readmitted to the crystallization vessel.

A number of experimental variables were studied with regard to their effect on overall concentration. These included rate of stirring, rate of lowering, crystallization-bath temperature, ambient temperature, and sample composition. It was found that no useful concentration was obtained without stirring. The optimum stirring rate depended, in turn, on other variables such as solute concentration, but the generalization was made that concentration efficiency increases with stirring rate up to about 33 rev s^{-1}. At lowering rates greater than 1.11×10^{-3} cm s^{-1}, supercooling and deviation from planar interface became troublesome. Annular bias heaters were used to avoid formation of a parabolic interface shape. It was also found that programmed lowering rates gave better results than constant lowering rate. It was asserted that a run starting at 1.67×10^{-3} cm s^{-1} and gradually diminishing should give a much better final result than a constant lowering rate of 0.83×10^{-3} cm s^{-1} (64). Variation of the bath temperature between 258 and 243 K did not much affect the concentration. Lower temperatures were found to increase the risk of supercooling, although lower temperatures did provide a steeper gradient which in turn should give better separation. It was observed that salts and other low-molecular-weight solutes, which are able to hydrogen-bond water molecules, lead to entrapment of the solute in the ice lattice. The entrapment, in turn, decreases the perfection of the solidifying cylinder and leads to a rough surface which diminishes the effectiveness of stirring. Experimentally, it was found that the upper limit for starting salt concentration is about 0.1% (0.02 M) at a lowering rate of 5.6×10^{-4} cm s^{-1} and a stirring rate of 25 rev s^{-1}. If continuous dialysis is added to the crystallization process, it is possible to increase the above-mentioned limit for salt concentration by a factor of 20.

Studies were carried out on the concentration of four enzymes: trypsin, chymotrypsin, dextranase, and cellulase. Table 4.3 summarizes the concentrations of protein and enzyme activity in the liquid that overflowed during solidification and in the liquid remaining unfrozen at the conclusion of solidification, for several of these systems. In general it can be said that about 90% of the protein and of the enzyme activity are recovered in about 15% of the original volume.

Isotope enrichment in water was examined by Posey and Smith (65). Spherical/radial crystallization was executed with shaking, until half the charge was frozen. The ice contained 18.33 mole% deuterium, while the liquid contained 17.89 mole%. The calculated distribution coefficient was 1.021. Natural directional crystallization, as manifested in the composition of glacial ice, showed much stronger segregation of D_2O. This effect could arise from one or more of the following factors:

• Different k for different concentrations
• Slower crystallization in the glacier
• Multiple effect through repeated freezing and thawing in the glacier

Mahler and Bechtold (66) have described a unique directional crystallization of an aqueous solution of polysilicic acid. As the water crystallizes, the polysilicic acid emerges from solution retaining the geometry conferred by the surrounding

Table 4.3. Recovery of Enzymes by Directional Crystallization of Aqueous Solutions (63)

Sample	Unfrozen Liquid			Overflow Liquid		
	Volume, %	Protein Recovery, %	Enzyme Activity, %	Volume, %	Protein Recovery, %	Enzyme Activity, %
Trypsin[a]	6.3	69.2	71.1	7.1	22.2	23.2
α-Chymotrypsin[b]	7.2	67.0	66.0	7.5	27.0	26.0
Dextranase	8.3	77.0	82.5	5.2	14.7	16.7

[a]50 μg mL^{-1} in 0.01 M potassium phosphate buffer, pH 7.
[b]10 μg mL^{-1} in 0.01 M potassium phosphate buffer, pH 7.2.

ice. Aqueous silicic acid at pH 5 was gelled in a plastic cylinder (2.7 cm $\times$ 24 cm), then was directionally crystallized at about 1.1×10^{-3} cm s^{-1} by lowering into a cold bath at about 200 K. The resulting ingot was thawed and all of the silica was found to be present as parallel fibers of polygonal cross section, about 50 μm in diameter and 15 cm long. Similar fibers have been obtained at freezing rates from 10^{-4} to 4.2×10^{-2} cm s^{-1}, at temperatures from 77 to 263 K. The resulting fibers were amorphous and porous. High-temperature firing (1300 K) led to incipient crystallization, but fiber geometry was unchanged.

2.3 Organic Systems

Matthews and Coggeshall used a two-stage preconcentration to quantify impurities in benzene (67). After a twentyfold enrichment in impurity concentration, it was possible to measure the concentrations of C_7 paraffins and C_8 and C_9 aromatics that were not detectable in the original material. The calculated concentrations were 0.06 wt% C_7H_{16}, 5×10^{-4} wt% C_8 aromatics, and 10^{-3} wt% C_9 aromatics. In addition, impurities that were detectable in the starting material were concentrated by about twenty-fivefold; these included toluene, C_6 and C_7 cyclics, and/or mono-olefins.

Sloan (68) described a device for stirring the remaining melt during directional crystallization, and applied it to impurity enrichment in benzene and dimethyl sulfoxide. In the former case, the number of impurities detectable by gas chromatography increased from 4 to 12; the total impurity content rose from 0.18 area% in the starting material to 1.59 area % in the concentrate.

3 DETERMINATION OF PHASE DIAGRAMS

The crystal-growth procedures based on directional crystallization are of great importance in establishing the phase relationships in binary and higher-order systems. Some of these applications are discussed in Section 2.4, Chapter 2 and others in Chapter 10.

4 GROWTH OF SINGLE CRYSTALS

The motivations for growing single crystals are as varied as science itself. Single crystals may be sought for use in structure determination by X-ray crystallographic means, for spectroscopic studies, and for the study of mechanical, optical, and electrical properties. Further, the availability of single crystals makes it possible to study the anisotropy of various physical properties. Of course, with the growth of single crystals, there is often the possibility of obtaining high purity, and this has already been discussed. Finally, it should be mentioned that single crystals are grown in large numbers for device applications utilizing the optical, electrical, and magnetic properties of many materials. These applications include solar cells, laser materials, and semiconducting devices.

Modern methods for preparing single crystals by directional crystallization spring from the work of Andrade (69) and of Tammann (70). Andrade grew mercury crystals by slow cooling of the liquid metal in capillaries. Tammann contributed the idea of drawing out the bottom of the growth vessel to a capillary extension. The melt was cooled in such a way that crystallization began in the capillary. Bridgman used vessels of the same type, and lowered them through a temperature gradient generated in a single furnace (71).

The techniques for growing single crystals by directional crystallization have evolved into two families. The first of these includes the Bridgman, Stockbarger, and Stöber methods, while the second includes Nacken, Czochralski, Kyropoulos, and Verneuil methods. All these methods will be discussed; they have been reviewed in detail elsewhere, with varying emphasis (72, 73, 74, 75, 76). An extensive discussion is available, relating thermodynamic and other properties to the selection of crystal-growth methods (77).

4.1 Bridgman–Stockbarger and Stöber Methods

Stockbarger's modification of the Bridgman technique involved use of a second furnace, to provide better control of the temperature gradient (see Section 2.1.1, Chapter 3). It is common practice to refer to the Bridgman and Stockbarger techniques jointly, in reference to a two-furnace assembly.

The Stöber technique, less widely used now, consists in placing a stationary vessel in a temperature gradient, which is then caused to move through the charge. A hemispherical vessel containing the melt is heated from above and cooled from below, while embedded in a material whose thermal conductivity is similar to that of the melt. At the outset, the upper heater is hot enough to melt the entire charge; if the charge has low thermal conductivity, care must be taken to avoid overheating its upper portion. The gradient is displaced vertically by gradually reducing the temperature of the upper hotplate. Stöber suggested that:

- Crystallization should proceed from a point
- The temperature gradient should be large
- The crystallization front should be planar

The Stöber apparatus is shown schematically in Figure 4.9 (78).

Some furnace assemblies suitable for Bridgman–Stockbarger growth have been described previously (Section 2.1.1, Chapter 3). Additional devices appropriate to various categories of materials are described below.

4.1.1 Containers

In growing organic and inorganic crystals from the melt, borosilicate glass is by far the most widely used container material. It offers easy fabrication, charging, evacuation, and sealing. Unfortunately, many organics adhere strongly to glass; beyond posing problems in the removal of the grown crystal, adhesion can cause plastic deformation, especially of soft organic solids. Coating

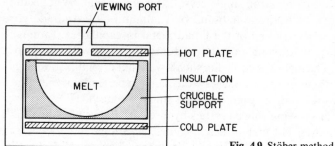

VIEWING PORT

HOT PLATE

INSULATION

MELT

CRUCIBLE
SUPPORT

COLD PLATE

Fig. 4.9 Stöber method of melt crystallization.

the inner walls of growth tubes with silicone fluids (Dri-Film SC-87, General Electric Co, Silicone Products Department) (79) can solve the problem without introducing any detectable impurity.

Metallic and inorganic crystals have likewise been grown in glass and silica vessels, with the same advantages and difficulties. Bridgman used an oil film to minimize adhesion. Graphite crucibles have been used for high-temperature growth, but it should be noted that they may require outgassing. Split crucibles of graphite have been used to facilitate removal of the grown crystal (see Figure 4.10) (80).

Interaction between strontium chloride and silica containers was reduced by using an internal coating of pyrolytic carbon. In the absence of such a coating, $SrCl_2$ crystals stuck to the ampoule with sufficient tenacity to break it (81).

The differences between directional crystallization for the purpose of

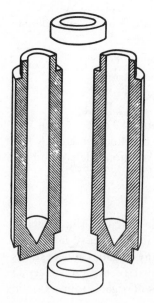

Fig. 4.10 Split crucible of graphite with locking end rings; from Wernick and Davis (80).

purification and for the purpose of growing single crystals reside primarily in the shape of the sample container and in the growth gradient. Early workers used vessels with conical bottoms or capillary tips for the initiation of crystal growth, with the intent of constraining growth to a single orientation and eliminating growth in other orientations. The seed whose maximum growth rate is parallel to the capillary axis will crowd out seeds growing in other directions. Sherwood and Thomson (82) have discussed the effectiveness of various tube designs in promoting growth of single crystals of anthracene from the melt. Figure 4.11a shows a conventional tube with a capillary nucleating tip, which was said to produce only polycrystals. The expanded capillary shown in Figure 4.11b produced somewhat better results, as did the bent capillary of Figure 4.11c. Figures 4.11d, 4.11e, and 4.11f gave successively better results in the growth of anthracene. Since the growth oven used in this study did not embody optimum temperature control, it may be that the primary function of the annular "jacket" in tube f lies in its role as an insulator in buffering temperature oscillations.

An earlier study of capillary design led to the use of an initiating sphere (83). Upon cooling the melt in the sphere (Figure 4.12), crystals grow radially toward its center and only those growing parallel to the capillary can enter and propagate. By angling the capillary with respect to the tube axis, other growth directions may be selected, such as angle α in Figure 4.12.

Datt and Verma (84) tried various growth-tube designs and said that their best results were obtained with helical nucleating sections in which the conical transition between the helix and the growth tube had a vertical angle (see Figure 4.11d) of 45° to 60° and in which the tip of the helix was not "visible" from any point in the body of the tube.

Growth of crystals for study of infrared absorption poses two major problems: the crystals must be very thin, and crystallographic orientation must be carefully specified. An elaborate apparatus and procedure have been described for growing crystals 6–25 μm thick, between salt plates, with an

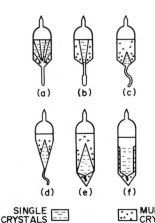

SINGLE CRYSTALS ▭   MULTI-CRYSTALS ▦

Fig. 4.11 Tube designs for single-crystal growth by the Bridgman–Stockbarger method; after Sherwood and Thomson (82).

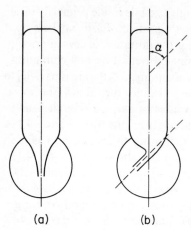

(a)   (b)

Fig. 4.12 Nucleation sphere for Bridgman–Stockbarger growth, with preferred orientation.

observable area about 0.7 cm × 1.5 cm. The device allows continuous observation of the growing crystal; observed flaws can be melted and regrown. A seed crystal is contained in a capillary whose orientation with respect to the cell may be varied, so that three crystals with mutually perpendicular orientation can be grown from a common seed crystal (85, 86).

It is often desirable to grow a number of crystals from a single portion of starting material of uniform high purity without exposure to air or other contaminants. This can be done in a vessel which allows a terminal purification by zone refining, followed by transfer of a portion of the ingot into crystal-growth tubes. Figure 4.13 shows such a "combination tube" for zoning and crystal growth. Note that the individual crystal growth tubes A have been provided with ports D through which measured portions of a dopant can be added. In some cases, it may be desired to carry out the zoning in an inert atmosphere (to suppress sublimation, for example) and to grow the crystals in vacuum. For this, the vessel should be provided with a breakseal through which the inert gas can be removed.

In practice prepurified material is loaded into the large upper tube B of the vessel, which is evacuated to remove air and other volatiles. The vessel is filled with an inert gas; the charge is melted and allowed to flow through the constriction C into the lower, central tube. After solidification, the vessel is evacuated and solid adhering to the constriction is sublimed away by careful local heating. The vessel may be sealed in vacuum or after readmission of an inert gas. Zone refining may be carried out by reciprocal motion of the tube or a heater battery. If the presence of an impurity with distribution coefficient greater than unity is suspected or known, then the top of the ingot may be sublimed or melted into a sidearm or bulb. The zone-melted ingot may be used in two ways:

• A portion of the ingot may be melted and homogenized in the main tube; liquid of constant composition may then be decanted into the several crystal tubes.

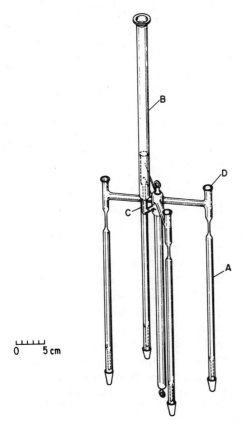

0   5 cm

Fig. 4.13 Combination tube for zone melting and crystal growth.

· The ingot may be melted in small fractions which are decanted serially into the crystal tubes. This procedure makes it possible to study the effects of solute gradient in the zoned ingot, using single-crystal methods. This procedure is especially useful if the solute has a distribution coefficient greater than unity.

Solid dopants of low vapor pressure can be weighed directly into the crystal tubes through ports D. If very small amounts (less than 5 μg, for example) are required, they may be added as weighed segments of a solid solution in the same host, or as aliquots of a solution in a volatile solvent.

Liquid dopants can be injected by syringe, down to quantities of tenths of a microliter. Another technique uses a calibrated capillary containing a measured column of liquid. After insertion into a suitable sidearm of the vessel, the liquid may be frozen to immobilize it during evacuation. Gaseous dopants can be admitted to a measured pressure in a vessel of known volume, and then transferred by condensation into the crystal tube. Figure 4.14 shows an apparatus in which such a dopant can be stored in sealed sidearm A while the host substance is refined in tube B. After a portion of the zoned material has

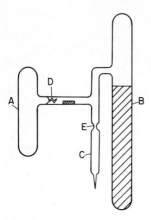

Fig. 4.14 Sealed apparatus for transfer of gaseous dopant to a host crystal after purification by zone refining.

been decanted into crystal-growth tube C, the breakseal D is broken and the dopant condensed into tube C, which is finally sealed off at constriction E.

A simpler method can be used for the Bridgman growth of one crystal from zone-refined solid. The starting material is charged into a straight tube with a nucleating capillary at its lower end (see Figure 5.36). After seal-off at point 1, the charge is transferred to the zoning section 2, and zoning is carried out toward tip 1. After completion of the zoning, a portion from the top of the ingot is transferred to crystal-growth section 3; the central constriction 4 is heated gently to sublime away adhering solid. Then section 3 is removed by flame fusion at 4. The vessel may be sealed under an inert atmosphere. If argon is used as the protective gas, it may be condensed in liquid nitrogen before sealing at 4, to give very low gas pressure during crystal growth. This may be desirable at least for some systems in view of the reported detrimental effects of residual gases on crystal quality (87).

4.1.2 Heating and Cooling

Many of the devices and techniques described in Chapter 3 are applicable to crystal growth by the Bridgman–Stockbarger method. Other assemblies, appropriate for various temperature regimes, are described here.

Lupien et al. (87) have described a simple device for maintaining a fairly stable interface between a hotter (upper) zone and a cooler (lower) zone, for crystals melting between 320 and 523 K. In their apparatus (Figure 4.15) a lower liquid layer A was contained in a jacketed beaker B, heated by a hotplate and a low-wattage heater controlled by a thermoregulator. The upper bath contained a liquid C in a container D consisting of three concentric glass tubes with immersed heater-coil E. Silicone oil was used in both baths and it was found desirable to seal the upper container to prevent access of air. Temperature was monitored by thermocouples or thermometers. Immersion of the outermost member of D in the lower oil bath minimized air turbulence at the crystal's solid/liquid interface. Anthracene crystals were grown with the upper bath at about 500 K and the lower at 480 K.

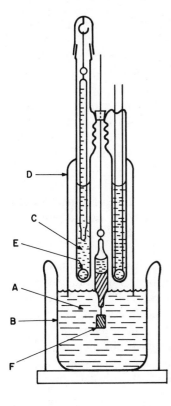

Fig. 4.15 Combined air bath/liquid bath used for Bridgman growth of low-melting materials; after Lupien et al. (87).

Hood and Sherwood (88) used a similar procedure to purify and crystallize cyclohexane. They used a vertical tube A of 2.6-cm diameter, sealed through the base of 2-L beaker B (Figure 4.16). The beaker was heated by an immersion heater to keep the temperature above the melting point of cyclohexane (279.7 K). The beaker rested on insulating board C which in turn covered beaker D, filled with acetone. Beaker D was thermostatted at low temperature in a 15-L Dewar flask. With bath A at 323 K and the lower bath at 195 K, the gradient at

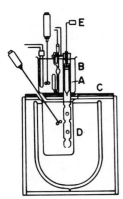

Fig. 4.16 Apparatus for low-temperature Bridgman growth using two separately thermostatted liquid baths; Hood and Sherwood (88).

the interface was $15\,\mathrm{K\,cm^{-1}}$. This apparatus is suitable for crystals melting between 220 K and room temperature.

Dahlgren et al. (89) have built a cryostat for crystal growth at very low temperature, cooled by a pulsed stream of liquid nitrogen. A mixture of liquid and gaseous nitrogen, from a pressurized storage Dewar, enters the annular channel of the stainless-steel cryostat (Figure 4.17) through tangential inlet F, and is then forced through six slits G, symmetrically distributed around the periphery, into the outer chamber. The coolant passes through perforated cylinder B and cools copper jacket C; it exits through two ports H. The coolant delivery is actuated by resistance thermometer A. The cryostat is filled with a liquid D (ethanol or isopentane) into which crystal tube E is lowered. The vessel is insulated with 5 cm of polystyrene foam (not shown). Temperatures between 273 and 123 K were maintained with a constancy of $\pm1\mathrm{K}$ over a period of $8.64 \times 10^4\,\mathrm{s}$. Crystals of n-alkanes, butadiene, and vinyl monomers were grown in this cryostat.

Datt and Verma (84) have also studied the effects of growth gradient and tube design on the growth of crystals of aromatic hydrocarbons. Low gradients (2.5 K $\mathrm{cm^{-1}}$ and $5.0\,\mathrm{K\,cm^{-1}}$) gave polycrystals at growth rates of $2.3 \times 10^{-5}\,\mathrm{cm\,s^{-1}}$; high gradients ($40\,\mathrm{K\,cm^{-1}}$) consistently gave good single crystals.

The accurate measurement of furnace gradients at high temperatures is a difficult task. Conventional methods use one thermocouple, moved stepwise through the central zone, or several stationary thermocouples at appropriate positions. The first method is perturbed by electrical leakage from heater

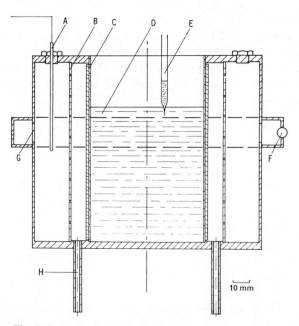

Fig. 4.17 Cryostat for crystal growth at very low temperature.

windings to the thermocouple, causing noise in the measuring circuits. The second method is perturbed by the leakage and by inhomogeneities among the thermocouples, so that an isothermal distribution does not necessarily produce uniform readings.

These difficulties are overcome by an ingenious stratagem, requiring only that each hot junction of the probe thermocouples be provided with an attached wire loop or a chip of a pure metal (90). To measure a gradient, the furnace is equilibrated slightly below the melting point of the metal, and is then heated slowly. When the temperature at each thermocouple reaches the melting point of the metal, there is an arrest in the time-voltage recorder trace, providing a precise internal calibration. At the end of the arrest, the thermocouple output rises rapidly to the furnace temperature at the end of melting. The true temperature differences among the thermocouples can be calculated from the elapsed time between successive melts and the rate of temperature increase as given by any of the thermocouples.

Better heat transfer can be obtained if the crystal tube, including its nucleating capillary, is of uniform outer diameter, so that close contact can be maintained with the furnace. A tube of this kind is shown in Figure 4.18 where A and C are separately controlled metal-block thermostats, B is a transparent insulating spacer, and D is a spindle by means of which the tube is lowered through the gradient with rotation. The use of this device in determining phase diagrams is discussed in Section 2.4.2, Chapter 2. It is noted there that several crystals can be grown simultaneously in such a device, over a wide range of temperature. This stratagem, aside from saving time, assures that each crystal has experienced the same growth conditions.

Ahmad et al. (91) devised an elegant method for growing small crystals in quartz tubes suited for optical measurements. They used a small tube (0.5 cm od × 2 cm long) with a polished quartz window at the bottom. Brass clips were clamped near the top and bottom of the tube and connected through copper

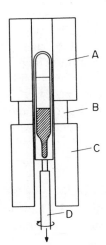

Fig. 4.18 Crystal-growth tube of uniform outer diameter, for improved heat transfer.

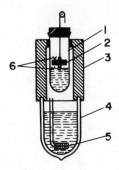

Fig. 4.19 Low-temperature Bridgman apparatus, using liquid nitrogen as coolant.

braids to thermoelectric modules that served as heat sinks. The temperature of the bottom of the cell was controlled by a feedback circuit involving a heater and a GaAs sensor attached to the lower clip. A second heater was attached to the top of the cell to provide a temperature gradient which was monitored by a differential thermocouple. The device was used to grow a crystal of cyclohexane, using a sample that was degassed in the cell before the cell was sealed off. The temperature was lowered slowly, while a temperature difference of about 1 K was maintained until solid appeared on the window. The spontaneously formed seed was annealed for 0.72 to 1.08×10^4 s; then the temperature was lowered at a rate that provided upward growth at about 1.4×10^{-5} cm s^{-1}.

A compact, easily assembled version of the Bridgman–Stockbarger apparatus has been described for growth of alkane crystals doped with aromatic hydrocarbons (92). Aliphatic/aromatic mixed crystals are required for measurement of Shpol'skii quasi line spectra at low temperature (Figure 4.19). The apparatus consists of a large glass tube 1 partially filled with propanol and containing a steel ring 2 with an internal heater, positioned about 0.5 cm above the liquid. Tube 1 is surrounded by foamed-plastic sleeve 3 with an observation window. The assembly rests on Dewar container 4, which contains liquid nitrogen and an immersed heater 5. The temperatures above and below the interface are monitored by thermocouples 6; gradients about 100 K cm^{-1} are attainable between 225 and 575 K with suitable coolants.

The device described above was used to grow crystals 1.0 to 1.5 cm long in quartz tubes of 0.4 cm od. Growth rates of 2.8×10^{-5} to 1.7×10^{-4} cm s^{-1} were tried; best results were obtained at 4.2×10^{-5} cm s^{-1} while at rates higher than 8.4×10^{-5} cm s^{-1}, the resulting solids were polycrystalline. It was possible to prepare clear crystals of n-octane (mp 216.7 K) and n-decane (mp 242.9 K) containing 10^{-5} to 10^{-3} moles L^{-1} of the solutes anthracene, pyrene, and coronene. It was not possible to prepare monocrystals of n-nonane with aromatic hydrocarbon solutes, at any rate of solidification.

4.1.3 Transport

Methods for moving Bridgman–Stockbarger crystal tubes through a temperature gradient are essentially the same as those described in Section 2.3, Chapter 2. In growing single crystals, it is necessary to ensure very uniform growth rate.

4.1.4   Mixing

The conditions of mixing in bulk and interface liquid during Bridgman–Stockbarger growth are critical in assuring successful growth. Various methods of agitation are described in Section 2.5, Chapter 3.

4.2   Nacken, Czochralski, Kyropoulos, and Verneuil Methods

The second family of melt-growth techniques based on directional crystallization is related to the names Nacken, Czochralski, Kyropoulos, and Verneuil.

The Czochralski and Kyropoulos techniques derive from earlier work of Nacken, who grew organic crystals from seeds immersed in melts. In Nacken's apparatus, the melt is contained in two interconnected vessels A and B (see Figure 4.20) within which it may be circulated by raising or lowering the pressure on one side. The seed crystal C is mounted on the hemispherical tip of copper rod D, which is surrounded by copper tube E. The upper portion of D is cooled by an air stream through jacket F and the seed is cooled by conduction so that it remains somewhat cooler than the melt. The entire assembly is immersed in a thermostat. In all of these methods the growing crystal is unconstrained, in that it grows without contacting the walls of the container.

Czochralski's method starts with a crystalline seed or a cold rod immersed in a melt. As the seed, or nucleating rod, is withdrawn from the melt, the crystal grows upwardly with a solid/liquid interface just above the level of the bulk melt, as shown schematically in Figure 4.21a. The diameter of the growing crystal may be increased by reducing the pulling rate (Figure 4.21b) (93). Solute re-

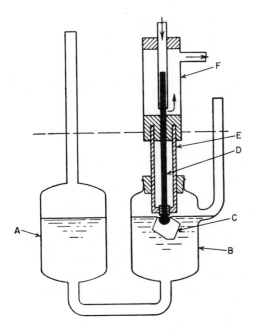

Fig. 4.20 Nacken's method of melt crystallization.

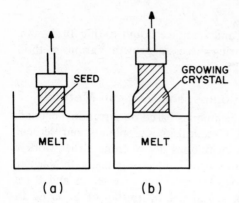

Fig. 4.21 Czochralski method of crystal pulling from the melt.

distribution in the Czochralski configuration is discussed in the classic BPS treatment, which is referred to in Chapter 3 (94). Since then, there have been many theoretical and theoretical/experimental works devoted to understanding the role of geometrical factors such as shape of the liquid surface and interface shape, and thermal factors such as static and dynamic heat flow. Kvapil et al. (95, 96) have provided a detailed theoretical analysis of temperature distribution during Czochralski growth, along with confirming experimental observations relating to growth of alumina and yttrium aluminum garnet (YAG). The surface aspects of Czochralski and certain other growth modes have been reviewed. There exists a three-phase boundary involving crystal, melt, and ambient fluid (gas or liquid). The geometric equilibrium depends on whether or not the crystal is completely wetted by its own melt (97).

The advantages of the Czochralski pulling technique are:

- High crystal perfection can be attained because the crystal is unconstrained during growth and cooling.
- The rotation normally imposed on crystal and/or crucible leads to a homogeneous melt, which in turn affords uniform distribution of dopants.
- Frequently only a small fraction of the melt is crystallized; as a result, the crystal shows only a minor axial concentration gradient.
- It is possible to view the growing crystal, and to program the pulling rate to achieve a desired shape.

A recent development in Czochralski growth allows continuous crystalline ribbons to be pulled by using either a melt-wettable die shaper (the "Edge-Defined Film-Fed Growth" method) (98) or a notwettable die shaper (the Stepanov method) (99). A variety of die shapers have been successfully used to pull continuous ribbons of silicon and sapphire (100) and of semiconductor, dielectric, and metal crystals (101). Applications of the Stepanov method have been reviewed recently (102).

An inverted Czochralski apparatus has been used for the continuous drawing of crystalline filaments from melts (103). The apparatus included a resistively heated platinum crucible with a nozzle in its base. A conical crucible was used to minimize heat transfer to the filament and thus make it possible to establish high-temperature gradient in the filament. A jet of cooling gas just below the nozzle stabilized the drawing of filaments from very fluid melts. In this procedure, the crystal front is stationary at constant distance from the nozzle at a point where the interface temperature is such that growth rate equals drawing rate. This condition is achievable only on the high-temperature side of the growth-rate maximum of Figure 4.22. This controlled fibrous crystallization requires a glass composition and a temperature that together give high rates of growth with little or no tendency to internal nucleation. In geometry, this assembly resembles the one used by Dewey for growth of aluminum ingots by directional crystallization (1). Metals however, solidify immediately below the liquidus temperature while silicate melts increase in viscosity over some temperature range before crystallizing.

To grow crystals of a volatile material, a liquid sealant can be used to inhibit evaporation from the free surface. This liquid-encapsulated Czochralski (LEC) method was introduced for growth of GaAs, with B_2O_3 as the encapsulating sealant (104). Gallium arsenide is of tremendous interest as a solid-state material. It is in use in light-emitting diodes, in solar cells, in optoelectronic devices, and in lasers, and has been grown in many ways. Ingot growth by a horizontal gradient method has been described recently. A melt is prepared in a stationary silica boat in a stationary furnace. Directional crystallization is brought about by gradual reduction of power to the furnace. Ingots of 15-cm length were solidified in 50 h (8.3×10^{-5} cm s^{-1}) (105). Silicon is usually used for preparation of highly doped n-type GaAs for optoelectronic devices, but this engenders a high concentration of crystal defects. Double doping has been used to overcome this problem, and the results of an extensive study have been

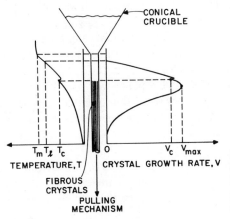

Fig. 4.22 Inverted Czochralski apparatus for continuous drawing of crystalline filaments. Melt emerges from the crucible at T_m, becomes glassy at T_i, and crystallizes at T_c; after Maries and Rogers (205).

reported (106). GaAs has also been grown by the liquid-encapsulation method under B_2O_3 (see Section 4.4). Crystals were pulled at 5.5×10^{-4} cm s^{-1} (2 cm h^{-1}) under 1.5×10^5 Pa of inert gas, with rotation at 0.17 rev s^{-1} (10 rev min^{-1}) (107).

Gallium phosphide crystals up to 4.2-cm diameter were grown by B_2O_3 encapsulation of melts using resistance and induction heating. In all cases, the growth direction was [111] (108).

Indium phosphide has been grown in a quartz crucible under 3.5×10^6 Pa of phosphorus pressure using LEC under B_2O_3 (109).

In recent years a variant of the Czochralski technique has been introduced for use with electrically conductive materials. The novel method, designated Electrochemical Czochralski Technique (ECT) combines electrochemical crystallization from a molten salt solution, with the crystal-pulling technique. Because ECT is a faradaic process, material is deposited at the interface by the passage of current. Consequently, the system is relatively insensitive to thermal fluctuations and only slight attention need be paid to heat flows within the system. The technique was initially used for growth of large crystals (up to 2.5 cm diameter by 11 cm length) of sodium tungsten bronze, Na_xWO_3, at 1023 K (110).

The Czochralski technique has been little used with organic compounds, in large measure because the high vapor pressures of many organics at their melting points make it difficult to retain them in a crucible. With some organics, of course, vapor pressure is not a problem; among these, long-chain aliphatics stand out. In fact, the softness of these compounds makes it hard to remove their crystals from the tubes normally used for Bridgman growth, while Czochralski growth is ideally suited to them.

Cook and Gwan (111) have grown large crystals of 12-tricosanone (laurone) by pulling from a melt contained in a jacketed crucible in an evacuable chamber, Figure 4.23. The pulling rod was insulated to prevent ice formation. The laurone was outgassed above its melting point at 10^{-4} Pa; purified argon was admitted to 50 kPa to suppress evaporation. At this pressure, bubble formation at the interface was not a problem. In the absence of a seed, solidification was begun by immersing the stainless-steel rod shown in Fig. 4.23, and withdrawing it at 10^{-5} cm s^{-1}. The resulting ingot was "necked" three times, by increasing the pulling rate, to promote growth of a single crystal, which grew with its cleavage plane parallel to the surface of the melt. A seed was cleaved from the spontaneously nucleated ingot, and reoriented so that the cleavage plane was parallel to the pulling direction. The diameter of the ingot grown on this oriented seed was made to increase by gradual reduction of the crucible temperature. When the diameter reached 1 cm, axial growth was continued at 3×10^{-6} cm s^{-1} for 8.6×10^5 s.

The Czochralski method has been applied with success to crystallization of benzophenone (112); single crystals up to 2 cm in diameter and several centimeters long were grown. Attempts to apply the method to other compounds, such as coumarin derivatives, were not successful (112). It should be

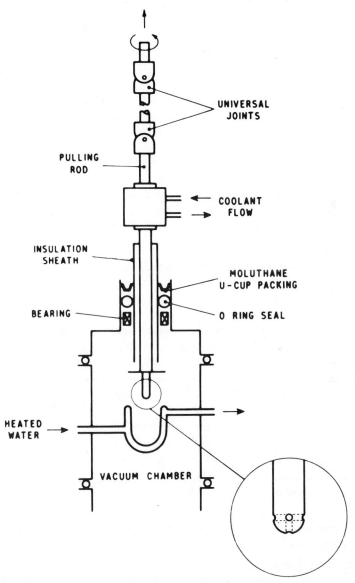

Fig. 4.23 Czochralski apparatus for pulling organic crystals from a melt; after Cook and Gwan (111).

noted that in this work, the method was called "Kyropoulos," but the continuous pulling and rotation of the growing crystal mark this as Czochralski growth.

Neither convection alone nor continuous rotation of the crucible provides adequate mixing during Czochralski growth. Cyclic acceleration and deceleration of crucible and/or crystal can produce enhanced, controlled mixing (113)

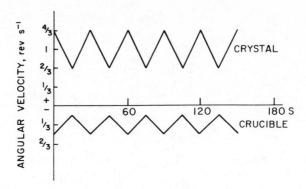

Fig. 4.24 Cyclic acceleration and deceleration of crystal and crucible rotation in Czochralski crystal growth; after Turner et al. (115).

(see Figure 4.24). Constant acceleration of the crucible about its vertical axis generates an outer region of liquid which promptly follows the change in crucible motion. The inertia of the central melt, however, keeps it at rest and a shearing action ensues between inner and outer liquid. In time, the inner liquid will rotate at the same rate as the crucible. Reversal of the sense of rotation provides a wide range of turbulence, on the basis of the frequency and amplitude of the acceleration cycles. Moreover, vertical and radial mixing result from Ekman flow. Flow velocities up to 10^2 cm s^{-1} relative to the crystal are easily attained; these reduce the thickness of the stationary boundary layer to about 10^{-2} to 10^{-3} cm. To grow crystals with only convective mixing requires a low cooling rate, with consequent slow growth (about 5×10^{-7} cm s^{-1}). The accelerated crucible rotation technique (ACRT), with flow at only 1 cm s^{-1} provides for stable growth at 10^{-6} to 5×10^{-6} cm s^{-1}.

Experimental and theoretical aspects of this method have been elaborated by Scheel and his co-workers in a long series of papers (113, 114). Recently an analysis was presented of three important problems in ACRT:

- Change in melt level and meniscus angle during acceleration and deceleration
- Development of surface waves
- Mechanical stability of the seed crystal

Theory predicted and experiment (in a simulated crystallization system) confirmed that increasing the crucible's rotation rate from one value to another causes a change in liquid height that is proportional to the difference between the squares of the rotation rates. The pulling rate must then be adjusted to this change of melt height. Mechanical stability of the crystal/melt meniscus was evaluated by just touching the crystal to the parabolic surface of the rotating liquid pool, then raising the crystal slowly until the meniscus broke. The break-

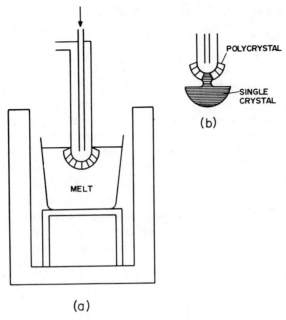

POLYCRYSTAL

SINGLE
CRYSTAL

(b)

MELT

(a)

Fig. 4.25 Kyropoulos method of crystal growth from the melt.

off height was found to depend only slightly on rotation rate of the crucible. It was concluded that meniscus angle will not change drastically and should not influence the diameter of a crystal grown by the ACRT. The viscous friction between the crystal and melt, and the inertial torque, were found to be negligible (114).

A versatile circuit has been described to drive a direct-current motor in ACRT growth (115). It consists of a timer and waveform generator which provide cycle times of 10–260 s and a wide variety of acceleration patterns.

The Kyropoulos method is similar to the Czochralski technique, but involves no axial movement of the seed during growth. A platinum tube is immersed in the melt and cooled by an axial flow of compressed air so that a polycrystalline hemisphere grows outward into the melt, as shown schematically in Figure 4.25. After the hemisphere's radius reaches about 4 × that of the platinum tube, the tube is raised slowly until the bottom of the hemisphere is barely in contact with the melt. The flow of coolant is increased, and a single crystal grows radially from the seed (Figure 4.25b). When the crystal approaches the crucible wall, it is raised and the remaining melt in the crucible is allowed to solidify. The crystal is then separated from its support and allowed to cool slowly in the covered furnace (116).

The Verneuil method was the earliest to avoid the use of a container completely. It is applied primarily to high-melting oxides, and involves gradual application of powdered starting material, through a flame, to a seed crystal which is slowly lowered (117, 118).

4.3 Summary of Crystal Growth Results

4.3.1 Noble Gases

In an experimental *tour de force*, Batchelder et al. have grown crystals of neon, argon, and krypton (119). Their method was essentially the Stöber technique carried out as follows.

Thin-walled Mylar® tubes of 0.3- and 0.6-cm diameter were used. The bottom of the Mylar® tube was closed with a copper plug having a spike 1 cm long; the tube was immersed in a cryostat at a temperature appropriate to the particular gas, which was introduced and condensed in the tube at its triple point. The temperature of the tube was then slowly lowered until a seed appeared at the tip of the copper spike. After an anneal of about 10^5 s at the temperature and pressure at which seeding took place, growth was started by slowly cooling the tube while still maintaining a temperature gradient of 1 K over the tube. The temperature difference was then very slowly decreased, causing the horizontal solid/liquid interface to move slowly up the specimen tube. Crystals were produced in lengths up to 10 cm. Because of the thermal isolation of the sample tube and the melt, the heat of fusion could be carried away only through the poorly conducting crystal; hence slow growth rates were necessary. In general, these were between 10^{-5} and 5×10^{-5} cm s^{-1}. At these growth rates, significant annealing of the crystals took place during growth and at the lower growth rates, crystals of higher quality were produced.

After growth, each crystal was annealed for several hours at the saturation vapor pressure. In order to prevent strain during cooling, the crystal was separated from the walls of the specimen tube by first filling the specimen chamber with exchange gas and then pumping slightly to reduce the vapor pressure above the crystal in its specimen tube. Evaporation was rapid at the Mylar®/crystal interface, and the separation of a few hundred micrometers was readily effected. After separation, the specimen chamber was cooled to liquid helium temperature in 3.5×10^3 to 1.5×10^4 s.

It should be noted that ^{4}He could be grown at rates up to 5×10^{-3} cm s^{-1} (120). Neon, argon, krypton, and helium were grown by this technique.

Because of their relative simplicity, crystals of the noble gases have been subjected to extensive theoretical analyses with respect to crystal-lattice dynamics and crystal-lattice defects. In spite of the extensive theoretical treatment available, these crystals have rarely been produced and manipulated in a form satisfactory for experimental studies. Growth of noble gases is beset with experimental difficulties resulting from low melting points, small latent heats of fusion, high vapor pressures, exceedingly large coefficients of thermal expansion, very small yield strengths at high temperature, and severe brittleness at low temperature.

4.3.2 Aliphatic Hydrocarbons

Methane (CH_4) and deuteromethane (CD_4) have been grown by the same method as was used for the noble gases. CH_4 and CD_4 have freezing points of 90.7 and 89.8 K, respectively. The CH_4 crystals had to be grown at rates lower than 2.2×10^{-5} cm s^{-1} because crystals grown at higher rates were visibly flawed. The interface was slightly convex toward the liquid and was very smooth during normal growth. A smooth, rounded interface agrees well with the predictions of Jackson that materials with low entropies of fusion will grow from the melt with nonfaceted interfaces. In order to maintain the grown crystals as strain free as possible, they were separated from the wall of the sample tube by admitting warm helium gas to the chamber surrounding the tube and pumping gently to remove the vapor above the crystal. The solid was then cooled slowly to 25 K, keeping the pressure slightly lower than the vapor pressure. It should be noted that CH_4 is a plastic crystal (see below). Higher-melting n-alkanes ($C_{32}H_{66}$, $C_{94}H_{190}$) have been grown by a modification of the Bridgman method (121), using two immiscible liquids, with the upper phase at a higher temperature than the lower.

Three medium-length alkanes ($C_{19}H_{40}$, $C_{20}H_{42}$, and $C_{22}H_{46}$) have been brought to high purity (about $4N$) by very slow zone melting (4.2×10^{-5} cm s^{-1}) in a low gradient, with effective stirring (250 zone passes required about one year!) (122). The purified materials were easier to convert to single crystals than were the starting materials; eicosane ($C_{20}H_{42}$) had the highest purity and gave the clearest crystals. Growth was carried out in an all-glass Bridgman apparatus (123) (see Section 2.1.1, Chapter 3), using acetone (bp 329 K) or methanol (bp 338 K) in the upper chamber and unheated air (293 K) in the lower. Each of the three compounds undergoes a phase transition a few degrees Kelvin below its melting point; for the most part the eicosane crystals passed through the transition and grew successfully in the low-temperature form in the growth gradient (15 K cm^{-1}). Occasionally crystals grew in the high-temperature form and underwent transition slowly, with evident deterioration of quality.

The cleavage and etching properties were studied; in spite of low etch-pit densities (10^4 to 10^5 cm^{-2}), it was not possible to obtain well contrasted X-ray topographs (122).

4.3.3 Polymers

Polymers do not normally yield single crystals in the sense used earlier in this chapter. The spherulites that grow are instead organized arrays of single crystals. These have been studied by the Bridgman–Stockbarger method. Polyethylene, for example, has been directionally solidified in a device consisting of two cylindrical copper blocks (similar to those described in Section 2.1.1, Chapter 3). The polymer was contained in an evacuated glass tube of about 0.2 cm id which was lowered through a steep temperature gradient between an upper block at 473 K and a lower block at room temperature. At high rates of crystallization (5.3×10^{-4} cm s^{-1}), essentially no orientation resulted. At much lower rates (7.8×10^{-6} cm s^{-1}) a relatively high degree of lamellar orientation was obtained, with the crystallographic b axis lying along the growth direction (124). Similar results were obtained by Fujiwara with isotactic polypropylene; oriented β-phase crystals resulted from directional crystallization at growth rates less than 2.8×10^{-5} cm s^{-1}. Intermediate rates (2.8×10^{-5} to

2.8×10^{-4} cm s^{-1}) gave mixtures of nearly unoriented α phase and oriented β phase, while higher speeds ($>2.8 \times 10^{-4}$ cm s^{-1}) gave nearly unoriented α phase (125).

Lam and Geil (126) have shown that single crystals of polyethylene grow from the glassy amorphous state when it is annealed slightly above its glass transition temperature. The effect is reminiscent of the crystallization of supercooled polar liquids which is brought about by repeated cycling through the melting range, in that restricted molecular mobility and reduced viscosity are at work in both cases. Polymer crystallization has been reviewed recently (127).

4.3.4 Laser and Nonlinear Optical Crystals

$LiNdP_4O_{12}$ (LNP) is a representative stoichiometric compound that has found use in continuous-wave solid-state lasers. Crystals up to 2 cm in diameter and 4 cm long were grown from seed by a pulling technique. The pulling rate was varied from 1.4×10^{-6} cm s^{-1} to 1.4×10^{-4} cm s^{-1}, and the rotation rate from 1 to 5 rev s^{-1}. A platinum crucible was charged under nitrogen with $LiPO_3$ and NdP_3O_9 in 3:1 ratio, giving a starting solution containing about 73 wt% LNP, with a saturation temperature of 1233 K. An oriented seed was immersed in the melt about 5 K above saturation, and its surface was melted back to remove fabrication damage. The melt temperature was reduced to the saturation value and pulling was carried out at 5.5×10^{-6} cm s^{-1}, with rotation at 3 rev s^{-1}. The melt temperature was reduced at 5.6×10^{-5} K s^{-1} to 1.1×10^{-4} K s^{-1} during growth. Low growth rate and high rotation rate were found to be essential to the growth of inclusion-free crystals (128).

Crystals of gadolinium molybdate (GMO) suitable for physical characterization have been grown by the Czochralski technique, which is applicable because GMO melts congruently at 1444 K and has negligible vapor pressure at its melting point. Feed material was produced by the solid-state reaction $Gd_2O_3 + 3MoO_3 \rightarrow Gd_2(MoO_4)_3$, and melted in an iridium container. Platinum or platinum/rhodium containers were found to be unsuitable since platinum is slightly soluble in the GMO melt. Boules of 1.3-cm diameter were pulled from the melt at about 2×10^{-4} cm s^{-1} while rotating the seed at 1.7 rev s^{-1} (129).

The closely related oxide $LiGd_3(MoO_4)_5$ can be considered a solid solution of the type $Li_2MoO_4 \cdot xGd_2(MoO_4)_3$, where $x = 3$ (130). This material was grown from the product prepared by reacting Li_2MoO_4, Gd_2O_3, and MoO_3 (7.7, 23.1, and 69.2 mole%, respectively) in a platinum–rhodium crucible at about 1110 K. The cooled product was ground and heated at about 1360 K and single crystals were prepared by seeding on a platinum wire and pulling at 0.2–0.3 cm h^{-1}, with rotation at 0.5 rev s^{-1} (130). The physical properties of these ferroelectric-ferroelastic rare-earth compounds are expected to be of great value in electrooptical and electroacoustic devices.

$Hg_{1-x}Cd_xTe$ has been used for IR detectors and spin-flip Raman lasers. The relationship between structural and chemical inhomogeneities in this material has been studied in crystals prepared by zone melting, high-temperature recrystallization, and Bridgman methods. All the crystals have mosaic structures, but the Bridgman method gave the best structure (131).

Single crystals of rare-earth garnets, primarily yttrium aluminum garnet (YAG) and yttrium iron garnet (YIG), have been used extensively in lasers and for microwave and magnetic applications. For the most part, flux methods (high-temperature solution growth) have been applied. However, YAG has been grown by the Czochralski technique. Great care was required in the synthesis to maintain stoichiometric composition. The starting material was heated to the melting point (2253 ± 10 K) in an iridium crucible, by induction heating; slow heating was required to prevent ebullition. A seed was formed on an iridium rod by withdrawing it slowly from the melt. Periodic changes in the withdrawal rate caused changes in diameter of the seed; the "necked" regions served to orient the growth. Typically, crystals were grown at 7×10^{-5} to 3×10^{-4} cm s^{-1}, with rotation at about 0.5 to 1 rev s^{-1}. Oriented seeds were later used to grow large crystals (1-cm diameter $\times$ 5 cm long) in any desired orientation (132). Similar conditions have been used for Czochralski growth of ruby crystals (133). Organic crystals have also been shown to function as lasers (134).

Theory, based on observations of many inorganic compounds, asserts that the nonlinear susceptibility of a material is the geometric sum of the nonlinear susceptibilities of its interatomic bonds. This theory has been extended to molecular solids to establish that intramolecular bonds make a stronger contribution to the nonlinear terms than do intermoelcular bonds. In suitably substituted compounds containing the benzene nucleus or another conjugated system, this fact is most evident. Molecules substituted with both donor and acceptor groups seem most likely to give large nonlinear effects. Tests on the second-harmonic generation of emission at 1.06 μm in meta-disubstituted aromatic compounds in powder form confirm the presence of strong nonlinear susceptibility.

A number of such crystals were grown by directional crystallization methods (135). Sample seeds of m-bromonitrobenzene were prepared by sublimation and immersed in a melt at 0.1–0.2 K below the freezing point. During a period of a week, a crystal about $2.5 \times 1 \times 2.5$ cm was obtained. The seed propagated only slightly along the original needle axis. The related compounds m-dinitrobenzene and m-dinitroaniline were grown by a Bridgman–Stockbarger technique in a gradient heater containing two immiscible liquids (see Section 2.1.4, Chapter 3). In this case, the liquids were glycerol upon mercury with temperatures of 408 and 323 K in upper and lower layers, respectively. The growth rate was 2×10^{-5} cm s^{-1}. Crystals up to 2-cm diameter and 7 cm long were grown. A number of closely related compounds were grown from solution.

One of the above-mentioned compounds, m-dinitrobenzene, has been grown in fibers of 2-μm diameter in an ingenious inverted Bridgman configuration. A number of tubes were placed in a vertical position in a crucible containing a quantity of mDNB. The crucible was placed in the lower zone of an apparatus similar to that described by McArdle et al. (123) (see Section 2.1.2, Chapter 3), with the upper and lower zones maintained at 353 and 373 K by boiling benzene and by boiling water, respectively. Surface tension causes the tubes (10 cm long) to fill with melt completely; raising the crucible at 3×10^{-5} cm s^{-1} into the colder, upper zone causes downward crystallization over the entire length of the tube. X-ray studies show that the c axis of crystals grown in this way lay within 2° of the fiber axis (136). The inverted method described here overcame difficulties occasioned by the greater density of the solid and the vapor pressure of the solid which collectively interrupted growth in the conventional Bridgman mode in small tubes.

Lithium tantalate (LiTaO$_3$) single crystals have been used in electrooptic, piezoelectric and pyroelectric applications. Recently, surface acoustic wave (SAW) properties were reported (137). The starting material for crystal growth was prepared by melting Li$_2$CO$_3$ and Ta$_2$O$_5$ in stoichiometric ratio, followed by pressing and calcining. Single crystals were pulled from melts contained in platinum–rhodium crucibles, at 2.8×10^{-5} to 4.2×10^{-4} cm s^{-1}, with rotation at 0.2 to 1.0 rev s^{-1}. Crystals were grown in diameters up to 7.5 and 12 cm long. They were yellow-brown in color, as a result of rhodium impurity (137).

Grazhulene et al. (138) have described the purification and growth of crystals of 2,4-diaminotoluene (mp 368.7 K), which is of interest as a frequency doubler of laser radiation. Crystallization from tetrahydrofuran and sublimation gave chromatographically pure material which was charged into glass ampoules (1–2 cm in diameter and 10–15 cm long) with helical nucleating tips. These were lowered through a crystallizing gradient in a two-liquid bath (see Section 2.1.4, Chapter 3) at 1.38×10^{-5} to 4.14×10^{-5} cm s^{-1}, annealed at 363 K, and cooled to room temperature at 1.67×10^{-3} K s^{-1}. Etching of the (101) and (001) planes of the crystals showed a marked increase of etch-pit density with increasing growth gradient (see Table 4.4).

Growth and properties of laser crystals have been reviewed in recent books (139, 140).

Table 4.4. Variation of Etch-Pit Density with Growth Gradient for 2,4-Diaminotoluene

Growth Gradient, $K\ cm^{-1}$	Pits cm^{-2}
14	3×10^2–2×10^3
15	1×10^3–9×10^3
17.6	5×10^3–3×10^4
22.6	1×10^5–5×10^5
34	8×10^5–6×10^6

4.3.5   Aromatic Hydrocarbons

NAPHTHALENE

While naphthalene (mp 353.4 K) has been less widely studied than anthracene (see below), its crystallization has been pursued, especially for spectroscopic work.

Gordon (141) grew naphthalene crystals for a study of their plastic flow in shear and compression, following the very early work of Kochendorfer (142). Growth at 2.9×10^{-3} cm s^{-1} by a simple version of the Bridgman technique, gave two or three large crystal grains in each tube, usually with [001] parallel to the tube axis. It was easily possible to establish the orientation of the [100] direction in a cleaved plate merely by rotating the plate on a paper bearing a drawn straight line; a double image of the line is observed except when [100] is parallel to the line, since the [010] orientation is the direction of the twofold symmetry axis.

Corke et al. (143) tried bromination, nitration, and oxidation as means of etching dislocations selectively on naphthalene (and anthracene) crystals, without success. Brief sulfonation with fuming sulfuric acid containing about 6% sulfur trioxide, followed by a water wash, proved satisfactory. Rhombic pits were observed on (001) with their long axis lying on [100]. Because of the high volatility of naphthalene, the pits became diffuse on standing. As-grown crystals showed 10^5 to 10^6 dislocations cm^{-2}, which could be reduced to 10^3 to 10^4 by annealing for 4.32×10^4 s at 351 K. Thomas et al. (144) obtained similar etching results on Bridgman-grown and vapor-grown crystals, using a totally different etching technique: they subjected the crystals to attack by an aqueous solution of a cell-free enzyme derived from pseudomonads.

Robinson and Scott (145) also grew naphthalene crystals by the Bridgman–Stockbarger method, in Pyrex tubes. With the upper zone at about 363 K and the lower zone at about 343 K, they grew crystals 0.3 cm in diameter and 12 cm long, at 5.5×10^{-6} cm s^{-1}. Since naphthalene adheres strongly to Pyrex glass, the tubes were removed from the crystals by dissolution in hydrofluoric acid. It should be noted that the dissolution generates enough heat to melt naphthalene, and the process is best carried out in an acid-resistant plastic tube in a cold-water bath. Tensile testing was carried out on crystals grown from naphthalene samples of varying purity. The temperature dependence of critical resolved shear stress (CRSS) for slip on the (001) [110] system was found to depend strongly on impurity content. The purest crystals deformed plastically to large strains, down to 170 K, with only moderate increase in the CRSS with decreasing temperature. Less pure crystals failed by brittle fracture at much higher temperatures, without plastic deformation. It is possible that the marked effect of impurities on the (001) [110] slip system results from production of stacking faults by [110] dislocations.

Corke and Sherwood (146) purified naphthalene for Bridgman growth, using recrystallization, sublimation, and zone refining. These steps and the crystal growth itself reduced impurity content to about 1 ppm, determined by spectroscopic and gas chromatographic methods. Experimental samples were annealed after they were cut from the grown crystals. Surface etching (see above) revealed the presence of about 10^6 to 10^7 dislocations cm^{-2} in the original crystals, while annealing at 340 K for 8.64×10^5 s and 1.73×10^6 s reduced this value by factors of 10 and 100, respectively. Plastic flow of the crystals was characterized by measurement of strain rate at constant temperature. For small stresses, strain rate is proportional to a power of the stress, and the value of this stress exponent is characteristic of the creep mechanism. The measured exponent was 5.3, while the values expected for creep by a vacancy migration mechanism and by dislocation climb are unity and 4.5, respectively. Hence dislocation climb appears to be operative in naphthalene. The same family of crystals was used for measurement of self-diffusion of tagged naphthalene.

Activation energies for creep and for self-diffusion were evaluated from the respective temperature dependencies, and were found to be 130.8 ± 14 and $178.5 \pm 6.7 \, kJ \, mole^{-1}$. The near equivalence is in accord with theory and with earlier measurements on other materials; it indicates that high-temperature creep is self-diffusion controlled. This conclusion is consistent with the observation that both of the above-mentioned activation energies are considerably higher than the latent heat of sublimation for naphthalene ($74.9 \, kJ \, mole^{-1}$).

Matvienko et al. have described a variation of the Bridgman–Stockbarger method for growth of naphthalene in which a single large tube (3.2-cm diameter by 11 cm long) is filled with a bundle of small (0.3-cm diameter) tubes. All of the small crystals were grown from a common nucleus and had uniform crystallographic orientation whose scatter did not exceed 30′; the individual crystals gave a dislocation density of $10^4 \, cm^{-2}$ (147).

Much work has been invested in the study of possible relationships between solid-state luminescence and defect structure of naphthalene crystals. In pure, relatively defect-free crystals of naphthalene, the relative quantum yield of the exciton luminescence is independent of temperature from 4.2 to 77 K. Crystals with high defect levels show an increase in the quantum yield of exciton luminescence which is attributed to structural defects (148). Similar observations were made by Vorob'ev and Mel'nik, who also suggested that in naphthalene crystals containing 2-methylnaphthalene, some of the impurity molecules are associated with defects (149).

ANTHRACENE

The crystallization of anthracene (mp 490 K) has been studied intensively because it has been used widely in investigation of energy transport, solid-state photochemistry, and photoconductivity. Sherwood (150) has reviewed the purity requirements for successful Bridgman growth of anthracene and described conditions for obtaining good crystals. An improved transparent oven was used with two pairs of concentric windings, one pair above, and another below the central baffle. The upper and lower heaters (see Figure 3.25) were controlled by thermistors placed just above and below the baffle, which fit closely to the crystal tube.

Karl, in a monograph on the semiconducting properties of organic crystals, has also reviewed the purification, analysis, and crystallization of anthracene (151). He described a modified growth apparatus in which a stationary crystal tube A (Figure 4.26) is placed in a uniformly heated, tubular oven B, while a gold-coated reflector C is raised around B. For the growth of good crystals, it was found that the solid surface at the growing interface had to be convex. Further growth had to proceed from a polycrystalline seed, rather than from the solidification of supercooled melt. The "olive-shaped" expansion of the capillary was found to be effective in selecting a single nucleus; growth generally proceeded with the *b* axis nearly parallel to the growth direction. The homogeneity of orientation of the anthracene crystals was evaluated by microscopic observation of cleavage surfaces in reflected light. The defects observed were thought to have arisen from deformation resulting from easy glide on the *ac* plane during cleavage. A similar purification and crystallization sequence was described by Saleh (152), along with measurements of charge-carrier yield and mobilities.

Several attempts have been made to grow anthracene crystals in other orientations than the one described above. For example, Nakada used oriented seeds inserted in Bridgman tubes to grow crystals with (001) perpendicular to the growth direction (153). Other workers have reported that the angle between the growth direction and nucleating capillary is decisive with respect to crystal

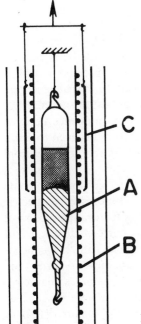

Fig. 4.26 Bridgman growth with interface movement controlled by a traveling reflector.

orientation (154); if the capillary/axis angle is greater than 45°, it is said that anthracene crystals grow with (001) perpendicular to the growth direction. Fisher (155) has observed that the dielectric polarizability of anthracene is sufficiently anisotropic that application of an electric field can orient the molecules parallel to the field. A field of $30 \, kV \, cm^{-1}$ is adequate to bring about the reorientation and hence to give rise to crystals oriented with (001) at 20° to the horizontal. The modified orientation was said to produce crystals with improved mechanical quality. It should be noted that a magnetic stirrer was levitated within the melt during growth (see Section 2.5.3, Chapter 3). Much smaller electric fields ($> 20 \, V \, cm^{-1}$) have been found to cause orienting effects in Bridgman growth of p-azoxyanisole, which is a liquid-crystalline substance (156).

The etching of emergent dislocations in anthracene by fuming sulfuric acid has been reported (157, 158). Since this work, the defect structural and mechanical properties of anthracene have been widely studied, especially as they relate to the solid-state photochemistry of the compound. Robinson and Scott (159) extended the etching study, and found that an etchant comprised of nitric acid, glacial acetic acid, and water (2:2:1) was better than fuming sulfuric acid in that the action is less rapid and hence more controllable. Moreover, the ratio of pit depth to width can be increased by saturating the etchant with silver nitrate. Generally, the axes of the pits are parallel to [100] and [010] directions, but the faces of the pits do not appear to be crystallographic.

Plastic deformation of Bridgman-grown anthracene crystals occurs by slip on (001) [101] or (001) [110], depending on the orientation of the tensile axis, and a yield point is observed when the crystal deforms on either system (160). Thomas and Williams reviewed studies of dislocation behavior in a wide variety of organic crystals including anthracene (161). They concluded, on geometrical grounds, that dislocation cores (especially those emergent at the basal surface) would facilitate the dimerization reaction. In fact it has been observed that when one-half of a cleaved crystal of anthracene was etched to reveal emergent dislocations, and the other half was illuminated by a medium-pressure mercury lamp to generate the photodimer, there was a close correspondence between etch pits and dimerization centers. The enhanced reactivity is believed to result from a reduction in the excitation energy of molecules that are displaced from their regular lattice sites by dislocations (162).

An entirely different category of physical phenomena has been found to relate closely to the defect structure of anthracene crystals. Fluorescence spectra, thermoluminescence, and conductivity all depend on the textural perfection of the test specimen. Thomas, Williams, and Cox (163) have correlated peak height in thermally stimulated conductivity (TSC) measurements with the concentration of specific dislocations introduced by mechanical deformation. Anthracene crystals, studied as grown, showed optical glow curves characterized by five well-defined peaks in the range 4–65 K. After they were annealed at 353 K for 1.8×10^4 s, the crystals gave no observable TSC maxima (164). A study of space-charge-limited currents has been made in anthracene crystals grown from melt, vapor, and solution; the effect of mechanical deformation was studied in those grown from the melt, by Thomas et al. (165). They found that the concentration of carrier traps corresponded well with the nonbasal dislocation density.

Williams and Clarke (166) have investigated the subtle inter-relationships between chemical dopants and textural defects in anthracene crystals. Their spectra show that the exciton traps introduced by structural imperfections (e.g., dislocations) are similar to those introduced by impurities. It was speculated that the traps may actually be "perturbed" (i.e., misoriented) anthracene molecules at internal surfaces where impurity has segregated. In later work (167), it was found that the triplet lifetime of melt-grown anthracene crystals was 27.1 ms in a sealed growth tube, but diminished following removal from the tube, or cleavage. It might be argued that these changes result from a chemical effect, namely adsorption of or reaction with ambient oxygen. However, other experimental results argue convincingly that purely mechanical differences can influence triplet lifetime. Thus Williams and Zboinski (167) prepared anthracene crystals from melt and vapor, using a single batch of starting material, and found that the vapor-grown crystals showed shorter triplet lifetimes. It has also been observed (Sloan, unpublished) that growth from melt and vapor in a single sealed vessel gave crystals whose lifetimes were inversely related to the growth temperature. The result implies that the growth-derived defects that are responsible for the diminished lifetime are traps consisting of misoriented molecules, which increase in concentration with increasing growth temperature.

The relationship between fluorescence spectra and defect structure has been more sharply defined by measurement of emission from microscopic regions of a crystal (168). Spectra were measured on "as-grown" and thermally or mechanically deformed crystals. The latter crystals exhibited an additional series of emission lines, red-shifted by 240 cm^{-1}. Deformed crystals were etched with oleum to reveal emergent dislocations as pyramidal pits. When emission spectra were measured from pitted regions, they were found to show the additional emission series, while pit-free regions showed the same spectra as undeformed crystals.

Earlier, Riehl and Thoma (164) had studied the thermoluminescence from anthracene crystals in which traps had been populated by polarized UV illumination at 4.2 K. They found a series of five discrete traps. The results showed a preferred direction of the light vector for optimal trap filling. In another experiment, a melt-grown crystal was first annealed for 1.8×10^4 s at 353 K then cooled to a 4.2 K and irradiated; on heating, no glow peaks were observed.

In all, these results show that the crystals contain physical as well as chemical defects, and that spectroscopic probes can reveal their presence and orientation. Of course, chemical defects can influence spectroscopic behavior of anthracene crystals as well as mechanical ones. A classic case involves deliberate or adventitious contamination with tetracene. Even at low concentrations (e.g., 10^{-4} mole mole^{-1}), UV light absorbed by the bulk crystal is nevertheless emitted almost completely from the tetracene impurity. The explanation lies in the facile transmission of the excitation energy through the anthracene until it is trapped by the impurity. Phenazine at 2×10^{-6} mole mole^{-1}, acridine at 10^{-5} mole mole^{-1}, and tetracene at 10^{-7} mole mole^{-1} have been shown to act as hole and/or electron traps in anthracene (169).

Anthracene crystals have been grown from the melt, with electron-accepting dopants. At concentrations of TCNQ, TCNE, chloranil, and bromanil below 100 ppm, good single crystals resulted at growth rates below 0.05 cm h^{-1}. At higher concentrations, polycrystalline ingots resulted (170).

PHENANTHRENE

Phenanthrene (mp 373 K), while chemically very similar to anthracene, poses quite different problems in crystal growth because of the phase transition which takes place at about 345 K. Naturally, attainment of high purity is basic to any attempt at crystal growth. Sherwood et al. (171) have explored several routes to pure phenanthrene, starting with commercial material of coal-tar origin. The following sequence was judged most effective:

- Treatment with maleic anhydride to remove anthracene
- Fusion with sodium to remove fluorene
- Chromatography and vacuum distillation
- Treatment with Raney nickel to remove sulfur-containing heterocyclics
- Sublimation and zone refining

Zone-refined product was transferred directly to tubes for Bridgman growth. The tubes used were of the type shown in Fig. 4.11f. A dual vapor-bath Bridgman oven (see Fig. 3.28) was used, with toluene (bp 388 K) and benzene (bp 353 K) as the refluxing liquids, giving a growth gradient of 9 K cm^{-1}. Identified impurities amounted to less than 1 ppm, and an unidentified impurity X, which formed when the melt was heated more than 20 K over the melting point, was present in larger amounts. Some evidence exists that X is dihydrophenanthrene.

If the grown crystals were cooled slowly through the transition temperature, optically clear boules resulted, up to 2 cm in diameter. It is especially noteworthy that the purest of crystals tended to fracture on passing through the transition. Anthracene-doped crystals showed an opaque region in the first portion grown, where the solid solubility (about 103 ppm) was exceeded as a result of preferential inclusion in the crystal (note: $k > 1$). The midportion of the crystals was optically clear and the uppermost portions (last solidified) were fractured.

Cleaved crystals were etched with oleum and with solvents (wet benzene and bromobenzene). Low concentrations of oleum gave triangular pits and higher concentrations gave rhombic pits whose short axis was parallel to [010]. Wet benzene gave similar rhombic pits. It was concluded that plastic deformation occurs by slip of dislocations of the type (001)[010] or (001)[100] and (201)[010] (171).

Phenanthrene crystals have been subjected to many of the measurements made on anthracene: electrical (172) and optical properties and lattice dynamics (173), and self-diffusion (174).

CHRYSENE

Another aromatic hydrocarbon that has been purified and crystallized is chrysene, mp 529 K. Purified chrysene was crystallized without exposure to air, and the crystals were characterized by gas chromatography, X-ray diffraction, and dislocation etching; finally, thermally simulated current was measured (175).

4.3.6 Plastic Crystals

Among organic crystals, considerable interest has developed in the class of substances which show a "rotator phase" in the solid below the melting point. In this phase, molecules show nearly free rotation about the lattice site as center. Materials showing this behavior are known as "plastic crystals" since they show plastic flow under application of small stress (176). Crystals consisting of nearly spherical molecules generally show this behavior; it is this property that eases the accommodation of the molecules in the growing crystal, since molecular reorganization is rapid near the melting point, and the orientation of the entering molecule is not critical. Plastic crystals can be grown at much higher rates than less-symmetrical organic molecules.

Succinonitrile is representative of this class of crystals. It is a cubic crystal that has been studied extensively.

- Because its entropy of fusion is low, it crystallizes in a nonfacetted, metal-like manner, making it a suitable transparent model for indirect study of the crystallization of metals (177).
- The thermal conductivities of solid and liquid are almost identical, simplifying the theoretical calculation of crystallization behavior (178).
- In common with other plastic crystals, melting is accompanied by a small volume change.

Succinonitrile has been purified by sublimation and zone refining. Impurities were present with $k > 1$ and $k < 1$; hence material for crystal growth was taken from the center of the ingot (179). Single crystals were grown in sealed Pyrex tubes by the Bridgman–Stockbarger method (180, 88). Cyclohexane, camphene, hexamethylethane, and pivalic acid have been grown in the same way (181). In summary, it may be said that organic single crystals, while less extensively studied than inorganic and metallic ones, have received considerable attention in recent years. As work on their mechanical properties has accumulated, certain generalization have emerged. Sherwood, for example (J. N. Sherwood, personal communication), has divided organic crystals into four classes on the basis of plasticity, and related dislocation density to this property and to mode of growth (see Table 4.5).

Table 4.5. Crystal Perfection as a Function of Growth Method and Crystal Plasticity (after Sherwood)

		Melt Growth		*Vapor Growth*		*Solution Growth*	
		F	NF	F	NF	F	NF
Brittle	d	Low	—	—	—	Zero	—
(benzophenone)	ms	10	—	—	—	10	—
Low plasticity	d	—	High	Medium	—	Zero	—
(pyrene)	ms	—	100	50	—	10	—
Medium plasticity	d	—	High	Medium	High	Medium	—
(anthracene)	ms	—	1000	120	1000	50	—
High plasticity	d	—	High	High	High	Medium	—
(adamantane)	ms	—	1500	50	200	10	—

F = facetted crystals.
NF = nonfacetted crystals.
d = dislocation density, cm^{-2}.
ms = mosaic spread, from full-width at half-maximum of the rocking curve of a thick crystal in X-ray topography, s (of arc).

4.3.7   Alkali Halides and Silver Halides

The alkali halides represent an important category of crystals in the midrange of melting points. They are widely grown for use as prisms and windows, especially for use in IR spectrophotometry and in scintillation counting. In addition numerous research groups are studying the effects of low temperature and IR and UV light on the optical absorption and thermal conductivity properties of these crystals.

Lithium fluoride (mp 1143 K) has been grown in crucibles of platinum or graphite, in diameters up to 15 cm, at about 2.78×10^{-5} cm s^{-1}. When grown in air, LiF crystals develop a yellow color, which has been studied extensively (182). Sodium fluoride (mp 1265 K) is grown in the same way as LiF. Sodium chloride (mp 1073 K) is grown in platinum crucibles.

Growth of sodium iodide (mp 935 K) poses severe problems because it is so hygroscopic, but its usefulness as a scintillator has led to much effort to find routes to good crystals. It is grown from dry, carefully purified material in conical platinum crucibles. In one procedure, the vessel is lowered at 2. 1×10^{-5} cm s^{-1} (1.8 cm d^{-1}) for a period of 2.5 to 3 weeks, to give crystals suitable for optical or scintillation use (183).

The potassium halides behave much as NaCl does, with chemical stability decreasing from chloride to bromide to iodide. Gründig (184) observed that oxygen-containing potassium salts in KBr (mp 1015 K) were converted to KBr by treatment with Br$_2$. Moreover, the bromine-treated melt did not attack quartz, even at 1675 K, and the resulting crystals did not adhere to the quartz surface. Gründig also described a simple furnace for growth of KBr, with a transparent central baffle of quartz. Later work on KCl (mp 1043 K) focused on the preparation of extremely pure starting material, by zone refining in graphite boats within quartz tubes under 5×10^4 Pa of chlorine. The zone-refined material was transferred to a graphite crucible for crystallization by the Kyropoulos technique. Throughout the work, much attention was directed to the cleanliness of utensils and purification of reagents, including the chlorine and inert gases. Schönherr (185) has used the Kyropoulos method to grow KCl crystals with very low dislocation densities by using properly shaped seeds of good perfection, by controlling the seeding process, and by avoiding thermal stress through temperature control and suppression of convection in the atmosphere above the melt. It was found that pressures $> 1.3 \times 10^4$ Pa cause temperature fluctuations by convection and lead to dense dislocation networks. Long, slender seed crystals (8 cm × 0.5 cm) were prepared by the Kyropoulos method. Selected portions were etch polished and thinned to a small tip (0.1 cm^2) to reduce propagation of dislocations. The [111] direction was favored because dislocations have a reduced tendency to propagate along it.

Large, nearly perfect crystals of potassium chloride and potassium bromide have been grown by the Czochralski method at pulling rates of $1-2$ cm h^{-1}. The seeds were rotated at 0.5 rev s^{-1} in copper holders, carefully designed to minimize wobble. In each case, the melt was contained in a ceramic crucible at temperature controlled to ± 1 K. The grown crystals were characterized by X-ray topography and by rocking curves (186).

Growth of cesium fluoride (mp 955 K) requires platinum crucibles, but the bromide (mp 909 K) and iodide (mp 919 K) may be crystallized in quartz vessels. Cesium chloride crystallizes in the NaCl structure, but undergoes a transformation at 742 K. Hence only small crystals can be grown from the melt, and these may be cooled slowly through the transformation temperature (187).

Strontium chloride and its solid solutions with PrCl$_3$ have been crystallized in vessels that were treated with carbon to prevent sticking (81).

Epitaxially oriented crystals of NaI, KBr, KI, and RbI have been grown on cleaved surfaces of mica (188). In each case, the (111) plane and the [110] direction of the overgrowth were parallel to the (001) plane and the [100] direction of the mica.

Progress in the growth of alkali halide crystals for optical purposes has been reviewed by Reed; related processes (cutting, grinding, polishing) were also covered (189).

The silver halides are of great interest in optical, electrical, and photochemical studies. Of course, it is desirable to investigate large, pure, unexposed crystals. Nail et al. (190) prepared pure and mixed halides (chloride, bromide, chlorobromide, and bromoiodide) as well as iron- and copper-doped silver chloride. Crystals were grown in Pyrex or quartz tubes at 1.4×10^{-4} cm s^{-1} in a Bridgman furnace. The halide charge was heated to about 20 K above the melting point in the crystal-growth tube and then treated with a stream of halogen-saturated nitrogen to convert traces of silver and silver oxide to the halide, and to remove volatile impurities from the melt. Excess dissolved chlorine was then removed in a stream of pure dry nitrogen. A slight excess pressure of inert gas or halogen was maintained over the sample during crystal growth. All of these operations were carried out in red safelight. The grown crystals were cooled slowly to room temperature. During the early stage of cooling, crystals adhered to the container, but at room temperature the glass broke readily from the crystal. Rapid cooling caused breakage of the sample tube. (190).

4.3.8   Metalloids

SELENIUM

The viscosity of Se is quite high at its melting point and it usually forms a glass at growth rates above 10^{-7} cm s^{-1}. Harrison and Tiller (191) crystallized Se at 5×10^8 Pa and found that the melting point was raised sufficiently to reduce the average chain length to such an extent that the viscosity fell to about one-fifth of its value at 10^5 Pa. Moreover the interface transport was accelerated at the higher temperature, making it possible to grow crystals at rates up to 10^{-5} cm s^{-1}. The pressure effects observed in this work may be quite general, and may facilitate crystallization of both organic and inorganic glass formers (e.g., hydrogen-bonded polyalcohols and silicates, respectively).

SILICON

Modern technology is heavily dependent on single crystals of silicon and the structures derived from such crystals. Extensive discussion of growth of silicon is beyond the scope of this volume. Specialist monographs describe the many approaches used in preparing single-crystalline silicon (192).

5   CRYSTALLIZATION IN LOW-GRAVITY ENVIRONMENTS

The occurrence of Space Shuttle flights makes it reasonable to think of long-lasting experiments on crystallization in near-Earth orbit, where the gravitational effect is 10^3 to 10^6 times smaller than at the Earth's surface. Solidification in microgravity environments can shed light on four aspects of crystallization:

- Containerless processing avoids reaction with and contamination by containers. Moreover, isolated melts can be highly undercooled, and the kinetic and thermodynamic characteristics of solidification of such melts can be studied.
- Convective flow brought about by thermally or compositionally induced density gradients can be greatly reduced.
- Sedimentation and Stokes flows are eliminated.
- Hydrostatic pressure is eliminated.

While larger crystals can be grown by containerless methods in microgravity environments than can be grown on Earth, there does not seem to be much practical advantage to growing such crystals. Nor is it certain that space-grown crystals will be of substantially higher quality than those grown on Earth (193, 194). Moreover, higher perfection of the starting crystals may not confer improved properties on fabricated devices because the fabrication processes can themselves introduce many defects. For these reasons, it appears that the major thrust of crystallization studies in space should be toward increased understanding of fundamental physical processes (195). It is self-evident that such intrinsically costly studies should be preceded by careful experiments in terrestrial laboratories.

In such a study of this type, microgravity environment was simulated for brief periods (1–6 s) by free fall of droplets in an evacuated drop tube, to study containerless solidification. Niobium droplets up to 0.5 cm in diameter were formed by omnidirectional electron bombardment at the top of a 32-m tube. Droplets of known size and temperature ($T_f = 2741$ K) were detached; they cooled radiatively until nucleation took place. Consequent release of the droplet's latent heat of fusion caused a flash of light more intense than the radiation emitted at the nucleation temperature. The time of this flash was monitored and the cooling rate was measured by silicon photovoltaic detectors installed at instrumentation ports along the drop tube. From the two measurements, undercoolings up to 525 K ($\Delta T/T_f = 0.19$) were achieved in droplets of various diameters (0.2–0.5 cm). This value of undercooling might be thought to indicate that homogeneous nucleation took place. However, the nucleation temperature is almost exactly the melting temperature of NbO, and it is thus possible that traces of NbO are responsible for heterogeneous nucleation of individual droplets.

Solid spheres formed from highly undercooled melts consisted of single grains with wrinkled skins. Spheres formed from slightly undercooled melts showed large internal cavities and smooth surfaces. Perhaps the most striking result of the study was the calculated rate of solidification of the highly undercooled droplets: 320 m s^{-1}, a value comparable to rates achieved in splat cooling. This finding implies that the mechanics of high-speed solidification can be studied without the stresses induced by impact (196).

Microgravity crystallization studies have mainly been directed toward taking advantage of the first two of the four aspects mentioned above, namely containerless processing and reduced convection. When buoyancy effects are suppressed, surface-tension-driven flow (Marangoni flow) may be observed as the dominant mechanism of convection. Both buoyancy and surface tension effects on convection have been considered in theoretical studies (197). Another theoretical study (198) has considered two-dimensional thermal convection in a cylinder with imposed circumferential temperature distribution. Solution of the relevant equations for arbitrary time-dependent accelerations makes it possible to calculate the transient convection resulting from motion of an astronaut, one source of "g-jitter."

Experiments on containerless growth require some positioning mechanism, because stable levitation is made difficult by residual accelerations and temperature gradients. In one approach, spheres of Bi, Te, InSb, and Bi_2Te_3 were centered on thin silica threads which traversed a silica tube. A single silica tube containing six samples was positioned in a gradient furnace so that each material was heated just above its melting point. In one case crystallization was achieved by lowering the furnace temperature and in another by transporting the sample tube to a cooler region of the furnace. The grown crystals were characterized by macrophotography, X-ray diffraction, sectioning, and etching. Control experiments were carried out in identical apparatus on Earth; in all cases the molten spheres fell from their holders owing to gravity (!). In general, the textural properties of the space-grown crystals were better than those of the Earth-grown ones (199).

Directional crystallization has been carried out in space in a variety of containers. $Bi_{1-x}Sb_x$ and PbTe were grown as single crystals (200). Germanium solidified with a free surface of crystallization, without wetting the walls of its quartz ampoule. Even under forced-contact conditions, there was no close contact with the container wall, because of mounds and ridges on the surface of the growing crystal; the noncontact led to improved crystal quality. Moreover, stable, convectionless growth conditions resulted in chemical microhomogeneity, both axially and radially (201). Cadmium mercury telluride crystals were grown in evacuated quartz ampoules, at rates from $5.6 \times 10^{-5}\,cm\,s^{-1}$ to $1.4 \times 10^{-3}\,cm\,s^{-1}$ (0.2 to $5\,cm\,h^{-1}$). Slow crystallization gave diffusion-controlled growth and homogeneous product. At rates above $8.3 \times 10^{-5}\,cm\,s^{-1}$ ($0.3\,cm\,h^{-1}$), spontaneous crystallization occurred, creating cavities and giving rise to inhomogeneous material (202).

Directional solidification of a saturated solution of NH_4Cl in water was carried out in a suborbital rocket at $10^{-5}\,g$, for comparison with laboratory results on Earth. Experiments at $1\,g$ showed extensive fluid flow with plumes carrying numerous crystallites vertically above the interface. The bulk interface advanced at the same rate as the individual dendrites. Dendrites growing ahead of the interface were asymmetric, with secondary arms on the side where no flow was apparent. Solidification at low g gave an interface which grew at a slower rate than the individual dendrites, which grew symmetrically. No crystal-carrying plumes of liquid were observed (203).

A composite eutectic structure of LiF fibers embedded in an NaCl matrix was produced both on Earth and in space. In the latter experiments the included fibers grew continuously, and this result was attributed to the absence of convection currents. The space-grown eutectics showed better optical properties as a result of the superior fiber alignment in the growth direction (204). This result could be an opening to actual technological practice in space, since the structures formed could not be duplicated on Earth.

An experimental apparatus has been constructed for zone refining in space, using radiant heating (see Section 2.1.4, Chapter 5) (205).

REFERENCES

1. J. L. Dewey, U.S. Patent 3,163,895, January 5, 1965.

2. M. Zief, Ed., *Purification of Inorganic and Organic Materials; Techniques of Fractional Solidification*, Dekker, New York, 1969.

3. L. Repiska, L. Komorov, *Zb. Ved. Pr. Vys. Sk. Tech. Kosiciach*, **1977**(2), 137–145.

4. A. N. Kirgintsev, S. I. Raspopin, and N. I. Lastushkin, *Deposited Doc.*, **1975** VINITI, 3223–3275; through CA **88**:57048t.

5. P. I. Artyukhin, Yu. L. Mityakin, and B. M. Shavinskii, *Izv. Sib. Otd. Akad. Nauk SSSR, Ser. Khim. Nauk*, **1981**, 87–93.

6. P. I. Artyukhin, B. M. Shavinskii, and Yu. L. Mityakin, *Izv. Sib. Otd. Akad. Nauk SSSR, Ser. Khim. Nauk*, **1979**(3), 102–106.

7. P. I. Artyukhin, Yu. L. Mityakin, and B. M. Shavinskii, *Izv. Sib. Otd. Akad. Nauk SSSR, Ser. Khim. Nauk*, **5**, 66–69 (1979) (Russ.).

8. A. N. Kirgintsev, V. I. Kabysheva, V. I. Kosyakov, and V. M. Sokolov, *J. Appl. Chem. USSR*, **45**, 2272–2274 (1972).

9. A. D. Pruett, *J. Phys. Chem. Solids*, **28**, 2346–2347 (1967).

10. A. Poczynajlo, *Nukleonika*, **19**(2), 11–26 (1974).

11. F. Feher and H. D. Lutz, *Z. anorg. allg. Chemie*, **334**, 235–241 (1965).

12. A. N. Kirgintsev and N. I. Lastushkin, *Izv. Sib. Otd. Akad. Nauk SSSR, Ser. Khim. Nauk*, **1971**(4), 47–57.

13. V. N. Vigdorovich, A. E. Vol'pyan, and V. V. Marychev, *Tsvetn. Met.*, **1978**(4), 52–54.

14. Sh. Movlanov and A. A. Kuliev, *Izv. Akad. Nauk Azerb. SSSR, Ser. Fiz. Mat. i Tekhn. Nauk*, **3**, 55–62 (1961).

15. P. Poeydomenge, G. Cizern, and P. Lacombe, *J. Nucl. Mater.*, **8**(1), 138–142 (1963).

16. A. Lunden, E. Svantesson, and H. Svensson, *Z. Naturforsch.*, **20a**, 1279–1282 (1965).

17. V. I. Kosyakov, *Izv. Sib. Otd. Akad. Nauk SSSR, Ser. Khim. Nauk*, **1971**(2), 3–10.

18. A. N. Kirgintsev and L. I. Isaenko, *Deposited Publ.*, **1973** VINITI, 6273–6273, 48 pp.; through CA**84**:187634j.

19. A. S. Aloi and A. N. Kirgintsev, *Izv. Sib. Otd. Akad. Nauk SSSR, Ser. Khim. Nauk*, **1977**(1), 59–61; through CA**86**:161937b.

20. A. N. Kirgintsev, N. I. Latushkin, and E. Ya. Aladko, *Deposited Publ.*, **1973** VINITI, 6131–6173, 53 pp.; through CA**85**:134531n.

21. V. I. Zharinov, E. E. Konovalov, Sh. I. Peizulaev, and T. V. Kayurova, *Russ. J. Phys. Chem.*, **48**(4), 572–574 (1974).

22. A. N. Kirgintsev, V. A. Isayenko, N. I. Krainyukov, and N. N. Smirnov, *Izv. Sib. Ot. Akad. Nauk SSSR, Ser. Khim. Nauk*, **1972**(6), 119–121.

23. P. Süe, J. Pauly, and A. Nouaillle, *Bull. Soc. Chim. France*, **5**, 593–602 (1958).

24. A. B. Blank, V. G. Chepurnaya, and V. Ya. Vakulenko, *Monokrist. Tekh.*, **1972**(7), 149–155; through CA**80**:53047d.

25. A. B. Blank and N. I. Komishan, *Fiz. i Khim. Kristallov*, **1977**, 11–23.

26. A. B. Blank and V. G. Chepurnaya, *Massov. Kristallizatsiya*, **1977**(3), 84–91.

27. A. N. Krigintsev and B. M. Shavinskii, *Deposited Publ.*, **1973** VINITI, 6137–6173.

28. H. Schildknecht and F. Schlegelmilch, *Chemie-Ing. Techn.*, **35**(9), 637–640 (1963).

29. P. Gaubert, *Comptes. rendus*, **194**, 109–111 (1932).

30. F. W. Schwab and E. Wichers, U.S. Department of Commerce National Bureau of Standards Research Paper RP1351, **25**, December 1940.

31. F. W. Schwab and E. Wichers, *J. Res. NBS*, **32**, 253–259 (1944).

32. M. V. Mayer, *Verfahrenstechnik (Mainz)*, **8**, 221–223 (1974).

33. T. E. Naida, E. A. Basistov, T. B. Dement, A. E. Golub, and G. Z. Blyum, *Vsesoyuznyi nauchno-issledovatel'skii institut khimicheskikh reaktivov i osobo chistykh khimicheskikh veshchestv*, **1974**, 248–254, no. 36.

34. H. Schildknecht, G. Rauch, and F. Schlegelmilch, *Chemiker Ztg.*, **83**, 549–552 (1959).

35. C. W. Bunn, A. J. Cobbold, and R. P. Palmer, *J. Polym. Sci.*, **28**, 365–376 (1958).

36. F. R. Anderson, *J. Appl. Phys.*, **35**, 64–70 (1964).

37. F. R. Anderson, *J. Polym. Sci., Part C*, **8**, 275–285 (1964).

38. P. H. Geil, F. R. Anderson, B. Wunderlich, and T. Arakawa, *J. Polym. Sci.*, **2**, Part A2, 3707–3720 (1964).

39. J. D. Moyer and R. J. Ochs, *Science*, **142**, 1316–1318 (1963).

40. P. D. Calvert and T. G. Ryan, *Polymer*, **19**, 611 (1978).

41. A. B. Blank, L. I. Afanasiadi, E. S. Zolotovitskaya, B. M. Fidel'man, and V. G. Chepurnaya, *Zh. Anal. Khim.*, **25**, 2291–2296 (1970).

42. A. M. Kirgintsev, *J. Anal. Chem. USSR*, **26**(9), 1719–1724 (1971).

43. A. R. McGhie and G. J. Sloan, *J. Crystal Growth*, **32**, 60–67 (1976).

44. A. B. Blank, L. I. Afanasiadi, and L. P. Eksperiandova, *Zh. Anal. Khim.*, **1972**, 27 (2), 221–224.

45. A. B. Blank, N. I. Shevtsov, and E. S. Zolotovitskaya, *Zh. Anal. Khimii*, **30**, 2036–2039 (1975).

46. E. Ya. Aladko and A. N. Kirgintsev, *Izv. Sib. Otd. Akad. Nauk SSSR, Ser. Khim. Nauk*, **1976**(6), 122; through CA**86**:14940q.

47. A. B. Blank, L. I. Afanasiadi, E. S. Zolotovitskaya, and B. M. Fidel'man, *Monokrist. Tekh.*, **1972**(7), 156–159 through CA**80**:41721f (1974).

48. A. B. Blank, N. I. Shevtsov, and I. T. Kogan, *J. Anal. Chem. USSR*, **31**, 833–835 (1976).

49. M. N. Matyukhina, *Zavod. Lab.*, **46**(9), 814–816 (1980).

50. A. B. Blank and V. G. Chepurnaya, *Zh. Anal. Khim.*, **29**(5), 1006–1008 (1974).

51. A. B. Blank, R. P. Pantaler, I. V. Pulyaeva, and L. P. Eksperiandova, *Zh. Anal. Khim.*, **33**(9), 1771–1773 (1978).

52. A. B. Blank, E. S. Zolotovitskaya, N. I. Komishan, Z. V. Shtitel'man, and L. I. Afanasiadi, *J. Anal. Chem. (USSR)*, **34**(5), 792–794 (1979).

53. A. N. Kirgintsev, S. B. Gryaznova, and Z. V. Zil'berfain, *Zhur. Anal. Khimii*, **28**(6), 1069–1075 (1973); *J. Analyt. Chem. (USSR)*, **28**, 954–957 (1973).

54. H. Schildknecht and A. Mannl, *Angew. Chem.*, **69**, 634–638 (1957).

55. H. Schildknecht, G. Rauch, and F. Schlegelmilch, *Chemiker Ztg.*, **83**, 549–552 (1959).

56. J. Shapiro, *Science*, **133**, 2063–2064 (1961).

57. J. Shapiro in M. Zief, Ed., *Purification of Inorganic and Organic Materials; Techniques of Fractional Solidification*, Dekker, New York, 1969, p. 147.

58. J. Shapiro, *Anal. Chem.*, **39**, 280 (1967).

59. S. Kobayashi and G. F. Lee, *Anal. Chem.*, **36**, 2197–2198 (1964).

60. N. Yonehara, M. Kamada, and K. Fukunaga, *Bunseki Kagaku*, **30**(9), 620–622 (1981).

61. T. E. Wilson, D. J. Evans, Jr., and M. L. Theriot, *Appl. Microbiol.*, **12**(2), 96–99 (1964).

62. R. A. Baker, *J. Water Pollution Control Federation*, **37**, 1164–1170 (1965).

63. J-C. Janson, B. Ersson, and J. Porat, *Biotechnology & Bioeng.*, **16**, 21–39 (1974).

64. T. H. Gouw, *Separation Sci.*, **2**(4), 431–437 (1967).

65. J. C. Posey and H. A. Smith, *J. Am. Chem. Soc.*, **79**, 555–557 (1957).

66. W. Mahler and M. F. Bechtold, U.S. Pat. 4,122,041, October 24, 1978.

67. J. S. Matthews and N. D. Coggeshall, *Anal. Chem.*, **31**, 1124 (1959).

68. G. J. Sloan, *Anal. Chem.*, **38**, 1805–1806 (1968).

69. E. N. da C. Andrade, *Phil. Mag.*, **27**, 869–870 (1914).

70. G. Tammann, *Lehrbuch der Metallographie*, L. Voss, Leipzig, 1923.

71. P. W. Bridgman, *Proc. Am. Acad. Arts Sci.*, **60**, 385–421 (1925).

72. N. B. Hannay, Ed., *Treatise on Solid State Chemistry*, Vol. 5, Plenum, New York, 1975.

73. I. Tarjan and M. Matrai, *Laboratory Manual on Crystal Growth*, Akademiai Kiado, Budapest, 1972 (UNESCO).

74. A. Smakula, *Einkristalle*, Springer-Verlag, Berlin, 1961.

75. P. Hartman, Ed., *Crystal Growth: An Introduction*, North Holland, 1973.

76. H. E. Buckley, *Crystal Growth*, Wiley, New York, 1951.

77. R. Roy, *North-Holland Ser. Cryst. Growth*, **2** (Cryst. Growth: Tutorial Approach), 67–90 (1979).

78. F. Stober, *Z. Krist.*, **61**, 299–314 (1925).

79. G. J. Sloan, *Mol. Cryst.*, **1**, 161–194 (1966).

80. J. H. Wernick and H. M. Davis, *J. Appl. Phys.*, **27**, 149–153 (1956).

81. R. J. Walker, *J. Crystal Growth*, **44**, 187–189 (1978).

82. J. N. Sherwood and S. J. Thomson, *J. Sci. Instr.*, **37**, 242–245 (1960060).

8 P. A. Polibin and A. I. Froiman, *Z. Krist.*, **85**, 322–325 (1933).

84. S. C. Datt and K. D. Verma, *Ind. J. Pure Appl. Phys.*, **6**, 96–97 (1968).

85. K. M. M. Kruse, *J. Phys. E.*, **3**(8), 609–614 (1970).

86. K. M. M. Kruse, *J. Phys. E.*, **8**, 592–595 (1975).

87. Y. Lupien, J. O. Williams, and D. F. Williams, *Mol. Cryst. Liquid Cryst.*, **18**, 129–141 (1972).

88. G. M. Hood and J. N. Sherwood, *Brit. J. Appl. Phy.*, **14**, 215–217 (1963).

89. T. Dahlgren, T. Gillbro, G. Nilsson, and A. Lund, *J. Phys. E.*, **4**, 61–62 (1971).

90. C. K. Ma, M. Ohtsuka, and R. E. Bedford, *Rev. Sci. Instrum.*, **51**(1), 52–54 (1980).

91. S. F. Ahmad, H. Kiefte, and M. J. Clouter, *J. Chem. Phys.*, **69**(12), 5468–5472 (1978).

92. Yu. Glushkov, I. Yarovskii, and B. M. Oreshin, *Kristallografiya*, **15**, 1273–1274 (1970).

93. J. Czochralski, *Z. Phys. Chem.*, **92**, 219–221 (1918).

94. J. A. Burton, R. C. Prim, and W. P. Slichter, *J. Chem. Phys.*, **21**, 1987–1996 (1953).

95. J. Kvapil, J. Kubelka, and B. Perner, *Kristal Technik*, **13**, 1369–1375 (1978).

96. J. Kvapil, J. Kubelka, and R. Vadura, *Kristal Technik*, **13**, 1357–1367 (1978).

97. D. T. J. Hurle, *Adv. Colloid Interface Sci.*, **15**, 101–130 (1981).

98. H. E. Labelle, U.S. Pat. 3,471,266, Oct. 7, 1969.

99. A. V. Stepanov, *Bull. Acad. Sci. USSR, Phys. Ser.*, **33**, 1775–1782 (1969).

100. T. Surek, B. Chalmers, and A. I. Mlavsky, *J. Crystal Growth*, **42**, 453–465 (1977).

101. P. I. Antonov, S. P. Nikanorov, and V. A. Tatarchenko, *J. Crystal Growth*, **42**, 447–452 (1977).

102. *Izvest. Akad. Nauk SSSR, Fiz.*, **44**(2), (1980).

103. A. Maries and P. S. Rogers, *J. Mater. Sci.*, **13**, 2119–2130 (1978).

104. J. B. Mullin, B. W. Straughan, and W. S. Brickell, *J. Phys. Chem. Solids.* **26**, 782–784 (1965).

105. P. D. Greene, *J. Crystal Growth*, **50**, 612–618 (1980).

106. M. R. Brozel, K. Laithwaite, R. C. Newman, and B. Ozbay, *J. Crystal Growth*, **50**, 619–624 (1980).

107. J. B. Mullin, A. Royle, and S. Benn, *J. Crystal Growth*, **50**, 625–637 (1980).

108. H. Kotake, K. Hirahara, and M. Watanabe, *J. Crystal Growth*, **50**, 743–751 (1980).

109. J. Matsui, K. Watanabe, Y. Seki, *J. Crystal Growth*, **46**, 563–568 (1980).

110. R. C. DeMattei, R. A. Huggins, and R. S. Feigelson, *J. Crystal Growth*, **34**, 1–10 (1976).

111. J. S. Cook and P. B. Gwan, *J. Crystal Growth*, **33**, 369–371 (1976).

112. E. A. Decamps, M. Durand, and M. M. Granger, *Bull. Soc. Sci. Bretagne*, **73**, 48 (Hors Serie) 103–114.

113. H. J. Scheel, *J. Crystal Growth*, **13/14**, 560–565 (1972).

114. H. J. Scheel and H. Muller-Krumbhaar, *J. Crystal Growth*, **49**, 291–296 (1980).

115. C. E. Turner, A. W. Morris, and D. Elwell, *J. Crystal Growth*, **35**, 234–235 (1976).

116. S. Kyropoulos, *Z. Anorg. Chem.*, **154**, 308–313 (1926).

117. A. Verneuil, *Compt. rend.*, **135**, 791–794 (1902).

118. A. Verneuil, *Ann. Chim. Phys.*, **3**, 20–29 (1904).

119. D. N. Batchelder, D. L. Losee, and R. O. Simmons, in "Crystal Growth," Proc. Int. Conf. on Crystal Growth, Boston, June 1966, pp. 843–847.

120. B. A. Grass, S. M. Heald, and R. O. Simmons, *J. Crystal Growth*, **42**, 370–375 (1977).

121. Publications of the National Bureau of Standards, July 1960–June 1966, Technical Notes 174, 197, 236, 251, 260, Growth of High-Melting Alkanes.

122. R. S. Narang and J. N. Sherwood, *J. Crystal Growth*, **49**, 357–362 (1980).

123. B. J. McArdle, J. N. Sherwood, and A. C. Damask, *J. Crystal Growth*, **22**, 193–200 (1974).

124. J. M. Crissman and E. Passaglia, *J. Res. Nat. Bur. Stds.*, **70A**, 225–232 (1966).

125. Y. Fujiwara, *Kolloid Z.Z. Polym.*, **226**, 135–138 (1968).

126. R. Lam and P. H. Geil, *Science*, **205**, 1388–1389 (1979).

127. A. J. Pennings, *J. Crystal Growth*, **48**(4), 574–581 (1980).

128. J. Nakano, T. Yamada, and S. Miyazawa, *J. Crystal Growth*, **47**, 693–698 (1979).

129. J. R. Barkley, L. H. Brixner, E. M. Hogan, and R. K. Waring, Jr., *Ferroelectrics*, **3**, 191–197 (1972).

130. R. K. Pandey, *J. Crystal Growth*, **48**, 355–358 (1980).

131. T. C. Harman, in M. Aven and J. S. Prener Eds., *Physics and Chemistry of II-VI Compounds*, North Holland, Amsterdam, 1967, p. 767.

132. M. Kestigian and W. W. Holloway, Jr., "Single-Crystal Growth and Optical Studies of Rare-Earth Alumina Garnets," in H. S. Peiser Ed., *Crystal Growth*, Pergamon, New York, 1967, pp. 451–456.

133. F. R. Charvat, J. C. Smith, and O. H. Nestor, "Characteristics of Large Ruby Crystals Grown by the Czochralski Technique," in H. S. Peiser, Ed., *Crystal Growth*, Pergamon, New York, 1967, pp. 45–50.

134. N. Karl, "Laser Emission from Organic Crystals," in *Lasers in Physical Chemistry and Biophysics*, Elsevier, Amsterdam, 1975, pp. 61–76.

135. A. Carenco, A. Perigaud, and J. Jerphagnon, *Mol. Cryst. Liq. Cryst.*, **32**, 101–104 (1976).

136. D. W. G. Ballentyne and S. M. Al-Shukri, *J. Crystal Growth*, **48**, 491–492 (1980).

137. T. Fukuda, S. Matsumura, H. Hirao and T. Ito, *J. Crystal Growth*, **46**, 179–184 (1979).

138. S. S. Grazhulene, L. A. Musikin, G. F. Telegin, and V. D. Shigorin, *Zhur. Prikl. Khimii*, **50**(11), 2473–2475 (1977).

139. P. A. Arsenjew, S. Bagdasarow, K. Bienert, E. F. Kustow, and A. W. Potjomkin, *Kristalle in der Modernen Lasertechnik*, Akademische Verlagsgesellschaft, Leipzig, 1980.

140. A. A. Kaminski, *Laser Crystals–Their Physics and Properties*, Springer-Verlag, New York, 1981.

141. R. B. Gordon, *Acta Metallurgica*, **13**, 199–203 (1965).

142. A. Kochendorfer, *Z. Krist.*, **97**, 263–299 (1937).

143. N. T. Corke, A. A. Kawada, and J. N. Sherwood, *Nature*, **213**, 62–63 (1967).

144. J. M. Thomas and J. O. Williams, *Nature*, **211**, 181–182 (1966).

145. P. M. Robinson and H. G. Scott, *Acta Met.*, **15**, 1230–1231 (1967).

146. N. T. Corke and J. N. Sherwood, *J. Mater. Sci.*, **6**, 68–73 (1971).

147. V. N. Matvienko, N. V. Pertsov, and E. D. Shuchkin, *Mol. Cryst. Liq. Cryst.*, **51**, 1–8 (1979).

148. N. I. Ostapenko, Yu. A. Skryshevskii, A. N. Faidysh, and M. T. Shpak, *Izv. Akad. Nauk SSSR., Ser. Fiz.*, **39**(9), 1786–1792 (1975).

149. V. P. Vorob'ev and V. I. Mel'nik, *Izv. Akad. Nauk SSSR, Ser. Fiz.*, **39**, 1877–1881 (1975).

150. M. Zief, Ed., *Purification of Inorganic and Organic Materials; Techniques of Fractional Solidification*, Dekker, New York, 1969, p. 157.

151. N. Karl, "High Purity Organic Molecular Crystals," in H. C. Freyhardt, Ed., *Crystals–Growth, Properties, Applications*, Vol. 4, Springer, New York, 1980.

152. M. Saleh, *Japan. J. Appl. Phys.*, **17**, 1031–1036 (1978).

153. I. Nakada, *J. Phys. Soc. Japan*, **17**, 113–118 (1962).

154. J. N. Sherwood and S. J. Thomson, *J. Sci. Instr.*, **37**, 242–245 (1960).

155. D. Fischer, *Mat. Res. Bull.*, **3**, 759–764 (1968).

156. E. N. Nikitin, I. P. Areshev, and N. V. Zaitseva, *Kristallografiya*, **25**(1), 205–207 (1980).

157. A. R. McGhie, P. J. Reucroft, and M. M. Labes, *J. Chem. Phys.*, **45**, 3163 (1966).

158. J. O. Williams and J. M. Thomas, *Trans. Faraday Soc.*, **63**, 1720–1729 (1967).

159. P. M. Robinson and H. G. Scott, *J. Crystal Growth*, **1**, 187–194 (1967).

160. P. M. Robinson and H. G. Scott, *Acta Metallurgica*, **15**, 1581–1590 (1967).

161. J. M. Thomas and J. O. Williams, *Prog. Solid State Chem.*, **6**, 199–254 (1971).

162. J. M. Thomas and J. O. Williams, *Chem. Comm.*, 432–433 (1967).

163. J. M. Thomas, J. O. Williams, and G. A. Cox, *Trans. Faraday Soc.*, **64**, 2496–2504 (1968).

164. N. Riehl and P. Thoma, *Phys. Stat. Sol.*, **16**, 159–169 (1966).

165. J. M. Thomas, J. O. Williams, and L. M. Turton, *Trans. Faraday Soc.*, **64**(9), 2505–2513 (1968).

166. J. O. Williams and B. P. Clarke, *J. Chem. Soc.*, **73**(10), 1371–1384 (1977).

167. J. O. Williams and Z. Zboinski, *J. Chem. Soc.*, **74**(3), 618–629 (1978); *Faraday Trans.* 629.

168. V. A. Lisovenko, M. T. Shpak, and V. L. Salo, *Phys. Stat. Sol. (a)*, **29**, K101–103 (1975).

169. K. H. Probst and N. Karl, *Phys. Stat. Sol. (a)*, **27**, 499–508 (1975).

170. B. Marciniak and W. Waclawek, *J. Crystal Growth*, **52**(2), 623–629 (1981).

171. B. J. McArdle, J. N. Sherwood, and A. C. Damask, *J. Crystal Growth*, **22**, 193–200 (1974).

172. R. A. Arndt and A. C. Damask, *J. Chem. Phys.*, **45**, 4627–4633 (1966).

173. A. C. Damask, R. A. Arndt, D. H. Spielberg, and I. Lefkowitz, *J. Chem. Phys.*, **54**, 2597–2601 (1971).

174. G. Burns and J. N. Sherwood, *Mol. Cryst. & Liquid Cryst.*, **18**, 91–94 (1972).

175. P. J. Reucroft, A. R. McGhie, E. E. Hillman, and V. V. Damiano, *J. Crystal Growth*, **11**(3), 355–357 (1971).

176. J. N. Sherwood, Ed., *The Plastically Crystalline State*, Wiley, New York, 1980.

177. K. A. Jackson and J. D. Hunt, *Acta Metal.*, **13**, 1212–1215 (1965).

178. R. J. Schaefer, M. E. Glicksman, and J. D. Ayers, *Phil. Mag.*, **32**, 725–743 (1975).

179. D. R. H. Jones and G. A. Chadwick, *Phil. Mag.*, **22**, 291–300 (1970).

180. H. M. Hawthorne and J. N. Sherwood, *Trans. Faraday Soc.*, **66**(7), 1792–1798 (1970).

181. M. Brissaud, C. Dolin, J. LeGuigou, B. S. McArdle, and J. N. Sherwood, *J. Crystal Growth*, **38**(1), 134–138 (1977).

182. J. J. Gilman and W. G. Johnston, "Dislocations in Lithium Flouride Crystals," in F. Seitz and D. Turnbull, Eds., *Solid State Physics–Advances in Research and Applications*, Vol. 13, Academic Press, New York, 1962.

183. R. W. Spurney, U.S. Pat. 3,926,566, May 18, 1973.

184. H. Grundig, *Z. Physik.*, **158**, 577–594 (1960).

185. E. Schönherr, "Growth of KCl Crystals of Extremely Low Dislocation Density," in H. S. Peiser, Ed., *Crystal Growth*, Pergamon, Oxford, 1967, p. 825.

186. K. Lal, A. Murthy, S. K. Halder, B. P. Singh, and V. Kumar, *J. Crystal Growth*, **56**, 125–131 (1982).

187. A. Smakula, *Einkristalle*, Springer, Berlin, 1961, p. 250.

188. C. D. West, *J. Opt. Soc. Amer.*, **35**, 26–31 (1945).

189. J. Reed, *Proc. Soc. Photo-Opt. Instrum. Eng.*, **109**, 2–5 (1979).

190. N. R. Nail, F. Moser, P. E. Goddard, and F. Urbach, *Rev. Sci. Instr.*, **28**(4), 275–278 (1957).

191. D. E. Harrison and W. A. Tiller, *J. Appl. Phys.*, **36**, 1680–1683 (1965).

192. J. H. Matlock, "Advances in Single Crystal Growth of Silicon," in H. R. Huff and E. Sirtl, Eds., *Semiconductor Silicon*, Electrochemical Society Proceedings, 77–82 (1977).

193. A. F. Witt, H. C. Gatos, M. Lichtensteiger, and C. J. Herman, *J. Electrochem. Soc.*, **125**(11), 1832–1840 (1978).

194. A. F. Witt, *Space Res.*, **19**, 503–506 (1979).

195. National Academy of Sciences, *Materials Processing in Space*, Report by Committee on Scientific and Technological Aspects of Materials Processing in Space, Washington, DC (1978).

196. L. L. Lacy, M. B. Robinson, and T. J. Rathz, *J. Crystal Growth*, **51**, 47–60 (1981).

197. V. I. Polezhayev, K. G. Dubovik, S. A. Nikitin, A. I. Prostomolotov, and A. I. Fedyushkin, *J. Crystal Growth*, **52**, 465–470 (1981).

198. R. F. Dressler, *J. Crystal Growth*, **54**, 523–533 (1981).

199. H. Rodot, M. Hamidi, J. Bourneix, A. S. Okhotin, I. A. Zoubridski, V. T. Kriapov, and E. V. Markov, *J. Crystal Growth*, **52**, 478–484 (1981).

200. H. J. Fischer, R. Kuhl, H. Oppermann, R. Herrman, K. Hilbert, A. S. Okhotin, G. E. Ignat'ev, V. T. Khryapov, E. M. Markov, and I. V. Barmin, *Adv. Space Res.*, **1**(5), 111–115 (1981).

201. E. V. Markov et al., Eur. Space Agency [Spec. Publ.] ESA SP, ESA SP-142, Mater. Sci. Space, Proc. Eur. Symp., 3rd, 17–23 (1979).

202. R. R. Galazka, T. Warminski, J. Bak, J. Auleytner, T. Dietl, A. S. Okhotin, R. P. Borovikova, and I. A. Zubritskii, *J. Crystal Growth*, **53**(2), 397–408 (1981).

203. M. H. Johnston, C. S. Griner, R. A. Parr, and S. J. Robertson, *J. Crystal Growth*, **50**, 831–838 (1980).

204. A. S. Yue, C. W. Yeh, and B. K. Yue, NASA (Spec. Publ.) SP-412, Apollo-Soyuz Test Proj., Summ. Sci. Rep. Vol. 1, N78-17088, 491–500 (Eng.).

205. T. Eyer, R. Nitsche, and H. Zimmerman, *J. Crystal Growth*, **47**, 219–229 (1979).

Chapter **V**

ZONE MELTING

Zone melting is a generic term introduced by Pfann (1) as a name for a family of crystallization techniques in which a molten zone is passed through a solid charge of material. The zone volume is usually much smaller than the total volume of the charge. The most important (and the most widely used) of these techniques is zone refining, in which the goal of the work is purification of the charge. The term "refining" derives from the early application of the method to metallurgy; chemists tend to use the term "zone purification" synonymously. Naturally, a technique that purifies a sample without destroying the impurities must, in turn, concentrate the latter. Experimentally, the procedure may be carried out so as to maximize the yield of pure product or the concentration of impurities. The technique can be carried out continuously or in batch mode.

If the intent of the work is controlled, homogeneous distribution of an impurity, related procedures (known collectively as zone leveling) may be used.

Although neither zone melting nor zone leveling necessarily generates a single crystal, both may be used to do so.

A family of zone-refining techniques has evolved in which a "foreign" substance is added to the sample whose purification is desired. Depending on the phase relationships between the sample and this added component, the techniques are called traveling-solvent method (TSM), eutectic zone refining (EZR), and zone-melting chromatography (ZMC). Yet another variant makes use of the fact that a molten zone of added impurity can be caused to move through an ingot by imposing a stationary temperature gradient over the length of the ingot; this procedure is known, appropriately, as temperature-gradient zone melting (TGZM). These "added-component techniques" (ACT) are discussed in Chapter 7.

In Chapter 1, we describe the origins of zone melting and introduce the concept of the distribution coefficient, which is used extensively in our discussion of directional crystallization in Chapter 3. Three zoning situations can be treated more or less readily by mathematical procedures:

1. A single zone pass
2. An infinitely large number of zone passes
3. Intermediate cases

All of the early theoretical treatments are based on several simplifying assumptions:

- Distribution coefficient independent of concentration
- Constant volume of the molten zone
- Uniform concentration profile along the charge at the start (except for the starting-charge-only case)
- Negligible diffusion in the solid
- No density change accompanies melting

There has been considerable examination of the validity of these assumptions and the results of such studies are discussed later in this chapter.

1 THEORY

1.1 Single-Pass Distribution

Interest in single-pass zone melting is primarily theoretical because useful purification does not normally result from one pass. In fact, a single directional crystallization, taking about the same length of time, yields product of higher purity. However, it is useful to know the distribution after one pass in order to compare theory with experiment and to provide a basis for more accurate calculation of multipass distributions. Single-pass distribution can start with a homogeneous ingot or with all of the impurity in the initial zone. Both cases have been treated mathematically.

1.1.1   Homogeneous Ingot

EUTECTIC SYSTEM

If a binary mixture of the simple eutectic type (see Figure 2.3) is quenched, it gives a macroscopically homogeneous ingot. Consider an ingot of unit cross section and length L and let a molten zone be formed from a length l of solid (as in Figure 1.3a). In this system,

- N_A is the number of moles of substance A in the zone.
- C_0 is the initial uniform concentration.
- C_l is the solute concentration in the molten zone.
- C_s is the instantaneous solute concentration in the freezing solid at position x.

If the zone moves through a distance dx a material balance for solute A requires that

$$dN_A = C_0 \, dx - C_s \, dx \qquad (5.1)$$

Here $C_0 \, dx$ is the number of moles of A entering the zone at its leading edge and $C_s \, dx$ is the number of moles of A deposited in the solid. At the start of the pass, the solid consists of pure A (i.e., $C_s = 0$) and the zone is enriched in solute; the enrichment can only proceed until the eutectic concentration is reached. If this concentration is reached before $x = (L - l)$, that is, before the end of the ingot, then the zone concentration must remain below the eutectic concentration, since the zone is being fed with solid of hypoeutectic concentration, namely C_0.

This circumstance is described mathematically by substituting $C_l = N_A/l$ into Equation 5.1, giving

$$dC_l = \frac{(C_0 - C_s) \, dx}{l} \qquad (5.2)$$

since C_l becomes constant, $dC_l = 0$ and $C_s = C_0$; that is, after a transition region of pure A, the zone deposits solid of solute content C_0 until the zone reaches $x = (L - l)$. Here the zone attains the eutectic composition since no more solid of concentration C_0 is entering, and solid of eutectic composition freezes. Thus the concentration profile after one zone pass shows two discontinuities, as seen in Figure 5.1.

The eutectic concentration of an unknown binary system can be simply estimated from the geometry of the zoned ingot. Integration of Equation 5.2

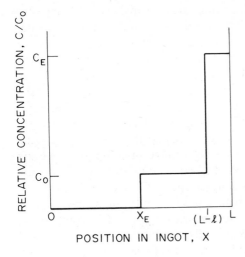

RELATIVE CONCENTRATION, C/C_0

POSITION IN INGOT, X

Fig. 5.1 Concentration profile after a single zone pass through a simple eutectic ingot with complete mixing in the liquid zone (schematic).

from the head of the ingot $(x = 0, C = C_0)$ to position $x(x = x, C = C_l)$ gives

$$\int_{C_0}^{C_l} dC_l = \int_0^x \frac{C_0 - C_s}{l} dx, \quad \text{yielding} \quad C_l - C_0 = \frac{(C_0 - C_s)x}{l} \quad (5.3)$$

The solute concentration in the molten zone reaches C_E at a point designated X_E, where C_s is still zero. Hence substituting $C_l = C_E$ and $X = X_E$ in Equation 5.3 gives

$$C_E - C_0 = \frac{C_0}{l} X_E$$

or

$$X_E = \frac{C_E - C_0}{C_0} l \quad (5.4)$$

From a knowledge of the starting concentration and the position of the transition from pure A to C_0, the concentration C_E may be computed. If C_0 is very low, C_l may not reach C_E until the zone has reached position $(L - l)$ in the ingot. In this case, the final ingot will contain only a zone of pure A and a terminal eutectic segment.

SOLID-SOLUTION SYSTEM

Let us consider next a cylindrical ingot of a binary solid solution of length L and unit cross section. If the heater is positioned at the top of the ingot, a length of solid l will yield a molten zone in which the solute concentration is C_0. As the heater moves through the ingot, the binary mixture is melted at the leading edge of the zone, taking into the zone additional solid of composition C_0. Simultaneously liquid at the trailing edge of the zone is crystallizing and depositing solid of concentration $k_{eff} \cdot C_0$ at position $x = 0$. If $k_{eff} < 1$, the solid will be purer than the starting material and consequently the liquid in the zone becomes richer in solute. Conversely, if $k_{eff} > 1$, the solid deposited will be richer in solute and the zone will be depleted. The former case is more common. In either case, enrichment or depletion continues until the concentration at the freezing interface reaches C_0/k_{eff}, after which the solid leaving the zone has the same concentration as that entering and no further change in composition occurs until the last zone length of the ingot solidifies. Here, to maintain mass balance, the impurity removed from the starting zone is deposited, for $k_{eff} < 1$. Figure 5.2 shows a plot of relative concentration against position in the ingot after passage of a single well mixed zone (a), for a system with $k_{eff} < 1$. For comparison, concentration profiles are shown for directional crystallization both with (b) and without (c) mixing of the liquid. It is evident that directional crystallization with mixing is considerably more effective than one zone pass. In the absence of mixing, both zone melting and directional crystallization give similar distributions with a sharp initial rise to C_0. The advantage of zone melting derives from the ready possibility of changing the distribution by multiple zone passes.

Pfann has pointed out (1) that the initial transient and the plateau can be described by a single equation

$$\frac{C_s}{C_0} = 1 - (1 - k_{eff}) \exp\left(-\frac{k_{eff} x}{l}\right) \tag{5.5}$$

The terminal transient, on the other hand, is described by the directional crystallization equation, Equation 3.2. Calculated distributions are usually presented as plots of relative concentration against distance (expressed in units

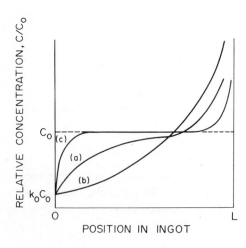

Fig. 5.2 Concentration profile after a single zone pass through an ingot of a solid solution with $k_{eff} < 1$ (curve a). For comparison profiles are shown for directional crystallization with liquid mixing (b) and without liquid mixing (c).

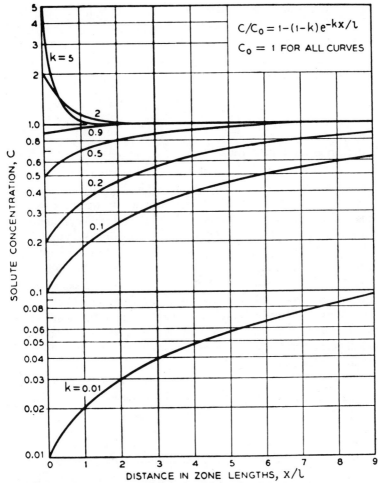

Fig. 5.3 Solute distribution after single-pass zone melting; C is the solute concentration in the solid as a function of distance from the head of the ingot, with distribution coefficient, k, as parameter, (after Pfann (1)).

of zone length, x/l), excluding the last zone length. Figure 5.3 shows such plots for several values of k_{eff}.

If the concentration profile is measured in an ingot after a single zone pass, a plot of the results gives an exponential curve approaching C_0 (see Figure 5.3). Schildknecht has pointed out that k_{eff} can be calculated from each point of the curve, using Eq. 5.5. Hence, this method affords a check on the constancy of k_{eff} with increasing concentration of solute.

1.1.2 Starting-Charge-Only

Another zoning configuration for which mathematical analysis is desired is that in which a quantity of solute is distributed from a small starting zone into an ingot of pure solvent. In this case, the single-pass profile is given by Equation

5.6, where C_s is the local concentration of solute at point x and C_0 is the concentration of solute in the first zone length $(0 < x < l)$. Here, if k_{eff} is independent of concentration, the logarithm of solute concentration varies linearly with x. Hence, a logarithmic plot of the solute profile gives a value of k_{eff} both from its slope and its intercept.

$$C_s = k_{eff}C_0e^{-k_{eff}x/l} \qquad (5.6)$$

1.2   Multipass Distribution

Early in our treatment of crystallization we referred to the single most important aspect of the zone-melting technique: its capacity for iteration. Even before the first zone passage is complete, another can start. Successive zones can follow one another at a spacing limited only by the thermal conductivity of the intervening solid (see Figure 1.3b). After a single directional crystallization, it is necessary to separate pure from impure fractions before attempting further purification; in zone melting, on the other hand, no manipulation is required. It is easy to see qualitatively what the effect of repeated zone passages will be: given an inhomogeneous ingot with a concentration profile generated by a single zone pass, a second pass will lower the solute concentration at the "head" of the ingot and raise it at the "tail" $(k < 1)$. The intermediate region of constant composition will become progressively smaller as impurity is removed from the head and deposited in the tail.

Another advantage of zone melting derives from the small volume of the molten zone. Each solidifying interface requires the presence of only a small amount of melt; hence zone melting minimizes exposure of the sample to high temperature. This is important if a material is only marginally stable at its melting point, or if it is somewhat reactive with its container.

The goal of theory is to relate solute distribution to the phase diagram and to give the experimenter an *a priori* idea of what the concentration profile will be in ingot after an arbitrary number of passes. Two cases can be considered:

- Ultimate distribution after a large number of passes
- Intermediate cases, in which the number of passes is greater than 1 and less than the ultimate

1.2.1   Ultimate Distribution

After a very large number of zone passes, a steady state is reached because the forward flux of solute which results from crystallization is exactly balanced by diffusive backflow resulting from the increase in concentration gradient at the tail of the ingot. Ultimate distribution is in fact the result sought in most zone-refining experiments, since it represents the best separation of impurity that can be achieved for a given chemical system under the experimental conditions imposed. This steady state or ultimate distribution has been calculated by Pfann

(2) as

$$C(x) = Ae^{Bx} \tag{5.7}$$

where $C(x)$ is the solute concentration at distance x from the head of the ingot and A and B are constants given by

$$k_{eff} = \frac{Bl}{e^{Bl} - 1} \tag{5.8}$$

$$A = \frac{C_0 BL}{e^{BL} - 1} \tag{5.9}$$

where l is the length of the molten zone, L is the total length of the ingot, and k_{eff} is the effective distribution coefficient. Figure 5.4 shows logarithmic concentration profiles for ultimate distributions in systems with various values of k_{eff}, with $L/l = 10$. Equation 5.7 does not hold in the last zone to freeze, as this is described by the directional crystallization distribution. More exact solutions have been derived which take this end effect into account (3, 4, 5). These treatments, however, differ little from the expression derived by Pfann.

Both Eqs. 5.8 and 5.9 have the form of the Einstein function, $\phi = W/(e^W - 1)$. Although it cannot be solved explicitly for W, there are published tabulations of its argument (6); of course it is a trivial exercise to find ϕ for an assumed value of

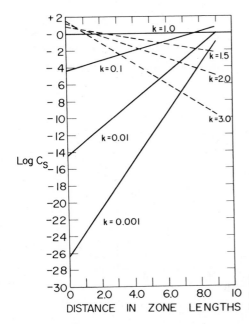

Fig. 5.4 Concentration profiles at ultimate distribution, with effective distribution coefficient as parameter.

W. Using these, it is possible to compute the constant B from a known value of k_{eff}; that is, if k_{eff} is equivalent to ϕ, then $W = Bl$. Figure 5.5 gives values of B for various values of k_{eff} with $l = 1$. Using this value of B in Eq. 5.9 one calculates A. Even in the absence of tabulated values of the Einstein function, one can get a rough idea of the value of B for k_{eff} in the range $0.75 < k_{eff} < 1.25$, since it has been shown that $B \simeq 2(1 - k_{eff})$ (7). Equation 5.7, in logarithmic form, reads

$$\ln C(x) = \ln A + Bx \qquad (5.10)$$

At the head of the ingot (i.e., where $x = 0$);

$$\ln C(0) = \ln A \qquad (5.11)$$

From Eqs. 5.10 and 5.11 it is clear that the concentration at the head of the ingot, at ultimate distribution, is given by

$$\ln A = \frac{L}{l} \ln k_{eff} \qquad (5.12)$$

Thus, knowing only k_{eff} and the geometry of an experiment, it is possible to

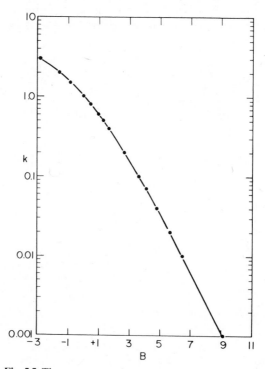

Fig. 5.5 The constant B of Eqs. 5.7–5.9, as a function of k.

calculate the highest purity attainable. Naturally this is a rough approximation because it is based on the assumption of constant k_{eff} and neglects the linear terms in Equations 5.8 and 5.9. Table 5.1 shows how markedly the ultimate distribution depends on L/l and hence how important it is to achieve narrow zones. On the other hand, it should be borne in mind that the number of passes required to reach ultimate distribution increases with L/l.

Braun (8) has shown that for $0.1 < k_{eff} < 0.3$, ultimate distribution is attained after about $2[(L/l) + 1]$ passes. For systems in which k_{eff} is near unity, ultimate distribution is reached after about $4(L/l)$ passes.

Pfann has introduced the concept of number of theoretical stages, s, for a system at ultimate distribution such that

$$k_{eff}^s = \frac{C(0)}{C_0} \tag{5.13}$$

where $C(0)$, the concentration at the head of the ingot at ultimate distribution, is equal to A from Eq. 5.9. He found empirically that the number of stages s required to attain the steady state is simply related to the geometry of the system by

$$s = \frac{fL}{l} \tag{5.14}$$

where f is approximately constant, having a value between 1 and 2; that is, the number of theoretical stages is proportional to the number of zone lengths in the ingot. This relationship makes it clear that maximum purification requires that the ratio L/l be as high as may practically be attained. However, narrow zones are less effective carriers of solute than wide ones, and large L/l implies longer purification time.

Table 5.1. Concentration A at the Head of a Zoned Ingot After Attainment of Ultimate Distribution for Various Geometries

	L/l		
k_{eff}	10	20	50
0.1	10^{-10}	10^{-20}	10^{-50}
0.2	10^{-7}	10^{-14}	10^{-35}
0.5	10^{-3}	10^{-6}	10^{-15}
0.9	0.35	0.12	5×10^{-3}
1.2	6.2	38	9×10^{-3}
1.5	57.7	3.3×10^3	10^9
2.0	10^3	10^6	10^{15}

1.2.2   Intermediate Distribution

Reiss (9) and Lord (10) derived a differential equation that relates the solute concentration $C_{s,n}(x)$ in the solid formed from the molten zone at coordinate x during the nth pass to the concentration during the $(n - 1)$th pass:

$$C_{s,n}(x) = \frac{k_{\text{eff}}}{l} \left[\int_0^{x+l} C_{s,n-1}(x) \, dx - \int_0^x C_{s,n}(x) \, dx \right] \qquad (5.15)$$

The zone length changes as the zone passes over the end of the ingot, and Eq. 5.15 is modified to account for this:

$$C_{s,n}(x) = \frac{k_{\text{eff}}}{L - x} \left[\int_0^L C_{s,n-1}(x) \, dx - \int_0^x C_{s,n}(x) \, dx \right] \qquad (5.16)$$

Early solutions of Equation 5.15 (8, 9, 10, 11) involved the assumption of an infinite or semi-infinite ingot; that is, they did not account for back reflection of solute from the enriched ingot end. Moreover they were mathematically cumbersome. Later Helfand and Kornegay (12) derived a very general equation relating concentration and position in a zone-melted ingot after an arbitrary number of zone passes. When n is small this treatment is equivalent to the earlier one of Braun and Marshall (11) while for large n, it gives the ultimate distribution of Velicky (5).

Hamming proposed a procedure (1) for calculation of concentration profiles by apportioning the ingot into cells of arbitrarily selected length and assigning some number of cells (e.g., 5 to 10) to each zone, with the zone moving in one-cell increments. Burris, Stockman, and Dillon (13) used Hamming's method to compute distributions in an ingot of unit length, comprising N increments of length Δx. They formulated the problem as a differential difference equation; Laplace transforms were used to handle the positional variable, and generating functions were used for the n index. Their treatment assumes constant zone length and k_{eff}, negligible diffusion in the solid, and uniform composition in the liquid. They solved Equations 5.15 and 5.16 numerically by use of the trapezoidal rule, which gave the expressions

$$C_{s,n}^0 = \frac{k_{\text{eff}}}{b} \sum_{j=1}^b \frac{1}{2} (C_{s,n-1}^{j-1} + C_{s,n-1}) \qquad (5.17)$$

and

$$C_{s,n}^i = \frac{2k_{\text{eff}}}{2b + k_{\text{eff}}} (S_{n-1}^{i+b} - S_n^{*i}) \qquad (5.18)$$

where i is the number of increments, counting from the head of the ingot, $\ell = b\Delta x$, and zone position is given by $x = i\Delta x$.

At $x = 0$ and in the region $0 \leqslant x \leqslant (1 - l)$, respectively, where

$$S_n^i = \sum_{j=1}^{i} \frac{1}{2} (C_{s,n}^{j-1} + C_{s,n}^j) \tag{5.19}$$

$$S_n^{*i} = S_n^i - \tfrac{1}{2} C_{s,n}^i \tag{5.20}$$

and

$$S_n^i = S_n^{i-1} + \tfrac{1}{2} C_{s,n}^{i-1} + \tfrac{1}{2} C_{s,n}^i \tag{5.21}$$

$$S_n^{*i} = S_n^{*i-1} + C_n^{i-1} \tag{5.22}$$

In the region $(1 - l) \leqslant x < 1$, the profile is given by

$$C_{s,n}^i = \frac{2k_{\text{eff}}}{2(N - i) + k_{\text{eff}}} (S_{n-1}^N - S_n^{*i}) \tag{5.23}$$

Equations 5.17, 5.18, and 5.23 afford the concentration profiles after various numbers of passes. These curves are useful, especially when k_{eff} is close to unity. This is because a single zone pass or a directional crystallization in such a system gives a barely detectable concentration gradient. Application of several passes will then give a concentration profile which can be compared with the calculated ones for various values of k_{eff}, assuming that the experimental geometry was close to that chosen in the calculation.

It is useful to know at what point in an ingot, after zoning to ultimate distribution, the concentration reaches C_0. This point defines the maximum yield of product that is purer than the starting material. Figure 5.6 shows this crossover point as a function of k_{eff} for various values of L/l.

Herington (7) has proposed several "rules of thumb" as guides to judging the completeness of zone refining. One of these relates $C(0)/C_0$, the relative concentration at the head of the ingot, to number of passes expressed in terms of the number of zone lengths contained in the ingot. Table 5.2 shows these concentrations for various numbers of passes. A second observation is that after $n = 4L/l$, additional passes become less effective. A good general procedure then, is to pass $(4L/l)$ zones through the ingot and to collect as product the first half of the ingot. An additional caveat, based on experience, is that the first zone length is often less pure than theory predicts. This deviation may be the result of the presence of bubbles, inclusion of sublimed material, and supercooling.

Soon after the basic treatments of solute distribution appeared, attempts were made to achieve less restrictive descriptions of solute distribution. Nelson et al. (14) and Brandt-Petrik et al. (15) derived and applied computer routines which permit calculation of concentration profiles in systems with variable distribution coefficients, for ingots with arbitrary initial solute distribution. Frenkel and

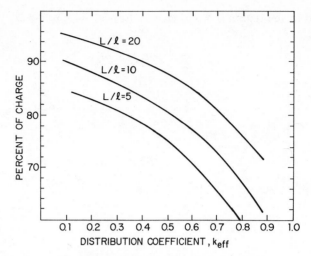

Fig. 5.6 Percentage of starting charge that is purer than the starting material, at ultimate distribution (as a function of k_{eff}) with L/l as parameter.

Table 5.2. Relative Solute Concentration at the Head of an Ingot After Various Numbers of Zone Passes

Number of Passes, n	Relative Solute Concentration at Head of Ingot $C(0)/C_0{}^a$
$\frac{1}{2}(L/l)$	log $A/4$
L/l	log $A/2$
$2(L/l)$	3 log $A/4$
∞	log A

$^a A$ is defined by Eq. 5.9.

Yaroshevskii provided a theory which gave concentration profiles after zoning with zones of variable width (16).

1.3 Zone Leveling

While most zone-melting operations are carried out with the intent of segregating impurities, this is not always the case. If, for example, it is desired to fabricate electrical devices from segments of a zoned ingot, it is necessary to assure that the impurity concentration will be as uniform as possible, in order to provide constant properties in the resulting devices. Another circumstance in which uniformity of concentration is desired is that arising when a single crystal is sought for some physical characterization. Pfann introduced a family of zone-melting techniques which do, in fact, provide homogeneous ingots and which he named "zone leveling."

1.3.1   Single-Pass Leveling

A starting ingot of nominally uniform concentration will show inhomogeneity at the microscopic level. As is pointed out in Section 1, passage of a single zone through such an ingot, containing solute at concentration C_0, yields a solute-depleted region at the head of the ingot, a region of nearly constant composition C_0, and a terminal, solute-enriched transient. If k_{eff} is sufficiently close to 1, the central region can be quite uniform and comprise a large fraction of the ingot. Even for $k = 0.5$, with $L/l = 10$, nearly three-quarters of the ingot varies by only $\pm 10\%$ from the average concentration.

At the other extreme, when k_{eff} is very small (e.g., $< 10^{-2}$), a uniform ingot results when solute is introduced into a pure ingot by a single pass. In practice the solute is dispersed in the first zone length of the starting charge, and this solute-rich zone is then moved through the ingot. The procedure is often called the starting-charge-only (SCO) method. It is discussed in detail in connection with the determination of distribution coefficients in Chapter 10 and in Section 2.

Jackson and Pfann have described a general method of producing a linear gradient of solute concentration in the SCO method by changing the volume of the molten zone (17). Zone leveling is a special case in which the length l of a molten zone of constant cross section is given by

$$l = l_0 - kx$$

The maximum value of x, for which $l = 0$, is l_0/k and the maximum value of k is unity.

Wang has derived a more general formula for the distribution of an impurity in an ingot from the passage of one molten zone. From this, the zone-leveling expression can be derived using either constant zone length with variable cross-sectional area, or constant area with variable zone length (18).

1.3.2   Multipass Leveling

If $|1 - k|$ is large, a single pass will not provide a useful length of constant composition. However, if the direction of zoning is reversed, then the solute originally carried to one end will be transported toward the starting point of the ingot, and after a sufficient number of passes, homogeneity ensues. Pfann has given an equation relating initial average concentration C_0 (not necessarily uniform) and final zone-leveled concentration:

$$\frac{C_f}{C_0} = \frac{1}{1 + \dfrac{l}{L}\left(\dfrac{1}{k_{eff}} - 1\right)} \tag{5.24}$$

where l is zone length, L is ingot length, and k_{eff} is the effective distribution coefficient (19). Naturally, this equation does not apply to the final zone length, where directional crystallization takes place, yielding solid of concentration C_f/k_{eff}.

1.4 Reconsideration of the Assumptions

Early in this chapter we mention that all the theories of solute redistribution in zone melting are underlain by several assumptions. These, not surprisingly, turn out to be only partly consistent with the behavior of real systems. Most of the "corrective" studies designed to reconcile theory with experimental fact have been experimental in nature, and will be treated in later chapters on technique. Nevertheless, we will describe here some attempts to amend theory so as to make it less dependent on arbitrary assumptions.

1.4.1 Negligible Diffusion in the Solid

Mathematical modeling of zone processes has assumed that solid-state diffusion is negligible. While this is generally the case, it is not universally correct. At this point it is useful to recall that diffusion can occur by several mechanisms. Intrinsic bulk diffusion can take place by vacancy or interstitial mechanisms. In imperfect crystals, extrinsic diffusion can occur along one- or two-dimensional defects, such as dislocations and grain boundaries. In most crystals, intrinsic diffusion is 10^4–10^6 times slower than liquid-state diffusion (10^{-15}–$10^{-16}\,m^2\,s^{-1}$ vs 10^{-8}–$10^{-9}\,m^2\,s^{-1}$). The rate of migration along defects, on the other hand, can be substantial. If the density of defects is high, then appreciable diffusive transport can take place through the imperfect solid (20, 21).

Intrinsic diffusion processes in most solids are much slower than liquid diffusion. There is, however, an important exception, namely "rotator phase" crystals. At or near their melting points, these have diffusivities of 10^{-11}–$10^{-12}\,m^2\,s^{-1}$, which put them about halfway between liquids and intrinsic solids (21).

Fischer (22) amended the treatment of Burris, Stockman, and Dillon (13) to include a term representing solid-state diffusion, giving rise to a time-dependent concentration profile. In an experimental study of solute transport in anthracene-doped carbazole, he measured concentration profiles which were in accord with a diffusivity of $3 \times 10^{-9}\,m^2\,s^{-1}$ at the melting point. This value is high enough to produce a substantial deviation of the concentration profile from the theoretical one described by Eq. 3.2, which assumes no solid-state diffusion. Fischer concluded that for diffusivities $< 10^{-12}\,m^2\,s^{-1}$ the influence of back diffusion becomes negligible.

1.4.2 Constant Distribution Coefficient

By far the weakest crutch of conventional solute-distribution theory is the assumption of constant distribution coefficient. Few phase diagrams show solidus and liquidus slopes which truly yield constant k_0. If one considers the effects of solute accretion at a solidification interface, it is clear that k_{eff} is even less likely to be invariant with concentration.

One of the earliest attempts to consider variation of k is due to Matz (23); the basis of his approach is set out in Chapter 3. Using the same notation, a material balance for solidification of a differential thickness dx by advance of a molten zone through an ingot yields

$$dN = -X_{A,s}\,dx + C_0\,dx \qquad (5.25)$$

or

$$l\,dX_{A,l} = (C_0 - X_{A,s})\,dx \qquad (5.26)$$

Relating $X_{A,l}$ and $X_{A,s}$ by Equation 3.5 and separating variables, one obtains

$$d\frac{\dfrac{\phi X_{A,S}}{1 + (\phi - 1)X_{A,s}}}{C_0 - X_{A,s}} = \frac{dx}{l} \qquad (5.27)$$

with the boundary condition that $X_{A,l} = C_0$ at $x = 0$. This equation was solved for constant separation factor ϕ, and was evaluated numerically to give concentration profiles under a variety of operating conditions. When $(\phi - 1)$ $C_0 \ll 1$, the solution of Equation 5.27 becomes Equation 5.5. Thus Eq. 5.5 is a special case in Matz's general treatment. The critical distinction between the two methods is that in Pfann's treatment the concentration profile is dependent on k but independent of C_0, while in Matz's treatment, the profile depends on both ϕ and C_0, as is seen in Figures 5.7a and 5.7b. In turn, it must be recognized that ϕ may also be concentration dependent, and the functional relationship was assumed to be exponential. If the exponential expansion is truncated after the

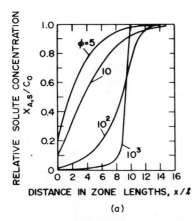

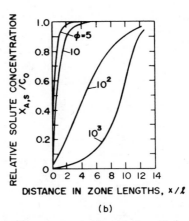

(a)   (b)

Fig. 5.7 Concentration profile along an ingot after single-pass zone melting (a) $C_0 = 10^{-1}$ and (b) $C_0 = 10^{-2}$ with separation factors $\varphi = 5, 10, 10^2$, and 10^3.

quadratic term,

$$\phi = \phi^0(1 - aX_{A,s} + bX_{A,s}^2) \tag{5.28}$$

where a and b are constants which can be obtained by curve fitting to an equilibrium phase diagram. Concentration profiles were computed numerically using such expressions for variable ϕ (23).

While the foregoing treatment faced the incorrectness of assuming constant k, it is based explicitly on the maintenance of solid/liquid equilibrium. Singh and Mathur (24) have considered mass transfer in zone refining under conditions approaching still more closely to experimental ones, namely variable k and no liquid mixing. They used a Laplace transformation to derive a relationship in which the distribution of solute between solid and liquid is assumed to be of the form

$$C_s = k_1 C_l + k_2 C_l^2 \tag{5.29}$$

Practical application of the derived relationship was said to be unlikely because of a dearth of reliable phase diagrams and diffusivity data for organic systems (24).

1.5 Matter Transport

Most solids show a change of density upon melting. Pfann observed that this fact results in the movement of an ingot during zone melting (25) and was able to describe the movement quantitatively. Consider an ingot of height h_0 in a horizontal boat of uniform rectangular cross section (Figure 5.8). Melting a length l of the ingot yields a liquid zone whose height $h = \alpha h_0$, where α is the ratio of solid density to liquid density. In most cases, α is greater than unity. As the zone advances from the end of the ingot, solid forms at height h; assuming that the ingot has unit width, a section dx units long has volume $h\,dx$. In the same advance, the heater melts a volume $(h_0\,dx)$ of solid. Since α is greater than unity, the zone loses height as it travels. A material balance during advance through distance dx leads to a change in solid volume given by $(h_0 - h)\,dx$ and a

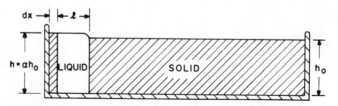

Fig. 5.8 Matter transport as a result of density change upon melting.

corresponding change in liquid zone volume dv given by

$$dv = l\,dh = \alpha(h_0 - h)\,dx \qquad (5.30)$$

The solution of this equation is

$$\frac{h}{h_0} = 1 - (1 - \alpha)\exp\left(\frac{\alpha x}{l}\right) \qquad (5.31)$$

It applies to all of the ingot except for the last zone length, where the following relationship holds:

$$\frac{h}{h_0^1} = (1 - g)^{\alpha - 1} \qquad (5.32)$$

and h_0^1 is the height of the molten zone at the start of terminal directional crystallization.

The equations describing matter transport are thus of the same form as those for solute distribution in zone melting (Equations 5.15, 5.16) and directional crystallization (Equation 3.2).

As with solute distribution, there is an ultimate height distribution in which relative height (h/h_0) replaces relative concentration and α replaces k (see Equations 5.7, 5.8, and 5.9). For many organic compounds, α has values around 1.15. Assuming that $L/l = 20$, then (h/h_0) is approximately 6 at $x = 0$ and 0.03 at $x = 0.9$. While an actual pileup of this magnitude may not be observable, the calculation shows that matter transport can indeed be substantial.

The validity of attributing matter transport solely to the density change that accompanies melting (26) has been questioned. Consideration of surface tension and the wetting of solid and container by the melt indicates that surface tension and wetting effects can prevail over density change. For water and aqueous solutions α is less than unity; consequently one expects matter transport to occur in the same direction as zoning. In the presence of surface–active agents, the reduced surface tension led to spreading over the adjacent solid. The result of this situation was retrograde matter transport. A number of organics (biphenyl, 3,4-dimethylphenol), however, showed the expected transport.

Pfann showed that matter transport may be overcome by tilting the ingot at an angle which makes α effectively unity. This critical angle θ_c is

$$\theta_c = \tan^{-1}\frac{2h_0(1 - \alpha)}{l} \qquad (5.33)$$

If a weir is installed in a zone-melting boat, matter transport will cause a small overflow if the boat is tilted at an appropriate angle. This effect can be used as a kind of ingot cropping or as a way of introducing voids into an ingot for continuous operation.

2 EXPERIMENTAL METHODS

2.1 Heating

Heating and cooling are of central importance in zone melting. In large measure, the design of the heating/cooling system determines the efficiency of the overall zoning apparatus, because the temperature gradient at the interface is an important determinant of solute distribution and because temperature fluctuations can lead to entrapment of impure liquid at the interface.

Every imaginable means of heating, and some unimaginable ones, have been used. Without a doubt, a wide variety of heaters will continue to be used. This is so because the requirements of zone melting are so disparate. One user may want to purify large (kg) amounts of one material at minimum cost; another may want a versatile system usable with materials of differing melting points, in various quantities; yet another will be interested in very close control of temperature for studies of solute distribution in small samples. We shall describe a wide variety of heaters and heating modes. Most of the heaters will be suitable for use to about 750 K, a range adequate for the majority of chemicals. Extension to higher temperatures will be possible in many cases by minor changes in construction.

2.1.1 Resistance Heating

By far the preponderance of laboratory-scale zone melting of chemicals has been carried out in cylindrical glass tubes. While other container shapes and materials have been used and will be used, the advantages of cylindrical glass tubes are so substantial that they will continue to be widely used. For this reason, we will first discuss heaters suited to this type of container.

INDIVIDUAL HEATERS

Coils of bare resistance wire have been used in zone melting, with considerable success but with some failures as well. Pfann et al. used single turns of Nichrome wire in order to achieve narrow zones (27, 28). They succeeded in establishing zones that were as short as one-fourth of the diameter of the tubular container. Since their experiments were carried out in the horizontal mode, convective heat flow causes the zone to be much broader at the top than at the bottom. This problem was overcome by rotating the sample tube slowly (about $0.017 \, \text{rev s}^{-1}$) about its long axis; the method was thus called the "rotation-convection" technique. Bare-wire heating suffers from sensitivity to air currents and variation in room temperature. Moreover, the low mass of the heaters makes them sensitive to voltage fluctuations as well. A more serious flaw emerges when one considers the situation that arises when a bare, unsupported wire is used to heat a tube of large diameter (>2 cm) or a high-melting substrate. In either case, the heater must be operated at a high temperature in order to generate the required amount of heat. Ultimately, the wire sags or is distorted until it makes contact with the tube. It is well known that application of a hot wire to a scratch is a good method of breaking or cutting glass tubes. Hence, any imperfection in the zone-melting container, once contacted by the heater, is the source of almost certain breakage of the container. Zone-melting tubes may also break because of expansion of the contents. (This phenomenon and ways of avoiding it will be discussed later.)

Spialter and Riley (29) made annular heaters by winding resistance wire on notched rings of high-silica glass. The rings were 1.9 cm to 3.2 cm in diameter and 1.2 cm to 1.9 cm tall; the notches were 0.16 cm deep, at 0.32-cm intervals (Figure 5.9a). Nichrome wire (No. 21) was wrapped snugly around the cylinder (Figure 5.9b), covered externally with asbestos tape or a glass-fiber tape, and spliced to a power cord (Figure 5.9c). Such heaters are efficient, because heat is transferred directly to the sample tube by conduction and radiation from the bare resistance wire. Their external insulation confers stability, uniformity, and ease of control. Nevertheless, their exposed inner windings pose some hazard and other designs may be preferable.

In one such design (30), a coil of bare resistance wire is slightly recessed in an insulating housing (Figure 5.10). Other sturdy heaters can be prepared simply, to make a very close fit to the sample tube. A rectangle of plastic film is wrapped around the sample tube (or a mandrel of like diameter) to form a sleeve;

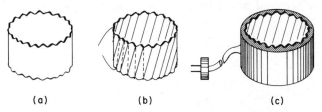

(a)   (b)   (c)

Fig. 5.9 Annular heaters made by winding a resistance wire on a notched ring of silica glass: (a) glass former; (b) former wrapped with wire; and (c) complete insulated unit.

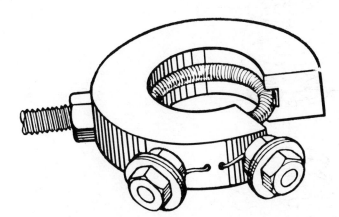

Fig. 5.10 Helical resistance heating element recessed in an insulated housing.

resistance wire is wound around the sleeve, then embedded in electrically resistive cement, and finally insulated thermally. A variety of suitable potting compounds is available (see Appendix II). The glass mandrel and the layer of film are removed, leaving a heater whose inner diameter is larger than the outer diameter of the sample tube by the thickness of the layer of film, which may be as small as 25 μm, although 75- or 125-μm films are preferable.

Rugged heaters are available in stainless-steel sheaths of 0.32-cm ($\frac{1}{8}$-in.) diameter; these can be wound in tight coils which give the advantages of bare wires, without the disadvantages (Watlow Electric Manufacturing Co.). Flat, annular heaters have been constructed (31) (Figure 5.11) by winding nickel–chromium wire (110 cm long and 0.23 cm in diameter) radially on thin, insulating formers A (2.2 cm od, 1.2 cm id, 0.2 cm thick). These were assembled in pairs with a mica spacer B. Each three-piece assembly was wrapped peripherally with glass–fabric tape and encased in insulating blocks C which were provided with a groove D for a thermocouple.

Annular heaters are available commercially, but it is hard to find them in the small diameters and heights normally needed for laboratory-scale zone melting. A very useful heater can be made by embedding two cylindrical heaters A in a metal plate which is centrally bored to pass the sample tube (Figure 5.12a). The heater bores connect with smaller, coaxial bores C, which are threaded. These serve to ease removal of the heaters when required. Appropriate heaters are

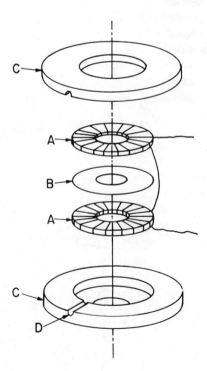

Fig. 5.11 Exploded view of an annular heater (31).

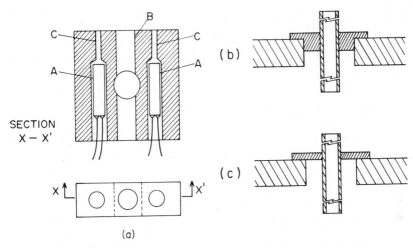

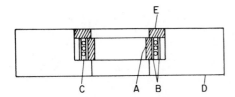

Fig. 5.12 (*a*) Plate heater for zone melting. (*b*) Close-fitting reduction adapter for plate heater. (*c*) Unconstrained adapter for plate heater.

available in diameters as small as 0.32 cm ($\frac{1}{8}$ in.). If a heater of 0.95 cm ($\frac{3}{8}$ in.) or larger diameter is to be used, it is possible to buy cartridge heaters with internal limiting thermostats (ITT Vulcan Electric, Kezar Falls, ME 04047). A bore B may be provided and fitted with glass or quartz plugs with polished ends, extending nearly to the central hole. These viewing ports make it possible to observe the solid/liquid interface without disturbing the system. The entire block may be encased in sheet insulation. Because of the high thermal conductivity of the plate, zones tend to be quite flat even without rotation of the sample tube. To accommodate samples of varying size, the heater plate may be provided with reducing adapters as shown in Figure 5.12*b*. In fixed mountings of heaters such as these, careful alignment is critical to avoid binding the sample tube. This difficulty may be avoided by using a close-fitting flat adapter which can move horizontally over the heater surface as shown in Figure 5.12*c*. In this way, standard Pyrex tubing may be used as disposable sample containers, rather than resorting to costly precision-bore tubing.

Metal rings may be used as cores of heaters as shown in Figure 5.13. Ring A is wrapped with one or two turns of insulating tape B (3M Company, Tape No. 27; see Appendix II) before application of the resistive winding C. An additional

Fig. 5.13 A metal ring as the core of an enclosed heater.

turn of tape serves to keep the winding in place. A block of insulation D protects the heater and can also be used to attach mounting hardware for assembly of a group of such heaters. An insulating ring E can be used to lock the heater in place. The insulating block D and ring E may be machined from sheet insulation (see Appendix II) but some of these materials contain asbestos and pose health hazards during machining. Moreover the machining can be costly. It is possible to avoid both problems by using soft insulation board such as K-Fac® (United States Gypsum Co.), which can be worked easily with hand tools.

Yet another way to avoid the health problems and machining costs is to embed each heater in a cast ceramic or foam which is thermally as well as electrically insulating. Foamed silicone rubbers are available with thermal conductivities as low as $0.043 \text{ W m}^{-1}\text{ K}^{-1}$ and usable to temperatures above 475 K (see Appendix II). Numbers of heaters can be made serially, to close dimensional tolerances, in inexpensive molds. Figure 5.14 shows a casting box in two stages of preparation of heaters. In Figure 5.14a, wooden box A has been filled with castable ceramic paste B, around former C. After the paste has hardened, heater D is emplaced and additional ceramic paste is added around a second, cylindrical former E. Box A and formers C and E should be coated with a silicone-containing grease to facilitate disassembly.

Vacuum-deposited films of indium antimonide, 9-μm thick, have been zone melted on glass substrates (32). The glass was pretreated with pure indium to ensure adhesion of the InSb and to compensate for In depletion during later oxidation. Regions as large as 2.5 cm $\times$ 2.5 cm were processed by passage over a single Nichrome wire (0.03-cm diameter) at a distance of 0.03 cm (wire-to-substrate). A motor-driven micrometer moved the substrate at 0.02 to 20 μm s^{-1}. A helium atmosphere was used to promote heat transfer.

A microscale zone-refining assembly has been described, making use of a single resistance wire to form a molten zone in a planar sample sandwiched between glass slides. The device, shown in Figure 5.15, was designed for use as a microscope stage, for detailed observations of the solidification of polymers (33). In Figure 5.15, A is a machined copper block having a central, rectangular aperture B through which the optical path of the microscope is established.

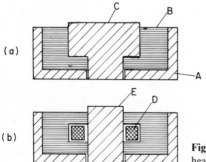

Fig. 5.14 Use of castable ceramic in fabrication of heaters for zone melting.

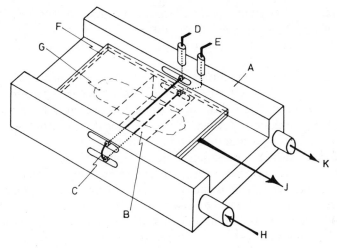

Fig. 5.15 Small-scale zone refiner for use on a microscope stage.

Bores H and K are connected by transverse bores on both sides of the aperture to allow flowing coolant to surround the aperture. A single loop of No. 26 Nichrome wire, C, provides heating; the wire is insulated from block A by ceramic spacers. Temperatures up to 575 K were attainable with local gradients as high as 400 K cm^{-1}. Sample G is sandwiched between heat-resisting slides F. The sandwich is moved horizontally by a cable J, attached to a syringe pump which gives speeds between 0.13 cm s^{-1} and 7.5×10^{-6} cm s^{-1}.

HEATER BATTERIES

Single heaters may be used in the measurement of distribution coefficients by single-pass zone melting (see Chapter 10). For zone refining, it is desirable to increase efficiency by using groups of heaters, suitably spaced. Many methods have been used to build such heater batteries.

Naturally, any of the heaters just described can be used in groups, by attaching them to rigid supports or inserting them in cooling blocks. When batteries of heaters are being assembled, it is necessary to assure that the individual heaters are accurately and stably mounted to allow smooth passage of the zone-melting tube. If the heaters are of uniform wattage, one of them can be fitted with a thermocouple or thermistor as a sensor for a controller, and all the heaters may be identically powered. If such a battery is operated vertically in an enclosure, a chimney effect is likely to produce higher temperatures in the upper heaters. This undesired effect may be overcome by "tapering" the wattages of the heaters so that progressively less power is applied to the upper heaters. Alternatively radiation/conduction fins may be applied to dissipate some heat externally from the upper heaters, or additional insulation may be applied to the lower ones. As zoning proceeds and impurity is deposited at the ingot end, the melting point of the charge at that end will decrease. Consequently less heat will be required to produce melting and adjustable heating is desirable. Of course each heater or pair of heaters may be provided with a controller.

Another method of mounting multiple resistance elements to assure alignment is to wrap them on an insulating tubular mandrel through which the sample tube can move. Pyrex, quartz, and Kapton tubes have been used as mandrels. Such heaters are easy to make but pose the problem that it is difficult to cool the interzone region through the additional layer of insulating mandrel. The individual heaters can be separated by lengths of glass tubing.

Heater batteries have been constructed in which a single heater (Figure 5.16a) heats a number of elements B and these, in turn, heat the sample container. The metal elements may be formed from wire or rod (Figure 5.16a) or machined from blocks (Figure 5.16b). Schildknecht and Vetter (34) used this approach to build a microrefiner in which coils consisting of five turns of copper wire were wrapped tightly on a single heater, to generate 18 zones in a 6-cm ingot. The copper wire, of 0.05-cm diameter, was brought from the heater to form two-turn coils of about 0.1-cm id, spaced about 0.3 cm apart (see Figure 5.16a). In one embodiment, the heater consisted of a tube through which hot oil circulated. Another version used an electrical cartridge heater. A cooler battery was made by wrapping identical copper-wire coils on a copper tube through which liquid coolant was circulated. Assemblies such as these have the advantages of simplicity and ruggedness; the spacing and alignment of the individual heaters are easily maintained. On the other hand, the single heat source makes it impossible to vary the temperatures of the individual elements, which is sometimes desirable. This deficiency may be overcome by applying additional external insulation to certain heaters as required.

A sturdy heater battery has been assembled from 18 annular heaters (see Figure 5.11) and 19 coolers. The heaters, connected in parallel, were controlled by using 18 thermocouples and three automatic controllers. The temperature of each heater was measured for 12 s out of every 72 s; the voltage applied to the heater was maintained until the next measurement interval, by a circuit memory system included in the controller. Heater temperature was constant within

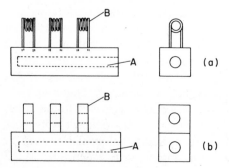

Fig. 5.16 Heater battery using a single cartridge heater, A, to heat individual heating elements, B; (a) individual heater formed from wire or rod; (b) machined block heaters.

±3 K. The sample tube was moved reciprocally, with rotation, through a distance of 10 cm (twice the heater spacing) and this pattern was said to even out differences in output of adjacent heaters. The battery was oriented at 50° from vertical, to allow gas bubbles to escape from the center of the tube to the container wall (31).

Karl has described a basically different approach to the use of resistance heating in zone melting. A stainless-steel tube A is machined to provide a stepped outer diameter (see Figure 5.17a). A low voltage is applied over the length of the tube; more heat is generated in the thin-wall segments than elsewhere. The profile may be sharpened by providing insulators B around the hot zones. Alternatively a heated tube A of constant wall thickness may be provided with insulating rings B (Figure 5.17b). If the assembly is immersed in a low-temperature thermostat, heat loss will be reduced by the insulators B, generating hot zones internally. If, on the other hand, highly conductive shunts are applied to the heated tube, colder zones will be generated. The two ideas are combined in Figure 5.17c in which insulators B alternate with conductive shunts C (35).

Many chemicals sublime at their melting points, creating problems during zone melting. These may be overcome to some extent by sealing the zone-melting tube under an inert gas, at as high a pressure as conveniently possible. A different approach is to modify the heater battery so that the top is maintained at a temperature that prevents condensation (36). A snug-fitting stopper of silicone rubber may be placed atop the ingot to impede evaporation (34).

The length of the molten zone does not usually need to be controlled very accurately in multipass zone melting. In zone leveling however, constant zone length is critical. With constant power applied to a heater, zone length may change during a pass because heat dissipation at the ends of the ingot differs from that in the middle. The problem is more serious with materials having high thermal conductivity than with those of low conductivity. According to Lastushkin et al. (37), the zone volume decreases parabolically as a zone enters a rotating horizontal ingot, and rises parabolically as the zone exits. In the case of

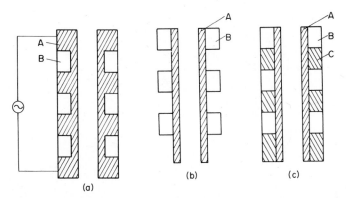

Fig. 5.17 Heater battery consisting of a profiled stainless-steel tube.

a long ingot, there may be a central region of constant zone volume. These authors devised a method of programming the heater voltage on the basis of heater position. The contact of a slide-wire potentiometer was attached to the heater carriage to generate a reference voltage which was used with a function generator to vary the voltage applied to the heater. Zone volume was held within $\pm 5\%$ of the average volume throughout the zone melting of bismuth and tin.

2.1.2   Immersed Heaters

Early in the history of zone melting of organic chemicals, Pfann proposed that alternate zones of solid and liquid might be produced by sequenced passage of heat-exchange fluids through tubes embedded in an ingot (38). Marrelli (39) has used this method for the separation of a test system, p-dichlorobenzene/p-chlorobromobenzene. He used a vertical column with a square cross section traversed by a network of heat-exchange tubes. Twenty-four such tubes were disposed through the height of the column (Figure 5.18a), each layer consisting of a serpentine with six traverses within the column (Figure 5.18b). Three independent fluid circuits were used, with control by solenoid valves. Circuit 1 fed layers 1, 4, 7, ... in series. Similarly circuits 2 and 3 fed layers 2, 5, 8, ... and layers 3, 6, 9, ..., respectively. In operation, two of the circuits were connected to heating fluid at a given time, while the third was filled with refrigerant, providing a total of eight molten zones, separated by solid zones about twice as thick. Zone movement was effected by sequenced switching of the fluid; a layer that is at one instant traversed by hot fluid and is consequently surrounded by melt, is at a later time traversed by refrigerant, bringing about solidification of the melt. Figure 5.19a shows the valving system in a schematic view of the layers in the column. Figures 5.19b and 5.19c show the layers, at two later times, to make clear the downward progression of zones. There are three groups of valves, each connected to one of the fluid circuits. Each group consists of four

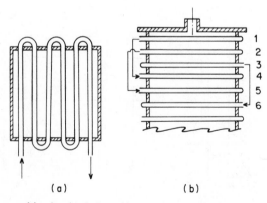

(a)   (b)

Fig. 5.18 Zone melting by circulation of heat-exchange fluids through immersed tubes.

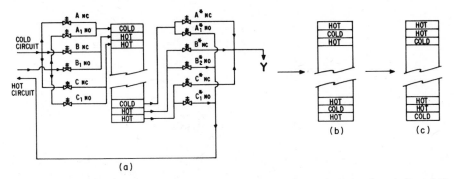

Fig. 5.19 Valving circuits for heat-exchange fluids. Cold fluid enters through valves A, B, and C (normally closed) and exits through A*, B*, and C* (normally closed). Hot fluid enters through A_1, B_1, and C_1 (normally open) and exits through A_1^*, B_1^* and C_1^* (normally open). Cold fluid is discharged at Y; (a), (b), and (c) show the sequencing used to provide zone motion.

valves, two at the inlet and two at the outlet. For example, circuit 1 is served by valves A and A_1 (inlet) and valves A* and A_1^* (outlet). Normally open (NO) and normally closed (NC) valves were used in the hot and cold circuits, respectively.

Rennolds described an extension of this method, in which finned heat-transfer elements are embedded in an ingot of rectangular cross section. Hot and cold fluids are circulated through the elements in sequence, with the timing of the hot-to-cold switching adjusted to generate an array of solid wedges extending into the liquid. The resulting corrugated solid/liquid interface provides significantly increased crystallization area in a charge of a given volume (40).

Pfann described a related method in which zones are moved through a stationary charge by sequentially energizing a series of electrical resistance wires embedded in the charge (38).

2.1.3 Augmented Heat Transfer

In discussions of the rotation-convection method of zone melting, it is said that direct contact of the liquid coolant upon the surface of the zone-melting tube between heaters resulted in improved efficiency of solidification because the usual air gap between cooler and the sample tube is eliminated (28). This analysis of the operation overlooks the low thermal conductivity of the organic material within the tube. Although the onset of solidification within the cooled zone takes place more rapidly than in the presence of an air gap, the rate of solidification diminishes steadily as solid builds up within the tube. Moreover, the rate of fusion brought about by the heaters is likewise limited by the low conductivity of the chemical sample being purified. In fact, the low thermal conductivity of organic chemicals has generally restricted the application of zone melting to cylindrical organic samples less than 3 cm in diameter. In larger tubes, convection in the molten zone leads to formation of a central "dome" of

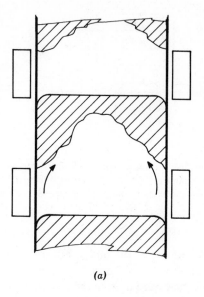

(a)

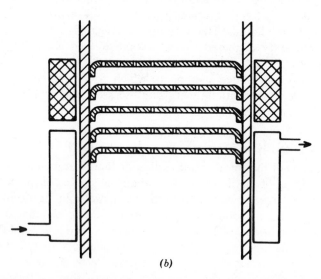

(b)

Fig. 5.20 (a) Schematic of "dome" shaped molten zone resulting from convection in material of low thermal conductivity. (b) Use of perforated metal disks to provide enhanced radial heat transfer in the charge. (c) Photograph of sharp, well-defined, narrow molten zone obtained by using heat-conducting disks (the vertical dark image is that of a central rod used to mount the disks).

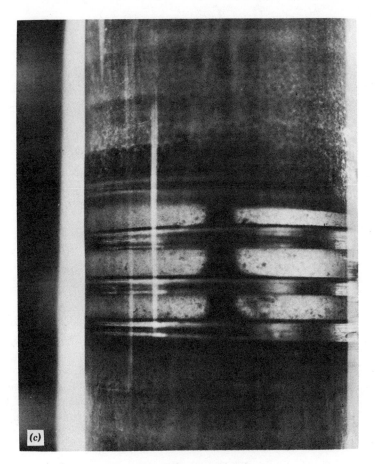

Fig. 5.20 (Continued).

liquid, which increases the minimum interzone spacing that is attainable (Figure 5.20a). A method has been proposed for overcoming this deficiency by including within the sample space a number of chemically inert, thermally conductive members such that heat conductivity in the radial sense is substantially augmented while axial conductivity remains essentially that of the chemical charge (41). Figure 5.20b shows schematically the manner in which perforated disks of high heat conductivity may be incorporated within a cylindrical sample container. Heating and cooling may be provided externally by any means, with the general result that the rate of penetration of the latent heat of fusion from the heater to the interior of the sample is increased. Likewise, the rate of withdrawal of the latent heat of fusion from the innermost portion of the sample to an external cooler is enhanced. The heat-conducting disks can also serve to

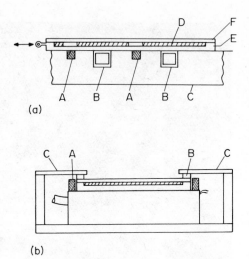

Fig. 5.21 (*a*) Cross section of a planar zone refiner for small samples. Components A–F are identified in the text. (*b*) End view of planar zone refiner. Components A–C are identified in the text.

augment matter transport and to improve mixing in the molten zone. A substantial temperature gradient exists in the vicinity of an included metal disk, so that the liquid in contact with the disk will be hotter than the bulk and in consequence a convective mixing cell is generated near the disk. This method makes it possible to zone materials with low thermal conductivity in containers of large cross section without the use of circulating heat-exchange fluids. Figure 5.20*c* shows the planar solid/liquid interface of a molten zone in a 4.5-cm Pyrex tube containing perforated conductive disks.

Most of the heaters described thus far are suited to cylindrical sample containers. With other container configurations, different heaters are useful. For example, with a helical container, a single bar or ribbon heater can generate multiple zones (see Figure 5.39). In an annular charge, a single helical heater can be used to generate a zoning effect. Such a heater can be used externally (38) or internally.

Sloan has used commercially available heaters with square cross sections to make a heater battery suitable for small samples in thin-film form (Figure 5.21*a*). Heaters A and coolers B are cemented in grooves milled into insulating board C. Sample D is contained in a flat boat E of Pyrex or Teflon®, covered by plate F. One such apparatus used five heaters and six coolers of 10-cm length, on 5-cm centers, in a board of 30-cm length. The sample area could be as large as 150 cm^2; for a thickness of 0.3 cm, this corresponds to a charge of 45 mL. However, samples were also used in narrow channels (about 0.1 cm wide × 0.1 cm deep × 5 cm long) in volumes as small as 0.05 mL. Several small samples having about the same melting point can be zoned simultaneously in separate channels. The flat boat is moved reciprocally over the heater/cooler battery between guides A (Figure 5.21*b*); good contact is maintained by leaf springs B mounted in the overhanging plates C.

2.1.4   Radiant Heating

Radiant heat from incandescent sources offers a number of advantages: it can be focused to a point or to a line image; many sources are commercially available and it provides rapid response to changes in demand. Except for an ingot in a slab configuration, it is difficult to envisage a multiple-zone battery based on radiant heating; that is, it would be difficult to achieve multiple zones in the usual laboratory configuration of a cylindrical charge in a glass container, by radiant heating. Moreover, it is required that the specimen being melted must absorb the radiation in order to be heated, or an absorbing body may be included within the ingot.

In one of the earliest applications of zone melting to organic compounds, Herington and Handley used a small projection lamp (Osram S14, 100 W) at the minor focus (5 cm) of an ellipsoidal mirror to concentrate the radiation on a specimen at the mirror's major focus (35 cm). Organic samples sensitive to the lamp's radiation were protected by a metal sleeve in which a small optical filter (Chance OX2) was provided, to transmit heat only. Temperatures up to 620 K were obtained (42). Baum used a circular array of six projection lamps to achieve higher temperatures (43) and still higher temperatures were attained by Kooy et al., who used a carbon-arc radiation furnace (44). The latter device provided high radiation density and rapid heating, but its carbon rods had a life of less than 2 h.

A zone refiner has been constructed for experiments in space (see Section 5, Chapter 4). It is based on a mirror furnace consisting of two adjoining rotary ellipsoids having a common focus. The ingot to be zoned is positioned at this focus while the radiation sources (two halogen lamps) are located at the other foci; about 95% of the total radiation is focused. The furnace, which can be operated in vacuum or with a protective gas, can reach temperatures up to 2300 K, depending on size and properties of the samples. Power consumption is very low (45).

Pfann proposed the use of radiant heating to zone a thin slab of a chemical by

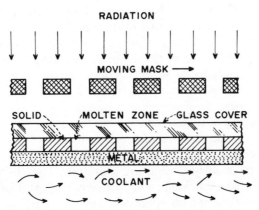

Fig. 5.22 Zone melting a thin layer of chemical by irradiation through a moving mask.

allowing the radiation to impinge upon the sample zones through a moving opaque mask (28) (Figure 5.22). Pfann has described a fair-weather refiner in which solar radiation provides zone heat and zone transport as well. The sun is imaged by a sequence of cylindrical lenses and the diurnal transit of the sun produces motion of the zones (28).

In considering radiant heating for zone-melting chemicals, it should be remembered that many compounds are sensitive to short-wavelength radiation. Pyrex containers, for example, are transparent into the near UV, a region in which many aromatic compounds absorb strongly and are subject to photo-chemical degradation. The difficulty can be turned to advantage, as in the removal of anthracene from phenanthrene by photochemical dimerization of the former, to facilitate its segregation (46).

2.1.5 Induction Heating

Induction heating may be applied directly to materials that have high electrical conductivity, such as metals and semiconductors. It can be applied indirectly to nonconducting substrates by using a conductive container or a susceptor. In the latter case, the susceptor can be an external sleeve surrounding a nonconductive sample in a nonconductive container or it can be an immersed conductive body.

Induction heating has not been widely used in the purification of organic and inorganic chemicals by zone melting. This may be due as much to the prejudices and limited experience of organic and inorganic chemists, as it is to any deficiencies inherent in the procedure, although there are a number of disadvantages that weigh against the manifest advantages. To begin with, induction heating for the zone melting of chemicals offers the efficiency associated with any mode of internal heating, that is, the heat required to melt the substrate is generated internally. It is also true that in the zone melting of electrically conducting materials such as metals, the efficiency of zone melting is further enhanced by eddy currents, which provide mixing in the molten zone. This advantage, however, is not present when organic chemicals are zone melted by induction heating. On the other hand, certain inorganic materials such as alkali salts become sufficiently conducting in the liquid state to be induction heated directly and here the additional advantage of eddy current mixing would in fact be present (see below). One of the major drawbacks to the use of induction heating is the relatively high cost of generators providing the necessary power output, and in addition, the critical tuning that must be attained in order to produce stable zones.

Induction heating does make it possible to achieve the extremely narrow zones which are required for multizone purification. This advantage gains increasing importance when attempts are made to process substantial amounts of material in large-diameter cylindrical containers. Thus, for example, Cremer and Kribbe (47) used induction heating with columns of 10–30 cm diameter for the purification of elemental phosphorus in quantities up to 30 kg. Magnetically supported iron susceptors were inductively heated; these were provided with many holes through which the liquid could pass (see also Section 2.5).

Plancher et al. (48) used induction heating with metal susceptors for the purification of a number of organic materials. Their experience showed that a 2.5-kW, 450-kHz generator was adequate to form 10 zones in samples of materials melting up to 713 K. In their work, a single susceptor was moved through an ingot by raising the container of sample through the load coil with the susceptor initially at the head of the ingot. This allowed the susceptor to descend from top to bottom of the ingot. The disadvantage of this procedure, of course, is that a new susceptor had to be added to the ingot for each pass.

Warren (49) chose induction heating and the float-zoning mode for purification of potassium chloride, after consideration of other methods. The mutual problems of sample contamination and corrosion eliminated methods using containers and/or immersed susceptors. KCl cannot be heated by induction at low temperatures because of its low electrical conductivity. Hence, a carbon-knife susceptor was used at the head of the ingot to initiate a liquid zone. After the zone was formed, the susceptor was withdrawn and the work coil was moved along the containerless ingot in a selected atmosphere (O_2, N_2, H_2, HCl) within a closed chamber. The containerless mode of zone melting known as float zoning has evoked a range of heating techniques particularly adapted to it. In the first float zoning, heat was provided by an incandescent cylinder of tantalum. In later work, heating has generally been by induction. Extensive theoretical analysis has defined the geometric limits for float zoning of cylindrical rods. The maximum stable zone length in a rod of radius r is given by $l_{max} = 3.46r$ for small rods and by $l_{max} = 2.84(\gamma/\rho g)^{1/2}$ for large rods, where γ is surface tension, ρ is density, and g is gravitational acceleration (50). This expression indicates that the method is best applied to substances with low density and high surface tension. The size limitation can be circumvented by using charges in shapes other than cylindrical rods: Pfann et al. (51) used plates and tubes. In the case of cylindrical rods, efficiency may be enhanced by rotating one end of the ingot to stir the zone.

Other methods of heating have been used to advantage with highly reactive materials. It is possible to introduce a molten zone into a free-standing ingot, without a container. The ends of the vertical ingot are firmly supported and a molten zone is formed near one end of the ingot. The zone is stabilized by surface tension or a combination of surface tension and electromagnetic forces, and then moved through the ingot.

Many metals have been float zoned by electron bombardment (52). The electron beam originates at a wire cathode and is focused on the sample, which serves as anode. If the operation is carried out in high vacuum, impurities may be removed by evaporation from the surface as well as by solid/liquid partitioning. Naturally the vacuum requirement precludes use of this method with volatile materials. Oxides such as alumina and other high-temperature conductors have also been float zoned by electron bombardment (53). Carbon-arc-image furnaces, like induction heaters and radiant heating, do not require vacuum operation. Laser heating has been described by Cockayne et al. and by Gasson and Cockayne (54, 55); power densities up to $50\,kW\,cm^{-2}$ were reported.

A modified floating-zone method has been described in which a filament is drawn from a glassy silicate melt into a platinum coil that heats and supports the melt. The filament cools as it emerges from the coil, but is held at a temperature that is too high for spontaneous nucleation. Crystallization was initiated by inserting a platinum-wire probe into the bottom of the molten zone, giving a stable, self-propagating crystal front (56).

A second method of zone melting without a container uses part of the charge to contain the molten zone ("cage zone refining"). The current induced in a prism of triangular cross section melts only the interior, and the solid "corners" contain the melt (57).

2.1.6   Dielectric Heating

Dielectric heating is in a sense complementary to induction heating, since the former is applicable to conductors and the latter to nonconductors. A nonconductor is placed between electrodes to which a high-frequency voltage is applied. Voltages are usually 1 to $2\,kV\,cm^{-1}$ and frequencies are 1 to $200\,MHz$. The phase lag of dipoles in the nonconductor gives rise to energy dissipation, and hence to internal heating.

Although dielectric heating offers the advantages of speed and specificity for organic and inorganic chemicals, large generators are required and costs are high.

2.1.7   Discharge Heating

High-melting nonconductors and metals may both be heated by discharge methods. An arc may be struck between an electrode (usually of carbon or tungsten) and the substrate in its container. The container is moved horizontally below the electrode, to produce zoning. Arc techniques are usually carried out in an inert-gas atmosphere. In contrast, electron beam methods require vacuum operation. Thermionic electrons from a tungsten cathode are accelerated and focused on the substrate, which functions as the anode.

2.2   Cooling

2.2.1   Heaters Immersed in Coolant

Many workers have zone melted chemicals with melting points as low as $250\,K$ by enclosing conventional zone-melting apparatus within a domestic refrigerator or freezer, appropriately modified for safety. Assemblies for carrying out zone melting within a refrigerator have been described by Plancher et al. (48).

For operation at temperatures below those attainable in domestic freezers, a zone heater may be immersed within an insulated container or Dewar vessel (Figure 5.23). A cryogenic fluid may be circulated through a coil in the vessel or

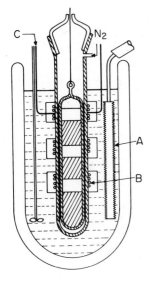

Fig. 5.23 Zone melter for low-melting materials.

maintained at constant level in it. Alternatively, low temperature may be maintained by immersing near the heater a probe A, through which a refrigerant is circulated from a closed-cycle compressor. Suitable refrigerator/probe combinations are available from many sources for operation at temperatures down to about 170 K (see Appendix II). The heaters B must be heavily insulated to minimize the heat load on the refrigerating system. Thermocouples may be installed in the heaters and in the bath to monitor temperature distribution. The physical state of the specimen in the inner tube may be monitored by fiber-optic lines C, carefully aligned on the diameters of the heaters (one pair is shown). All of the input lines may be illuminated by a single source and the output lines may be inserted in numbered holes in a board for convenience in checking the conditions of the several zones. The heater tube is closed to minimize entry of atmospheric moisture. A lateral flow of inert gas is used to assure a moisture-free environment.

H. Röck (58) purified benzene in an annular container within which a heater was lowered, while the exterior was cooled by ice water in a Dewar. A related method was used later by Ball et al. (59).

Dugacheva and Anikin (60) extended the technique of Röck to still lower temperatures, using a similar experimental configuration (Figure 5.24): an annular sample chamber A surrounded by a jacket B which could be partially evacuated (for higher-melting materials) or filled with hydrogen to promote rapid abstraction of heat by the surrounding bath of liquid nitrogen C. A single heater D was embedded in a copper ring that made a close fit within the central well; it was lowered at 2.1×10^{-3} cm s^{-1} (7.5 cm h^{-1}) from spindle E. The sample was charged into the vessel through three-way stopcock F, which allowed evacuation of the sample chamber or admission of an inert gas. It must be noted that this assembly does not provide uniform temperature along its

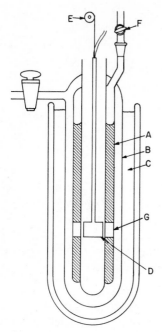

Fig. 5.24 Low-temperature zone refiner using an annular sample charge.

vertical axis. To maintain constant height of zone G, it was necessary to increase the heater current during each zone pass.

2.2.2   Circulating Coolant

A zone-melting heater battery may be modified for application to lower-melting materials by inserting individual cooling devices between the heaters. Figure 5.25 shows coolers of several types. A jet of compressed air is the simplest such device, and a commercial zone refiner is available in which this method is used (Enraf Nonius; Kratos Scientific Instruments, Westwood NJ 07675). Maire and Delmas (61) used copper coils interposed in a heater battery, to provide better definition of the cooled region than is offered by open jets of air. Often water is used in such coils, but of course other refrigerated liquids may be used. Coils of metal tubing are easy to make and use, but they make only slight contact with the enclosed zone-melting tube. This deficiency may be overcome by soldering the coil to a smooth metal sleeve whose inner diameter is slightly larger than the zone-melting tube. A cross section of such a soldered coil is shown in Figure 5.31, as parts A' and B'. The cooler (and associated heater C') are housed in insulating blocks D'. Heat-exchange tubing is available with semicircular cross section, so that maximum contact is made on the flat surface. Maeda et al. (62) used U-shaped coolers which contacted only about half the periphery of the zone-melting tube; since the tube was rotated, uniform cooling resulted.

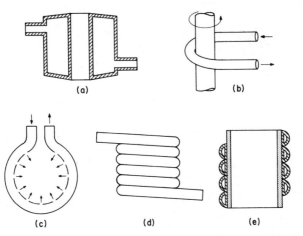

(a)   (b)

(c)   (d)   (e)

Fig. 5.25 Individual cooling units for use between heaters of a zone-melting heater battery.

Individual coolers may also be cooled by immersion in a constant-temperature medium in an insulated container. Süe et al. (63) used metal plates passing through the wall of such a container; each plate was provided with a bore through which the sample tube passed (Figure 5.26). In this construction, the coolant may consist of a cryogenic liquid such as boiling nitrogen, a eutectic bath for higher temperatures, or it may be a liquid cooled by an immersed probe from a refrigerator. If the coolers are exposed to air while they are below 273 K, atmospheric moisture will freeze on the cooling elements and ultimately impede the motion of the sample tube through the cooler. Consequently if such a device

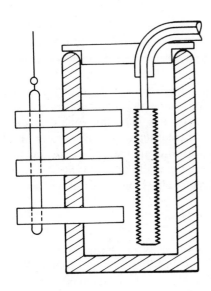

Fig. 5.26 Individual coolers immersed in a cold bath. The bath may consist of liquid nitrogen or a slurry of solid carbon dioxide and a solvent. A refrigerated probe (as shown) is more covenient.

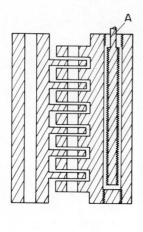

Fig. 5.27 Cooling and heating by conductive blocks.

is to be used, it must be enclosed within a sealed housing to avoid contact with atmospheric moisture, or a dry gas may be passed through the system to prevent entry of atmospheric moisture. However, this introduces the possibility of disturbing the stable thermal conditions in the heater/cooler elements. If the apparatus is enclosed in a vertical housing, the working section may be protected by a dynamic seal in which a dry gas is admitted laterally and exits upwardly through a narrow gap, thereby impeding the access of atmospheric moisture to the working apparatus below (see Figure 5.23).

2.2.3   Heat-Exchange Blocks

The individual elements described by Süe may be cooled by conduction from a common metal block as well as from a liquid bath. Figure 5.27 shows a cross section of such a unit, which may be cooled by circulation of a liquid coolant or by insertion of a refrigerated probe A. One convenient aspect of this kind of cooler is that two such units may be "nested," one within another, to serve as heater and cooler, respectively. The conducting fins may have one or more bores, and the bores may be of the same or different diameters. The heater and cooler may be enclosed in a box and the space around and between them may be filled with castable ceramic or foam insulation; the working bore is filled with a lubricated rod during the casting. Upon its removal, a continuous smooth-bore heater/cooler is available.

2.2.4   Thermoelectric Cooling

Thermoelectric elements offer advantages of easily localized cooling, easy control of cooling capacity, and simplicity in use. Schildknecht and Schnell (64) described two zone melters incorporating commercial thermoelements. One of these used 10 elements accommodating a 1-cm od zoning tube and the other used a single thermoelectric element to cool 12 metal plates through which a 0.2-cm tube traveled. In spite of the evident convenience of this mode of cooling, it appears not to have been used recently.

2.3 Zone Transport

To move molten zones through a solid charge, there must be relative motion between the heater and the charge. The motion may be brought about by simple or complex mechanisms. The choice depends on the scale of operation, the degree of reliability desired, and considerations of cost. Even when fairly large amounts of material are being purified, the rates of transport in general are sufficiently low that only small amounts of power are required and mechanical components of rather simple construction may suffice. Thus, for example, mechanical components normally thought to be "toys" may be used: Erector Sets and Meccano components have been used, and clock motors may be used to move heaters or zone-melting tubes.

A number of passes may be carried out simply by moving the charge repeatedly through a single heater (Figure 5.28a). This approach is least costly with respect to apparatus and laboratory space, but it is very time consumimg: ten passes through a 40-cm ingot at 2.78×10^{-4} cm s^{-1} (1 cm h^{-1}) require 1.44×10^6 s (2.4 weeks). If the heater moves over a stationary ingot, the

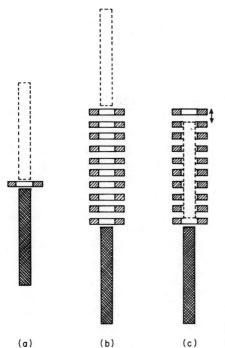

(a)   (b)   (c)

Fig. 5.28 Three modes of carrying out multiple zone passes.

experiment requires only a 40-cm space. If the ingot is moved through the heater, the space requirement increases to 80 cm. If several heaters are used, an ingot can be subjected to multiple zone passes during one excursion through the heater battery. In our 10-pass example, a single passage at 1 cm h^{-1} through 10 heaters spaced 5 cm apart requires 4.5×10^5 s (0.74 week) and 125 cm of working space (Figure 5.28b). A third mode of transport is feasible, in which the ingot is drawn upward into the heater battery and is then moved cyclically through one interzone spacing (Figure 5.28c) (see Section 2.3.3). The reciprocating motion is slow in the zoning direction and much faster in the "return" step. To avoid disturbing the zones, the return should be at least 100 times faster than the zoning speed. This reciprocal motion (65) reduces the space requirement to about twice the ingot length. In this mode, the top of the ingot always receives more passes than the bottom, since the top of the ingot must pass through the battery before the bottom reaches zoning position. The time required for the bottom of the 40-cm ingot to receive 10 passes is 3.4×10^5 s (0.57 week). The ingot cannot be inserted directly into the heater battery because (for substances that expand upon melting) an attempt to form a molten zone between solid segments almost inevitably leads to tube breakage. If the ingot is premelted, then it may be inserted in a heater battery and zones allowed to form by cooling the interzone spaces. The same result can be achieved by placing the solid ingot in the heater battery and energizing the heaters sequentially, each time the uppermost zone moves through one interzone spacing. The total time required in this case is 1.8×10^5 s (0.30 week) and less heater energy is consumed because all of the ingot is being zoned all the time. Further, the space requirement drops to 45 cm. These considerations have been evaluated in general terms by Pfann (28).

2.3.1   Single Motors

A simple clock-type motor with a metal cable or a thread wound on its shaft, may be used to raise or lower a zone-melting tube through a heater or battery of heaters. At the completion of one pass, the tube must be repositioned manually. Wilman suggested a clever mechanism for obtaining automatic reversal of travel, using only one motor (Figure 5.29) (30). Motor A was fitted with a gear B, from which several teeth were removed. For a short time during each revolution, the driving and driven gears are not enmeshed and the weight of the tube is sufficient to rotate pulley C until pin D contacts stop E. The drop distance was adjusted to equal the interzone spacing of the heater battery. If the tube is suitably counterweighted, the Wilman mechanism can be used for upward zoning. This travel mechanism may also be used for horizontal motion if the "return" is powered by a spring rather than by gravity.

A simple threaded shaft may be used to provide reciprocal motion: the tube is raised as the shaft is rotated. When the filament reaches the end of the screw thread, the tube will drop approximately one screw circumference, and then reengage.

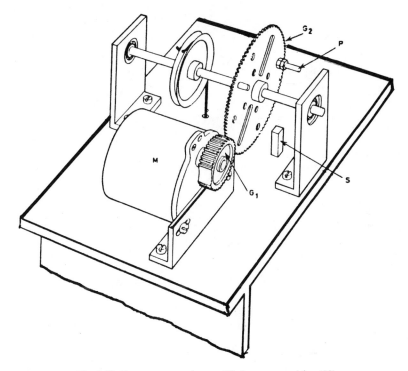

Fig. 5.29 Zone transport by modified spur-gear drive (30).

A cam can be used to drive a tube upwardly or downwardly through a fixed distance, followed by rapid return (Figure 5.30). The cam is in the form of an Archimedean spiral, which is the locus of a point whose distance r from the origin O is proportional to the angle θ between the radius vector and a fixed initial position; $r = a\theta$ where a is a constant defined by $r_0 = 2\pi a$.

A lead-screw drive (Figure 5.31, E) or a rack-and-pinion may be used to drive either an ingot or a heater/cooler battery. Connecting plate F, mounted on a captive nut attached to the rotating screw, propels heater/cooler battery A through a distance defined by limit switches G. Motors are available that provide reversibility and a very wide range of speeds. It is possible to use such a motor (Figure 5.31, D) for both advance and return steps of reciprocating motion (Electrocraft Corp).

A clutch motor (Hurst Manufacturing Corp., Princeton, IN 47670) may be used to raise a tube (Figure 5.32) until switch A is contacted, deenergizing the clutch. The freely turning rotor of the motor allows the tube to drop to a second switch B which reenergizes the clutch and starts another pass.

Pfann has described a number of devices in which a helical or annular charge is exposed to a single linear heating source such that continuous rotation of the charge with respect to the heat source provides effective motion of a molten zone (38).

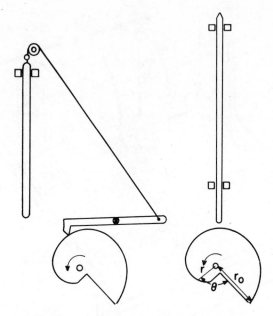

Fig. 5.30 Cam-actuated zone-transport mechanism.

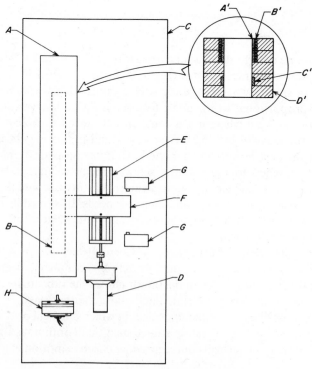

Fig. 5.31 Lead-screw transport mechanism.

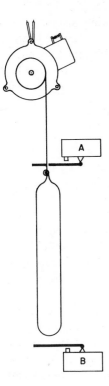

Fig. 5.32 Clutch motor transports tube upward until contact with switch A disengages clutch; tube drops to switch B and reengages clutch.

2.3.2 Hydraulic Drives

Knypl and Zielinski (66) described a transport mechanism in which water is slowly introduced into a chamber, causing the level to rise within it. A zone-melting tube rests on a float on the rising water. When the water level attains a critical height, a siphon drains the reservoir, causing the float to fall to a predetermined level, whereupon the cycle begins anew. Maire and Delmas used a similar technique in which the motion of the hydraulically propelled float was transmitted through a network of pulleys and cables to an external tube so that the tube itself was not immersed in the hydraulic bath (67). Herington has described a method for raising or lowering a zone-melting tube by the gradual deflation of a rubber balloon through a capillary leak (7).

2.3.3 Two-Motor Assemblies

Small differential-drive assemblies are available commercially in which two motors of greatly differing speed and opposite sense of rotation drive a common output shaft. Sloan and McGowan (65) described the use of such an assembly to perform reciprocating motion in a heater battery. Because insertion of an entire solid ingot into the heater would lead to breakage of the container, the drive was constructed so that the ingot started its travel from a position completely below the heater battery. In this way, each zone is introduced at the top of the ingot. After all the desired zones are formed, the stops that control reversal are "captured" between two switches that define the reciprocating distance. An alternative way to avoid tube breakage is to insert the molten ingot in the heater battery, and allow solid zones to form in place before start of ingot transport. Such differential-drive assemblies, if driven by small instrument motors, can be used to move a sample container (gross mass about 0.5 kg), but not a more massive heater/cooler battery; somewhat greater lifting capacity may be attained by counterweighting the sample container.

An alternative to the differential drive may be made by using two high-torque clutch motors. A fast motor and a counter-rotating slow motor are mounted to drive a common output shaft. While the clutch of the slow motor is energized, it drives the output shaft and the fast motor idles. At the end of the zone pass, the clutch of the slow motor is deenergized and that of the fast motor is energized, providing rapid return to complete a zoning cycle. Figure 5.33 shows the mounting of two such motors with a central output gear. A flanged adapter on the output shaft allows attachment of a pulley on which a drive cable is wrapped. The pulley is located in place by a nut (not shown) on the threaded end of the adapter. Suitable motors are available from Hurst Manufacturing Corp. (Princeton, IN 47670, Model AR-DA).

Adjustable zoning speed may be attained in either one-motor or two-motor drives by using a variable-speed motor or by using an on–off cyclic timer to vary the fractional duty cycle of the slower motor. Because the speeds involved in zoning are generally quite low, the difference between continuous motion and step-wise motion is not critical if the switching interval is only a few seconds.

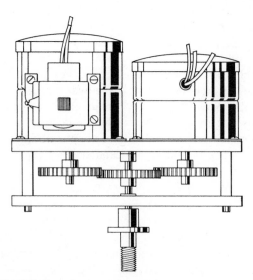

Fig. 5.33 Two clutch motors driving a single output shaft.

2.3.4   Complete Transport Assemblies

As in the case of directional crystallization, it is possible to buy transport mechanisms that provide variable speed and adjustable travel distance. Such devices are available from a number of commercial suppliers (see Appendix II).

2.3.5   Miscellaneous Transport Methods

It was pointed out earlier that effective zone motion can be brought about without moving parts, by sequential passage of heat-exchange fluids through a network of embedded tubes. In like fashion, a series of closely spaced parallel resistance heaters may be energized sequentially to give the effect of zone motion.

An ingot may move past an external radiant source; the charge may be stationary in the laboratory frame of reference, but in relative motion with respect to the sun as the radiant source.

2.4   Containers

A wide variety of shapes and materials can be used in zone-melting containers. It is only necessary that the material to be zoned shall not attack the container and not be contaminated by it. It is also necessary that the container be impermeable to atmospheric or other contaminants. Additionally, it is desirable that the thermal conductivity of the container should be low, if well-defined zones are to be achieved.

The relevant properties of a number of container materials are assembled in Table 5.3.

The commonest container in use for the zone melting of organic and inorganic chemicals is a cylindrical tube of borosilicate glass. The container may be simply a length of tubing, closed with a stopper at one end and filled by decanting the molten material into the tube. For various reasons, somewhat more sophisticated sample containers may be used. Consideration of chemical stability of the sample to be zone melted as well as considerations of safety of the experimenter may dictate that the melting be carried out in a closed system. Because most chemicals, as synthesized or purchased, have a relatively large volume compared to that of a solidified ingot, it is necessary that the container have a section of relatively large volume to accommodate the powdered crystalline material before it is melted into the zone melting tube. Figure 5.34a shows a typical configuration for such a tube, in which the section of large volume is a spherical flask with the zone-melting tube appended below. These tubes may be cleaned easily, and dried by baking in vacuum. On occasion it is necessary that the tube be loaded in an inert atmosphere if, for example, the material to be purified is air- or moisture sensitive even at room temperature. The adapter shown in Figure 5.34b (with an O-ring-sealed valve) makes it possible to connect the zone-melting tube to a vacuum line to bake it, fill it with an inert gas, and then transfer it to a glove box, within which the sensitive compound may be charged into the tube. The adapter is connected to the tube

Table 5.3. Properties of Container Materials

Material	Maximum Working Temperature, K^c	Thermal Conductivity,[f] $Wm^{-1} K^{-1}$	Coefficient of Linear Expansion, $K^{-1} \times 10^6$	Useful With
Glasses				
Borosilicate (Pyrex®)	725	1.13	3.2	Organics, some metals
Vycor®	1375	1.36	0.75	Organics, metals, alkali halides
Vitreous silica	1600	1.36	0.55	Organics, metals
Metals				
Gold	1150	315.0	14.2	Alkali metals, halides
Stainless steel, 304	1100	30	17.3	Organics, alkali metals
Nickel	1600	91	13	Fluorides
Platinum	1900	73.4	9	Alkali metals, cyanides, halides
Iridium	2500	148	6	Oxides, corrosives
Tantalum	3000	57.4	6.5	AlSb
Molybdenum	2300	138	5	Halides, cyanides
Silver	1100	427	19	
Aluminum	750	237	25	
Ceramics				
Alumina	2200	1.7	4.6	Al, Mg, AlSb
Magnesia	2900	37	12.7	Fe
Beryllia	2600	170	8.4	U
Boron nitride	2000	5^e	0.2, 3^a	B, GaAs
Silicon nitride	1800	9.6	6.4	Si, AlSb
Carbons				
Graphite	2850	150^b		
Pyrolytic graphite	3300	1960 Layer		
		5.7 Layer		

Plastics[d]				
Teflon®	525	0.33	100	All chemicals except molten alkali metals
Polyethylene, low density	340		150	(2)(3)(5)(6)(9)
high density	375	0.50	120	(2)(3)(5)(6)(9)
Polyvinyl chloride (PVC)	340	0.15	100	(2)(3)(9)
Polypropylene	410	0.24	90	(2)(3)(5)(6)(7)(9)

[a] Anisotropic.

[b] Average.

[c] Depends on pressure and ambient gas.

[d] The use ranges of these plastics are described in terms of compatibility with several classes of compounds:
1. alkyl halides and polyhalides
2. alcohols
3. dilute mineral acids
4. concentrated mineral acids
5. aldehydes and ketones
6. esters
7. aromatic hydrocarbons
8. dilute alkalis
9. concentrated alkalis

[e] "Isotropic" BN shows a maximum conductivity of 6–9 $Wm^{-1} K^{-1}$ at 500–600 K, decreasing to about 2 $Wm^{-1} K^{-1}$ at 100 K. Crystalline BN gives a maximum 25 $Wm^{-1} K^{-1}$ at 235 K, parallel to the deposition plane and a gradual decrease from 2.5 $Wm^{-1} K^{-1}$ at 100 K to 1.7 at 800 K, perpendicular to it.

[f] The values cited here refer to temperature near 300 K.

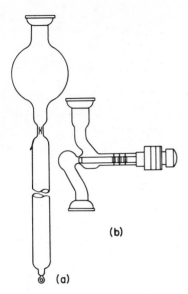

Fig. 5.34 Flask for charging zoning tube (*a*), with adapter (*b*).

by an O-ring joint. The tube is then again closed with the adapter and returned to the vacuum line.

After a period of evacuation suitable to the compound in question, an inert gas is admitted to the zone-melting tube so that the ensuing melting does not produce extensive sublimation, which is often a problem with organic chemicals. After the molten material has flowed into the zone-melting tube and solidified, the tube is reevacuated and the constriction between the upper and lower sections is heated gently to remove any adhering chemical by sublimation. At this stage, the sample may be outgassed by freeze/pump/melt cycling, and finally the zone-melting tube is removed from the charging flask by flame fusion at the constriction.

Solid chemicals generally retain entrapped and/or adsorbed gases very tenaciously and even prolonged evacuation does not remove such impurities. If the chemical will be damaged by being melted in the presence of these gases, it can be sublimed in vacuum before being melted (see "sausage-tube" technique in Section 2.6.2, Chapter 3). If the amount of retained gas is small, it may be allowed to react with a portion of the charge to form a product that will be segregated during the later zoning.

Tubular containers may not be used in arbitrarily small diameters. In general, it is difficult to introduce melts into bores smaller than about 0.3 cm. This is a result of high viscosity and/or surface tension. Even if a sample can be charged into a small-bore tube, surface tension tends to produce voids or gaps in the ingot. Once such voids form, solute segregation ceases. This problem may be alleviated to some extent by inserting an inert elastomeric plug at the head of the ingot to avoid entry of gas bubbles and ingot losses by sublimation.

Small samples are more readily zoned in channels machined in plates (see Section 2.1.3) or in microboats. Schildknecht has described an ingenious way of making a microboat from a length of glass tubing. One end of the tube is closed, and a spot is heated intensely with a very small flame; careful application of suction to the tube causes formation of a "dimple" about 1 cm long. The tube is quickly pulled at both ends, causing the dimple to elongate, giving a very narrow channel (68).

On occasion, it is necessary to transfer zone-melted material after processing, in the absence of air. This may be achieved conveniently by use of the break-seal tube shown in Figure 5.35. The charging procedure is exactly as described above. After the zone-melting tube is sealed off at the constriction, it is inverted and the charge is melted away from the break-seal A. B is a small hole that avoids pressure buildup during zoning. The zone melting is carried out in this configuration so that most impurity will be transported to the end that was sealed off, away from the break-seal. The break-seal end of the tube may then be attached by flame fusion to other apparatus; the break-seal is opened and the upper portion of the ingot may then be transferred into an appropriate receiver, either by sublimation or melting.

Fig. 5.35 Zone-melting tube with break-seal.

It is possible to use one vessel for zone purification and crystal growth; Figure 5.36 shows a tube for this sequence. A prepurified sample is charged into the tube as described above and the tube is sealed off at constriction 1. The sample is then melted away from the capillary and zone melted in section 2. After the zoning is complete, the upper portion is melted toward the capillary, constriction 4 is freed of adhering solid by gentle heating, and the tube is sealed off. The pure material in the capillary end may be crystallized by the Bridgman method (see Section 4.1, Chapter 4) or by sublimation.

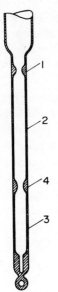

Fig. 5.36 Combination zone-melting and crystal-growth tube.

More sophisticated vessels may be used for terminal zone melting followed by crystal growth. One of these is described in Section 4.1.1, Chapter 4.

The "sausage tubes" described in Section 2.6.2, Chapter 3, can, of course, be used to charge tubes for zone melting. The same kind of vessel can be used to distill liquids into tubes for zone melting; the process is the same, regardless of the freezing point of the sample. If fine fractionation is desired, a zone-melting tube may be used as one receiver attached to the fraction collector of a multiplate distillation column.

Most of the container designs discussed thus far imply single usage: the purified product is removed from the zone-melting tube by breaking the tube. This procedure, while convenient in many cases, may become costly or inconvenient in others. Reusable tubes of precision-bore glass or other material may be closed with suitably inert plugs, inserted first at one end, and then the other, after the sample has been charged. Figure 5.37 shows such plugs A, machined from Teflon®, sealed to the container B by O-rings C of an appropriate elastomer. The closures may be provided with bores so that a thermocouple may be included, passing wholly or partly through charge D. This arrangement makes it possible to monitor the temperature of any point within the tube or to use a thermocouple for control of the heaters. Plug-type tube closures have been used by a number of workers to help avoid the breakage of the zone-melting tube from expansion accompanying melting. Smit et al. used an internal plug of Teflon® for this purpose (69). A related approach has been used, in which a movable Teflon® plug, backed by a spring, allows limited displacement of the plug as melting takes place in each pass (70, 71).

The design of Figure 5.37 leads naturally to the idea of an annular tube. The

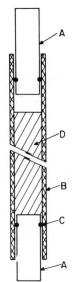

Fig. 5.37 Zone-melting tube with removable closures.

centrally bored closure makes it possible to center an inert rod or tube within the outer cylinder, defining an annular sample space. By reducing the radial distance through which heat must be fed and abstracted, it becomes possible to process larger amounts of material with reduced zone lengths. If the central cylinder is hollow, it may be used for internal cooling or passage of an internal heater. All-glass annular containers can be made, resembling Liebig condensers.

Maire and Delmas (67) have described a tube of 1-m length and 5-cm diameter which was provided with a central tube of Teflon$^{®}$. The inner tube defined an annular sample space and also served to take up stresses resulting from melting of the contents, by deforming flexibly. Figure 5.38 shows a method of mounting such a tube A; if the Teflon$^{®}$ tube is reasonably straight, it may be cemented to glass closures at top and bottom. If it is bent, the closures may be joined by a rigid glass rod B. Teflon$^{®}$ must be treated with an alkali–metal etchant before it can be bonded with epoxy adhesives.

Although the stratagems described above do make it possible to zone larger amounts of material, the experimenter is cautioned to use great care in scaling up laboratory zone melting. Borosilicate tubing of diameter larger than 5 cm is especially prone to breakage under the high gradients required for zoning. Breakage may involve more than inconvenience, since the toxicological properties of many materials are not known. Wrapping the tube, as described below, is a prudent measure, serving to contain the contents of a tube in case of breakage.

The annular configuration, with its reduced heat path, lends itself to formation of a single helical zone. Pfann proposed such an arrangement in which the annular container would rotate within a stationary helical heater;

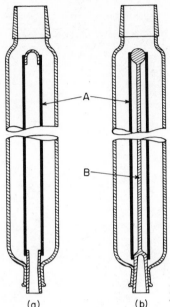

Fig. 5.38 Zone-melting tube with a central tubular insert of "Teflon."

(a)   (b)

McGhie et al. have described an improvement on this approach (72). The annular configuration is not alone in providing reduced heat path. Containers with rectangular cross section, one of whose dimensions is small, confer the same advantage.

For operation at temperatures above about 675 K, borosilicate glass can be replaced with a high-silica glass such as Vycor[R] or with quartz.

Metal tubes have been used, but they are troublesome in two respects:

1. Their high thermal conductivity makes it hard to maintain sharp, narrow zones; this is especially true when organic chemicals are being processed, because the thermal conductivity in the radial sense is low.

2. They are opaque. Temperature profiling with an internal thermocouple is helpful, but if the temperature/composition relationship is unknown, it becomes impossible to say whether a given temperature inside the vessel corresponds to solid or liquid. A better solution to the problem of opacity is the incorporation of an optical probe.

A number of workers have used zone-melting tubes containing a porous closure at or near the lower end of the tube, so that low-melting impurity can flow through the closure and be physically removed from the ingot. The intention of this device is to produce higher overall purity by preventing back diffusion of impurity from the bottom of the ingot into the purified upper portion. In continuous zone refining (that is, in operations involving a continuous supply of starting material and simultaneous withdrawal of product and waste) this problem does not arise.

Tubular containers have been used in a number of shapes other than right-circular cylinders. Smith and Thomas used a helical coil of Tygon® tubing A (shown schematically in Figure 5.39) for the zone-melting of water (73). The tubing was wound on a cylinder of wire mesh of 3.8-cm diameter (not shown). The coil was mounted with its axis horizontal and about half its diameter was immersed in a bath B at 263 K. The end of the tubing at which zones originated was folded inside the helix and remained frozen, to provide a constant seed for continuing crystallization. The 40-turn spiral was heated from above by radiant heater C through slotted cover D, which limited melting to a length of about 3.8 cm, or one-quarter of the circumference.

Plastic tubing may be used as a liner inside a glass container for the processing of chemicals which would attack glass, such as strong bases. Moreover, the product may be recovered from the plastic tube by cutting it. Plastic sheet or tape can be applied to the exterior of glass zone-melting tubes as a safety measure, against the possibility of tube breakage. Thin, strong, transparent tapes of Kapton® are available with adhesive backing, able to withstand long exposure to 475 K (Connecticut Hard Rubber Co. 407 East St., New Haven, CT 06509).

Schildknecht used open, ring-shaped containers of glass or silicone rubber in horizontal operation. This design has the advantage that simple rotation about a vertical axis provides zone motion (74). If the ring is interrupted, zone refining takes place; if the ring is continuous, zone leveling results.

Open boats and trays are usable in horizontal zone melting. Graphite boats and crucibles have been used for zoning and for crystal growth at temperatures to 3500 K. The use of glass or Teflon® trays for zoning of small samples is described earlier in this section.

Zone melting may be effected without a container, in several ways. In the first of these, the ends of a vertical ingot are supported and a single narrow zone is formed near one end and then caused to move slowly toward the other. The molten zone is held in place by surface tension. This method was introduced by Keck and Golay (75) in connection with purification of silicon, and has come to be known as "float zoning." Methods of heating the floating zone, critical to success of the process, have been discussed earlier in this section.

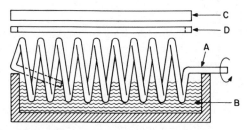

Fig. 5.39 Zone melting by rotation of a helical container in a cold bath. A radiant heater warms part of each turn, through a slotted cover.

Other container-free methods have been described. These include levitation, magnetic suspension, two-liquid zone melting, and cage zone melting (28).

Glass tubes, for all their advantages, suffer from the drawback of breakability. An attempt to introduce a molten zone anywhere but at the top of an ingot is almost certain to cause breakage of a glass container because the confined zone, increasing in volume as it melts, exerts great pressure. Under certain conditions, the problem may be overcome and zones may be introduced at the closed end of an ingot. In one approach, the solid segments separating the molten zones were not allowed to cool much below the freezing temperature, to keep the solid in a somewhat plastic state. The expansion that accompanies formation of the molten zone, in this arrangement, merely causes the sequence of solid and liquid to slip axially in the tube (76). Yet another solution to the problem is to include a displaceable bottom closure below the ingot. The expansion that accompanies melting, in this case, leaves the ingot unmoved, but causes downward slippage of the close-fitting closure. Travel of the closure may be limited by a spring between it and the end of the tube (69, 70, 77).

Another method of avoiding tube breakage is applicable to horizontal refiners operating in the rotation-convection mode. By filling the zoning tube only partially, an axial hole is produced in an annular ingot. The liquid zone fills less than half the cross section of the tube, so that free expansion can take place (78).

The cleaning of containers warrants consideration. Solutions of sodium dichromate or potassium dichromate in sulfuric acid have long been used for the cleaning of glassware, especially for removal of organic residues, but it is known that chromium ions are adsorbed on glass and are not readily removed (79). Moreover, interaction of "chromic acid" with halides can yield volatile chromyl compounds, which are said to be carcinogenic. Strongly oxidizing sulfuric acid preparations are now available with no chromium content (Nochromix® Godax Laboratories, 6 Varick St., New York, NY 10013). Other acid formulations have been used, including mixtures of nitric and sulfuric acids, and nitric and hydrofluoric acids with a detergent (80). Less aggressive methods using dilute acids, alkalis, and sequestering agents have been found effective (81).

Ultrasonication in organic solvents or in aqueous detergents is a generally effective cleaning procedure for glasses and metals. The effectiveness can be enhanced by heating the solvent under reflux in an ultrasonic bath.

2.5 Mixing

No experimental aspect of zone melting has received greater attention than the question of mixing in the molten zone. Theory predicts and experiment confirms that solute distribution depends sensitively on the thickness of the layer of melt in which matter transport is diffusion controlled. Since mixing reduces this thickness, it improves the overall efficiency of solute partitioning and makes it possible to zone at higher rates than can be used with quiescent liquids. Two families of stirring methods have been used, depending on the electrical conductivity of the sample being zoned.

Nonconductive materials are mixed by mechanical means. The obvious stratagem of inserting a stirrer in the zone suffers from several drawbacks. It is applicable only to horizontal ingots, and it is difficult to seal an immersed stirrer against atmospheric contamination. If it is in fact possible to work with an open ingot in the form of a tray, then mixing can be achieved by pumping the melt out of the molten zone and returning it. The melt may be heated in the pumping circuit, and this heating may provide the heat of fusion required for advancement of the zone (28). These ideas do not seem to have been used and are probably of interest only in larger-than-laboratory operations.

Cremer and Kribbe (47) purified kilogram batches of phosphorus in glass tubes of 4-cm diameter. The low thermal conductivity of phosphorus made it impractical to heat charges of this size externally. An ingenious application of induction heating made it possible to do so, and stir the molten zone at the same time. In operation, tube A (Figure 5.40) was provided with a number of silver-plated soft-iron cores B, one of which is shown. The cores were perforated to allow easy passage of liquid. An external alternating-current magnet system C (Figure 5.40b) was provided to heat the core by eddy currents and hysteresis loss and generate a zone of 1-cm height. This technique is applicable because of the low melting point of phosphorus (317.3 K). The system was improved by notching the edge of the core to produce turbulence as a rotating external field caused the core to rotate.

The same effect was achieved with single-phase alternating current, using the split-pole stator configuration shown in Figure 5.40c. Effective impurity segregation required a low zoning speed of 1.4×10^{-4} cm s^{-1} (0.5 cm h^{-1}). Since the core is "captured" at the bottom of the ingot after a zone pass, a new one must be introduced for each successive pass. For this purpose, a supply of cores was stored in an appendage above the ingot and single cores were maneuvered into

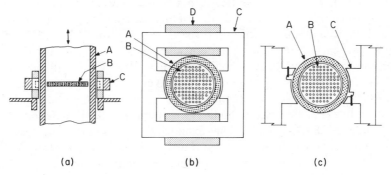

(a)   (b)   (c)

Fig. 5.40 Zone melting a nonconductive material with induction heating of a traveling susceptor. (a) Vertical cross section:

A—Container tube.

B—Perforated susceptor.

C—Magnet system.

(b) Horizontal cross section through zone. (c) Horizontal cross section through zone, showing magnetically driven rotation.

place by an external permanent magnet. The number of zone passes was thus limited to the number of stored cores.

Avramenko et al. used a glass-covered steel rod to stir a single molten zone in ingots of stearic acid (82). The internal magnet was vibrated axially by a mechanically driven external permanent magnet, at rates of 0, 0.8, 1.7, 3.3, 5.3, 7.0, and 8.3 oscillations s^{-1}. Zoning was carried out at constant rate of 8.3×10^{-4} cm s^{-1} (3 cm h^{-1}) and the ingots were sectioned for analysis after five passes. The effectiveness of zoning was greatest for agitation at about 4 oscillations s^{-1}. The worsening of the results at higher rates was attributed to enhanced nucleation. Apparently, excessive agitation disrupted the crystallization interface and led to entrapment of impurity. Since this effect is substance specific, the optimum rate of agitation by this method would have to be determined for each material. Tube rotation is by far the most widely used method of mixing the molten zone in nonconductive materials. Ordway zoned in glass tubes at rates of 2.8×10^{-5} cm s^{-1} to 1.7×10^{-3} cm s^{-1}) (0.1 to 6 cm h^{-1}), with rotation at 0.17 to 0.5 rev s^{-1} (71). This measure was not very effective in promoting efficient partitioning because unidirectional rotation leaves stagnant liquid in the center of the tube, remote from the inner wall. A reverse rotation, vertical-tube method was used by Sloan for directional crystallization (83) and has been applied extensively to zone melting. Smit and his co-workers (69, 77, 84) rotated tubes at speeds of 25 rev s^{-1}, with 2 reversals s^{-1}. The bottom of the zoning tube, which had an inner diameter of 0.94 cm, was closed by insertion of a Teflon® stopper to the extent of 20 cm. Raising a heater battery around the tube caused upward zone motion. To avoid tube breakage, the tube/motor coupling allowed the Teflon® stopper to be displaced as each molten zone was introduced. The enhanced mixing made it possible to zone at speeds up to 0.033 cm s^{-1} (120 cm h^{-1}) but normally zoning was carried out at 0.0167 cm s^{-1} (60 cm h^{-1}). Iida and Tachikawa (85) used apparatus similar to Smit's: vertical tubes of 0.35–0.9 cm id were rotated at speeds up to 100 rev s^{-1} while zoning a test mixture consisting of Sudan Red in naphthalene. Runs were carried out at speeds ranging from 1.3×10^{-2} cm s^{-1} to 2.6×10^{-2} cm s^{-1} and several patterns of reversal were used at each speed: (1) constant rotation at 100 rev s^{-1}, (2) rotation at 100 rev s^{-1} with a change of about 5 rev s^{-1} every 2 s, (3) 100 rev s^{-1} stopped every 2 s, (4) reversal after every rotation; (5) reversal twice per second. Modes 1 and 3 gave the best separation.

Utsonomiya et al. (86) used a similar device to purify alkali–metal nitrates in vertical tubes of 0.8-cm diameter. They explored the effects of rotation speed and reversal interval on the efficiency of separation of sodium chromate from sodium nitrate. All the experiments were carried out at a zoning rate of 1.67×10^{-3} cm s^{-1} (6 cm h^{-1}) with a zone length of 2.5 cm ($L/l = 10$). Increasing the rotation speed from about 8 rev s^{-1} to 50 rev s^{-1} did not cause a demonstrable change in segregation. Changing the reversal interval had a dramatic effect, however, as shown in Figure 5.41. Intervals shorter than 0.5 s were not used because of mechanical considerations. In any case, the time required for reversal was about 0.25 s, so that much shorter reversal intervals were not accessible.

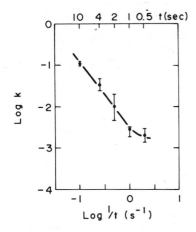

Fig. 5.41 Variation of distribution coefficient with frequency of reversal of rotation.

Not all reversible motors are suited to long-term use with frequent reversal. Light-weight "pancake" motors (PMI Motors) have been found to be applicable. A stepping motor was used by Wolter et al. to provide reversing rotation with extremely short reversal periods (down to 1 step of 1.8° at 10^3 steps s^{-1}). Optimum separation efficiency was attained with an interval of about 0.03 s between reversals. This report describes in detail an elegantly flexible apparatus allowing wide variation of rate of rotation and interval between reversals.

Lastushkin and Kirgintsev used the reverse-rotation method with horizontal or nearly horizontal tubes in a cold box, for the purification of aqueous eutectic solutions of copper sulfate and potassium nitrate (87). Lastushkin also carried out a theoretical study of the effects of changing zone volume in a rotating horizontal container and described a method for stabilizing the zone length. Yeh and Yeh made a similar study and describe methods for optimizing the zone length in a multipass operation (88). Yoshida and Ueno (89) concluded that the effect of stirring is maximized when the half-period of rotation (i.e., time between reversals) is equal to the time required to reach maximum speed. When naphthalene containing azobenzene was zoned at 0.024 cm s^{-1} (86.4 cm h^{-1}) with rotation at 35 rev s^{-1} and reversal time of 0.5 s, k_{eff} was 0.01. Zoning under these conditions for 1.8×10^3 s (0.5 h) removed all of the solute from 80% of a 25-cm ingot.

A six-sample vertical zone refiner has been described in which stirring is provided by an oscillation system that consists of a sprocket gear fitted to the base of each zoning tube, and driven by a variable speed motor (90).

The rotation-convection method uses a rotating horizontal container, but the rotation is quite slow and is not intended to produce mechanical mixing. The intent is merely to generate a planar interface and narrow zones (27, 28).

Convection, driven by temperature or concentration gradients, should also be considered as a means of mixing. While the density differences resulting from concentration gradients are not usually large enough to provide mixing at the interface, temperature gradients can be effective. This is especially true if heat transfer is promoted by inclusion of conductive bodies in an ingot. The difference in heat conduction between the charge and the conductive body causes large local temperature gradients even if the entire zone is not heated greatly above the melting temperature of the solid.

Of course, the conductive bodies can be made to oscillate if they are mounted loosely on a central shaft (41). Further, each conductive disk can be fitted with a magnet which in turn can be driven by an external magnet; only the disks in the molten zones will be rotated by the magnetic drive. In these ways, mechanical mixing can be combined with thermally driven convection.

A stationary axial conductor, with diameter small in comparison to the inner diameter of the tubular container, can enhance convection in molten zones. A mixture of naphthalene and 2-naphthol was charged into a glass tube of 1.8-cm id, with a central stainless-steel tube of 0.32-cm id and 0.04-cm wall thickness. A single zone pass was carried out with and without the conductive tube; calculations showed that the diffusion-controlled interface layer was reduced from 0.15-cm thickness to 0.05 cm by the conductor (91).

Yet another way of promoting convective mixing at the interface is to use periodic regrowth. Jackson and Miller (92) demonstrated that in thin-film crystallizations, the quality of the crystal improved if a fraction of the newly grown solid was remelted. The meltback was repeated cyclically so that eventually all of the charge was multiply crystallized. Although this stratagem seems not to have been used in bulk zone melting, it seems likely that it would lead to a temperature inversion in the melt near the interface. Especially when zoning vertically downward, convective mixing should be promoted at this stage.

Conductive materials can be agitated by any of the mechanical methods just described. In addition, some methods are applicable only by virtue of the conductivity of the sample. Induction heating, for example, produces eddy currents which in turn produce hydrodynamic flow in the zone.

Pfann and Dorsi (93) described a method of mixing based on interaction between an impressed current and a magnetic field. The current is passed axially through a horizontal ingot; simultaneously an orthogonal magnetic field is applied near the end of the zone. The current and field interact to produce a force normal to both. If the magnetic field is nonuniform, the resultant force varies and stirring results.

Braun et al. (94) described another method suited to conductive materials in horizontal containers. A zone is produced by an electric-resistance furnace and an array of three coils of copper wire is disposed symmetrically around the furnace with their axes horizontal. The coils are energized by three-phase alternating current. The result is that magnetic lines of force thread sectors of the molten zone sequentially, at a rate determined by the frequency of the alternating current.

Normally, convective mixing is strongly desired in zone refining, to overcome the accretion of impurity in a layer at the solid/liquid interface. There is,

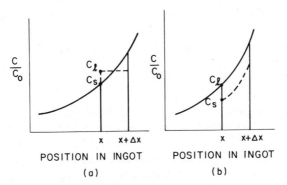

Fig. 5.42 Use of fibrous packing to impede back-mixing in zone melting.

however, a circumstance in which convective mixing should be minimized. To grasp this need, one must consider the concentration profile at the end of an ingot where impurities with $k_{eff} < 1$ accumulate. When the molten zone enters this high-impurity "tail," the trailing edge of the zone (position x in Figure 5.42a) has a higher impurity concentration C_l than the solid from which it formed. The solid later formed from this zone can have, at best, the same concentration, C_s, as before. It is this backflow that leads to the existence of an ultimate distribution. To overcome this phenomenon, it has been suggested that the tail section be packed with fibrous material to impede mixing (95). Especially if the fibers are aligned parallel to the direction of zoning, it is possible to maintain in the molten zone, the concentration gradient that existed in the solid. The solid forming at the trailing edge of the zone will have a lower solute concentration C'_s than existed in the solid after the prior pass. In this way, the terminal gradient is shifted toward the end of the ingot and a more favorable ultimate distribution ensues (dashed curve, Figure 5.42b).

2.6 Orientation of Apparatus and Direction of Zoning

Zone-melting apparatus has been operated vertically, horizontally, and in intermediate orientations. Early work with metals was largely carried out in open horizontal boats protected by tubular enclosures. This orientation, while it takes up relatively much laboratory space, makes it easy to use induction heating and to gain the resulting stirring effect. If the material being zoned is not volatile, there is no problem of evaporation from the heated zone. The boat-in-tube method allows use of opaque boats, since the surface of the ingot is visible through the tube.

Matter transport (see Section 1.5) can be troublesome in horizontal operation, but the problems can be overcome. Effects of volatility and surface tension are more serious. Organic compounds, especially, may evaporate completely from open boats; completely filling the cross section of the container is a remedy. In an early study of the zoning of organics, it was observed that the impure melt underran the purified solid as the latter contracted from the container (96). The result is not surprising, given the fact that the surface tensions of organics are at least an order of magnitude smaller than those of metals. As a result, most zone melting of chemicals has been carried out in vertical containers.

Downward zone transport in a vertical container affords a desirable situation for mixing in the zone. Since melting is taking place at the bottom of the zone, the bottom will be hottest. Thermal convection will tend to homogenize the zone and thin the barrier layer at the crystallizing interface, in which matter transport takes place by diffusion. Unfortunately, the contraction that takes place as most materials freeze, results in the formation of voids at the crystallizing interface; these can rapidly become large enough to reduce the effective crystallization area. Voids may be allowed to escape from an ingot by using a rotating container in an oblique orientation (31). In the downward/vertical mode, zones must be introduced at the top of the ingot, where expansion can compress a buffer volume of inert gas, to avoid tube fracture.

The problem of void formation at the crystallizing interface can be overcome by zoning upwardly. This mode, however, surrenders the convective-stirring advantage of downward zone transport; to overcome this lack of convective mixing, it is imperative that some form of mechanical mixing be provided. Moreover, there is greater risk of tube fracture during introduction of zones at the closed bottom of the container. Two methods have been used to cope with this problem. One of these merely uses a displaceable plug between the end of the ingot and the end of the tube. As a zone is formed, the plug moves down to accommodate the increase in volume. The second method of preventing breakage is to maintain minimum lengths of solid between the molten zones and to keep the solid close enough to its melting point to retain some plasticity. Fischer (97) and Benz et al. (98) have described horizontal refiners operating on this principle. The former provided zones of 2-cm length separated by 4-cm segments of solid, in a 40-cm ingot; the ingot and heater were surrounded by a cooling jacket that maintained the desired temperature in the solid. The latter used an aluminum heat-exchange block with discrete heaters inserted into pockets in the block. A 35-cm ingot in a tube of 1.5-cm od contained zones of 0.7-cm length separated by 2.8-cm segments of solid. In both cases, the entire ingot was able to slide in the container when a new zone was introduced at the closed end (see Section 2.1, Chapter 8).

In the context of zoning direction, a word is in order about nomenclature. Because of the diversity of practice, "top" and "bottom" do not have clear, unambiguous meanings as designations of position in an ingot. Consequently, we have used "head" and "tail" to denote the ends of an ingot at which each zone pass starts and finishes, without regard to the orientation of the ingot or the direction of zoning, referred to external coordinates.

2.7 Related Methods

If a gradient of an intensive property is imposed on a substance, matter transport can be promoted; the two gradients most widely used are temperature and electric field.

2.7.1   Temperature-Gradient Zone Melting (TGZM)

If a disk of substance B is placed between elongated rods of substance A, it is possible to move a thin $A-B$ zone through the structure. The mechanism is clarified with the help of Figure 5.43, which shows that the distribution coefficient of B in A is less than unity and that the composite structure is in a temperature gradient whose lowest temperature is above that of the AB eutectic, and whose highest temperature is below the melting point of A. Some B will dissolve in A at both faces of the disk, elongating it until saturation is reached at both ends. The concentration of A is higher in the melt at $(x + \Delta x)$ than at x, and A will diffuse along the concentration gradient, depositing solid at x. This solid contains B at concentration given by kC_x. The interface at $(x + \Delta x)$, now depleted of A, moves to the right to reattain saturation. In this way, gradient-driven diffusion moves the zone through the structure. The distinguishing feature of the process is the absence of a moving heat source, as is used in the related traveling-solvent method (see Section 3.4, Chapter 1). In principle, the zone does not have to be molten; the same result would ensue in an entirely solid system if the diffusivity were high enough. Thus, the process might be applied to organic rotator-phase solids.

Pfann has discussed the factors that control the length, shape, and composition of the zone, as well as its travel rate (99). It is possible to generate two-dimensional ("sheet") zones and one-dimensional ("wire") zones and these have been used to grow single crystals, with a number of advantages:

• The apparatus is simple, with no moving parts.
• The molten zone is so small that no container need be used.

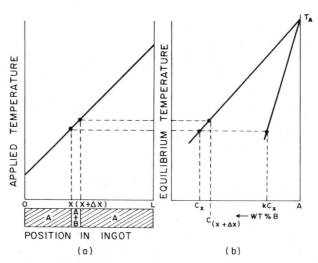

Fig. 5.43 Schematic representation of TGZM: (*a*) temperature gradient applied to ingot; (*b*) partial phase diagram of A/B system.

- Spontaneous nucleation is unlikely in a small, quiescent zone.
- The small surface-to-volume ratio minimizes evaporation and contamination.
- Constitutional supercooling is suppressed (100).

Keezer has used TGZM to grow single crystals of selenium, with thallium as the added component (101). Calcium molybdate crystals were grown by Parker and Brower (102), through the eutectic with lithium sulfate. Boah (103) used IR heating to dope semiconductor wafers.

TGZM has been related to microsegregation in succinonitrile. Observation of dendritic solidification in this material showed that the side arms migrated through several secondary arm spacings up the temperature gradient (104).

2.7.2   Solid-Solid Transformation

Solid-solid phase transformations can be used to distribute impurities in certain systems. Iron containing 0.4 wt% carbon, after 10 passes of a zone at 1473 K, showed a range of concentration from 0.1 to 0.8 wt% at the tail of the ingot. The 1-cm zone was passed along a rod of 20-cm length and 0.8-cm diameter at 2×10^{-3} cm s^{-1} (7.2 cm h^{-1}) (105). In this experiment the hot zone was above the α-to-γ transition temperature, and although the diffusivity of carbon in the high-temperature phase is 800 times smaller than in liquid iron, notable segregation took place. Similar results were obtained with nitrogen in chromium. After 10 passes of a zone at 1373 K through an ingot containing 0.4 wt% nitrogen, a profile was generated, with 0.12 and 1.28 wt% at the head and tail of the ingot.

2.7.3   Field Freezing

The eutectic composition is not separable by crystallization alone, unless pressure is used as a variable. However, if an electric field is applied to a conductive eutectic, a separation can be effected at constant pressure. This is so because the difference in the ionic mobilities of the two species will result in an accretion of one species at the anode and its depletion at the cathode. The melting temperature of the enriched material at the electrodes will rise to values higher than ambient, and solidification will ensue. The respective solid solutions of the two phases will grow isothermally toward the center of the cell, until all of the eutectic has been consumed.

Application of an electric field can also perturb solute distribution during directional crystallization of conductive melts. Depending on the direction of the electric current through the cell, the matter flux evoked electrically can add to or oppose the flux produced by crystallization. Considering a solute with $k_0 < 1$ it is possible to achieve $k_{eff} < k_0$ or even $k_{eff} > 1$, by application of a suitably directed current.

Patents (106, 107) have appeared in which electrolysis is used during zone melting of hydrates of alkali–metal hydroxides. In one of these (107) a rotating anode is immersed in the molten zone of a horizontal ingot whose container is the cathode. Starting with an iron concentration of $6 \times 10^{-3}\%$, 10 passes with electrolysis followed by 8 passes without electrolysis gave a 44% yield of product containing $10^{-6}\%$ Fe. The same procedure, without electrolysis, gave a 19% yield containing $5 \times 10^{-6}\%$ Fe. In another patent, the electrodes are situated at the ends of the container.

2.7.4 Vapor–Liquid–Solid (VLS) Crystal Growth

Wagner and Ellis (108, 109) described a mechanism for the growth of crystalline whiskers, which is of both theoretical and preparative importance. Qualitatively, growth proceeds as follows. A small droplet of substance B is placed on a surface of A, and the specimen is heated above the A/B eutectic temperature (Figure 5.44). If now a source of A is provided in the vapor surrounding the specimen, A will enter the droplet of eutectic, diffuse through it to the substrate, and freeze there. The accretion of A at the interface causes the original droplet to rise intact, with growth continuing at the interface between the droplet and the rising column of substance A. The original work was done with gold/silicon eutectic on silicon crystals.

Nittono et al. (110) carried out a detailed electron-microscopic study of the growth of copper whiskers by vapor-phase reduction of cuprous iodide. Questions about the growth mechanism, specifically whether growth took place at the root or at the tip of the whisker, were answered in favor of tip growth. Selenium crystals have been grown by Keezer et al., both with and without added thallium (101), with which it forms a eutectic. Crystals grown in the presence of thallium were terminated by a characteristic solidified droplet.

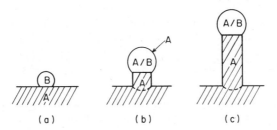

Fig. 5.44 Schematic representation of VLS growth: (a) droplet of B on substrate A; (b) eutectic formed by heating, fed from vapor source; (c) grown whisker.

2.7.5 Zone Leveling

The basic ideas of zone leveling are given in Section 1.3. The experimental requirements do not differ greatly from those of zone melting for purposes of purification. It is necessary to assure greater constancy of zone length, however.

The earliest applications of zone leveling were in the preparation of homogeneously doped ingots of semiconductors such as germanium and silicon. The SCO method (see Section 1.1.2) is applicable if the dopant has a very small

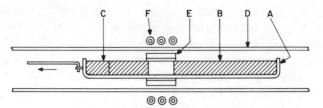

Fig. 5.45 Apparatus for zone leveling.

k_{eff}. Although zone leveling does not necessarily imply growth of a single crystal, the two processes have often been combined, and Pfann has provided a detailed description of the basic method. A quartz boat (Figure 5.45), cleaned by sandblasting, is coated with soot from a candle flame to prevent adhesion of the crystal. An ingot of germanium B, typically 30 cm long, is shaped to fit the boat. A seed crystal C, about 5 cm long, is placed in contact with the ingot and a pellet of dopant (antimony or indium in germanium) is added to the starting zone. The boat is placed in a tube D of clear silica, within a graphite susceptor E. The susceptor, an annulus 2–7 cm long with 0.6-cm wall thickness, is provided with a slot at its top. The induction coil F outside tube D is able to heat the melt directly through the slot in the susceptor, and hence to agitate the liquid. At the same time, the susceptor is heated and contributes to stable length and shape of the zone. The boat travels along quartz rods (not shown), lying in the furnace tube. A small portion of the seed is melted along with the dopant and the zone is then moved into the ingot at $2 \times 10^{-5}\,\mathrm{cm\,s^{-1}}$ to $10^{-3}\,\mathrm{cm\,s^{-1}}$, in an inert atmosphere (111).

Yim and Dismukes (112) and Dismukes and Ekstrom (113) described the application of this method to the growth of homogeneous bismuth/antimony single crystals. Their experiments were carried out in sealed ampoules under 5×10^4 Pa ($\frac{1}{2}$ atm) of hydrogen pressure. Even without the insertion of a seed crystal, the zone-leveled ingots were largely single crystals, especially for alloys containing less than 15 atom% antimony.

Kolkert (114) has applied multipass zone leveling to the preparation of homogeneous ingots of an organic solid-solution system, naphthalene/2-naphthol. His apparatus consisted of a modified rotation-convection system designed to facilitate microscopic observation of the freezing and melting interfaces (Figure 5.46). Ingot A is contained between outer tube B (1.27-cm id) and inner tube C (0.4-cm od), which are separated by spacers D of Teflon[R]. A thermopile (not shown) is provided within the inner tube, to monitor temperatures in the zone and ingot. Heater E consists of a single loop of resistance wire able to provide temperatures of 298–500 K, with a constancy of 0.06 K. Direct-contact cooling is provided by rings F, through which coolant is circulated at 250–300 K, with a constancy of 0.04 K. The container assembly is rotated through shaft G. Flat freezing and melting interfaces are achieved by adjustment of heating, cooling, and rotation speed; zones as narrow as 0.3 cm are attainable.

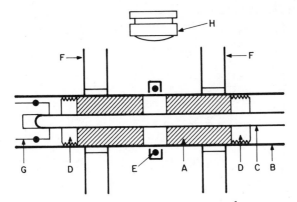

Fig. 5.46 Apparatus for multipass zone leveling of an organic material in an annular ingot.

The ingot is scanned through the heater/cooler assembly at 1.1×10^{-5} cm s^{-1} (0.4 cm h^{-1}), constant to $\pm 5\%$. The direction of travel is reversed automatically by photoswitches.

Kolkert's apparatus gave ingots that showed only 0.01 mole% variation over most of a 4-cm ingot. Distributions, after one and two passes, were compared with those calculated according to the method of Kirgintsev and Kudrin (115) and good agreement was found. The melting behavior of zone-leveled samples was compared with that of quenched samples. The results, which are striking, are of importance to the determination of phase diagrams by thermal analysis, and are discussed in Section 2.1.2, Chapter 2.

Van Genderen et al. (116) used an apparatus similar to Kolkert's to prepare homogeneous batches of 1,4-dichlorobenzene/1,4-dibromobenzene. The melting behavior of these solid solutions, monitored by differential scanning calorimetry, resembled that reported by Kolkert.

Multipass zone leveling can be achieved without mechanical reversal of zone transport, by the simple expedient of "bending the ingot back upon itself" to form a ring. Starting the zone at an arbitrary point will produce a depleted region whose solute concentration starts at C_0/k_{eff}. As soon as the first cycle is complete, solute-enriched material is fed into the depleted region and homogenization commences. After a sufficient number of passes, solutes of any k_{eff} will be distributed homogeneously, except in the final liquid zone. Kirgintsev and Kudrin (115) showed that the cyclic method requires only one-fourth as many passes as the reciprocating method.

Schildknecht applied the ring technique to the SCO method (see Section 1.1.2), with a circular ingot of stearyl alcohol through which a portion of ^{14}C-tagged cetyl alcohol was zoned. The container was an open ring of V-shaped cross section, so that the distribution could be monitored periodically during operation. Level concentration was attained after 58 passes.

Jackson and Pfann (17) described a means for generating a solid with a linear solute gradient by moving a solute-containing zone into an ingot of solvent. The

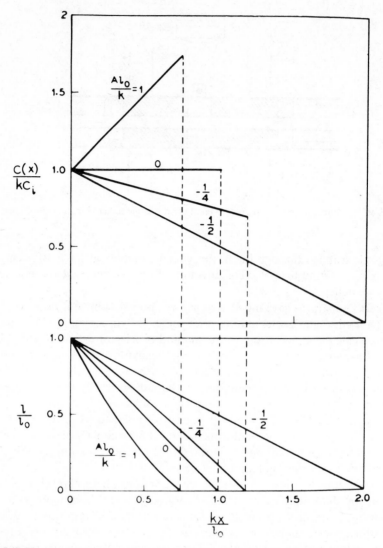

Fig. 5.47 The lower set of curves shows the variation in zone length (l/l_0) with position of freezing interface (k_x/l_0) necessary to produce the linear changes in composition [$C(x)/kC_i$] shown in the upper curves.

desired gradient is generated by varying the volume of the molten zone, that is, by varying the zone length at constant cross section. Figure 5.47 relates the variation in zone length (l/l_0) to the position in the ingot, and the position to concentration. Zone leveling is seen to be a special case of this procedure.

2.7.6 Helizone

McGhie et al. (117) have described a novel zone melter in which a slender helical heater rotates within an annular sample space (Figure 5.48). A film of

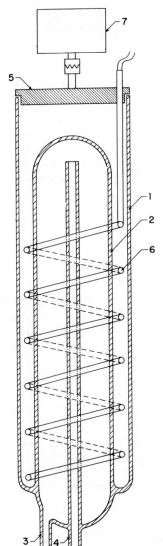

Fig. 5.48 Cross section of a zone melter with an immersed helical heater.

liquid is formed on the heater surface and heater rotation causes melting on the leading surface and crystallization on the trailing surface, to yield effective axial transport. The sense of rotation determines the direction of zoning. If the molten layer is thin and the ratio of helix diameter to helix pitch is large, then diffusive and convective mixing along the single molten zone are minimized and a large concentration gradient can be maintained.

This configuration confers several important advantages:

1. The immersed heater gives efficient heat transfer so that the energy requirement is small.

2. Container breakage is essentially eliminated because the single zone is always free to expand at its upper end, in case of fluctuating voltage or ambient temperature.

3. The assembly is mechanically simple, containing only one moving part.

4. Operation at low temperature is simplified because heat does not flow across air gaps and icing is avoided.

In Figure 5.48, the sample occupies the space between tubes 1 and 2. Coolant enters and exits through ports 3 and 4, respectively. Cap 5 supports heater 6 and is rotated by motor 7. Other versions of this device have been described (117) in which the sample chamber is sealed and the heater is powered through slip-rings; see also Chapter 6.

2.7.7 Vapor Zone Refining (VZR)

Although VZR is not a melt-crystallization technique, we describe it here because it relates so closely to zone melting in its experimental aspects.

Figure 5.49 shows schematically how the vapor/solid transition can be carried out in the zone mode. The compact charge A is confined in tube B between inert plugs C, as shown. Movement of heater D toward the right vaporizes the end of the ingot and allows the plug at the left to cool. Vapor condenses on the plug, leaving the most volatile components in the vapor space. Figure 5.49b shows the situation after some solid has condensed, and Figure 5.49c shows the completion of a pass. To start a new pass, the confining plugs are moved to the left, restoring the situation of Figure 5.49a. Since the ingot moves one zone length per pass, the number of passes is limited by the length of the container.

Weisberg and Rosi (118) used VZR for the purification of arsenic. Sloan and McGhie applied it to tetracene, an organic compound that cannot be zone

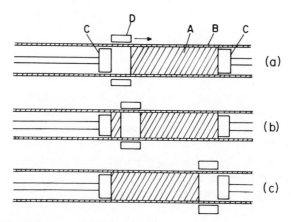

Fig. 5.49 Schematic of vapor zone refining: (*a*) beginning of pass; (*b*) start of condensation; (*c*) end of pass.

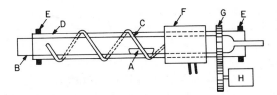

Fig. 5.50 Repeated sublimation by a rotating helical heater (Rotasub).

melted because it is unstable at its melting point (119). They described an apparatus in which a metal evacuation yoke may be connected to a length of disposable glass tubing as the sample container.

Dugacheva (120) has described a simple zone sublimation apparatus which was applied to the purification of benzoic acid. After five downward passes of a 1-cm zone through a 10-cm column, notable color segregation took place. The uppermost, colorless portion, comprising about one-fourth of the ingot, contained 0.7 mole% impurity. The purest, central, fraction, comprising 40% of the charge, contained $\sim 4 \times 10^{-3}$ mole%. The lowermost material, brown in color, contained 1.8 mole%. The same apparatus was used with several other compounds: salicyclic acid, 4-hydroxyacetophenone, 2-methoxynaphthalene, and chloroacetic acid were all purified by one to five passes of a vapor zone (121).

VZR is useful where solid/liquid methods are inapplicable, but it suffers from several drawbacks. Most basic is the low density of the moving zone, which limits its capacity to carry "solute" through the ingot. As an experimental matter, the process is slow and hard to automate.

2.7.8   Rotasub

McGhie has described (122) a related technique in which repeated sublimation is possible without manual adjustments (see Figure 5.50). The sublimand is contained in a boat A near one end of a tube B, which may be evacuated or purged with an inert gas. Tube B is centered within helical heater C, wrapped bifilarly on a glass or quartz tube D centered loosely within nylon rings E, which serve as bearings. The ends of the heater are connected to slip rings in housing F. A ring gear G is driven by motor H.

Rotation of the external helical heater subjects each point on the sample tube to a sinusoidally varying temperature. Suitably adjusted, the heater causes periodic evaporation and condensation at each point, producing multiple sublimation. If the pitch of the helical heater is varied, an overall temperature gradient can be imposed along the length of the tube.

2.8   Ancillary Techniques

Many of the procedures ancillary to directional crystallization (Section 2.6, Chapter 3) are naturally usable before zone melting, and those procedures should be consulted in connection with zone melting as well.

2.8.1   Product Recovery

One question almost universally overlooked in the literature of zone melting, but which arises in every case, is that of removal of purified material from zone-melting containers. In the simplest and most favorable case, if the purification has been carried out in a glass tube, and if the purified solid does not adhere to the glass, the tube may be broken and the ingot may simply be pressed out of its segment of glass tubing. If this cannot be done, each segment of the ingot may be dissolved in a suitable solvent, after the exterior of the glass has been cleaned. Alternatively, the segment may be placed in a sublimer and the contents removed by sublimation, or the contents may be melted and filtered to assure the absence of broken glass.

Materials that are suitably inert, such as hydrocarbons, may be obtained by dissolution of the glass container in hydrofluoric acid. Care must be taken to dissipate the heat of solution of the glass if the content of the tube is a low-melting solid. The dissolution may be carried out in an inert plastic container immersed in a coolant. This method often makes it possible to recover the ingot in one piece, which naturally must be washed carefully with water to remove the acid. Matvienko et al. (123) used this method to remove the glass from an ensemble of naphthalene crystals grown simultaneously in separate tubes.

The adhesion of chemicals to glass containers may be modified or prevented entirely by appropriate treatment of the glass surface before the sample is charged into the container. Typical of such materials is Dri-Film® SC87 (General Electric Company), which is a chlorotrimethylsilane. In use, the chlorine atom is hydrolyzed by reaction with the hydroxylic surface of the glass,

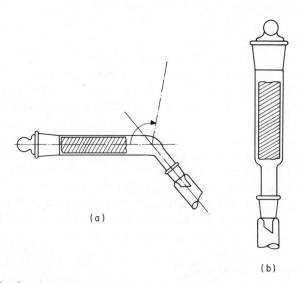

(a)

(b)

Fig. 5.51 Transfer of an air-sensitive chemical after zone melting. In version (*a*) the ingot is melted in the angled tube, which is rotated to effect transfer.

producing a chemically bonded trimethylsilyl surface. Excess reagent is rinsed from the glass with a dry hydrocarbon solvent; after drying of surface solvent, the treated vessel may be evacuated and baked at about 500 K.

Tubes containing air-sensitive chemicals may be segmented in a glove box and each segment inserted in an angled tube which is attached to a receiver. The ingot segment is melted and decanted into the receiver by rotation of the bent tube (Figure 5.51a). The adaptor may also have a constriction on which the vertical segment rests while it is melted (Figure 5.51b).

If zone melting is carried out in a tube fitted with a standard-taper joint or in a plugged length of precision-bore tubing, the tube may be coupled to a modified McCarter sublimer. The contents of the tube may be sublimed fractionally by using a series of heaters of increasing length so as to heat progressively lower portions of the ingot (see Section 2.6.4, Chapter 3). In larger-scale operation, the contents of the tube may be melted in fractions and pumped through a heated transfer line into a receiver, either under inert-gas pressure or by using a small positive-displacement pump.

If zone melting is carried out in plastic tubing, the product ingot may be sectioned with a scalpel or razor blade.

2.8.2   In Situ Analysis

It is very desirable to monitor the progress of purification and/or impurity enrichment within a zone-melting container. The best approach would provide continuous or intermittent assay during the zoning; a useful second-best approach requires interruption of the zoning for an analysis that does not require disruption of the ingot. A number of useful methods have been developed and others can be envisaged.

Optical spectroscopy comes immediately to mind as a method applicable to samples in sealed cells, assuming that the cell is transparent to the radiation to be used. Absorption spectra can be measured, in principle, but the highly scattering nature of polycrystalline solids makes it difficult to obtain reproducible results. If suitable optics are available, the measurement can be made through the molten zone, but even here, there are problems in maintaining constant geometry.

Emission spectra of various kinds are more reliably measured, either during or after zoning. In a study of purification and crystal growth of p-terphenyl, Sloan and Ern (124) examined delayed fluorescence of zone-melted ingots and found that the decay of emission from the head of the ingot did not conform to the pure exponential expected from a single annihilation process. This indication of an impurity or impurities with $k_{eff} > 1$ was confirmed by mass spectrometry. Two such impurities, of masses 306 and 304, were identified as 2-phenyl terphenyl and 2-phenyl triphenylene and their concentration in the starting material was estimated at 6–60 ppm. These traces might have been overlooked in the absence of the delayed-fluorescence measurements.

Numerous other workers have used the triplet lifetime of delayed fluorescence from aromatic hydrocarbons as an external probe of purity. Lupien and Williams (125) applied this method to anthracene purified in various ways. Figure 5.52 shows triplet lifetime as a function of position in the ingot after various numbers of zone passes. The anthracene used in the zone refining had been prepurified by chromatography and sublimation. The optical probe provided clear evidence of the attainment of ultimate distribution, at least with respect to impurities that affect triplet lifetime.

Singlet lifetime has been measured in microcrystalline fluoranthene samples purified in various ways (126). Unpurified, commercial fluoranthene gave a room-temperature singlet lifetime of 23 ns at 440 nm. Zone-refined fluoranthene gave a value of 37 ns at 440 nm. Chemical treatment with maleic anhydride and with chloranil, followed by zone refining, gave a value of 45 ns. Addition of small amounts of anthracene brought about drastic reduction in singlet lifetime.

A word of caution is required in connection with external "sampling" of solid ingots. If the optical method samples only the exterior of the ingot (strongly absorbed radiation), misleading results may be obtained because of radial segregation (see Section 1.4, Chapter 3). If an axial distribution is to be measured externally, it should be established that the probe samples the bulk, or that there is no substantial radial concentration gradient.

Brown and Aftergut (127) studied the electrical conduction of imidazole, using measuring electrodes within a zone-melting tube. The platinum electrodes, of 1.3-cm diameter, were separated by about 0.2 cm and mounted on platinum wires supported in glass sleeves (Figure 5.53) within a glass zone-melting tube of 1.85-cm od. A number of problems affected the measurements and made it impossible to arrive at absolute specific resistivities:

- The absence of voids between the electrodes could not be assured.
- The orientation of the anisotropic crystals between the electrodes could vary from measurement to measurement.
- The distance between the electrodes could vary slightly.

On the other hand, the measurements were easy to make and not subject to effects of air or moisture. To measure the temperature dependence of the resistance, the ingot was removed from the zoner and brought to a temperature 10 K below the melting point, then cooled slowly, while further measurements were made.

MacFarlane et al. (128) have studied the refractive index of succinonitrile, which shows plastic-crystal behavior. Crystals were grown in a sealed quartz cell, at $8.3 \times 10^{-6}\,\text{cm s}^{-1}$ ($0.03\,\text{cm h}^{-1}$). The cell included a thermocouple, which made it possible to measure the temperature dependence of refractive index.

Another method of nondestructive analysis of an ingot during zoning has been described by Kern (129), who studied distribution of several gamma-emitting radioisotopes in gallium trifluoride, using a scintillation detector. Zoning was carried out both horizontally and vertically. In the latter case the observed distributions agreed closely with theory.

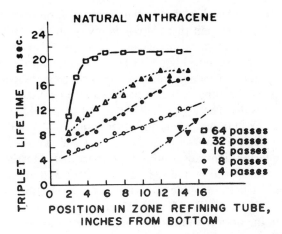

Fig. 5.52 Variation of anthracene triplet lifetime in a zone-melting tube as a function of the number of zones passed.

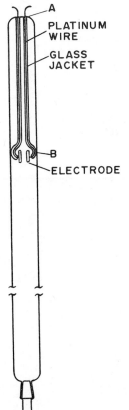

Fig. 5.53 Zone-melting tube containing electrodes for in situ measurement of resistivity, after Brown and Aftergut (127).

It was pointed out earlier that metal containers pose problems in zone melting because of their high thermal conductivity and opacity. Plastic tubes, while less conductive than metal, are generally opaque. Fiber optical devices can surmount the latter problem. Light can be introduced into an ingot through rigid or flexible guides, to act as a transverse internal probe. Figure 5.54 shows a cross section of a 6-cm tube in which two Pyrex rods of 0.4-cm diameter are inserted in Pyrex wells. The upper end of one rod is coupled to a flexible light guide from a 150-W illuminator and the other is similarly coupled to a sensitive detector. The lower ends of the rods are ground and polished to 45° prisms, directing the entering light across the 2-cm space between the rods into the detector system. In operation, the rigid guides are held in an aluminum clamp which also holds a thermocouple in a stainless-steel sheath (not shown). The assembly, 1.2 m long, is raised through the ingot at 8.3×10^{-3} cm s^{-1} (30 cm h^{-1}) and the optical and thermocouple outputs are displayed on a two-pen recorder. The optical probe offers two advantages. First, it denotes the solid/liquid interface much more reliably than the temperature probe (Figure 5.55). As impurity accumulates at the tail of an ingot, the liquidus temperature of the system changes in a way that is not usually known accurately. Hence, it is not known what temperature in the ingot corresponds to the interface. The optical probe, measuring the difference in transmission between transparent

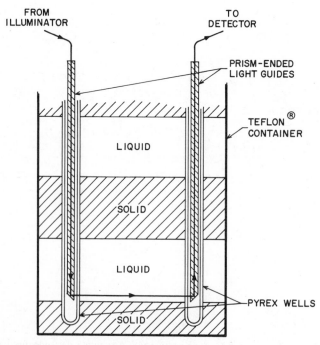

Fig. 5.54 Assembly for simultaneous temperature and optical probes of molten zones.

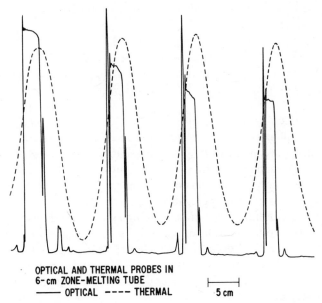

OPTICAL AND THERMAL PROBES IN
6-cm ZONE-MELTING TUBE
—— OPTICAL ---- THERMAL |——————|
                                    5 cm

Fig. 5.55 Temperature probe (----) and optical probe (————) of zone position in an opaque tube. (Note: zone positions do not coincide because both traces were recorded simultaneously with a two-channel recorder.)

liquid and relatively opaque solid, is temperature insensitive. Moreover, correlation of the two traces can provide insight into the slope of the liquidus. A second benefit of the optical probe is that transmission through the liquid at a given level in the ingot can be monitored during successive zone passes. Using monochromatic light, the accumulation or depletion of a given solute can be followed, and zoning continued only until a steady state is reached.

2.8.3   Scavenging (Use of Getters)

It is possible to add a reagent to a zone-refining charge, selected to react specifically with one or more minor components of the system. In this way, an impurity with an unfavorable distribution coefficient ($k \sim 1$) can be converted to another, more readily segregated substance. Sloan studied metallic impurities in anthracene (83) and found that some metals were not removed by zone melting or by sublimation, apparently because of the small particle sizes of the contaminants. Addition of small amounts of phthalonitrile to the starting charge led to reduction of iron and chromium levels. It is known that molten phthalonitrile reacts with those metals to form blue phthalocyanines, which were visually observable in the tail of the ingot. Evidently the dispersed metallic contaminants in anthracene are reacted and solubilized in the melt, facilitating their segregation by zoning. Other metals, less prone to formation of phthalo-cyanines, were not removed by phthalonitrile. Addition of a powerful electron acceptor (7,7,8,8-tetracyanoquinodimethane) led to a general reduction in metal contamination comparable with that effected by sublimation through a trapping layer of graphite.

In the realm of metals, it has been found that the removal of carbon from nickel is facilitated by the presence of oxygen. The carbon oxides produced during float zoning of the nickel, while not segregated in the zoning, are volatilized and removed from the ingot. Similar results were obtained with gold (130).

In the zone refining of zinc chloride, cupric chloride is not readily segregated. Addition of small amounts of zinc oxide convert the copper to the oxide, which is segregable, and zinc chloride ($ZnO + CuCl_2 \leftrightarrows CuO + ZnCl_2$). Another approach to the same problem merely requires that the zoning of copper containing zinc chloride be carried out in a hydrogen atmosphere; reduction to metallic copper ensues, and segregation is rapid (131).

Scavengers are useful as well in cases of high solid-state diffusivity. If a solute diffuses rapidly into a region from which it has been segregated, an improved result ensues if a getter is dissolved near the tail of the ingot, to react with and immobilize the diffusive impurity. Davies (132) has given a theoretical analysis of the effects of such solid-state diffusion, and recommended getters appropriate to silicon and germanium.

A similar procedure was used to augment the effect of zone melting tin; the metallic impurities concentrated by zoning were simultaneously extracted into stannous bromide or stannous chloride. With a salt/tin ratio of 1/100, detection limits were improved by one or two orders of magnitude for Na, Mg, Ca, Al, Fe, Zn, Ga, Cd, In, Tl, and Pb (133).

A related study was described by Jones and McDuffie (134), who examined the segregation of Fe^{3+} and Mg^{2+} from urethane. Zone refining removes Mg, at concentrations of 2–80 ppm, with $k = 0.31$. For iron, k is smaller, and has a strong concentration dependence. Iron in naphthalene showed a still smaller k. Addition of 0.1% β-quinolinol facilitated removal of both metals, giving k about 30% smaller in every case. The improvement was related to the size of the complex, as compared with that of the original impurity. Thus, provided that solubility in the melt is not a problem, larger chelating agents should be more effective than smaller ones. Another advantage of chelation as an adjunct to zone refining is that the products are often intensely colored, and removal of certain impurities can be followed visually down to the ppm range.

Photochemical transformation has been used to convert a nonsegregable impurity to a segregable one. Anthracene shows a distribution coefficient greater than unity in phenanthrene and is removed very slowly. While most of the anthracene can be removed by treatment with a dienophile such as maleic anhydride, traces (about 1 ppm) remain. Tschampa (46) illuminated the molten zone in a phenanthrene ingot with a high-pressure xenon lamp, through a quartz lens. The liquid became yellow at first and then brown; the reaction products collected at the tail of the ingot. The colored products might have arisen from reaction of the anthracene with residual oxygen, to form anthraquinone and/or hydroxyanthraquinones. In addition, anthracene dimers were probably formed. All of the products differ substantially from phenanthrene and segregate rapidly.

It is not surprising that lasers have come into use in photochemical purification, since they offer high fluence, monochromaticity, and tunability. While most laser photochemical purifications have been carried out with gaseous samples, separations in liquids are also known. In any case, gaseous intermediates are often purified to obtain important solid endproducts. For example, silane (SiH_4), which can be a source of pure silicon, has been freed from arsine (AsH_3), and phosphine (PH_3) by laser photolysis. The impurities are selectively fragmented and the resulting radicals form an easily separated polymeric solid (135).

Most laser-driven photochemical purifications are based on UV excitation. Nevertheless, selective IR multiphoton dissociation has been used for purification of $AsCl_3$ and BCl_3 (136, 137).

2.9 Efficiency of Zone Refining

2.9.1 Procedural Optimization

Most practitioners of laboratory-scale zone melting have not been greatly interested in optimizing the process. Rather, they have concentrated their efforts on achieving some target purity, regardless of the cost in time. Perhaps the extent of utilization of the method will increase if greater effort is devoted to reducing the time required for significant purification.

The degree of purification desired may be specified in terms of the reduction of solute concentration at the head of the ingot, $C(0)/C_0$. Plots of the number of passes required to achieve such specified reductions, for various values of k_{eff} are given by curves 86–94 of Appendix I, for varied geometries. One such plot, for $L/l = 20$, is given in Figure 5.56a. For example, the numbers of passes required to attain $C(0)/C_0 = 10^{-3}$ with $L/l = 20$, is given by the 10^{-3} ordinate in Figure 5.56a; these are shown below the graph. From Figure 5.56b one can obtain the normalized zoning rates corresponding to these k_{eff}'s assuming that k_0 is 0.1. The numbers of passes, zoning rates, and their quotients are given in Table 5.4. The time required to achieve the desired purification increases with n and decreases with increasing zoning rate. Hence, time for attainment of the desired purity is minimized when the quotient $nD/V\delta$ is minimized (28). Table 5.4 shows that this condition is attained when k_{eff} is 0.3. This value is achieved at a zoning rate substantially greater than that required to make $k_{eff} \simeq k_0$. This result is consistent with the generalization proposed by Harrison and Tiller, that optimum zoning is provided by operating at $V\delta/D \simeq 1$ (138). Hence, if the thickness δ can be reduced by effective stirring, V can be increased substantially without loss of efficiency, in accord with experimental results.

Using Equations 5.5 and 5.17, it is possible to calculate concentration profiles achieved after n zone passes, for any value of k_{eff}. Each k_{eff} in turn corresponds to a particular rate, V. The amount of impurity removed from a segment at the head of an ingot is given by the area under the appropriate curve, up to the

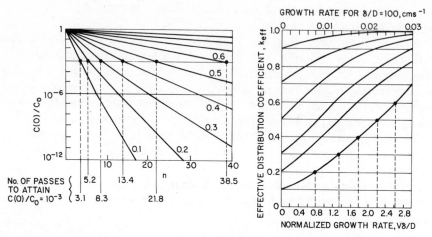

Fig. 5.56 (*a*) Relative solute concentration at the head of an ingot after *n* zone passes. (*b*) Effective distribution coefficients for a range of normalized growth rates, with k_0 as parameter.

Table 5.4. Optimum Zoning Speed Manifested in Minimum Value of $nD/V\delta$ for Reduction of Solute Concentration to $10^{-3}C_0$.

k_{eff}	0.1	0.2	0.3	0.4	0.5	0.6
n	3.1	5.2	8.3	13.4	21.8	38.5
$V\delta/D$	0	0.80	1.33	1.80	2.21	2.61
$nD/V\delta$	∞	6.5	6.2	7.4	9.9	14.8

coordinate of the end of the segment. In Figure 5.57, the area *ABCD* represents the amount of impurity removed by passing a zone as far as coordinate *B* at rate V_1. If $V_2 = 2V_1$, then in the same time interval, a zone moving at rate V_2 can reach coordinate *B'* and remove an amount of impurity *AB'C'D'*. Likewise, traversal at V_3 reaches *B''* and removes an amount corresponding to *AB''C''D''*. Pospisil (139) has suggested that a plot of such "equal-time areas" against zoning rate *V* will give a curve whose maximum indicates the optimum zoning rate V_0 (Figure 5.58).

It is not necessary to measure concentration profiles over a wide range of zoning rates in order to apply the "equal-time" method. If k_{eff} is measured at two or three rates, then a plot of $\ln[(1/k) - 1]$ against *V* gives a straight line of slope $-\delta/D$ and intercept $\ln[(1/k_0) - 1]$ (see Chapter 10). With these values, it is possible to calculate k_{eff} for any rate *V*, and hence to prepare plots similar to Figure 5.58 without extensive measurements.

Pospisil also suggested a simpler method for estimating f_0. The method is based on the fact that the slopes of the concentration profiles (Figure 5.57) at equal times are measures of purification rate. This arises because the position

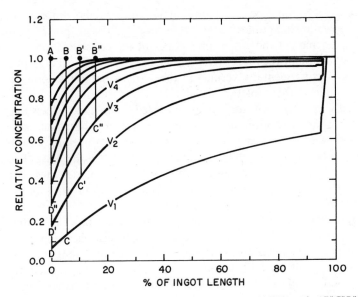

Fig. 5.57 Optimization based on equal-time areas. *ABCD*, *AB'C'D'*, and *AB"C"D"* represent amounts of impurity removed in a fixed time by zoning at progressively higher speed.

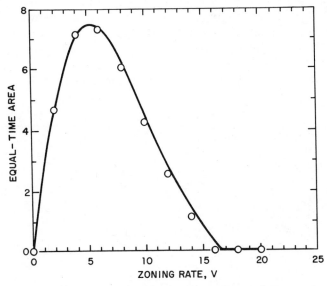

Fig. 5.58 Equal-time area against zoning speed, V.

coordinate can be replaced by (zoning rate) × (zoning time). Since the optimization is referred to equal times, the position coordinate is proportional to zoning rate. Hence a plot of the equal-time slopes of the concentration profiles will show a maximum for the rate corresponding to fastest removal of impurity. This analysis was experimentally verified in the systems naphthalene/methyl violet and acetanilide/methyl violet.

Pospisil's optimization study was extended to include effects of stirring. Figure 5.59 shows the optimum rates for zoning the naphthalene/methyl violet system (A) without stirring, (B) with continuous rotation, and (C) with oscillating rotation. From the figure it can be seen that V_0 increases from about $4 \, \text{cm h}^{-1}$ to $11 \, \text{cm h}^{-1}$ and finally to about $30 \, \text{cm h}^{-1}$, under these conditions. Table 5.5 shows the equivalence of determining V_0 from plots of equal-time areas or from the slopes at equal times.

If time is not the prime consideration, but instead product of the highest purity is sought, without regard to yield, then other measures are required. First of these is minimization of the zone length l as a fraction of ingot length L. The ingot is zoned to ultimate distribution at such a rate that k_{eff} is equal to k_0. This requires good mixing of the liquid and moderate rates. Further improvement, beyond ultimate distribution, can only be achieved by cropping the ingot and recasting a portion of it with a new geometry having longer L. Pfann has calculated the results attainable by this procedure (1). For example, taking half of an ingot of a system with $k_{\text{eff}} = 0.58$ and recasting into a new ingot of length equal to that of the original, gives an ultimate concentration about seventyfold lower than the first. Removal of the impure end of an ingot, followed by repeated zoning without recasting does not lead to substantial increase in purity. Vigdorovich and Selin (140) have reviewed these and other optimization procedures.

Matsui and Ishii (141) have discussed tactics for minimizing the time required

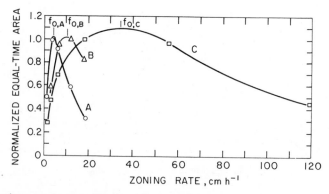

Fig. 5.59 Optimum zoning rate determined from equal-time plots. (A) no rotation; (B) continuous rotation; (C) oscillating rotation.

Table 5.5. Optimum Zoning Rate Determined from Area or Slope Plots at Equal Times

	Optimum Rate, f_0, cm h^{-1}		
	A	B	C
Naphthalene/methyl violet			
Area plot	4.5	10	39
Slope plot	5.0	11.5	34
Acetanilide/methyl violet			
Area plot	8.5	12	29
Slope plot	7.0	11	25

to obtain a maximum fractional yield of purified product; they evaluated both zoning rate and zone length as determinants of efficiency in the removal of tetracene from anthracene.

Iwano and Yokata (142) studied the efficiency of zone refining in several organic binary systems, as a function of zoning rate and degree of mixing. They defined effective distribution coefficients in terms of the fraction p of interface liquid occluded in the advancing crystal front under varied growth conditions. A modified BPS equation was derived, which included a constant, α, expressing stirring effectiveness and another (r) giving the rate of stirring

$$k_{\text{eff}} = \frac{k_i}{k_i + (1 - k_i) \exp\left[-\dfrac{\delta V}{D} \dfrac{\rho_s}{\rho_l} \exp(-\alpha r) \right]} \tag{5.34}$$

where $k_i = k_0 + p(1 - k_0)$, and ρ_s and ρ_l are the densities of solid and liquid, respectively. Purification efficiency, E, for one zone pass was expressed as follows.

$$E = \frac{C_l - C_s}{C_s} \frac{\delta}{D} V \frac{\rho_s}{\rho_l} = \left(\frac{1}{k_{\text{eff}}} - 1 \right) \frac{\delta}{D} V \frac{\rho_s}{\rho_l} \tag{5.35}$$

Substitution of Equation 5.34 in Eq. 5.35 gave a relationship between purification efficiency and normalized zoning rate, shown in Figure 5.60.

2.9.2 Apparatus Performance

Our discussion of efficiency has dealt with zoning rate as a variable, with the zone-refining apparatus remaining unchanged. Naturally, at a given zoning rate, refiners of different design will provide differing results. Among the decisive

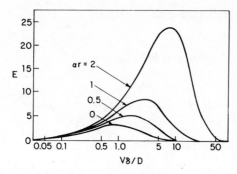

Fig. 5.60 Relationship between purification efficiency, E, and normalized zoning rate, $V\delta/D$, for one zone pass, with stirring rate, r, as parameter.

parameters, one may list

- Zone length relative to ingot length (l/L)
- Constancy of zoning rate
- Constancy of temperature in heated and cooled zones
- Temperature gradient at the crystallizing interface
- Mixing in the molten zone

Karl and Probst (143) have considered this question and noted the properties required of an appropriate test system. It is desirable that k_{eff} be about 0.1 so that concentration can be reduced from 0.5 to $10^{-4}\%$ in a small number of passes. Further, k should be independent of concentration, to facilitate mathematical analysis of measured concentration profiles. Another desideratum is that the solute should be easily measurable over a wide range of concentration. The system anthracene/phenazine meets these requirements and was used to "calibrate" several zone refiners.

For a given ingot length, the integral of the average relative concentration was determined over m zone lengths after n zone passes:

$$E_m(n) = \frac{1}{m} \int_0^m \log \frac{C(x)}{C_0} \, dx \qquad (5.36)$$

The quantity $E_m(n)$ is a measure of the average depletion achieved in the purest segment of the ingot, m zone lengths long, after n zone passes. One unit, operating horizontally with zones introduced at the closed end of the ingot, gave $E_8(4) = 3.1$, while to vertical refiners gave $E_8(4) = 1.9$ and 2.3, respectively.

Anthracene in fluorene ($k_{eff} = 1.3$) was proposed by Karl (35) for calibration with respect to solutes with $k_{eff} > 1$.

2.10   Commercial Zone Refiners

Several complete units are available commercially, embodying different modes of construction, Desaga offers a device based on a design of Schildknecht, incorporating integral heating ad cooling blocks, nested one within the other

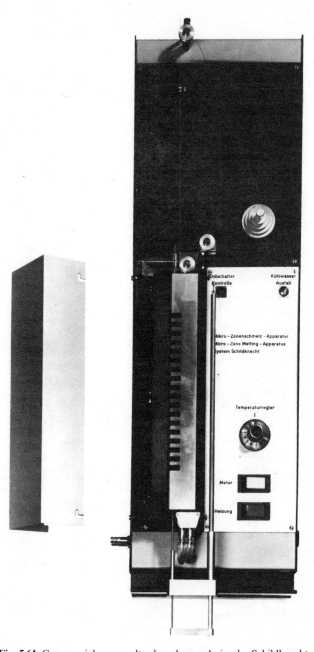

Fig. 5.61 Commercial zone melter based on a design by Schildknecht.

(see Figure 5.61). The blocks have 18 and 17 fins, respectively. They are capable of handling charges from 0.01 to 3 g at zoning rates of 1.4×10^{-5} to 1.4×10^{-4} cm s^{-1} (0.05 to 0.5 cm h^{-1}). A larger unit is also offered for charges of 10–25 g. Gallard-Schlesinger offers the Instrolec-200 zone refiner for quantities of 20–160 g of materials melting between 250 and 675 K, in a battery of nine heaters and eight coolers.

Enraf Nonius manufactures a zone refiner based on the design of Smit, marketed in the United States by Kratos Scientific Instruments. It handles charges of 0.2 to 4 mL at zoning rates up to 0.039 cm s^{-1} (140 cm h^{-1}). Addresses of these firms are given in Appendix II. Other firms offering specialized equipment for zoning and crystallizing metals are also mentioned there.

REFERENCES

1. W. G. Pfann, *Zone Melting*, 2nd ed., Wiley, New York, 1966, pp. 29–30.
2. Ibid., p. 42.
3. L. W. Davies, *Phil. Mag.*, **3**, 159–162 (1958).
4. B. A. Volchock, *Sov. Phys.–Solid State*, **4**, 789–790 (1962).
5. B. Velicky, *Phys. Stat. Sol.*, **5**, 207–212 (1964).
6. J. Sherman and R. B. Ewell, *J. Phys. Chem.*, **46**, 641–662 (1942).
7. E. F. G. Herington, *Zone Melting of Organic Compounds*, Wiley, New York, 1963.
8. I. Braun, *Brit. J. Appl. Phys.*, **8**, 457–461 (1957).
9. H. Reiss, *Trans. A.I.M.E.*, **200**, 1053–1059 (1954).
10. N. W. Lord, *Trans. A.I.M.E.*, **197**, 1531–1533 (1953).
11. I. Braun and S. Marshall, *Brit. J. Appl. Phys.*, **8**, 157–162 (1957).
12. E. Helfand and R. L. Kornegay, *J. Appl. Phys.*, **37**, 2484–2488 (1966).
13. L. Burris, Jr., C. H. Stockman, and I. G. Dillon, *Trans. A.I.M.E.*, **203**, 1017–1023 (1955).
14. E. T. Nelson, M. S. Brooks, and A. F. Armington, *Anal. Chem.*, **36**, 931–932 (1964).
15. E. Brandt-Petrik, L. Cser, and J. Nagy, *Acta Chimica (Budapest)*, **88**, 217–221 (1976).
16. M. Ya Frenkel and A. A. Yaroshevskii, *Dokl. Akad. Nauk. SSSR*, **175** (2), 403–406 (1967).
17. K. A. Jackson and W. G. Pfann, *J. Appl. Phys.*, **36**(1), 320 (1965).
18. E. Y. Wang, *J. Electrochem. Soc.*, **121**, 1671–1672 (1974).
19. W. G. Pfann, *Zone Melting*, Wiley, New York, 1966, p. 200.
20. J. N. Sherwood and D. J. White, *Phil. Mag.*, **15**, 745–753 (1967).
21. J. N. Sherwood, *Surface Defect Prop. Solids*, **2**, 250–268 (1973).
22. D. Fischer, *J. Appl. Phys.*, **44**, 1977–1982 (1973).
23. G. Matz, *Chemie-Ing. Technik*, **36**, 381–394 (1964).
24. D. C. Singh and S. C. Mathur, *Separation Science*, **7**(3), 243–248 (1972).
25. W. G. Pfann, *Trans. A.I.M.E.*, **197**, 1441–1442 (1953).
26. H. Schildknecht, *Zone Melting*, Academic Press, New York, 1966.
27. W. G. Pfann et al., *Rev. Sci. Instr.*, **37**(5), 649–652 (1966).
28. W. G. Pfann, *Zone Melting*, Wiley, New York, 1966, p. 97.
29. L. Spialter and J. R. Riley, *Rev. Sci. Instr.*, **30**, 139 (1959).
30. W. G. Wilman, *Chemistry and Industry*, 1825–1826 (1961).

31. Y. Ishizuka, *Bull. Chem. Soc. Jap.*, **50**(3), 563–565 (1977).

32. A. R. Clawson, *Thin Solid Films*, **12**, 291–294 (1972).

33. A. J. Lovinger, J. O. Chua, and C. C. Gryte, *J. Phys. E., Sci. Inst.*, **9**, 927–928 (1968).

34. H. Schildknecht and H. Vetter, *Angew. Chem.*, **71**, 723–726 (1959).

35. N. Karl, "High Purity Organic Molecular Crystals," in H. C. Freyhardt, Ed., *Crystals, Growth, Properties Applications*, Vol. 4, Springer Verlag, New York, 1980.

36. F. H. Horn, Jr., *J. Appl. Phys.*, **32**, 900–901 (1961).

37. N. I. Lastushkin, A. N. Kirgintsev, and A. F. Neermolov, *Deposited Document* VINITI, 555–574 (1974).

38. W. G. Pfann, *Zone Melting*, Wiley, New York, 1966, p. 92.

39. L. Marrelli, *Quaderni Dell'ingegnere Chimico Italiano*, **11**, 95–99 (1975).

40. P. J. Reynolds, U.S. Patent 4,255,390, March 10, 1981.

41. G. J. Sloan, U.S. Patent 3,844,724, October 29, 1974.

42. E. F. Herington, *Zone Melting of Organic Compounds*, Wiley, New York, 1963, p. 45.

43. F. J. Baum, *Rev. Sci. Instr.*, **30**, 1064–1065 (1959).

44. P. Kooy and H. J. M. Couwenberg, *Philips Tech Rev.*, **23**, 161–166 (1962).

45. T. Eyer, R. Nitsche, and H. Zimmerman, *J. Crystal Growth*, **47**, 219–229 (1979).

46. A. Tschampa, *Z. Naturforsch.*, **22a**, 112–118 (1967).

47. J. Cremer and H. Kribbe, *Chemie-Ing. Technik*, **36**, 957–959 (1964).

48. H. Plancher, T. E. Cogswell, and D. R. Latham, "Zone Melting Chromatography of Organic Mixtures," in M. Zief, Ed., *Purification of Inorganic and Organic Materials*, Dekker, New York, 1969.

49. R. W. Warren, "Purification of KCl by RF Heating," in M. Zief, Ed., *Purification of Inorganic and Organic Materials*, Dekker, New York, 1969.

50. W. Heywang, *Z. Naturforsch.*, **11a**, 238–243 (1956).

51. W. G. Pfann, K. E. Benson, and D. W. Hagelbarger, *J. Appl. Phys.*, **30**, 454–455 (1959).

52. A. M. Calverley, M. Davis, and R. F. Lever, *J. Sci. Instr.*, **34**, 142–147 (1957).

53. J. C. Brice, *The Growth of Crystals from Liquids*, American Elsevier, New York, 1973, p. 271.

54. B. Cockayne, D. B. Gasson, and N. Forbes, *J. Mater. Sci.*, **5**, 837–838 (1970).

55. D. B. Gasson and B. Cockayne, *J. Mater. Sci.*, **5**, 100–104 (1970).

56. A. Maries and P. S. Rogers, *Nature*, **256**, 401–402 (1975).

57. P. H. Brace, A. W. Cochardt, and G. Comenetz, *Rev. Sci. Inst.*, **26**, 303 (1955).

58. H. Rock, *Naturwiss.*, **43**, 81 (1956).

59. J. S. Ball, R. V. Helm, and C. R. Ferrin, *Petrol. Engr.*, **30**, C36–C39 (1958).

60. G. M. Dugacheva and A. G. Anikin, *Russ. J. Phys. Chem.*, **38**(1), 110–112 (1964).

61. J. C. Maire and M. A. Delmas, Purification of Kilogram Quantities of an Organic Compound," in M. Zief, Ed., *Purification of Inorganic and Organic Materials*, Dekker, New York, 1969.

62. S. Maeda, H. Kobayashi, and K. Ueno, *Talanta*, **20**, 653–658, (1973).

63. P. Süe, J. Pauly, and A. Nouaille, *Bull. Soc. Chim. France*, **5**, 593–602 (1958).

64. H. Schildknecht and J. Schnell, "On the Construction of An Ice Zone Melting Apparatus with Peltier Elements," in H. Schildknecht, Ed., *Symposium on Zone Melting and Column Crystallization*, Karlsruhe, 1963.

65. G. J. Sloan and N. H. McGowan, *Rev. Sci. Inst.*, **34**, 60–62 (1963).

66. E. T. Knypl and K. Zielinski, *J. Chem. Ed.*, **40**, 352 (1963).

67. J. C. Maire and M. A. Delmas, *Rec. des Trav. Chim. Pays Bas*, **85**, 268–274 (1966).

68. H. Schildknecht, *Zone Melting*, Academic Press, New York, 1966, p. 102.

69. N. J. G. Bollen, M. J. Van Essen, and W. M. Smit, *Anal. Chim. Acta*, **38**, 279–284 (1967).

70. N. I. Wakayama, Y. Nakano, and Y. Mashiko, *Bull. Chem. Soc. Japan*, **46**, 2277–2279 (1973).

71. F. Ordway, *Anal. Chem.*, **37**, 1178–1180 (1965).

72. A. R. McGhie, P. J. Rennolds, and G. J. Sloan, *Anal. Chem.*, **52**, 1738–1742 (1980).

73. H. A. Smith and C. O. Thomas, *J. Phys. Chem.*, **63**, 445–447 (1959).

74. H. Schildknecht, *Zone Melting*, Academic Press, New York, 1966, pp. 67, 103, 110, 131.

75. P. H. Keck and M. J. E. Golay, *Phys. Rev.*, **89**, 1297 (1953).

76. W. G. Filby and K. Gunther, *Lab. Pract.*, **22**, 570 (1973).

77. W. M. Smit, *TNO-Nieuws*, **20**, 593–594 (1965).

78. C. J. Doherty and W. G. Pfann, *Separation Science*, **8**(5), 593–597 (1973).

79. E. P. Laug, *Ind. Eng. Chem., Anal. Ed.*, **6**, 111 (1934).

80. R. H. A. Crawley, *Chem. Ind.*, **45**, 1205–1206 (1953).

81. D. W. Margerum and R. K. Steinhaus, *Anal. Chem.*, **37**, 222–228 (1965).

82. N. V. Avramenko, G. M. Dugacheva, and A. G. Anikin, *Russ. J. Phys. Chem.*, **42**(5), 673 (1968).

83. G. J. Sloan, *Mol. Cryst.*, **1**, 161–194 (1966).

84. N. J. G. Bollen, M. J. Van Essen, W. M. Smit, and W. F. Versteeg, U.S. Patent 3,490,877, January 20, 1970.

85. Y. Iida and T. Tachikawa, *Seikei Daigaku Kogakubu Kogaku Hokoku, Tokyo*, **15**, 1207–1208 (1973).

86. T. Utsunomiya, T. Yasukawa, and Y. Hoshino, *Report of the Research Laboratory of Engineering Materials Number 1*, 119–127 (1976).

87. N. I. Lastushkin and A. N. Kirgintsev, *Izv. Sib. Otd. Akad. Nauk SSSR, Ser. Khim. Nauk*, 111–115 (1976).

88. H. Yeh and W. H. Yeh, *Separation Sci. Techn.*, **14**(9), 795–803 (1979).

89. I. Yoshida and K. Oeno, *Chemistry* (*Kyoto*), **31**, 142–145 (1976).

90. K. K. Saraf and A. P. Srivastava, *Indian J. Technol.*, **14**(7), 337–339 (1976).

91. W. R. Wilcox, U.S. Pat. 3,189,419, 15 June 1965.

92. K. A. Jackson and C. E. Miller, *J. Crystal Growth*, **42**, 364–369 (1977).

93. W. G. Pfann and D. Dorsi, *Rev. Sci. Instr.*, **28**, 720 (1957).

94. I. Braun, F. C. Frank, S. Marshall, and G. Meyrick, *Phil. Mag.*, **3**, 208–209 (1958).

95. N. E. Hamilton, U.S. Patent 3,190,732, June 22, 1965.

96. E. F. G. Herington, R. Handley, and A. J. Cook, *Chem. Ind.*, 292–295 (1956).

97. D. Fischer, *Mat. Res. Bull.*, **8**, 385–392, (1973).

98. K. W. Benz, H. Port, and H. C. Wolf, *Z. Naturforsch.*, **26a**, 787–793 (1971).

99. W. G. Pfann, *Zone Melting*, Wiley, New York, 1966, pp. 255–259.

100. Ibid., p. 263.

101. R. Keezer, C. Wood, and J. W. Mody, "Use of Impurities to Facilitate Growth of Hexagonal Selenium Single Crystals," in H. Peiser, Ed., *Crystal Growth*, Proc. Intern. Conf. on Crystal Growth, Boston, 20–24th June 1966, Pergamon, New York, 1967.

102. H. S. Parker and W. S. Brower, "Growth of Calcium Molybdate Crystals by a T. G. Z. M. Technique," in H. Peiser, Ed., *Crystal Growth*, Proc. Intern. Conf. on Crystal Growth, Boston, 20–24th June 1966, Pergamon, New York, 1967.

103. J. K. Boah, German Patent 2,621,418.

104. D. J. Allen and J. D. Hunt, *Solidification Cast. Met., Proc. Int. Conf. Solidification, 1977* (Pub. 1979) 39–43 (Eng) Met. Soc., London.

105. D. S. Kamenetskaya, I. B. Piletskaya, and V. I. Shiryaev, *Soviet Physics—Doklady*, **13**, 245–247 (1968).

106. K. Eckschlager and J. Veprek-Siska, Certificate of Invention #158,898, July 15, 1975, Czechoslovak Socialist Republic.

107. K. Eckschlager and J. Veprek-Siska, Certificate of Invention #158,897, July 15, 1975, Czechoslovak Socialist Republic.

108. R. S. Wagner and W. C. Ellis, *Appl. Phys. Letters*, **4**, 89–90 (1964).

109. R. S. Wagner and W. C. Ellis, *Trans. Met. Soc. A.I.M.E.*, **233**, 1053–1064 (1965).

110. O. Nittono, H. Hasegawa, and S. Nakagura, *J. Crystal Growth*, **42**, 175–182 (1977).

111. W. G. Pfann, *Zone Melting*, 2nd ed., Wiley, New York, 1966, pp. 207–209.

112. W. M. Yim and J. P. Dismukes, "Growth of Homogeneous Bi-Sb Alloy Single Crystals," in H. Peiser, Ed., *Crystal Growth*, Proc. Intern. Conf. on Crystal Growth, Boston, 20–24th June 1966, Pergamon, New York, 1967.

113. J. P. Dismukes and L. Ekstrom, *Trans. Met. Soc. A.I.M.E.*, **233**, 672 (1965).

114. W. J. Kolkert, *J. Crystal Growth*, **30**, 213–219 (1975).

115. A. N. Kirgintsev and V. D. Kudrin, *Soviet Phys. Solid State (Eng. Trans.)*, **8**, 2169–2172 (1967).

116. A. C. G. Van Genderen, C. G. DeKruif, and H. A. J. Oonk, *Z. Phys. Chemie*, **107**, 167–173 (1977).

117. A. R. McGhie, P. J. Rennolds, and G. J. Sloan, *Anal. Chem.*, **52**, 1738–1742 (1980).

118. L. R. Weisberg and F. D. Rosi, *Rev. Sci. Instr.*, **31**, 206–207 (1960).

119. G. J. Sloan and A. R. McGhie, *Mol. Cryst. Liq. Cryst.*, **18**, 17–37 (1972).

120. G. M. Dugacheva, A. G. Anikin, and B. S. Pokarev, *Zh. Fiz. Khim.*, **39**(10), 2620–2622 (1965).

121. G. M. Dugacheva and N. V. Avramenko, *Russ. J. Phys. Chem.*, **50**(9), 1350–1352 (1976).

122. A. R. McGhie, U.S. Pat. 3,840,349, October 8, 1974.

123. V. N. Matvienko, N. V. Pertsov, and E. D. Shchukin, *Mol. Cryst. Liq. Cryst.*, **51**, 1–8 (1979).

124. G. J. Sloan and V. Ern, *Chem. Eng. Prog. Symp. Series*, **65**, 122–125 (1969).

125. Y. Lupien and D. F. Williams, *Molecular Crystals*, **5**, 1–7 (1968).

126. T. Kobayashi, S. Iwashima, S. Nagakura, and H. Inokuchi, *Mol. Cryst. Liq. Cryst.*, **18**, 117–127 (1972).

127. G. P. Brown and S. Aftergut, *J. Chem. Phys.*, **38**, 1356–1359 (1963).

128. R. M. MacFarlane, E. Courtens, and T. Bischofberger, *Mol. Cryst. Liq. Cryst.*, **35**, 27–32 (1976).

129. W. Kern, "Investigation of Zone Melting Purification of Gallium Trichloride by a Radiotracer Method," in M. Zief, Ed., *Purification of Inorganic and Organic Materials—Techniques of Fractional Solidification*, Dekker, New York, 1969.

130. P. H. Schmidt, *J. Electrochem. Soc.*, **112**(6), 631–632 (1965).

131. A. Gaeumann, *Chimia*, **18**(9), 300–305 (1964).

132. L. W. Davies, *Solid-State Electronics*, **7**, 501–504 (1964).

133. S. G. Gryaznova and L. A. Kondratenko, *Deposited Doc.*, **1974**, VINITI 2157–74, 26 pp.; through CA **86**:182537m.

134. L. N. Jones and B. McDuffie, *Anal. Chem.*, **41**, 65–70 (1969).

135. A. Hartford, Jr., E. J. Huber, J. L. Lyman, and J. H. Clark, *J. Appl. Phys.*, **51**, 4471–4474 (1980).

136. R. V. Ambartsumyan, Yu. A. Gorokhov, S. L. Grigorovich, V. S. Letokhov, G. N. Makarov, Yu. A. Malinin, A. A. Puretskii, E. P. Filippov, and N. P. Furzikov, *Soviet J. Quantum Electronics*, **7**, 96 (1977).

137. J. A. Merritt and L. C. Robertson, *J. Chem. Phys.*, **67**, 3545–3548 (1977).

138. J. D. Harrison and W. A. Tiller, *Trans. Met. Soc. A.I.M.E.*, **221**, 649 (1961).

139. P. A. Pospisil, Optimization of Zone Refining Rates, Ph.D. Thesis, Kansas State University, 1972.

140. V. N. Vigdorovich and A. A. Selin, *Teor. Osn. Khim. Tekhnol.*, **5**(6), 814–822 (1971).

141. A. Matsui and Y. Ishii, *Jap. J. Appl. Phys.*, **6**, 127–134 (1967).

142. T. Iwano and T. Yokota, *J. Chem. Soc. Japan (Chem. and Ind. Chem.)*, 21–26 (1974).

143. N. Karl and K.-H. Probst, *Mol. Cryst. Liq. Cryst.*, **11**, 155–171 (1970).

Chapter VI

CONTINUOUS ZONE REFINING

Zone purification of materials for laboratory use has been carried out largely in the batch mode. This approach will no doubt continue to be followed, because it offers simple operation and ease of sealing the sample against contamination. However, a substantial effort has been devoted to design and construction of continuous zone refiners. Continuous operation obviously offers the potential for increased production capacity. But continuous methods have other inherent advantages, which are often overlooked.

A comment on nomenclature is in order here. Devices designed to carry out zone melting on a continuous basis have purification as a universal goal. The other goals of zone melting, such as single-crystal growth or zone leveling, are not amenable to continuous operation. For this reason, this discussion refers to continuous zone refining, rather than to continuous zone melting.

In batch zone refining, purification efficiency decreases with increasing number of passes; that is, the fraction of the remaining impurity that is removed by each zone pass decreases as ultimate distribution is approached. Moreover, collection of a finite fraction of an ingot as product means that the product displays a concentration profile (see Figure 5.2). Hence the average overall purity can never be as high as that at the head of the ingot. In a continuous refiner, product and waste are collected at the extreme compositions prevailing at $g = 0$ and $g = 1$. The intervening solute profile can readily approach the ultimate distribution, since the time to attain this steady state need only be invested at the start of the operation.

In general terms, a continuous purification process requires an apparatus having a section in which the impurity is removed from the feed and another section in which the impurity concentration becomes correspondingly higher. Chemical engineers refer to these as "stripping" and "enriching" sections, respectively. Product and waste are removed at the ends of these sections and feed is introduced at some intermediate point. This description applies equally to purification based on any phase change, and continuous zone refining may be looked on as an analog of continuous distillation. Severe restrictions are imposed on continuous zone refiners, however, by the relative incompressibility of the fluid phase and by the fact that the denser phase does not flow. These properties require the introduction of methods quite different from those used in distillation. In vapor/liquid systems, interphase matter transport reaches equilibrium very rapidly and system efficiency is raised merely by increasing the area available for equilibration. In solid/liquid systems, interphase matter transport does not reach equilibration because of the slowness of diffusion in solids. An advancing molten zone acts as an impurity transporter, and multiple effect is possible precisely because a zone of intervening solid effectively isolates each interface from others present in the system.

Figure 6.1a shows schematically a continuous zone refiner and Figure 6.1b shows the relevant concentration profile. The profile cannot be exactly the

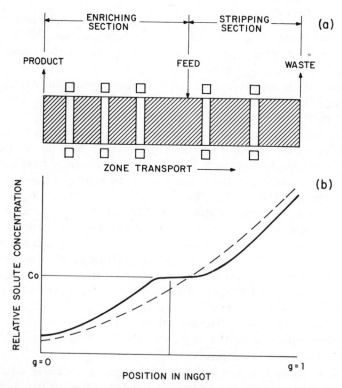

Fig. 6.1 Continuous zone refiner. (a) Schematic view of a continuous zone refiner. (b) The actual concentration profile is shown by the solid line and the ultimate distribution is shown by the dashed line, for a binary system with $k_{eff} < 1$.

ultimate distribution because the introduction of feed of concentration C_0 at a finite rate must produce a region of finite extent throughout which $C = C_0$. At ultimate distribution in batch operation, concentration C_0 prevails at only one point. The ultimate distribution would provide somewhat lower solute concentration at $g = 0$ and somewhat higher at $g = 1$.

Several experimental approaches have been practiced for continuous zone refining:

1. Zone-void method
2. Matter-transport/zone-transport methods
3. Helizone

In addition, two methods have been proposed, but evidently not carried out:

4. Reflux cross-flow method
5. Floating-zone method

1 ZONE-VOID METHOD

The zone-void method requires the introduction of a "bubble" or vapor space adjacent to each moving zone. Voids are generated at the ends of the enriching and stripping sections, by removal of quantities of melt as product and waste, respectively. Zone transport moves this bubble through the enriching section to a central point where the bubble is filled with feed. Heaters moving along the stripping section likewise cause periodic ejection of a quantity of melt as waste, causing formation of bubbles which move through the stripping section toward the feed point. The process is made clearer in Figures 6.2, 6.3, and 6.4, which refer to a vertical apparatus in the form of an inverted U-tube. The legs of the U are the enriching and stripping sections and the junction between them contains the feed mechanism.

Figure 6.2a shows the lowermost heater of the enriching section at the lower extremity of its travel. The material in the constricted region has melted and flowed downward into a collector. Because of the reduced diameter here, the volume of melt collected is less than that of a zone in the larger tube above. As the heater moves slowly upward to position b, solid is melted at the top of the heater, the melt drips through the void and freezes in the bottom of the column, sealing it as shown in Figure 6.2b. At a later time the entire zone and its associated void have reached the full diameter of the ingot (Figure 6.2c). At the top of the enriching section, the void is filled with fresh feed. After the battery of heaters has traversed one interzone spacing, d, it is returned rapidly to its original position (Figure 6.2a).

In the stripping section, zones are introduced in a similar way, but with somewhat different effect. In Figure 6.3a the lowermost heater of a battery of

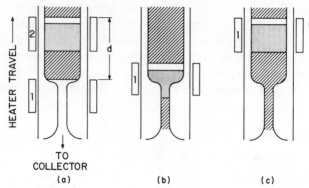

Fig. 6.2 Void generation in the enriching section of a continuous zone refiner.

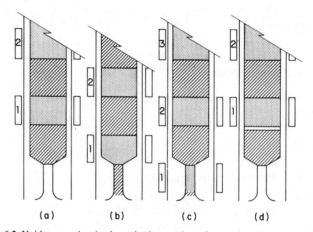

Fig. 6.3 Void generation in the stripping section of a continuous zone refiner.

heaters is shown in its highest position. The reduced-diameter section of the column is empty and the molten zone is flanked by solid. In Figure 6.3b, the zone has moved downward slowly and liquid has flowed from the bottom of the zone into the reduced-diameter section and solidified there. The loss of liquid from the zone has caused formation of a void at its top, of volume equal to that of the reduced-diameter section. Continued downward travel (Figure 6.3c) brings the lowermost heater nearly to the point of "breakthrough." After breakthrough, the liquid in the reduced-diameter section is discharged into the receiver and the heater battery moves rapidly upward through one interzone spacing to the position shown in Figure 6.3d. The void formed near the bottom of the column remains there. When the bottom of the heater reaches it during the next zone pass, liquid drips through the void and the void appears at the *top* of the zone. Hence, downward zone motion in the stripping section causes periodic upward motion of the voids. Ultimately voids in the stripping section rise to the feed point and are refilled.

The principles just described may be embodied in a device as shown in Figure 6.4 (1). The material to be zoned is charged into flask 1, which is provided with a charging port 2, an inert gas inlet 3, and stirrer 4. The charge is melted by heater 5 and flows into the stripping and enriching sections. The heater battery around the enriching section moves upward slowly through one interzone spacing and then reverts rapidly to its original position. The voids move upward with the zones. The heater battery around the stripping section moves downward slowly through one interzone spacing (which need not be the same as the spacing in the enriching section), and then reverts rapidly to its original position. The voids move upward discontinuously.

In the apparatus shown in Figure 6.4, the voids were generated around spindle-shaped plugs in the stripping and enriching sections. These plugs provide small annular apertures for discharge of melt and offer an advantage over a central capillary outlet in that the external annular passageway is closer to the surrounding heater. The central feed tank, 8-cm od $\times$ 10 cm tall, communicates with the vertical stripping and enriching sections (1.3-cm od $\times$ 37 cm long), each of which is provided with a 3-cm discharge section. Separate heater batteries traverse the two vertical tubes. The events taking place in void generation are the same as those described previously (see Figures 6.2 and 6.3). The device was tested on a mixture of naphthalene with 10^{-2} or 10^{-3}

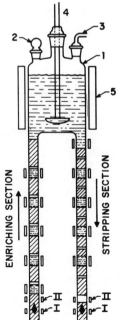

Fig. 6.4 Cross section of a continuous zone-void refiner. The arrows show the direction of zoning.

mole/mole of anthracene by zoning at speeds of $3.9 \times 10^{-4}\,\mathrm{cm\,s^{-1}}$ and $9.7 \times 10^{-4}\,\mathrm{cm\,s^{-1}}$ (1.4 and 3.5 cm h^{-1}) with zone lengths of 2 cm.

Product purity was determined by measuring anthracene fluorescence. It was found that product purity reached a constant value after 10 passes. Table 6.1 shows measured distribution coefficients for zoning at two rates, along with the separation ratios (C_p/C_0 and C_w/C_0 where C_p and C_w are the solute concentrations in the product and waste, respectively). Figure 6.5 shows the impurity concentration in product and waste as a function of number of zone passes.

Extensive mathematical modeling of the zone-void process is available (2). Probably because of its mechanical complexity and problems of breakage, no work on the zone-void method has been reported recently.

Early work on zone-void refiners was carried out by Pfann (2). Moates (3) and Buford and Starks (4) applied the technique to purification of silicon tetraiodide.

Table 6.1. Distribution Coefficients and Separation Ratios for Continuous Zone Refining of Anthracene in Naphthalene (1)

Initial Concentration C_0, mole/mole	Zoning Speed, cm s^{-1} $\times$ 10^4	Separation C_p/C_0	Ratio C_w/C_0	Distribution Coefficient, k_{eff}
10^{-3}	3.9	0.72×10^{-3}	7.5	0.468
10^{-3}	9.7	0.85×10^{-3}	6.0	0.472
3×10^{-3}	9.7	0.43	3.4	0.720
10^{-2}	3.9	0.47	3.2	0.735
10^{-2}	9.7	0.57	3.2	0.780

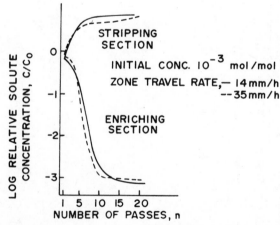

Fig. 6.5 Concentration profiles in the stripping and enriching sections of a continuous zone-void refiner.

Although their devices were operable, they were complex in structure and subject to breakage. Moreover, the purification that was attained fell short of expectation. Kennedy has described extensions of the application of the zone-void method (5). In an attempt to overcome the problems of breakage, Rozin et al. (6) proposed a modified zone-void method in which the stripping section is eliminated. Further modification was introduced to allow zone-void refining in orientations other than vertical (7); a horizontal device was used for the zone-void purification of aluminum (8).

Although zone-void devices are called "continuous," the feed, product, and waste move intermittently. The same is true of all the other modes described in this chapter except for the Helizone.

2 ZONE-TRANSPORT METHOD

Zone-transport devices use horizontal containers in which there is a difference in level between the solid entering a zone and that freezing out of the zone. The difference in level is brought about by the controlled introduction of feed at a central point. The product and waste are collected by overflow at the ends of the charge.

The container can have a square, rectangular, or other cross section with close-fitting external heaters as shown in Figure 6.6. Zoning in the direction shown moves solutes with $k < 1$ to the right and causes concentrated waste to overflow. The amount of overflow is indicated by the dashed triangle at the lower level of the exit port.

Kennedy and Moates (9) have described a horizontal hybrid continuous zone refiner that operates without a stripping section. The operation may be understood with the help of Figure 6.7. With only heaters 2 and 5 switched off, the feed is charged until a solid plug forms at the constriction in the vertical waste tube. Then the entire container is filled until liquid reaches the product-overflow level. Heaters 3 and 4 are then deenergized until the melt within them solidifies, making it possible to fill the vertical feed tube to the level shown in the figure. The zone heaters are set into motion toward the product overflow, with

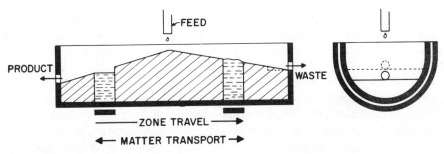

Fig. 6.6 Schematic diagram of continuous zone refining by the zone-transport method.

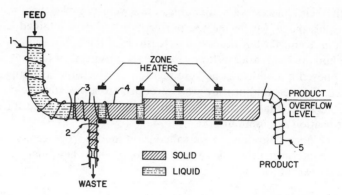

Fig. 6.7 Hybrid continuous zone refiner with no stripping section, after Kennedy and Moates (9).

heaters 1–5 under control of automatic timers. When the right-hand heater reaches the product exit, a quantity of melt overflows which is determined by the volume expansion accompanying melting. The heater battery then moves slowly to the left. When it reaches its extreme position, as shown, heater 2 is energized by its timer and the entire leftmost zone drains into the waste container. The heater battery then reciprocates to the right and the void formed by drainage to waste is refilled by deenergizing heater 2 and energizing heater 4. A timer energizes heater 3 allowing melt to fill the void within heater 4. Heaters 3 and 4 are deenergized, causing solidification of a solid barrier between the feed and zoning sections.

In the apparatus built by Kennedy and Moates, the zoning section was of 2.2-cm od tubing and the horizontal feed section was of 1.7-cm od tubing. The zoning section contained an empty displacement tube to equalize the zone volumes in the two sections. Thus, each 2-cm zone contained about 3.8 g of melt, and this amount was discharged to waste after each pass. At the same time, 0.65 g of product was collected, so that the recovery of purified product amounted to only 15%.

3   HELIZONE

In Section 2.7.6, Chapter 5, systems are described in which an annular charge is partially melted by a helical heater rotating within the charge. One such device is easily adapted to continuous introduction of feed and simultaneous removal of product and waste. The conversion from batch to quasicontinuous mode requires only that an access to the annular sample space be provided at some point between product and waste exits. Ideally, the point selected would be that at which the steady-state solute concentration equals the solute concentration in the feed. This arrangement cannot provide truly continuous operation, because the feed line can communicate with the helical molten zone only once per revolution. This defect is simply overcome by providing a small stationary horizontal zone as shown in Figure 6.8. This zone communicates with the rotating helical zone at all times and allows continuous introduction of feed.

Figure 6.9 shows a cross section of a vessel for continuous operation. It was used for purification of 1-nitro naphthalene from which the 2-nitro isomer and other impurities were readily removed as indicated in Table 6.2 (10).

Devices of similar geometry have been proposed by Anderson in connection with centrifugal zone refining (11). In these, an annular charge is rotated about a vertical central axis, and immersed heaters or external radiant heaters are rotated at a slightly different speed to provide slow motion of the heaters relative to the container. In these devices the centrifugal force field is vital to enhanced matter transport at the solid/liquid interface and serves to hold the charge in place. In principle, feed can be introduced to a fixed point in such a device (Figure 6.10a) and zone heaters may be arranged parallel to the external surface of the container. The container is tapered so that its walls form an angle with the vertical equal to the angle that a free liquid surface would assume at a given rotational speed. The nature of the zoning may be better understood from Figure 6.10b, showing the sample "unfolded" from its tapered, cylindrical form to a planar equivalent.

Feed is introduced through the central zone. The individual zones are formed by multiple external heaters whose motion relative to the charge is provided by the difference between the rotational speeds of ingot and heaters, respectively.

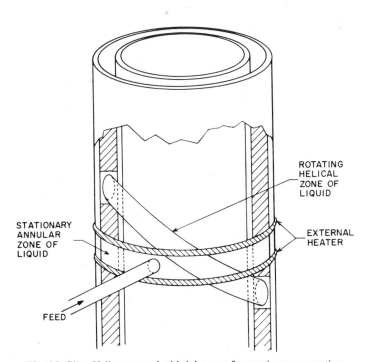

Fig. 6.8 Glass Helizone vessel with inlet port for continuous operation.

Table 6.2. Continuous Operation of Helizone; Purification of 1-Nitronaphthalene (10)[a]

Operating Condition No.	Hours at Steady State	Zoning Speed cm h^{-1}	Heater Wattage	Feed Stream, g day^{-1}	Product Stream, g day^{-1}	Waste Stream, g day^{-1}	Area % Impurity Product	Area % Impurity Waste
1	71	1.2	61	219	128	91	0.31	0.76
2	120	1.3	61	39	27	12	0.04	1.46
3	97	1.3	61	92	66	26	0.05	1.59
4	65	0.4	61	82	59	23	0.08	1.67
5	66	3.1	74	95	70	25	0.00[a]	1.58
6	64	1.3	74	92	68	24	0.00	2.03
7	73	1.2	74	105	70	35	0.00	2.31
8	73	1.2	74	81	35	46	0.00	1.73

[a]Limit of detection approximately 10^{-3} area %, by gas chromatography.

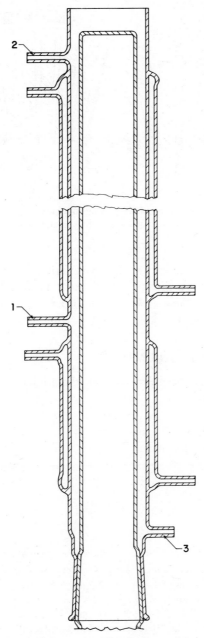

Fig. 6.9 Cross section of glass vessel for continuous operation of a Helizone refiner. 1. Feed port 2. Product port 3. Waste port.

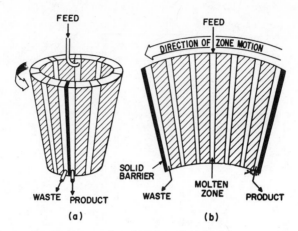

Fig. 6.10 Continuous centrifugal zone refiner. (*a*) Schematic view of apparatus; drive mechanism and heaters not shown for clarity. (*b*) Unfolded view of the annular charge showing zone motion.

In Fig. 6.10 the heaters are shown moving from right to left, and impurities with $k < 1$ are moved in this direction to a waste takeoff. Similarly, product is removed from the right side. In the actual device, the product and waste ports are adjacent to one another, but separated by an impermeable barrier.

4   OTHER CONTINUOUS MODES

Other methods have been proposed for the attainment of continuous zone refining. These include:

* Reflux cross-flow (2)
* Continuous floating zone (5)

Since these have not been exploited experimentally, they will not be discussed here.

REFERENCES

1. S. Hayakawa, T. Nakamura, M. Ohtani, and K. Abe, *J. Appl. Phys. (Japan)*, **36**, 977–981 (1967).

2. W. G. Pfann, *Zone Melting*, 2nd ed., Wiley, New York, 1966, pp. 10–12.

3. G. H. Moates, "Continuous Multistage Purification of Silicon Tetraiodide by Zone Melting," in M. Desirant and J. L. Michiels, Eds., *Solid State Physics in Electronics and Telecommunications*, Vol. 1, Academic Press, New York, 1960, pp. 1–8.

4. J. T. Buford and R. J. Starks, "Development of a Method for the Removal of B from SiI_4 by Zone Melting," in M. S. Brooks and J. K. Kennedy, Eds., *Ultrapurification of Semiconductor Materials*, Macmillan, New York, 1962, pp. 25–33.

5. J. K. Kennedy, Rev. Sci. Inst., **35**, 25 (1964).

6. K. M. Rozin, V. N. Vigdorovich, and A. N. Krestovnikov, *Izv. Akad. Nauk SSSR Otd. Tekhn. Nauk Met. i Toplivo*, **1961**(6), 56–73.

7. V. N. Vigdorovich and I. F. Chernomordin, Russian Pat. 161,488 (1963).

8. V. N. Vigdorovich and I. F. Chernomordin, *Izv. Akad. Nauk SSSR Met.*, **2**, 82–87 (1965).

9. J. K. Kennedy and G. H. Moates, "Continuous Zone Refining of Benzoic Acid" in *Purification of Inorganic and Organic Materials*, M. Zief, Ed., Dekker, New York, 1969, pp. 261–268.

10. A. R. McGhie, P. J. Rennolds and G. J. Sloan, *Anal. Chem.*, **52**, 1738–1742, (1980).

11. E. L. Anderson, "Purification of Naphthalene in a Centrifugal Field," in *Purification of Inorganic and Organic Materials*, M. Zief, Ed., Dekker, New York, 1969, pp. 269–284.

Chapter VII

ADDED-COMPONENT TECHNIQUES IN ZONE REFINING

---

1 Traveling-Solvent Method (TSM); $k_{eff} = 0$
2 Eutectic Zone Refining (EZR); $k_{eff} = 1$
3 Zone-Melting Chromatography (ZMC); $k_{eff} \neq 1$
4 Scavenging
References

In spite of the many unquestioned successes of zone refining, there are many systems and problems to which it is applied with difficulty if at all. As examples:

- Many chemicals decompose at their melting points or melt at temperatures so high as to be reached only with difficulty.
- Polymers, even if they do not decompose at their melting points, form entangled melts which do not yield well defined solid/liquid interfaces.
- Some binary mixtures have distribution coefficients very close to unity and are not practically separable.
- It is difficult to maintain a well defined solid/liquid interface in materials with very high vapor pressures at their melting points.
- Some materials are very reactive with the atmosphere or with containers.

To cope with these problems, a congeries of techniques has evolved. These are variously named and executed, but they have in common the addition of a "foreign" substance to the system whose separation is desired. The amount of the added substance may be small or large, relative to 'the mass being purified. The added substance may be mixed homogeneously or added only in portions large enough to form a single molten zone. Several related, added-component techniques will be discussed.

In conventional crystallization from solvents, one usually seeks a liquid solvent that is capable of dissolving a considerable amount of the material to be purified (the solute) and in which the solute has a large temperature coefficient of solubility. In seeking solvents for use in ACT, it is useful to consider the solubilities reciprocally; that is to say, the added component might be considered as a solute in the material being purified, since it is present in lesser amount. However, because the added component does not crystallize during the zoning, it will be referred to as the solvent. In zone crystallization from a solvent, three situations may be discerned (1):

1. The added component has distribution coefficient $k_{eff} = 0$.
2. The added component has $k_{eff} = 1$.
3. The added component has $k_{eff} \neq 1$.

The first two cases have been described fairly extensively. Case 1 has been called "the traveling-solvent method" (TSM) or "zone crystallization." Case 2 is called "eutectic zone refining"; the eutectic is by definition the composition such that solid and liquid, in equilibrium, have the same composition. Hence, no partitioning accompanies solidification of a eutectic, and $k_{eff} = 1$. Other circumstances can give rise to situations in which $k_{eff} \simeq 1$; for example, a system of solid solutions with a minimum melting point (Roozeboom type III) or a system in which a stable compound is formed. Case 3 is embodied in the process known as "zone-melting chromatography," wherein two or more substances are separated from one another on the basis of their differing distribution coefficients in a solid matrix.

Figure 7.1 shows schematically how TSM works. A small amount of

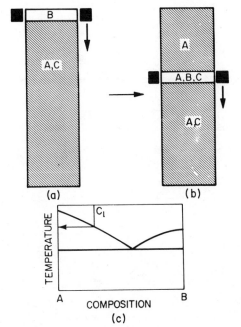

Fig. 7.1 Traveling-solvent method; a zone of solvent B is moved through an ingot of mixture A, C. A, B, and C form a single liquid phase, but A is not soluble in solid B.

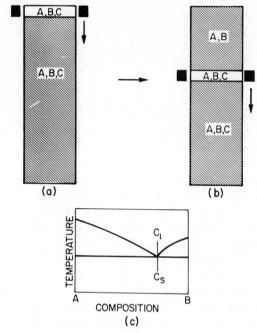

Fig. 7.2 Eutectic zone refining: enough B is added to the A, C mixture to form the A, B eutectic, which deposits in pure form as the molten zone descends, carrying impurity C with it.

$$0 < k_{eff} < 1 \text{ OR } k_{eff} > 1$$

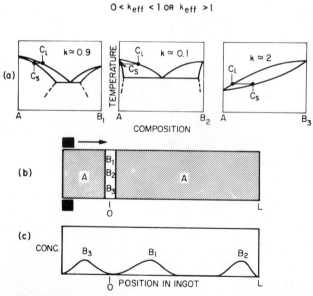

Fig. 7.3 Zone-melting chromatography: solutes B_1, B_2, and B_3 are deposited in an ingot of host A; B_1 and B_2 move with the zone, at different rates, while B_3 moves counter to the zone.

substance B (the solvent) is added to the head of an ingot consisting of substance A contaminated with C (Figure 7.1a). In Figure 7.1b a solvent zone has moved into the ingot carrying with it the impurity C and leaving behind pure A; Figure 7.1c shows that solvent B is not included in the crystallizing deposit of A.

Figure 7.2a shows the formation of an ingot of eutectic by addition of substance B to substance A contaminated with C. In Figure 7.2b, the molten zone has been moved into the ingot, carrying with it the impurity C and leaving behind the pure A/B eutectic. Figure 7.2c shows that passage of the molten zone produces a composition gradient in the eutectic ingot.

Figure 7.3a shows the phase relationships of components B_1, B_2, and B_3, with a selected host A. Their differing distribution coefficients determine their rates of migration during zoning. Figure 7.3b shows the zone at its starting position. Note that in this case, the mixture is not introduced at the head of the ingot, as it would be in conventional chromatography. Figure 7.3c shows a hypothetical distribution of solutes B_1, B_2, and B_3 after several zone passes. The fact that B_3 has $k > 1$ is responsible for its retrograde movement.

1   TRAVELING-SOLVENT METHOD (TSM); $k_{\text{eff}} = 0$

Nicolau (2) has discussed the nomenclature of purification methods based on passing a relatively small amount of added component with $k_{\text{eff}} = 0$ through a solid mixture. He distinguishes between traveling-solvent method (TSM) and the traveling-heater method (THM) on the basis of zone size and temperature gradients. But it appears to us that nearly all zone-melting methods could be called "THM" and that this name is not very descriptive or informative. Nicolau argues that TSM is a "renaming of the temperature-gradient zone melting process (TGZM)." In TGZM, however, the zone is not moved by the transport of a heater, but rather by the imposition of a stationary temperature gradient over the entire sample (3). Moreover, the zones are said to be very narrow. It is difficult to imagine the zoning of polycrystalline chemical systems taking place under these conditions. Hence, we feel that TSM is an appropriate designation for purification involving application of an added component at the beginning of a zone passage and removal of the added component at the end of the passage.

Nicolau (2) has derived the impurity profile along a semi-infinite ingot of a binary mixture after n passes of a zone of solvent. The derivation is based on the simplifying assumptions that (1) impurity diffusion and solvent inclusion in the crystallizate are negligible, (2) phase equilibrium obtains in the solution zone, and (3) the zone is homogeneous. The added component is called the solvent; the major component of the original mixture is called the solute, and the minor component of the original mixture is called the impurity.

For a parent mixture that forms solid solutions, the variation of impurity content in the solution zone as a function of heater position is given by

$$-\left(\frac{R_0 - kM}{M - R_0}\right)\frac{z}{l'} = \ln\frac{(R_0 - kM)Y - R_0Y}{(R_0 - kM)Y_0 - R_0Y} \tag{7.1}$$

where

R = the instantaneous weight ratio of impurity to solute in the crystallizate

R_0 = its value at the start

k = the distribution coefficient of the impurity between the crystallizate and the solution zone

$M = -Y/X$, the slope of the dissolution line

z = position of the heater

l = heater length

l' = reduced heater length = $\dfrac{l}{1 - \dfrac{1}{\alpha}}$

x = concentration of the saturated solute in the solution zone, by weight, at the crystallization temperature

X = concentration of the saturated solute in the solution zone, by weight, at the dissolution temperature; $\alpha = X/x$

Y = concentration of the saturated impurity in the solution zone, by weight, at the dissolution temperature

Y_0 = the value of Y at $z = 0$

In the derivation of Equation 7.1 it was assumed that the length of the solution zone is the same as the length of the zone of added solvent and that the solubilities of solute and impurity are independent.

A further relationship was derived from Equation 7.1, giving the profile of the impurity ratio in the crystallizate as a function of heater position:

$$R = kM + \frac{R_0 - kM}{1 + \dfrac{R_0(1 - k)e^{-hz/l'}}{k(R_0 - M)}} \tag{7.2}$$

where h is a constant equal to $(R_0 - kM)/(R_0 - M)$. This function is plotted in Figure 7.4 for various values of k.

The theoretical treatment leads to the conclusion that the most favorable results ensue with $k < 1$, while $k > 1$ is unfavorable. In TSM, impurity segregation depends on M and k. These quantities, in turn, depend on solute/impurity/solvent relationships and on temperatures chosen for dissolution and crystallization. Evidently the use of a solvent introduces greater flexibility than one encounters in binary melt systems and correspondingly makes more difficult the attainment of optimum conditions. In general, however, it can be said that the dissolution temperature T should be close to the boiling point of the solvent while the difference between T and t (the crystallization temperature) should be small.

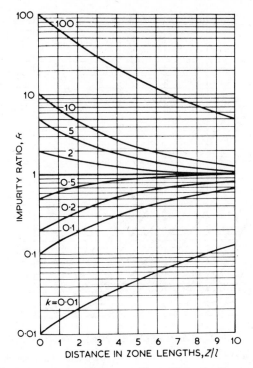

Fig. 7.4 Profiles of impurity ratio, R, in the crystallizate after a single passage of a solution zone, with distribution coefficient k as parameter.

Nicolau also derived an iterative formalism for calculating the impurity ratio after an arbitrary number of solution-zone passes. In this treatment, the solute zone at the start of each pass is assumed to contain both solute and impurity. Experimentally, however, it is common to introduce a fresh portion of pure solvent at the start of each pass; hence, the calculated results must be viewed as an approximation. Nevertheless, they show convincingly that a small number of zone passes should produce large diminutions of impurity content (see Figure 7.5).

Wolff and Mlavsky (4) have reviewed traveling-solvent techniques with emphasis on the growth of crystals of metals, alloys, and semiconducting compounds. For example, passage of a liquid gallium solution through solid GaP or GaAs removed a considerable amount of impurity. Similarly growth of SiC by passage of a zone of molten chromium produced substantial purification. Broader applications of TSM have emerged, as in the growth of large, optically clear crystals of calcite (5).

TSM has also been used for the growth of single crystals of lead–tin telluride, a semiconductor in use as a photodetector. A "sandwich" was prepared, using a source crystal, a substrate crystal, and an intermediate solvent layer. The source and substrate crystals were Bridgman-grown plates of $Pb_{1-x}Sn_xTe$, of about

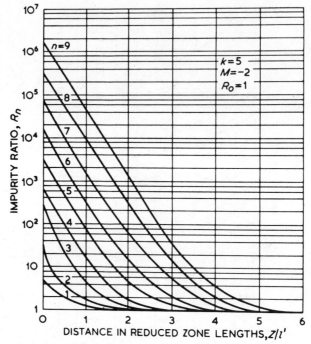

Figure 7.5 Profiles of impurity ratio, R, in the crystallizate after one through nine passages of a solution zone, with number of passes n as parameter.

0.1-cm thickness with cross-sectional areas of $0.5–1$ cm². The solvent layer consisted of a lead–tin foil about 200 μm thick, and the growth temperature was 858 K with a gradient of 10 K cm^{-1} (6).

TSM has been applied to the growth of single crystals of II–VI compounds such as HgTe (7), ZnTe (8), and CdTe (9, 10, 11). CdTe has been purified and crystallized by a modified TSM, in which an ingot of Te (solvent) was deposited in the bottom of a vertical quartz tube and an ingot of CdTe was placed above it. The heater was positioned just below the Te/CdTe interface to melt a 1-cm layer of Te. The heater, at 1023 K, was raised at 5.79×10^{-6} cm s^{-1}, (0.5 cm day^{-1}) until the Te was deposited at the top of the CdTe ingot. The heater was set for a second pass at a level 1 cm lower than for the first pass, and so on, for a total of five passes. The resulting crystal was of better quality and higher purity than crystals prepared by single-pass TSM (12).

Two devices have been described for large-scale purification by TSM (13). These are shown in Figures 7.6 and 7.7. The rotating, closed tube of Figure 7.6 advantageously provides a radially symmetrical solid/liquid interface, which leads to a more homogeneous solid crystallizate. Furthermore, the impressed temperature gradient is steeper in the glass tube than in the steel one, yielding better solute transport. In both cases, the tilt of the tube promotes drainage of the entrapped solution from the interface of the crystallizate. In the open tube,

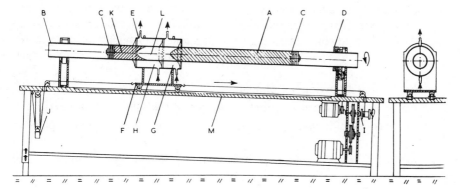

Fig. 7.6 Cross sections of a closed-tube apparatus for TSM zone refining.

A—solute/impurity system to be purified
B—glass sample tube
C—rubber seals
D—supporting gear or ring
E—glass surrounding sleeve
F—moving carriage
G and H—heating and cooling compartments, respectively
I—cable drum
J—counterweight
K—purified solute
L—solution zone
M—support table
N—stirrers
O—contact thermometer.

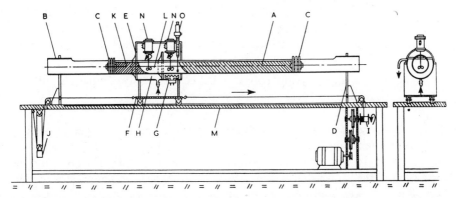

Fig. 7.7 Cross sections of a stainless-steel, open-tube apparatus for TSM zone refining. The components are identified by the legend of Fig. 7.6, with stirrers N and contact thermometer O in addition; in this case tube B and sleeve E are of stainless steel.

333

crystals that are detached from the dissolving interface may reach and deposit upon the crystallizing interface; this does not occur in the closed tube. In summary, these advantages lead to the attainment of higher purities in the closed-tube system. The closed-tube does have some important disadvantages:

- Filling and emptying are difficult.
- Air must be excluded to avoid disruption of the solidification interface.
- There is risk of breakage of the glass tube as a result of thermal expansion of solution.
- Samples can only be removed from the tube end, upon completion of zoning.

While TSM appears to be very broadly applicable, it is important to recognize the stringent criteria for selection of the solvent:

- It must dissolve both major and minor components of the system to be purified, but must not itself dissolve in the crystalline solute.
- The temperature coefficient of solubility should be positive.
- If a minor component forms a solid solution with the major component, k should be as far as possible from 1.
- If the minor components form eutectics with the major component, they should contain substantial amounts of the former.
- The solute should form a compact crystal mass to minimize solvent entrapment.
- The solvent should be readily removed from the solute.

In addition, two experimental conditions must be met:

1. The intrinsic crystallization rate should be high enough to allow a practical rate of zoning. To be sure, crystallization rate can be increased by increasing the difference between dissolution and crystallization temperatures, but it should be remembered that efficiency of purification is reduced if this difference is too large.
2. The solute, impurity, and solvent must be chemically compatible with the materials of construction, and not mutually reactive.

To test these ideas, a synthetic mixture of potassium dihydrogen phosphate with potassium chloride was prepared in the weight ratio 100:7; 5 kg of this mixture and 0.5 L of water filled 1.15 m of the refining tube. The solvent zone consisted of 0.3 L of water, which gave a 12-cm zone. The temperature in the solution zone was controlled at 363 K and the zone was moved through the slurry at 5.6×10^{-4} cm s^{-1} (2 cm h^{-1}). The parent KH_2PO_4/KCl system forms a eutectic and does not show solid solution or compound formation. The impurity ratio was determined readily by volumetric analysis and the resulting values, after five zone passes, are given in Table 7.1.

A similar experiment was carried out with the open stainless-steel tube shown in Figure 7.7. The solvent zone consisted of 0.4 L of water and the resulting solution zone was maintained at 363 K. As the starting water zone was heated, part of the solute mixture dissolved and the solution volume increased until saturation was attained. The zone was moved at 8.3×10^{-4} cm s^{-1} (3 cm h^{-1}) through the mixture of salts. As before, the entire mixture was dissolved and successively crystallized. After completion of each zone pass, the solution zone was decanted and a fresh portion of water introduced at the starting end. The profiles of the impurity ratio after one, two, and three passes are shown in Figure 7.8. The values after the third pass, namely 0.56 at the beginning of the crystallizate and 0.275 at the end of the crystallizate, compared with the starting value of 0.33, show that purification was effected but less markedly than in the closed-tube system (see Table 7.1).

In a related experiment starting with technical-grade disodium hydrogen phosphate, 10 zone passes in the closed-tube system lowered the concentrations of most impurities by about two orders of magnitude. These included chloride, nitrogen compounds, sulfate, arsenic, heavy metals, and iron.

Table 7.1. TSM of KH_2PO_4/KCl with Water, in the Closed-Tube System

Zone Pass No.	Impurity Ratio at the Beginning of the Crystallizate	Impurity Ratio at the End of the Crystallizate
0	7×10^{-2}	7×10^{-2}
1	2×10^{-3}	4×10^{-3}
2	4×10^{-4}	6.6×10^{-4}
3	1×10^{-4}	1.6×10^{-4}
4	2×10^{-5}	3×10^{-5}
5	8×10^{-6}	1×10^{-5}

The system nickel ammonium sulfate doped with iron (II) ammonium sulfate, which forms solid solutions, was also examined. The mixture of double sulfates, in weight ratio 10:1, was charged into the closed-tube system in the amount of 3.8 kg of mixture with 0.4 L of water at 373 K. The deaerated, cooled paste filled 1 m of the tube. A 20-cm water zone, containing 0.5 L of water, was introduced and passed through the polycrystalline ingot at a constant temperature of 343 K and a travel rate of 3.3×10^{-4} cm s^{-1} (1.2 cm h^{-1}). At the end of each zone pass, the solution zone was decanted and replaced at the starting end of the ingot with fresh water. Figure 7.9 shows the impurity ratios along the length of the ingot

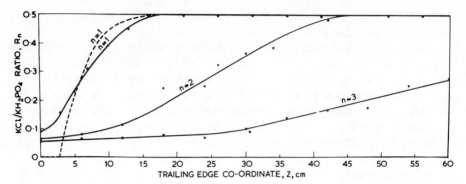

Fig. 7.8 Profiles of impurity ratio, R, in the crystallizate after one, two, and three passages of an aqueous solution zone through a charge of KH_2PO_4/KCl, in the open-tube apparatus.

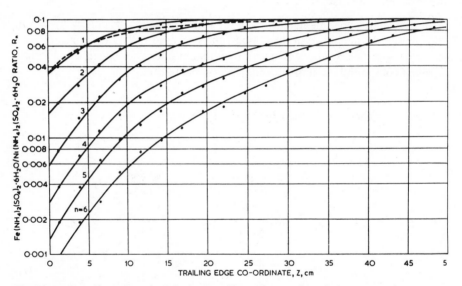

Fig. 7.9 Profiles of impurity ratio, R, in the crystallizate after one through six passages of an aqueous solution zone through a charge of $(NH_4)Ni(SO_4)_2 \cdot 6H_2O/(NH_4)_2Fe(SO_4)_2 \cdot 6H_2O$, in the closed-tube apparatus.

after passes 1 through 6. As a practical matter, it should be noted that the tube had to be broken for the sampling shown in Figure 7.9, so that each curve of the figure represents a separate charge of the same starting composition, carried to successively higher numbers of zone passes. It is evident that the system described by Nicolau produces substantial purification; but it is equally evident that considerable improvement is necessary before this technique becomes readily adaptable for broad laboratory use.

Eldib has described the attempted purification of microcrystalline waxes, which consist of complex mixtures of aliphatic hydrocarbons. The low diffusitivities in these systems hindered effective separation. The wax mixtures were, however, separated by the addition of various solvents followed by the passage of molten zones through the resulting ingots of solution. The situation here differs from the experiments described above in that the added foreign substance is at the outset uniformly distributed in the sample to be purified, rather than applied as a discrete zone (14).

It is relevant to describe an isothermal procedure for recrystallization from a solvent, which makes use of an advancing solvent front and may be carried out in a columnar configuration. If a suitable solvent is allowed to flow through a bed of finely powdered solid, the solvent will become saturated with respect to small particles while remaining unsaturated with respect to large ones. As the liquid solvent advances through the bed of solid, recrystallization takes place continuously, slightly behind the solvent front. Since fresh solvent is added continuously, the surviving crystals are washed and impurity collects at the liquid front (15).

Essentially the same procedure was "rediscovered" more recently, and was applied to the purification of dibenzofuran (16). The solid was packed into a jacketed column and coolant was passed through the jacket. Ethanol was added slowly to the column at a rate such that no liquid accumulated at the top. A black solution was collected, and the material remaining in the column was tan. Only after complete elution of the black material from the column did dibenzofuran begin to crystallize from the eluting drops. Using this procedure, triphenyl phosphine was said to have been removed from triphenyl phosphine oxide, with methylcyclohexane at 298 K. Likewise, impurities were removed from fluorene, with alcohol at 283 K. This column elution method with microparticulate solids has not been widely used and merits further study.

Smith and Downing (17) have used a related extraction procedure and described it in terms of phase solubility analysis (PSA). In PSA, different amounts of a crystalline sample are equilibrated with a fixed amount of a selected solvent. Solution compositions are measured and plotted against system composition (amount of solid per unit weight of solvent). In such a plot (Figure 7.10), point X_1 indicates saturation of the solution with respect to the main component. In the presence of soluble impurities, the solution composition continues to rise with increasing total solute concentration. The extrapolated solution concentration Y_0 is the solubility of the main component of the original solid sample. If a portion of solid is agitated with an amount of solvent insufficient to dissolve it (system composition X_2), then the enrichment of impurity in the solution is given by i_2/i_1, where

$$i_1 = \frac{\Delta Y}{X_2}$$

and

$$i_2 = \frac{\Delta Y}{Y_0 + \Delta Y}$$

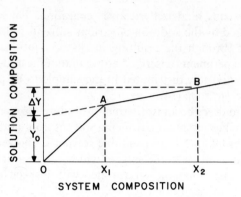

Fig. 7.10 Phase solubility analysis.

where i_1 represents the fraction of impurity in the original sample, i_2 represents the fraction of impurity in the solute after equilibration with solvent, and ΔY is the change in solution composition corresponding to the change in system composition from X_1 to X_2. Thus the solute recovered from the solution contains i_2/i_1 times more impurity than the original sample, assuming that at equilibrium the undissolved solid contains only the main component.

2   EUTECTIC ZONE REFINING (EZR); $k_{eff} = 1$

In one of the earliest papers on zone melting of chemicals, Süe et al. (18) suggested that zoning of chemical compounds might offer a route to some pure metallic elements without having to cope with the high melting points and/or volatilities of the elements themselves. As a logical next step, they zoned eutectic mixtures as a means of circumventing the instabilities or volatilities of the desired compounds themselves. Their results pointed to the general possibility of separating substances A and B by adding a third component C which forms a eutectic with the major component of the AB mixture and in which the minor component segregates.

The first unstable material zoned by Süe et al. as a eutectic was lead nitrate. This compound, which cannot be melted without decomposition, was mixed with potassium nitrate in the ratio 47 wt% lead nitrate, giving a eutectic mixture of mp 480 K. Radioactive nitrates of Na, Cs, Sr, and La were added in the ratio of 1.0 wt% with respect to the lead nitrate. After four zone passes, the concentrations of Na and Cs were reduced by factors of four and five, respectively, at the head of the ingot. The other added impurities were not substantially segregated. The same tactic was used for preparation of pure barium nitrate by zoning its eutectic with potassium nitrate (27 wt% $Ba(NO_3)_2$, 560 K) or with barium chloride. In both cases, alkaline earth cations were removed (19). Another study of the barium nitrate/potassium nitrate eutectic showed that after five passes the calcium content was reduced from 0.18 to 0.0012%. In related work, calcium nitrate was purified by zone refining its eutectic with potassium nitrate (20). The eutectic, which contains 40.5 wt% $Ca(NO_3)_2$ and melts at 435 K, was zoned at 0.7 cm h^{-1} in a glass tube in a horizontal apparatus, using a 3-cm zone and 12-cm ingot. Traces of aluminum, manganese, and copper were added to the eutectic and these were found to segregate effectively, along with two native trace impurities, magnesium and iron. The distribution coefficients of the major impurities (barium and strontium) were measured before and after addition of the trace contaminants. The substantial increase that was observed (Table 7.2) was attributed to formation of a difficultly soluble compound with $Ca(NO_3)_2$ (20).

Strontium and potassium nitrates form a eutectic containing 55 wt% of the strontium compound and melting at 480 K. EZR was carried out on a batch containing 0.016% calcium; after three passes of a molten zone through an ingot at 0.7 cm h^{-1}, the calcium content in about half the ingot was reduced to 4×10^{-3}% (21).

The eutectic of $InCl_3$ with NaCl (21.5 wt%) was zoned to remove Sn and Pb, present in amounts of $10^{-4}–10^{-2}$ wt%. The effective distribution coefficient of Sn increased in this range, while that of Pb was constant and equal to 0.21 (22). The $CoCl_2$/NaCl eutectic has been zoned and subjected to homogeneous reduction to give Co containing $< 10^{-5}$% impurity and an electrical resistivity ratio of 290 at 294 K with respect to that at 4.2 K (23).

EZR has been suggested as a means of purifying silver iodide by formation of eutectics with rubidium iodide or potassium iodide. Distribution coefficients of Ni, Fe, Pb, Mg, Zn, Sb, and Al were determined by directional crystallization of both eutectics and the effectiveness of EZR was established (24).

Eutectic cryohydrates have been used for purification of both organic and inorganic compounds. Thiamine bromide, thiamine chloride, and ascorbic acid were all zone refined as aqueous eutectics. The thiamine bromide system showed a stable eutectic at 11 wt% thiamine bromide and 272.5 K and a metastable eutectic at 41 wt% and 268.2 K. The thiamine chloride system similarly showed a stable eutectic (31.5 wt%, 268.7 K) and a metastable eutectic (44 wt%, 265.6 K).

Table 7.2. Effective Distribution Coefficients of Impurities in $Ca(NO_3)_2$/KNO_3 Eutectic

Impurity	Concentration wt%	k_{eff}
Al	1×10^{-3}	0.33
Mn	4×10^{-3}	0.33
Cu	2×10^{-3}	0.14
Mg	2×10^{-3}	0.25
Fe	1×10^{-3}	0.33
Sr	0.06	0.46, 0.56[a]
Ba	0.76	0.60, 0.78[a]

[a] After addition of trace impurities.

Ascorbic acid/water showed a eutectic at 11 wt% ascorbic acid and 269.7 K. In the absence of forced mixing in the molten zone, only slow zoning (5.5×10^{-5} cm s^{-1}) was effective (25). The eutectic cryohydrate of potassium chloride was purified by both directional crystallization and multipass zone refining. Various radiotracers were added at concentrations of 0.1 wt% referred to the KCl. The concentrations of sodium and strontium chlorides at the head of the ingot were reduced by factors of 28 and 10, respectively, after 10 zone passes (18). Potassium nitrate and copper sulfate have also been purified as eutectic cryohydrates in a nearly horizontal apparatus with a rotating tubular container (26). Calcium nitrate hydrate was effectively purified by zoning at 0.7 cm h^{-1} and 315 K. Five passes reduced the concentration of barium from 0.9 to 0.4 wt% ($k_{eff} = 0.69$). The distribution coefficient of strontium in the hydrate was found to depend strongly on its concentration: k_{eff} was found to be 0.11 for 0.05 wt% and 0.56 for 0.07 wt%. Trace impurities were efficiently segregated from the hydrate (20).

Eutectic zone refining has been applied to the anthracene/carbazole system (27). Although the components are stable at their melting points, segregation of carbazole is slow because its effective distribution coefficient is nearly 1. However, carbazole was removed quite rapidly from the anthracene/benzoic acid eutectic (~ 10 wt% anthracene, mp 390 K) in which its k_{eff} was found to be about 0.6. It was suggested that the enhanced segregation may result from the high degree of crystal perfection which results from solidification of a lamellar eutectic.

The eutectic method was also applied to tetracene (27), which is substantially more difficult to purify than anthracene because it is unstable at its melting point and is photochemically unstable. The eutectic composition of tetracene with 2-naphthoic acid was readily established by zone melting an arbitrary mixture and analyzing the terminal region of constant composition; it was found to contain about 7 wt% tetracene. A synthetic mixture of the eutectic with 2 wt% of 2,3-benzofluorene (referred to the tetracene) was prepared and subjected to 25 downward zone passes at 6.9×10^{-4} cm s^{-1} (2.5 cm h^{-1}). Analysis of the resulting ingot by liquid chromatography showed that nearly all of the ingot was free of 2,3-benzofluorene (the limit of detection was about 0.05%); the bottom contained essentially all of the impurity.

For effective eutectic zoning, the added component must:

- Form a thermally stable, nonreactive, single-phase liquid solution with the substance to be purified, with a melting point well below that of the compound of interest.
- Give distribution coefficients that are less than 1 for impurities.
- Give a eutectic containing a high concentration of the compound of interest.
- Be readily removable from the mixture.
- Be readily available.

Another ACT is which $k_{eff} = 1$ results from deliberate formation of a molecular compound to facilitate removal of impurities. This procedure was adopted to enhance the purification of aluminum bromide, which is itself not readily purified by zone melting. Addition of a mole of lithium bromide yields LiAlBr$_4$, from which impurities are readily segregated, and from which pure AlBr$_3$ is readily recovered (28). In similar fashion, TiCl$_4$ was converted to an addition compound with propionitrile; impurities were more readily zoned from the addition compound than from TiCl$_4$.

Haarer and Karl (29) attempted to purify both components of the complex between anthracene and pyromellitic dianhydride. The latter could not be zone refined because of decomposition at its melting point (556 K). The 1:1 complex itself, however, melted with slower decomposition, at 513 K. It was possible to purify the complex by zone refining and then to grow single crystals from the vapor. In later work, it was found that pyromellitic dianhydride can in fact be zone refined if it is free from pyromellitic acid. The acid decomposes at the melting point of the anhydride, with elimination of CO$_2$ and H$_2$O. Merely heating the mixture below the melting point results in gradual conversion of the acid to the anhydride, which can then be zoned without difficulty (30).

3 ZONE-MELTING CHROMATOGRAPHY (ZMC); $k_{eff} \neq 1$

In their classic 1958 paper, Süe et al. (18) anticipated the zone-melting mode later named zone-melting chromatography* (ZMC). They described a method for measuring distribution coefficients by applying an impurity to the head of an ingot and distributing it by passage of a single molten zone through the entire ingot. In one experiment, they distributed two impurities (rubidium and cesium nitrates) simultaneously through a single matrix ingot of potassium nitrate and found that the distribution coefficients differed by a factor of 2.5.

This result was said to imply that "the operational mode being described makes it possible to obtain a certain separation between two salts A and B introduced at the head of a cell containing an appropriately chosen substance C." Pfann (31) has defined zone-melting chromatography as "... placing a mixture of solutes at a point in a long column of solid solvent and passing many molten zones along the column to separate the components of the mixture into individual bands." Both definitions imply that each solute will migrate through the host at a rate determined by its effective distribution coefficient. The more k_{eff} differs from 1, the more rapidly it migrates. Since k_{eff} can be greater than 1 as well as less, migration can take place in both directions from the starting point. In this respect, ZMC differs from conventional chromatographies.

The mathematical description of ZMC originated with Reiss and Helfand (32) and Pfann (33), under the usual simplifying assumptions of constant k, no

---

*Both "zone chromatography" and "zone-melting chromatography" have been used as names for the technique under discussion here. Since almost all chromatography can be said to involve zones, we feel that the latter term is more descriptive.

diffusion in the solid, complete mixing in the liquid, constant zone length, and negligible change in density upon freezing. Assuming further that each solute originally occupies a delta distribution centered at coordinate l on the x axis (i.e., the ingot axis), and that index $m = (n - 1)$ where n is the number of passes, then

$$\frac{lC_n(x)}{kK} = \frac{C_n(x)}{kC_i} = \frac{1}{m!} \, [k(x_l + m)]^m \, \exp[-k(x_l + m)] \tag{7.3}$$

where $C_i = K/l$ is the concentration in the zone at $x = 0$ in the first pass, that is, when the front of the zone has reached the delta function of the solute at $x = l$ (31). This equation is essentially that of a Poisson distribution, whose properties are used extensively in the theory of gas chromatography. In Figure 7.11 such distributions are simulated by triangles in a plot of distributions for various k's and numbers of zone passes, as a function of relative concentration. The figure shows clearly that peaks move and spread more rapidly with decreasing k, and that peak movement is retrograde for $k > 1$ and zero for $k = 1$.

Maeda and his co-workers (34) have modified the Pfann/Reiss-Helfand (PRH) treatment in several respects. They made the realistic assumption that no solutes would be present with $k_{eff} > 1$, so that only the positive x axis needs to be considered; the treatment further assumes that the solute charge occupies only one zone length of the column at $x = 0$. An iterative solution of Equation 7.3 was derived and evaluated by numerical means, using a digital computer. The resulting distribution profiles differ only slightly from those calculated according to PRH. In particular, the distributions according to Maeda et al. are displaced to somewhat more positive values of x for small numbers of zone passes when k is near 1, that is, when the solute peak has not moved far from the origin. Another minor difference occurs in the predicted dependence of peak position upon number of zone passes; the relationship is linear according to PRH, while slight deviations from linearity result from the Maeda theory.

In a further extension of the PRH theory, Maeda et al. considered the case in which it is impractical to deposit all of the solute in a single zone length of the column. They derived a solution for the case in which the solute is deposited in an arbitrary length of the column, and established that the resulting distributions do not change much as the starting length varies from one to five zone lengths. This result is experimentally significant, since it implies that solutes that are only slightly soluble in a host column can be handled in dilute solution at the outset. The effect of zone length upon separation efficiency was considered next. An iterative procedure was derived for maximizing the amount of solute deposited in a given bottom-fraction of the ingot, by stepwise reduction of zone length. An optimum sequence of zone lengths can be calculated and in each case, the number of passes required to attain a desired "compression" of solute is less than for multipass zoning with constant zone length.

The behavior predicted by PRH was qualitatively confirmed by a model experiment in which a single solute, copper acetylacetonate, was distributed in a column of 2-methoxynaphthalene (35) (Figure 7.12). The position of the peak

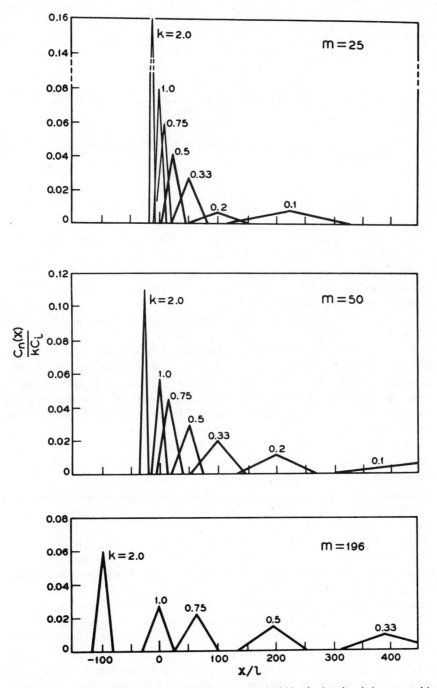

Fig. 7.11 Schematic representation of height, location, and width of solute bands in zone-melting chromatography, for various values of the distribution coefficient k, after $m = 25$, 50, and 196 zone passes (with permission of John Wiley & Sons).

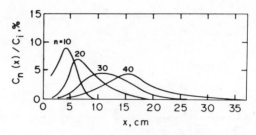

Fig. 7.12 Distribution of copper acetylacetonate in an ingot of 2-methoxynaphthalene, after 10, 20, 30, and 40 zone passes.

maximum varied linearly with number of passes according to the expression $x_{max}(n) = 0.40n - 0.45$. Moreover, it was found that the position of the peak maximum was independent of solute concentration over the range 340–1500 ppm. This too is in accord with Pfann's early theoretical predictions (33). To achieve these results, it was necessary to form small molten zones of constant size. This condition was attained by using circulating air for both heating and cooling, with baffles separating hot and cold zones. In a 0.6-cm tube, zones of 0.5-cm height were maintained with a relative standard deviation of 4.1%, with zoning at 1.4×10^{-3} cm s^{-1} (4.9 cm h^{-1}).

Even before the work of Maeda, Maruyama (36) had applied ZMC to the separation of acetylacetonates (AA) and thenoyltrifluoroacetonato chelates in a column of benzene. It was concluded that the technique offered promise for removal of iron from thorium. The acetylacetonate of Cu(II) was later separated from Co(II) (AA) and Ni(II) (AA) by zone melting in mercaptoacetanilide or in a mixture of mercaptoacetanilide and biphenyl (37).

ZMC has been applied, for example, to the separation of potassium and rubidium nitrates. These very similar substances give, with ammonium nitrate, distribution coefficients of 0.61 and 1.36, respectively. Hence, they should zone to opposite ends of an ingot of ammonium nitrate. Directional crystallization of ternary mixtures containing various amounts of K and Rb (1.32 and 5.40 mole%, 3.22 and 3.48 mole%, and 5.60 and 1.20 mole% of the two salts) confirmed that each was distributed independently of the other (38).

A different approach was followed by Plancher et al. (39) in seeking smaller, more stable molten zones. These workers used metal susceptors to concentrate energy from an external induction heater. Plate-type load coils were used to concentrate the radiofrequency energy from a 2.5-kW generator operating at 450 kHz into a shallow, horizontal plane. Each coil consisted of a two-turn, planar spiral of copper tubing ($\frac{3}{16}$-in. od), within the inner turn of which the $\frac{1}{2}$-in. od zoning tube traveled upward at 1.3×10^{-3} cm s^{-1} (4.7 cm h^{-1}). The susceptors were gold-plated disks of 18-mesh, galvanized-wire screen which were 0.038 cm smaller in diameter than the id of the tube.

The operation was carried out in a freezer at 243 K. The model system selected was a mixture containing coronene and benzo(ghi)perylene, 4.7 mg of which was applied to a column of solid benzene in the $\frac{1}{2}$-in. tube. Additional solid benzene was added above the sample to allow for possible retrograde transport of materials with $k_{eff} > 1$. A susceptor was applied to the top of the ingot and the tube was suspended so that the susceptor was 2.5 cm below the coil. As the tube was raised, the heated susceptor melted the ingot and sank to a position at which it received just enough energy to melt the benzene. The molten zone was about 0.1 cm long. This value is about one-tenth of the tube diameter and hence, an order of magnitude smaller than the zones formed by external heating. At the end of the pass, the susceptor and the melt adhering to it dropped from the open bottom of the tube into the receiver. Additional benzene was added to the bottom of the tube. A new susceptor was added and a second pass was begun. The distribution of the two solutes in a benzene ingot after 40 such passes is shown in Figure 7.13. Clearly, k_{eff} for coronene in benzene is less than that of benzo(ghi)perylene, and some separation was achieved. Equally clearly, a large amount of manual work was required to achieve this result and there is room for considerable improvement in efficiency.

Doron and Kirschner (40) tried to separate two-component salt systems and to resolve diastereoisomeric mixtures by zone melting through ice. They used a horizontal glass tube of 2-cm diameter in a chamber at 213 K. A single electric heater was advanced at 1.7×10^{-3} cm s^{-1} (6 cm h^{-1}), giving a zone that fluctuated between 1.5 and 2 cm in length (i.e., about one tube diameter). The heater was followed by a freezer containing solid CO_2 and ethanol. In some experiments, the after-cooler was omitted and the zone length was 5 to 6 cm.

Three pairs of salts with common anions were studied, namely copper/barium chlorides, calcium/copper acetates, and copper/barium nitrates. The salts were dissolved in a minimum amount of water at room temperature and the resulting solution was diluted with an equal amount of water. The solutions were frozen and subjected to six zone passes, sectioned, and analyzed. Final concentrations at the start and end of each ingot are given in Table 7.3.

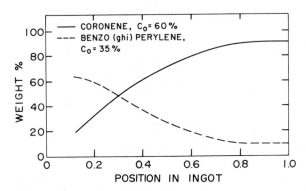

Fig. 7.13 Concentration profiles of coronene (——, $C_0 = 60\%$) and benzo[ghi]perylene (---, $C_0 = 35\%$) after ZMC. Note: The sample contained about 5% other materials.

Table 7.3. Separation of Two-Component Salt Systems in Aqueous Solution (40)

Salt Pair	% Decrease in Concentration^a at Start of Ingot				% Increase in Concentration^a at End of Ingot			
	Cu	Ba	Ca	E_l^b	Cu	Ba	Ca	E_m^b
Ca(OAc)$_2$·2H$_2$O/Cu(OAc)$_2$·2H$_2$O	4.0		4.7	0.70	17.0		17.4	0.41
CuCl$_2$·2H$_2$O/BaCl$_2$·2H$_2$O	61.0	49.0		12.0	89.5	30.5		59.0
Cu(NO$_3$)$_2$·3H$_2$O/Ba(NO$_3$)$_2$	47.6	26.2		21.4	78.3	4.0		74.3

^{a}With respect to C$_0$.

^{b}E = (% change of more soluble component) − (% change of less soluble component). Subscripts l and m refer to less soluble and more soluble component, respectively.

346

In Table 7.3 the three pairs are tabulated in order of increasing difference in the solubilities of the partners of the respective pairs. Clearly, separation took place, to an extent that increased with increasing solubility difference. The success with these systems suggested that diastereoisomeric pairs and racemic mixtures might be separated in the same way, and three such systems were explored: D,L-(Co(en)$_3$)Cl$_3$, racemic propylenediamine dihydrogen tartrate, and tris(acetylacetonato)cobalt(III). Even with a diastereoisomeric pair showing a large difference in solubility ($\Delta_{soly} = 36.3$ g/100 g H$_2$O), only a 0.5% resolution was achieved. This was taken by Doron and Kirschner to indicate limited applicability of the method. It should be noted, however, that the experiments were not carried out with small zones of constant size. If experimental conditions were optimized, better results might be expected.

Polymer fractionation is an area to which ZMC methods may be fruitfully applied: in fact, the earliest efforts along this line antedate the codification of added-component techniques. The assumption underlying this application is that in a polymeric material containing molecules having a range of molecular weights, each molecular species will have its own effective distribution coefficient. Peaker and Robb (41) appear to have been first to apply zone melting to polymer fractionation. They zone refined naphthalene in a 2.2-cm diameter × 30 cm tube and added to the pure end of the ingot 200 mg of polystyrene ($\bar{M}_n = 8 \times 10^4$; $\bar{M}_w = 2 \times 10^5$). The polymer was zoned through the ingot at 6.9×10^{-4} cm s^{-1} (2.5 cm h^{-1}) and its progress could be followed visually since its presence caused the transparent naphthalene ingot to become opaque. After eight passes, the polymer was distributed throughout the ingot, with 73% in the bottom fifth of the ingot. The ingot was sectioned and the naphthalene was sublimed away, leaving fine fibers of polymer. Almost three-fourths of the polymer was in the lowest fifth of the ingot, with the remainder distributed fairly uniformly in the upper four-fifths. The polymer fractions were dissolved in benzene and aliquots were titrated turbidimetrically with methanol. The material recovered from the zoning showed a narrower range of precipitation than did the starting material. Further, the initial precipitation points changed in regular fashion from fraction to fraction. This simple experimental sequence demonstrated that zone melting can be applied to polymer fractionation.

The zoning in this work was much too fast for the attainment of equilibrium. Had equilibrium prevailed, the distribution coefficients of all molecular weights doubtless would have been zero; all the polymer would have been transported to the bottom of the ingot and no fractionation would have been observed. Thus it was fortuitous that nonequilibrium zoning produced a range of effective distribution coefficients. These ideas were confirmed in the more extensive study reported by Loconti and Cahill (42), who found that polymer fractionation depended not only on freezing rate but on molecular weight and concentration as well. In an examination of monodisperse and polydisperse polystyrene in benzene ingots, they found that effective distribution coefficients increased toward unity with increasing molecular weight. Directional crystallization experiments indicated that for each molecular weight there is an ideal rate of crystallization which produces an intermediate rate of solute (polymer) removal from the host. For molecular weight fractionation, then, a programmed rate of crystallization would be required. At low rates, high molecular weights would be preferentially retained, since they show k_{eff}'s closer to unity. At high rates, lower molecular weights would be retained. Calculation of the best program would require knowledge of the molecular-weight distribution, which is generally unknown. However, an experiment in which directional crystallization was carried out at a linearly varying rate (1.4×10^{-4} cm s^{-1} to 1.7×10^{-3} cm s^{-1}) did, in fact, show enhanced separation.

The same polymer (Dow Styron 666-K27 $M_w = 3.05 \times 10^5$) was subjected to 10 zone passes in benzene, starting with 5% polymer concentration in the first 5% of the charge. The final ingot showed fairly uniform concentration (about 0.25%) and smoothly varying molecular weight from 1.0×10^5 to 4.0×10^5. The molecular-weight distribution is no sharper than that afforded by directional freezing. This appears to be the result of broad, irregular zones with inadequate mixing in the liquid.

Polystyrene was fractionated by zone melting its solutions in cyclohexane, in which it is only slightly soluble (43). At a zoning rate of 1.1×10^{-3} cm s^{-1} (4.0 cm h^{-1}), polymer molecules of all sizes showed retrograde motion, with k_{eff} ranging from 1.2 for molecular weight 35,000 to 9.0 for molecular weight 350,000. At a slightly higher rate of zoning (4.8 cm h^{-1}), the sense of solute transport was reversed; again the higher-molecular-weight material moved most rapidly. At a still higher rate (9.7 cm h^{-1}), decreased redistribution was observed. Loconti (44) investigated the behavior of the polystyrene/cyclohexane system, and found that near the freezing point, particles of gel formed. These gradually coalesced and precipitated in a discrete gel phase. Hence, at sufficiently low zoning rates, there is time for this process to go to completion, with the result that the gel adheres to the advancing interface and causes k_{eff} to be greater than unity. At higher rates, the gel particles do not coalesce and are rejected from the advancing interface into the melt, and show $k_{eff} < 1$.

Maeda et al. (45) have carried out extensive studies of ZMC of polystyrene in various hosts. To make the polymers in the host column visible, they introduced a "color tag" by chemical modification of the polymer. Nitration, reduction, diazotization, and coupling with N,N-dimethyl-2,4-dimethoxyaniline gave red polymers of molecular weight 2.0×10^2, 2.04×10^4, and 2.00×10^5. The availability of these tagged polymers made possible a direct spectrophotometric determination of polymer concentration. It was established by gel permeation chromatography measurements that the tagged polymers had essentially the same molecular weight as the initial polymers. ZMC indicated that the tagged polymers distributed in the same way as the initial ones. The availability of these tagged polymers, whose distribution in colorless hosts could be followed easily, also made possible a survey of potential hosts: eighteen aromatic compounds with melting points between 343 and 400 K were studied.

Dramatic differences were observed in the distribution coefficients of the polymers of lowest molecular weight in these hosts. After 30 passes, the polymer did not move at all in a column of 2-methoxynaphthalene. The same polymer moved 4 cm down a 33-cm ingot of biphenyl and was swept completely to the bottom of a column of durene. The intent of these studies was to find a relationship between ZMC behavior of polystyrene and some property of the host, so that chromatographic transport could be predicted. No such correlation was observed, but durene was selected as the best "compromise" host.

Distribution coefficients of monodisperse polymers were measured by the methods of Süe (18) and Sorenson (46) and were found to increase slightly (toward unity) with increasing concentration, evidently as a result of increasing entanglement at the solid/liquid interface. Model experiments were carried out in which single polymeric solutes were distributed in durene columns by multipass zoning. The results showed a marked dependence of k_{eff} upon molecular weight.

$\bar{M}_w$	k_{eff}
1.0×10^4	0.43
2.04×10^4	0.81
2.0×10^5	0.94

These values certainly lead one to think that molecular-weight fractionation can be achieved by ZMC. Further model experiments were done with mixtures of two monodisperse polymers in a single column of host, a 1:1 mixture of a polymer of molecular weight 2.1×10^3, and another of molecular weight 6.0×10^2 (peaks A and B in Figure 7.14). After only three passes, the top of the column contained 71% A/29% B and the bottom contained 9% A/9% B.

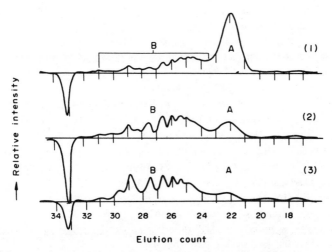

Fig. 7.14 Gel-permeation chromatograms of fractions obtained by ZMC of a 1:1 mixture of polymers with $\bar{M} = 2.1 \times 10^3$ and 6.0×10^2, at 11.8×10^{-4} mole g^{-1}, in a 32-cm ingot. Curves (1), (2), and (3) refer to top, middle, and bottom fractions, respectively. Peaks A and B correspond to polymers of $\bar{M} = 2.1 \times 10^3$ and 6.0×10^2.

The overall conclusions reached on the basis of these pair-wise distributions in durene were that:

- No solute/solute interaction takes place (i.e., the solutes are distributed independently of one another).
- Good chromatographic separation required that the molecular weights differ by a factor of 10.

With a polydisperse polystyrene having $\bar{M}_n = 1.6 \times 10^4$ and $\bar{M}_w/\bar{M}_n = 1.79$, the average molecular weight of the top fraction after ZMC was about two times higher than that of the bottom fraction. Correspondingly, the polydispersities of the fractions were lower than that of the starting material, namely 1.2 to 1.4 at the top and 1.4 to 1.7 at the bottom. In similar fashion, the polydispersity of a commercial sample of "monodisperse" polystyrene (2.04×10^4) was decreased by 10-pass ZMC from an original 1.053 to 1.047 and 1.044 at top and bottom of the column, respectively.

Maeda and Kobayashi have reviewed work on ZMC through 1976 (in Japanese) (47).

4 SCAVENGING

The addition of an extraneous substance to a zone-refining system to convert a nonsegregable impurity to a segregable or volatile one is discussed in Section 2.8.3, Chapter 5.

REFERENCES

1. W. G. Pfann, *Zone Melting*, 2nd ed., Chapter 5, Wiley, New York, 1966.
2. I. F. Nicolau, *J. Mater. Sci.*, **5**, 623–639 (1970).
3. W. G. Pfann, *Zone Melting*, 2nd ed., Chapter 10, Wiley, New York, 1966.
4. G. A. Wolff and A. I. Mlavsky, "Travelling Solvent Techniques," in C. H. L. Goodman, Ed., *Crystal Growth, Theory and Techniques, Vol. I*, Plenum Press, London, 1974, pp. 193–232.
5. C. Belin, *J. Crystal Growth*, **34**(2), 341–344 (1976).
6. R. Link, N. Noetzel, and W. Ermisch, *Krist. Tech.*, **13**(12), 1391–1397 (1978).
7. R. Triboulet, D. Triboulet, and G. Didier, *J. Crystal Growth*, **38**, 82–84 (1977).
8. R. Triboulet and G. Didier, *J. Crystal Growth*, **28**, 29–35 (1975).
9. R. O. Bell, N. Hemmat, and F. Wald, *Phys. Status Solidi (a)*, **1**, 375–387 (1970).
10. R. Triboulet, Y. Marfaing, A. Coronet, and P. Siffert, *J. Appl. Phys.*, **45**, 2759–2765 (1974).
11. J. C. Tranchart and P. Bach, *J. Crystal Growth*, **32**, 8–12 (1976).
12. R. Triboulet and Y. Marfaing, *J. Crystal Growth*, **51**, 89–96 (1981).
13. I. F. Nicolau, *J. Mater. Sci.*, **6**, 1049–1060 (1971).
14. I. A. Eldib, "Zone Precipitation and Allied Techniques," in M. Zief and W. R. Wilcox, Eds., *Fractional Solidification*, Dekker, New York, 1967.
15. W. M. Smit, *Chemisch Weekblad*, **57**, 143–147 (1961).
16. W. P. Trompen and J. Geevers, *Rec. Trav. Chim.*, **95**, 106–108 (1976).

17. G. B. Smith and G. V. Downing, *Anal. Chem.*, **51**, 2290–2292 (1979).

18. P. Süe, J. Pauly, and A. Nouaille, *Bull. Soc. Chim. France*, **5**, 593–602 (1958).

19. V. N. Vigdorovich, T. T. Got'manova, M. V. Mokhosoev, and V. G. Khudaiberdiev, *Izv. Vyssh. Ucheb Zaved. Tsvet. Met.*, **1**, 86–90 (1973).

20. M. V. Mokhosoev, T. T. Got'manova, and I. F. Kokot, *Russ. J. Inorg. Chem.*, **9**, 1360–1364 (1964).

21. M. V. Mokhosoev and T. T. Got'manova, *Izv. Vysshikh. Uchebn. Zav., Tsvetn. Met.*, **9**(3), 81–84 (1966).

22. P. I. Fedorov and N. S. Sitdykova, *Dokl. Akad. Nauk SSSR*, **153**(1), 126–128 (1963).

23. Yu. Kusaev, A. N. Zhukov, and R. K. Nikolaev, *Izv. Akad. Nauk SSSR Met.*, **3**, 33–36 (1978).

24. V. N. Zagorodnev, Yu. V. Korneenkov, and N. V. Lichkova, *Izv. Akad. Nauk SSSR Neorg. Mater.*, **17**(9), 1699–1701 (1981).

25. R. V. Fedorova, P. I. Fedorov, and N. Yu. P. Shvedov, *Zh. Prikl. Khim* (*Leningrad*), **44**(9), 2039–2044 (1971).

26. N. I. Lastushkin and A. N. Kirgintsev, *Izv. Sib. Otd. Akad. Nauk SSSR Ser. Khim.*, 111–115 (1976).

27. G. J. Sloan and A. R. McGhie, *Mol. Cryst. Liq. Cryst.*, **18**, 17–37 (1972).

28. E. Sirtl, German Patent 1,077,433, September 8, 1960.

29. D. Haarer and N. Karl, *Chem. Phys. Letters*, **21**, 49–53 (1973).

30. R. Anthonj, N. Karl, B. E. Robertson, and J. J. Stezowski, *J. Chem. Phys.*, **72**, 1244–1312 (1980).

31. W. G. Pfahn, *Zone Melting*, 2nd ed., Wiley, New York, 1966, pp. 58–59.

32. H. Reiss and E. Helfand, *J. Appl. Phys.*, **32**, 228–232 (1961).

33. W. G. Pfann, *Anal. Chem.*, **36**, 2231–2234 (1964).

34. S. Maeda et al., *Bull. Chem. Soc. Japan*, **46**(10), 3128–3133 (1973).

35. S. Maeda, H. Kobayashi, and K. Ueno, *Talanta*, **20**, 653–658 (1973).

36. K. Maruyama and M. Mashima, *Bull. Chem. Soc. Japan*, **44**, 2147–2149 (1971).

37. S. Fisel and C. Simion, *Rev. Roum. Chim.*, **22**(2), 305–311 (1977).

38. A. N. Kirgintsev and A. S. Aloi, *Izv. Akad. Nauk SSSR, Ser. Khim.*, 2808–2809 (1971).

39. H. Plancher, T. E. Cogswell, and D. R. Latham, "Zone Melting Chromatography of Organic Mixtures," in M. Zief, Ed., *Purification of Inorganic and Organic Materials*, Dekker, New York, 1969, p. 285.

40. V. F. Doron and S. Kirschner, *Inorg. Chem.*, **1**, 539–544 (1962).

41. F. W. Peaker and J. C. Robb, *Nature*, **182**, 1591 (1958).

42. J. D. Loconti and J. W. Cahill, *J. Poly. Sci., Part A*, **1**, 3163–3173 (1963).

43. A. M. Ruskin and G. Parravano, *J. Appl. Polymer Sci.*, **8**(2), 565–580 (1964).

44. J. D. Loconti, "Bulk Purification," in M. Zief and W. R. Wilcox, Eds., *Fractional Solidification*, Dekker, New York, 1967, p. 527.

45. S. Maeda, H. Kobayshi, and K. Ueno, *Talanta*, **21**, 1099–1107 (1974).

46. P. Sorensen, *Chem. Ind.*, 1593–1595 (1959).

47. S. Maeda and H. Kobayashi, *Kagako No Ryoik*, **30**(6), 479–488 (1976).

Chapter **VIII**

APPLICATIONS OF ZONE MELTING

---

1   PURIFICATION AND IMPURITY ENRICHMENT

By far the greatest success of zone melting has been in the area of purification, or zone refining. Elements and compounds, from noble gases to steroids, have been brought to purities unattainable in any other way. The results have been documented even in popular publications (1).

Purity is sought for a host of practical reasons:

* Pure difunctional monomers yield polymers of highest molecular weight.
* The measurement of intrinsic physical properties of materials requires highly purified samples.
* Impurities can alter the course of a chemical reaction.
* Crystal growth can be adversely influenced by impurities.
* Electronic properties of solids depend sensitively on purity.

In some cases high purity can be attained by direct application of zone refining to commercially available materials. More commonly, it is necessary to consider several important factors before zone refining. First, of course, it is necessary to establish that the material under study is thermally stable (see Section 2.6.1, Chapter 3 and Section 3). It is desirable to consider the likely range of impurities in the light of the history of the sample. Reaction by-products whose structures are very close to those of the major component may not be readily removed by zoning. They may, however, be removed chemically or they may be converted chemically to new materials with more favorable distribution coefficients. The conversion may be carried out concurrently with the zone refining by adding an appropriate scavenger (Section 2.8.3, Chapter 5).

Since zone refining partitions impurities between solid and liquid phases, the impurities are segregated and enriched in one or both ends of the sample ingot. There may be as much interest in the impure as in the pure product. Naturally, qualitative analysis is facilitated in the impurity-enriched end(s); in addition, quantitative analysis may be carried out with greater accuracy in a concentrated fraction.

In an early application of zone refining to organics, commercially available materials were subjected to a systematic sequence of operations. An initial portion was zoned, and melting points were measured on samples taken from several positions along the ingot. The segment of constant melting range was cropped from the ingot, recast, and rezoned. After each zoning, the melting-point distribution was remeasured until nearly the entire ingot melted at a uniform temperature (2). Samples from the head of the ultimate and penultimate ingots were analyzed by thin-film cryometry (see Section 2.2.4, Chapter 2 and Section 7, Chapter 11). Results of application of this methodology to azobenzene, benzophenone, benzil, stilbene, and naphthalene are given in Table 8.1.

Application of zone melting to purification is known as "zone refining," in accord with old metallurgical practice. No other zone-melting technique has

Table 8.1. Purity Achieved by "Fractional Zone Refining" (2)

Substance	Number of Times Zoned	Cumulative Number of Passes	Mole Fraction Impurity, $\times 10^{4(a)}$
Azobenzene	3	33	3.5
	5	57	2.9
	6	127	1.0
Benzophenone	4	78	2.3
	5	90	2.0
Naphthalene	2	23	6.0
	3	31	2.9
Benzil	2	45	2.7
Stilbene	2	23	1.8

[a] At head of ingot.

been so broadly or fruitfully applied. We will discuss several individual compounds, partly because of their intrinsic interest, and partly to show how the formation of an impurity profile in an ingot provides, at one time, a sample of purified material and a sample of impurity-enriched material in which qualitative detection of impurities is facilitated. In turn, more effective purification procedures can be applied to the previously purified product, and distribution coefficients may be measured for known impurities. Their concentrations may then be calculated at levels below analytical detectability.

1.1 Elements

ALUMINUM (mp 933.53 K)

Aluminum continues to be one of the most intensively studied metals. The oxide film that normally coats the melt has been removed by filtration through graphite. Further prepurification has been carried out by double electrolytic refining (3). Graphite and alumina boats have been used as containers for zone refining Al by resistance and induction heating. When graphite boats are used, care must be taken to keep the zone temperature close to the melting point because carbide formation takes place above 1275 K (4).

The effectiveness of electron-beam heating has been compared with that of induction heating. The latter was preferred because with electron-beam heating the ingot was contaminated with cathode material (5). Continuing efforts have been directed toward the identification of residual impurities and the attainment of maximum resistivity ratio. Material containing 0.5–5 ppm of metallic impurities showed resistivity ratios of 4.5×10^4 to 4.8×10^4. This work was carried out on material prepared by electrolysis of $NaF \cdot 2AlEt_3$ or by zone refining, or a combination of both (6). Zone refining in a vacuum of 0.33×10^{-4} Pa (10^{-6} torr) using metal of 5N purity gave an ingot of which $>80\%$ showed resistivity ratio greater than 3×10^4; the highest resistance ratio observed was 5×10^4 (7). Later work was devoted to the identification of trace elements in the refined material: Li, B, Na, Mg, Si, Ca, Sc, Mn, Fe, Cu, Zn, Sb, La, Ce, Sm, Dy, and Au were identified. The principal impurities contributing to residual resistivity were Si and Sc. Sc had $k_{eff} \simeq 1$, while Ta and W had $k_{eff} > 1$. Neutron activation analysis was used to detect the residual impurities Na, Sc, Cr, Ga, Gd, and Th (8). Not surprisingly, in single crystals grown by zone melting, the distribution of impurities was found to be inhomogeneous, with three to four times more at the crystal ends than in the middle. The etching behavior of crystals prepared by zone refining was different from that of Bridgman-grown crystals. In the central area, of highest purity, no etching of grain boundaries took place. In the crystal ends there was agreement between the observed etch pattern and dislocations and sub-boundaries (9).

A horizontal zone refiner for purification of metals has been described and applied to the purification of aluminum. A cylindrical ingot was extracted from an inductively heated melt through a sealed port, and then fed directly into a second melt chamber, from the opposite end of which the product ingot was extracted. Effective distribution coefficients were measured for tin and zinc, using radiotracer techniques (10).

ANTIMONY (mp 904 K)

Antimony is of interest as a dopant in silicon and germanium and also for the preparation of semiconducting compounds such as InSb. It is prepared by reduction of the trioxide and is generally zone refined in inert atmospheres in horizontal graphite or Alundum boats. Arsenic and bismuth are the most difficultly removed contaminants. Arsenic forms a continuous series of solid solutions with a minimum melting point and shows $k_{eff} = 0.8 \pm 0.1$ in the concentration range 0.009 to 0.8%. It has been found that addition of 1% of aluminum to the antimony charge serves as a getter and facilitates the removal of arsenic. The distribution coefficient of bismuth has been found to be 0.4 to 0.65. Lead and tin are also rather difficult to remove, with k for Sn about 0.6 to 0.7. Sulfur and selenium pose particular difficulties as contaminants because they yield two-phase liquids with antimony; nevertheless, k_{eff} has been measured for selenium and found to be about 0.6 for concentrations between 4.5×10^{-4} and 2.5×10^{-3}%. Most other elements are very insoluble in antimony and show k_{eff} much less than unity, so that removal is easy (11, 3). Large single crystals of antimony have been grown by horizontal zone melting.

ARSENIC (mp 1090 K, at 2.8×10^6 Pa)

Arsenic has not been successfully zone refined because of its high vapor pressure at the melting point, but it has been purified by vapor zone refining (see Section 2.7.7, Chapter 5).

BERYLLIUM (mp 1551 K)

Pure beryllium is required for growth of single crystals for use as neutron monochromators. The purification and conversion to single crystals can be effected by zone refining. Vertical ingots are purified by floating-zone induction heating in inert atmospheres. BeO, C, Al, Cr, Fe, Mg, Ca, Cu, and Si are all effectively segregated by zone refining. Nickel is also segregated, but with $k_{eff} > 1$ (3).

Crystals up to 2.5 cm in diameter have been grown, and the purest of these show striking ductility and tensile extensions up to 220% (12, 3).

BISMUTH (mp 544.5 K)

High-purity bismuth is of interest as a neutron monochromator and for fabrication of thermoelectric devices and semiconducting compounds. Preparation of material suitable for zone refining requires prepurification. In one approach the starting material was treated with chlorine to decrease the lead concentration from 2×10^{-2} to 4×10^{-4}%. Oxide impurities were reduced with hydrogen at 975 K, followed by vacuum degassing and filtering. After nine zone passes, material of this provenance showed no detectable Pb, Cu, Fe, or Ag (13). In another prepurification, anodic refining was the first step, followed by oxidation (14). In other work, starting with material of 99.98% purity, five zone passes gave a product of 4N5 to 5N purity (15). In general, bismuth is zone refined in vacuum, in boats of graphite, Pyrex, or fused silica; oxidized aluminum boats have also been used. Much work has been devoted to the measurement of distribution coefficients of various metallic impurities in bismuth. Values of k_0 for Pb, Cu, Ag, and Ni were reported to be 0.3, 0.4, 0.03, and 0.25, respectively (16). Antimony is apparently the only element having a distribution coefficient greater than unity, and a value of 11.6 has been reported (17). The least favorable distribution coefficient is that of selenium, 0.7 (17). The electrical resistivity ratio of bismuth single crystals has been found to be 570–650 in the [111] direction and 750–1000 in the [110] direction (13).

Recent work indicates that the effectiveness of zone refining can be enhanced by simultaneous application of electric and magnetic fields during the processing. A current density of 12 A cm^{-2} and a magnetic field intensity of 0.08 T gave an increase of efficiency by a factor of 1.4 or more (18).

BORON (mp 2573 K)

Boron has important functions both as a semiconductor and as an acceptor dopant in silicon. The extremely high melting point of boron contributes to the difficulty of purifying it by zone refining. Purified boron has been prepared by thermal dissociation of BBr_3 and by electrolysis of a melt of KBF_4 and KF (19). Zone refining is carried out on sintered or compressed powdered B and, in one case, rods 10–13 cm long by 0.5–0.8 cm diameter were float zoned in a solar furnace (20). Electron-beam heating has also been used in the float zoning of rods pressed from B powder with H_3BO_3 as a binding material in 5–40 wt% concentration (21).

CADMIUM (mp 594.1 K)

The volatility of cadmium lends itself to distillation as a prepurification method. Tantalum containers have been used for zone refining cadmium in helium containing 10% hydrogen, or in pure hydrogen. Quartz boats have also been used; these were etched and lined with graphite or pyrolytic carbon. Fifteen passes at 10^{-4} cm s^{-1} (1.5 cm h^{-1}) gave product of 6N5 purity (22). The resistance ratio of purified cadmium has been found to be 3.8×10^4 (3). Cadmium is used primarily in the preparation of cadmium telluride for fabrication of gamma-ray detectors.

CERIUM (mp 1072 K)

Cerium has been zone refined in vacuum by induction heating in a cooled copper vessel. The concentration of iron was reduced to 7×10^{-8} wt% as measured by neutron activation analysis (23). In the zone refining of cerium and other rare earth metals, it was found that metallic impurities had k_{eff} less than unity while interstitial O and N had k_{eff} greater than unity (24).

CHROMIUM (mp 2130 K)

Attempts to zone refine chromium were long foiled by its high volatility at the melting point. Recently success has been attained in preparation of high purity Cr by float zoning under an atmosphere of 90 kPa of hydrogen. The purified material showed a resistance ratio of 243 as compared with a starting value of 35 (25). In other work on chromium, an impurity level of 5×10^{-3}% was obtained (26).

COBALT (mp 1786 K)

Striation-free single crystals of cobalt have been grown reproducibly by single-pass electron-beam float zoning (27). In other work, it was observed that cobalt single crystals prepared in this way grew along the [0001] direction. As the growth deviated from this direction, striations resulted (28). Very pure cobalt, prepared by zone refining and homogeneous reduction of $CoCl_2$—NaCl eutectic, contained $<10^{-5}$% impurity and gave a resistivity ratio of 290 (29).

COPPER (mp 1357 K)

Copper single crystals of 5N purity have been grown at 1.7×10^{-4} cm s^{-1} by horizontal zone melting (30). The work was carried out in a graphite crucible in a nitrogen atmosphere, with induction heating. S, Se, Ca, and As were identified in the concentrated ingot end (3).

GADOLINIUM (mp 1596 K)

Gadolinium has been purified by zone refining in an argon atmosphere in water-cooled boats, by induction heating. Oxygen and nitrogen showed k_{eff} greater than unity; Cu, Ni, Fe, Ti, Si, and Al showed k_{eff} less than unity (31); the starting end of the ingot showed very low concentrations of metallic impurities. The concentration of tungsten in gadolinium has been reduced to <1 ppm by zone refining in vacuum (6.6×10^{-4} Pa) with induction heating (32). Gadolinium has also been zone refined by the floating-zone method, and it was found that Fe and Cu showed k_{eff} less than unity, while La, Nd, Tb, Eu, and Yb were barely segregated (33).

GALLIUM (mp 302.9 K)

Efforts at purification of gallium by zone refining were not very successful although growth of single crystals from the melt had been shown to distribute impurities effectively. The inconsistency was resolved by the observation that a monocrystalline interface rejected impurities more effectively than a polycrystalline interface; hence, successful zone refining would require a monocrystalline interface (3). Removal of silver from gallium by zone refining has been studied, using isotopes ^{65}Zn and ^{110m}Ag (34). Gallium of 5N to 6N purity has been further purified by zone refining in an ampoule to ultimate purity of 7N (35).

GERMANIUM (mp 1210.6 K)

After silicon, germanium appears to hold second place in importance to semiconductor technology. In fact, the need for ultrapure semiconductor-grade germanium was responsible for the development of zone refining. Commercially, germanium is purified as the oxide GeO_2, which is then reduced with hydrogen to provide starting material pure enough for some electronic purposes.

While liquid germanium dissolves all elements, the solid shows only limited capacity to dissolve other elements. It forms a continuous series of solid solutions only with its periodic table neighbor, silicon, with k_{eff} 5.5. Boron, likewise, forms extensive solid solutions with germanium with $k_{eff} > 10^{-1}$. Even at these levels however, the distributions of such foreign elements may be of interest and a rather exhaustive study is available (36).

The early zone refining of germanium was generally carried out in graphite or fused-silica boats (3). More recently, refining has been carried out in horizontal, rotating quartz tubes (37), and in nearly horizontal tubes (12–15' from horizontal). The tilted-tube method has been applied to the preparation of shaped single crystals (38). The chemical properties of germanium have not provided a strong impetus for the application of float zoning, but it has been applied in growth of crystals in diameters up to 3 cm (39). Single crystals have also been prepared by vertical zone leveling (40).

INDIUM (mp 430 K)

Indium is generally obtained as a by-product of the preparation of other nonferrous metals and commercial grades are contaminated with Ag, Cu, Cd, Sn, Pb, Fe, Zn, and Ni in small concentrations. Of these, Ag, Cu, and Ni are most detrimental to solid-state applications. It has been observed that indium does not form a continuous series of solid solutions with any element (11). Distillation is widely used for prepurification, especially for removal of cadmium and mercury. The zone refining of indium has been carried out in a boat at 2.2×10^{-3} cm s^{-1} (8 cm h^{-1}); the product was analyzed by mass spectrometry. The following elements were found to have distribution coefficients less than unity: Ag, P, S, Co, Ni, Cu, Se, Sb, Te, Ba, Tl, Bi, Sn, and Pb. Mn was found to have $k > 1$ (41). Other studies have indicated that Pb has $k > 1$. The inconsistency may be the result of the presence of other interacting impurities in the samples that were studied. Equilibrium distribution coefficients for Cu, Ag, and Ni have been reported to be approximately 0.07, 0.06, and 0.01, respectively (11). The distribution of the isotope ^{65}Zn in indium has been studied, and it was found that k was less than unity; in one study transverse segregation was greater than axial (34). The purification effect was maximum when the ingot had no surface oxide film, toward which most of the Zn migrates (42). The combined effects of distillation and 18-pass zone refining have been reported, and it was found that the resistance ratios and impurities in ppm were as follows: top of the ingot 1.53 (12), middle 1.61 (14), and bottom 4.2 (186) (43).

IRON (mp 1808 K)

Early zone refining of iron was carried out in horizontal boats.(3). More recently, float zoning has been used, and the resulting ingot was characterized by neutron activation analysis. The iron used in the study was obtained from a soluble salt which was prepurified by ion exchange (8).

A study of the Fe/Be and Fe/C systems has been reported, along with measurements of the distribution coefficients of these two elements. A periodic dependence of the distribution coefficient on atomic number was observed for a number of elements (44). High-purity sponge iron was consolidated by electron-beam melting and the resulting ingot was then float zoned to yield a product of 4N6 purity (45).

Iron has been prepared by zone refining for use in astrophysical and nuclear research; concentrations of Ra and U were reduced to $< 10^{-8}\%$ (46).

Multipass zone refining of iron in ultrahigh vacuum gave an ultimate resistance ratio of 4990 (47). Earlier work on the purification of iron has been reviewed (3).

LANTHANUM (mp 1194 K)

Lanthanum has been zone refined in an atmosphere of purified argon by induction heating in water-cooled boats. The conditions were the same as were used for purification of gadolinium (31).

LEAD (mp 600.7 K)

Lead may be prepurified by electrolytic refining; zone refining has been carried out in graphite boats within a quartz furnace. An exhaustive discussion of zone refining procedure and equipment has been provided by Tiller and Rutter (3). Tin, antimony, bismuth, indium, thallium, cadmium, and mercury have high solid solubility in lead, while lithium, silver, calcium, and barium show lower solubility. Only thallium and calcium appear to have $k_{eff} > 1$. Other elements are negligibly soluble in solid lead and in many cases the solubilities are below the limits of detection. In a study of the distribution coefficients of various elements in lead by directional crystallization, it was found that the degree of refining of the lead decreased when the total impurity content was greater than 0.25%. Strong mixing during zone refining of lead caused one hundredfold decrease in copper and silver concentration and fortyfold and fifteenfold decreases in cadmium and bismuth concentrations, respectively, after 15 passes. After 36 passes the amount of bismuth was decreased by a factor of 40 (48).

Solutes with $k_0 < 1$ and $k_0 > 1$ affect grain-growth rates in qualitatively different ways. In the presence of small amounts of Ag ($k_0 < 1$), grain boundary migration rate was proportional to impurity concentration, while in the presence of Ca ($k_0 > 1$), the rate was inversely proportional to concentration (49).

LITHIUM (mp 453.7 K)

Lithium has been zone refined in a molybdenum boat in a helium atmosphere, using electrical resistance heating. Zone refining produced an increase in resistance ratio from 25 to 5×10^2 (3).

MAGNESIUM (mp 922 K)

The high vapor pressure of Mg at its melting point poses problems in zone refining, but it has been zoned by covering the surface with a skin of $MgSO_4$ and MgO and carrying out the operation in an SO_2 atmosphere (3).

MOLYBDENUM (mp 2890 K)

The purification of molybdenum has been reviewed (50). Because of its high melting point, zone refining of molybdenum is carried out by either plasma-arc or electron-beam heating. Electron-beam zone refining has been said to remove Al, Si, Fe, Mn, Ni, Mg, and Ti in one zone pass at a speed of 1.3×10^{-3} to 2.6×10^{-3} cm s^{-1}. Removal of SiO_2 and MgO required either a greater number of zone passes or lower refining speed. Removal of C, W, Zr, Nb, Ta, Re, and ZrO_2 was said to be practically impossible. The refining in this work was carried out at a pressure of 7×10^{-4} Pa with a molten zone about 0.5 cm wide, at a temperature of about 3000 K. The volatility of molybdenum at its melting point is always a problem in this kind of work, and losses were about 30% when more than one zone pass was carried out at 6.7×10^{-4} cm s^{-1}. Experimental values of the effective distribution coefficients for Zr, Mg, and Si differed substantially from the calculated values, indicating that these elements are present as oxides (51).

In a study of electron-beam zone refining of molybdenum single crystals, it was found that passage of a single molten zone produced a single crystal with the [110] direction parallel to the rod axis. The resulting ingot consisted of two single crystals of the same orientation. When the upper crystal was rotated during the zoning, two single crystals were formed with a common [110] axis but with different orientation connected by the molten zone (52).

The structural perfection of zone-refined Mo single crystals has been studied by dislocation etching, X-ray diffraction, and microfractography (53). Molybdenum bicrystals have been prepared in controlled orientation and diameters of 1.0–1.5 cm, by vertical electron-beam zone refining (54).

Another study of distribution of impurities indicated that the k_0's for transition metals show a periodic variation and that the k_0 values depend on factors having to do with valence, size, and phase stability. From this dependence, k_0 values were determined for impurities whose phase diagrams with Mo are not known (55). The effects of vacuum, number of passes, and zoning rate on electron-beam zone refining of Mo were studied with respect to the impurities O, C, K, Al, Fe, and Cr. It was found that zoning was ineffective in removing C, but caused substantial reduction of the other impurities. The electrical resistivity ratio was increased from 44 in the starting material to 1650–2850, and the improvement was practically independent of the number of zone passes (56). Still higher purity was claimed in earlier work, with a resistivity ratio of 1.4×10^4 (57).

The plasma-arc method has been used for the growth of large Mo crystals (1.5- to 3.5-cm diameter and up to 25 cm long) in argon and helium atmospheres. The purity attained in these crystals was practically the same as that in electron-beam zone melting (58).

NICKEL (mp 1726 K)

Nickel is generally prepurified by electrolysis. In the floating-zone method, Ni single crystals grew preferentially along the $\langle 111 \rangle$ direction and were striation free. Striations resulted as the growth direction deviated from $\langle 111 \rangle$ (28, 59). The distribution of Ir, Os, and Ru in the zone refining of Ni has been discussed (60).

In more recent work on vacuum float zoning of nickel, it was found that Na, Sc, Cr, Zn, As, Ag, Sb, and Hg were removed by vaporization, while Se, Sb, Ta, Sm, and Tb were segregated. The main impurities, Fe and Co, were not removed by float zoning. Their removal required prior anion exchange (61).

The highest resistance ratio reported for zone refined Ni is 3.3×10^3 (3).

Preoxidation of Ni rods at about 1200 K before zoning allows for the removal of carbon as CO or CO_2 during zoning (3).

NIOBIUM (mp 2741 K)

Niobium has been exhaustively prepurified, starting with Nb_2O_3. After removal of tantalum by chemical extraction, the Nb_2O_3 was reduced to Nb vapor, which was deposited on resistance-heated Nb wires; these were then subjected to electron-beam zone refining (62). Nb has also been prepurified by electrodeposition from a molten salt. The initially high carbon level of the product was reduced by using a high partial pressure of oxygen during the zoning. After a high-temperature anneal, high resistivity ratio (6.5×10^3) and low flow stress (460 gm mm^{-2}) were attained. The impurity content, determined by chemical analysis, did not correlate with the resistivity ratio, but flow stress did (63).

Bicrystals of Nb have been prepared with a special seeding procedure, using a Y-shaped seed. Large bicrystals of specified orientation (0.63-cm diameter and more than 10-cm long) were grown by this method (64). Distribution coefficients for impurities in Nb were calculated from thermodynamics and known phase diagrams, as a function of temperature (65).

OSMIUM (mp 3318 K) AND RUTHENIUM (mp 2583 K)

These metals were purified in a two-stage procedure. The first stage involved degassing and electron-beam melting of the starting powder of 99.8% purity, followed by electron-beam zone refining at 8×10^{-3} to 0.7×10^{-3} cm s^{-1} in water-cooled copper boats. Spectroscopic and chemical analyses as well as measurement of the resistivity ratio were carried out on the products. The resulting single crystals showed resistivity ratios of about 2.5×10^3, and impurity contents of about 10^{-5} wt% (Fe, Ca, Cu, Mo, Pd, Rb, C, and P) in Ru and about 10^{-4} wt% in Os after 18 and 8 zone passes, respectively (66).

A three-stage zone refining procedure has been described for the preparation of purified single crystals of ruthenium. The zoning was carried out in vacuum at about 10^{-3} to 10^{-4} Pa. In the first stage, zone transport was carried out at 5×10^{-3} cm s^{-1}, and in the later stages, at considerably lower speed (67). In other work, vacuum-melted Ru, with a resistivity ratio of 5×10^2, was zone refined in an electric field (0.15 V cm^{-1}), and a product with a maximum resistivity ratio of 1.9×10^3 was attained (68).

PHOSPHORUS (mp 317.3 K)

Phosphorus has been zone refined in a vertical glass tube containing perforated, conductive disks which were rotated by an external electrical drive (see Sections 2.1.5 and 2.5, Chapter 5). Optimum separation, in 4-cm tubes, was obtained at 8×10^{-5} to 1.4×10^{-4} cm s^{-1} (0.3 to 0.5 cm h^{-1}). Starting with prepurified P containing 0.03% inorganic and 0.05% organic impurities, transparent ingots were obtained which after three zone passes contained < 10 ppm of both inorganic and organic impurities in the upper two-thirds of the ingot. After five passes, both categories of impurities were undetectable (69).

PLATINUM (mp 2045 K)

Fe, C, and Pd have been identified as the major impurities limiting the resistivity ratio in commercial platinum of 5N purity. The Pd content can be substantially reduced by vacuum melting and zone refining, while the Fe and C content can be reduced by zone refining. The limiting value of resistivity ratio is significantly improved by hydrogenation (70).

RHENIUM (mp 3453 K)

Powdered Rh was vacuum melted and electron-beam zone refined at 5×10^{-3} cm s^{-1}. The resulting purification was studied by mass spectrometry, chemical analysis, and measurement of electrical resistivity. Generally, impurities showed $k_{eff} > 1$ and purity improved with increasing number of zone passes until about 15 passes. The residual impurities detected were Ta, Si, and C. Carbon content was decreased by zone refining in a hydrogen atmosphere at 6.65×10^{-3} Pa.

The purity of the zone-refined single crystals was higher than that of those prepared by electrolysis (71). The resistivity ratio of the zone-refined single crystals was about 3×10^4 (72). Resistivity ratios as high as 5.5×10^4 have been reported (3). Maximum impurity displacement was attained by zone refining along the c axis (73).

SILICON (mp 1683 K)

There is little question that the purification of silicon has been more intensively studied than that of any other material. The work in this area has been reviewed in a recent book (74). Only lead and bismuth, among the elements, fail to dissolve in molten Si. However, only near neighbors in the periodic table show substantial solid solubility in Si and only Se forms a continuous series of solid solutions (75). B, O, P, Ge, and As show $k_{eff} = 0.8$, ~ 0.5, 0.4, 0.33, and 0.3, respectively.

Attempts to zone refine silicon in various containers before the mid-1950s were generally unsuccessful because of reactivity with and breakage of containers. Introduction of the float-zone method (Sections 2.1.5 and 2.4, Chapter 5) made possible the preparation of ingots of steadily increasing purity, textural perfection, and size. For 2 decades, a competition has taken place between the ability of crystal growers and the demands of fabricators of silicon-based solid-state devices. While the contest is certainly not over, the current state of affairs is represented by the work of Collins, who described the growth of dislocation-free, float-zoned crystals of 8-cm diameter, up to 24 cm long. Success in the work required optimization of the diameter of the polycrystalline starting rod, seed orientation, zone length, rotation rate, zoning speeds, finished crystal length, and crystal support mechanisms. Highly sophisticated crystal growth devices are commercially available for float-zoning of silicon and other high-melting materials, with features such as automatic diameter control (see Appendix II).

Early zoning of Si was carried out in quartz crucibles or boats, but recent work has generally used the float-zone technique (3). An Ar—H_2 plasma has been used for preparation of photovoltaic Si (76).

SILVER (mp 1235.09 K)

Ag has been zone refined in a graphite boat in an argon atmosphere, with induction heating. The resistance ratio was increased from 3.5×10^2 to about 8×10^2 (77). Higher resistance ratios (to about 2×10^4) and reduced impurity ($< 10^{-5}\%$) were attained in single crystals grown by the Bridgman method from metal prepared by reduction of zone-refined silver nitrate (78).

STRONTIUM (mp 1042 K)

Strontium has been prepared by a combination of vacuum distillation and crystallization methods. Distributions of Mg, K, and Fe have been reported (79).

SULFUR (mp 386 K)

Zone refining of sulfur is troubled by the β-to-α phase transition, which produces a fine-grained solid in which carbon-containing impurities are retained. A special refiner was built to avoid the transition, while keeping temperature below 423 K. Ingots were sealed in tubes of 1.2-cm diameter, 50 cm long, and 2.5-cm zones were passed at speeds of 0.028 to 0.108 cm s^{-1} (1.7 to 6.5 cm h^{-1}). Under these conditions, carbon concentrations as low as $2.6 \times 10^{-3}\%$ were attained. Some inorganic impurities were effectively removed at zoning rates of 0.017 cm s^{-1} (1 cm h^{-1}), but removal of organic impurities and some inorganics of low molecular weight (sulfides) required rates that were an order of magnitude lower. C, Si, Mg, and Fe were reduced to $<2.4 \times 10^{-4}$, $<1.3 \times 10^{-6}$, $<4.0 \times 10^{-6}$, and $<1 \times 10^{-4}\%$, respectively (80, 81). In more recent work, adsorbent carbon and silver have been used in the zone refining of sulfur (82).

TANTALUM (mp 3269 K)

Purification of tantalum by electron-beam float zoning and growth of single crystals have been reported by several groups (3). Spectroscopic analysis and electrical resistivity data have also been reported (3). Ta crystals have been grown by electron-beam zone melting using a transverse hearth. Methods of producing large single crystals with controlled orientation were devised (83, 84).

TELLURIUM (mp 725 K)

Tellurium has been zone refined in graphite-lined quartz boats at 2×10^{-4} cm s^{-1} in a helium atmosphere containing 10% hydrogen. Na and Si concentrations remain constant in the zoned ingots while Ag, Al, As, Au, Cu, In, Mg, Fe, P, Se, Zn, and O decreased; ultimate purity was 6N5 (22). High-purity Te single crystals with carrier concentrations of 10^{13} cm^{-3} have been prepared by zone refining in hydrogen or argon atmospheres (85).

Vigdorovich has applied modified zone-refining methods to Te, in which impurities were removed mechanically from the surface of the molten zone; experiments were carried out in both argon and hydrogen atmospheres. In the case of selenium, starting with 3×10^{-3} wt%, refining in argon gave slow reduction of solute concentration, to an ultimate level of 3×10^{-4} wt%. In hydrogen, the concentration fell to $<10^{-5}$ wt% after 20 passes. Charge-carrier concentrations approaching 10^{13} cm^{-3} were attained. Indirect evidence indicated that remaining metallic impurities were present as finely divided oxide inclusions (86). Purities to 6N have been attained by preliminary vacuum distillation and zone refining. The distillation was carried out in quartz at 3×10^{-4} Pa and temperatures of about 800 K. After three distillations, zone refining was carried out with induction heating in a hydrogen atmosphere. Residual impurity contents of the zone-refined ingots were measured mass spectrometrically. Distillation was found to remove many elements and zone refining removed Na, K, Cr, Au, Fe, Zn, Ge, Ag, In, Sn, Pb, Bi, and Cd (87).

TERBIUM (mp 1633 K)

Terbium was zone refined under 10^5 Pa (1 atm) of argon at 1073 K by 25 zone passes at a zoning rate of 10^{-3} cm s^{-1} (3.75 cm h^{-1}). Oxygen, nitrogen, and hydrogen, present originally at 740, 10, and 10 ppm, were reduced in concentration to 210, 0, and 10 ppm (88).

THALLIUM (mp 577 K)

Thallium containing a number of metal impurities has been zone refined at 5×10^{-4} cm s^{-1}. Ag and Cd were effectively segregated while Sn and Si migrated slowly. Pb, Cu, Al, and Bi were found to show $k_{eff} > 1$. After 40 passes, purity of the product was 6N in the central zone (89).

In another study, Tl of 99.9% purity was dissolved and TlCl was then extracted, and the Tl was recovered by electrolysis with a graphite cathode and a platinum anode. Zone refining of the electrodeposit gave Tl of 5N7 purity (90). An extensive report of the distribution of silver in thallium (91) is available. Ca, Bi, and Pb are all reported to have $k_{eff} \geqslant 1$. Sn, Sb, and Cs have $k_{eff} = 0.7, 0.85$, and 0.6, respectively (91).

THORIUM

See Molybdenum.

TIN (mp 505.0 K)

With its low melting point and low reactivity, tin has been zone refined readily in boats of silica, Pyrex, or graphite and in closed tubes. It has also been zone refined in rotating, vertical Pyrex tubes. Starting with tin doped to 10^3 ppm with Pb, Fe, Cu, Cd, and Bi, 20 zone passes reduced the Pb and Cu concentrations at the head of the ingot by about 10^2 and 10^3, respectively. In this work, the following values of k_{eff} were reported for Cd, Cu, and Pb: 0.5, 0.12, and 0.4 (92).

In other work, k_{eff} was found to be 0.6, <0.1, and 0.3 for Bi, Cu, and Pb. In continuation of this work, three different grades of tin containing different impurity distributions were subjected to 55–66 zone passes and the resulting ingots were analyzed. Zone refining was found to be particularly effective for As, Cu, Fe, and Bi (93, 94). Still purer tin (5N5–6N5) was further purified by zone refining at 0.9×10^{-4} cm s^{-1} (2.5 cm h^{-1}) in a rapidly rotating container with a 3-cm zone width. Impurity content was reduced by a factor of 2 after 40–60 passes. The resistivity ratio of the starting material was 4×10^4 and that of the product 1.06×10^5 (95). The effects of dc and ac fields on the zone refining of tin have been studied; the ac field produced a greater increase in efficiency than a dc field of either polarity (96).

Single crystals of extremely pure tin have been prepared from technical grade tin by preliminary electrolytic refining in molten $SnCl_2$ at 513 K, or by zone refining and directional crystallization. The final product had purity of 6N and a resistivity ratio of $1–1.1 \times 10^5$ (97). In related work, trace impurities were concentrated in tin by zone refining, with simultaneous extraction into molten $SnBr_2$ or $SnCl_2$ (98). Detection limits for metals extracted into the salt were 10–100 times lower than without preconcentration. Emission spectrography, applied to a concentrate of tin impurities, made it possible to determine impurities with a detection limit of 10^{-10} to $10^{-6}\%$ (99). Tin has also been purified by high-speed zone refining in a vertical tube with periodic reversal of rotation about the tube axis. After 20 zone passes, the concentration of major impurities was reduced by about two orders of magnitude (100).

TITANIUM (mp 1933 K)

Single crystals of Ti have been produced by zone refining in vacuum, both of pure Ti and Ti containing $<0.7\%$ impurities (Fe, Mn, Al, Cr, Zr, B, and Ni) (101). Distribution of several impurities in electron-beam zone-refined Ti has been studied. Solid solutions with $k_{eff} > 1$ were formed with Nb, Ta, Mo, W, Re, C, N, and O, while Zr, Hf, V, Cr, U, and Sc gave systems in which $k_0 < 1$. Solid solubility and $k_0 < 1$ were shown by systems with Cu, Ag, Au, Mn, Fe, Co, Ni, Sn, Pd, Ir, Bi, Ga, Ge, Al, Pb, Os, Rh, and Ru. Finally, variable solid solubility was shown by B, Be, Ce, Nd, Y, P, and Pu (102). Titanium ingots containing 800–1500 ppm of iron were purified by float zoning, and k_{eff} was found to be 0.34. A novel experimental method was used, in which the viscous frictional torque induced in the liquid zone by rotation of the ingot ends was measured and used to control the heat input to the zone. Rotation rates of about $2.5 \, \text{rev s}^{-1}$ gave maximum stability with a floating zone of 0.3-cm thickness (103).

TUNGSTEN (mp 3683 K)

W single crystals of $\langle 110 \rangle$ orientation have been grown by electron-beam zone refining. The crystals were studied by X-ray topography and by electrical resistivity measurements (104). Other crystals have been grown by electron-beam melting and their impurity distribution determined by mass spectrometry (105). W crystals, in diameters up to 1.6 cm, were grown in different orientations. X-ray diffraction, microhardness, and dislocation etching showed that the crystallographic orientation of grown crystals was the same as for the original crystals (106). A study of the effects of orientation, growth rate, and solid/liquid interface shape on the effectiveness of the floating-zone growth of W crystals has been reported (107). Bicrystals of 1.0–1.5 cm diameter have been prepared by electron-beam zone refining (108).

URANIUM (mp 1595 K)

Zone refining of U by electron-beam methods and other techniques has been discussed (3). In recent work, floating-zone refining was applied to uranium; after two passes, the impurity content was reduced to 0.4 atom%.

VANADIUM (mp 2163 K) (50)

The volatility of V at its melting point is such that under zone-refining conditions, impurity removal results equally from evaporation and from matter transport produced by the zoning. The relative rates of loss and transport are naturally dependent on the rate of zoning. Moreover, the predominant orientation of single crystals produced by zoning changed as a function of zoning speed. Single crystals with $\langle 111 \rangle$ orientation grew only from highly pure material (109). A more striking change of behavior takes place when high purity is attained in vanadium crystals, in that an allotropic transformation that normally takes place at high temperatures fails to occur in samples of high purity (110).

Compressed, electrolytic vanadium, after four zone passes in a vertical apparatus at 3.33×10^{-3} cm s^{-1} (12 cm h^{-1}), was found to be free of Mg, Ti, Mo, Cu, Fe, Al, and Si (111).

Single crystals of vanadium have been prepared by electron-beam float zoning, with resistivity ratios in the range $1-2 \times 10^3$ (112).

YTTRIUM (mp 1795 K)

The zone refining of Y has been carried out with application of an electric field. Metallic impurities were successfully concentrated, but gaseous impurities and carbon were not significantly segregated. After 3, 6, and 10 zone passes, metallic and interstitial impurities were moved toward the anode. Analysis showed that the initial part of the sample after 10 passes was a single crystal while the central portion consisted of large crystals and the end of the ingot was polycrystalline (113). The crystallographic orientation of single crystals resulting from zone refining has been related to the crystallization rate, melting temperature, and the form and size of the molten zone as well as the diameter of the ingot. For low crystallization rates, growth took place mainly in the [0001] direction, while in rapid crystallization the [1010] direction is preferred. Copper, iron, and hydrogen were segregated by the zoning (114).

ZINC (mp 692.8 K)

Zinc has been zone refined with the application of a dc field, and it was found that most of the Pb, Cd, Fe, Sn, and Bi had $k_{\text{eff}} < 1$ while Ag and Cu had $k_{\text{eff}} > 1$ and Al had $k_{\text{eff}} \simeq 1$ (115). Reduced zoning speed was relatively ineffective in the first stages of refining, when impurity concentrations were relatively high ($\sim 2 \times 10^{-4}\%$ Al), but the efficiency of zoning improved substantially with reduced rate at lower impurity concentrations (116).

Troitskii has examined the combined effects of electric fields and zone refining on isotope abundances in metals. Zone refining alone was first applied to zinc and cadmium. In the former, the ^{64}Zn isotope concentration at the end of the ingot increased with increasing number of zone passes; that is, $k_{\text{eff}} < 1$. After 70 passes the enhancement was nearly 3% (117). Enhancement was defined as $100 \Delta N/N$, where N is the ^{64}Zn-isotope concentration in the natural material, namely 48.9%. When an electric field was applied to a Zn wire during zone refining, a still-larger isotope enrichment ensued. After only four passes, the increase was 3–4%. More recently the experiment was carried out with melting produced by an electron beam and an enrichment of 9% was obtained after four passes (118). The metal was used in the form of wires 35 cm long and 0.2 cm in diameter; the zone was 0.63 cm long and was moved at 0.03 cm s^{-1} (120 cm h^{-1}). The sense of the electric field was found to have a profound influence on isotope migration; when the "head" of the ingot was positive, the field enhanced the fractionation resulting from zone refining; conversely when the "head" of the ingot was negative, the fractionation was reduced.

It has been reported that ultrasonication had a beneficial effect on the distribution coefficient of cadmium in zinc (119).

ZIRCONIUM (mp 2125 K)

Zirconium has been purified as the iodide and the metal recovered from this process was subjected to electron-beam zone refining at various rates. Segregation of metallic impurities was observed at zoning rates of 1.7×10^{-4} to 3.3×10^{-4} cm s^{-1} (0.6 to 1.2 cm h^{-1}) during the first two passes, along with a tenfold decrease in metallic impurity. Removal of C, O, and N was much less effective; concentrations at the head of the ingot after six passes decreased from 0.03, 0.06, and 0.004 to 0.011, 0.025, and 0.0015%, respectively (120). A high-vacuum pumping system has been used in the zone refining of Zr to generate an atmosphere essentially free from carbon in the residual gas (121).

1.2   III–V Semiconducting Compounds

Binary semiconductors prepared from elements of the third and fifth columns of the periodic table have provided interesting transistors, rectifiers, lasers, and transducers. While these have recently been largely displaced by germanium and silicon in device fabrication, they retain considerable interest. In general, III–V compounds are prepared by reacting stoichiometric quantities of the elements in graphite or graphite-lined quartz crucibles. In the case of aluminum antimonide, reaction in graphite yields aluminum carbide; reaction in alumina containers avoids this, but introduces oxygen into the product.

Indium antimonide is widely studied because of its conveniently low melting point (803 K), thermal stability, and low reactivity. After zone refining, the head of an ingot of InSb is p-type and the tail is n-type, because k_{eff}'s for acceptor and donor impurities are >1 and <1, respectively (122). Recently, InSb has been zone refined as a thin film (123). It has also been zone refined in a dual-boat apparatus in which a small portion of the molten zone is allowed to flow out of the main boat through a spillway. The removal of a portion of the impure molten zone at the end of each pass allows attainment of lower impurity levels at ultimate distribution. The purified product had lower majority carrier concentration and higher mobility than were attained in a conventional refiner (124).

Gallium antimonide is likewise p-type after purification by zone refining, because k_{eff}'s of acceptors are close to unity. Indium arsenide poses severe problems because of the high vapor pressure of arsenic at the melting point of the compound (1215 K). After zone refining, InAs is n-type, because of S and Se,

Table 8.2. Distribution Coefficients of Various Dopants in III–V Compounds (after Vigdorovich)

Dopant	Group	InSb	InAs	InP	GaSb	GaAs	GaP[a]	AlSb
Na	Ia	2.5×10^{-3}						
Cu	Ib	6.6×10^{-4}	<0.05			2×10^{-3}		0.02–0.1
Ag	Ib	4.9×10^{-5}				0.01		
Au	Ib	1.9×10^{-6}				0.1		
Mg	IIb	<1	0.7	1–1.4		0.3		0.1
Zn	IIb	~10, 3	0.8		0.3	0.1–0.9		
Cd	IIb	2.4–4.1, 0.26	0.1		0.02	0.02–0.2		
B	IIIb							0.01–0.02
Ga	IIIb	2.4						
Tl	IIIb	5.2×10^{-4}						<0.1
Si	IVb	0.04–0.1	0.4	0.05	~1	0.1		0.1
Ge	IVb	0.01, 0.02	0.07	0.3	0.2	0.02		
Sn	IVb	0.01, 0.057	<0.05	0.24–0.4		0.03		
Pb	IVb	0.16	<0.05			<0.02		<0.1
As	Vb	5.4						
V	Vb							
S	VIb	0.1	~1	0.8		0.5		<0.1
Se	VIb	0.17–0.35	0.9	0.6	0.013[b], 0.4	0.5		
Te	VIb	0.5–1, 4.2	0.4		0.3[b], 0.4	~0.1		
Cr	VIb						10^{-3}	
Mn	VIIb						0.015	
Fe	VIII	0.04	~5			3×10^{-3}	10^{-4}	<0.1
Ni	VIII	6×10^{-5}				<0.02		0.01–0.1
Co	VIII					<0.02		<0.1

[a]Ref. (209).
[b]Value for k extrapolated to zero growth rate (210).

whose k_{eff}'s are close to 1. The same problems are encountered with GaAs (mp 1513 K) and with InP (mp 1343 K). Gallium phosphide has been purified by the floating-zone method, and single crystals resulted, with specific resistance of $10^4 - 10^{10}$ ohm-cm (125).

Distribution coefficients of many elements in III–V compounds are assembled in Table 8.2.

1.3   Inorganic and Organometallic Compounds

1.3.1   Water and Aqueous Systems

Water has been zone refined to remove solutes and to achieve separation of the hydrogen isotopes. Schildknecht concentrated numerous organics from water by zone refining. Quinones and aldehydes were successfully enriched from very dilute solutions ($4 \times 10^{-4}\%$) (126), and materials of biological interest, such as enzymes, plankton, bacteria, and bacteriophage were isolated from water by zone refining. It was found that mechanical damage is caused by cyclic freezing and thawing; the plankton was unrecognizable after a few passes (126). Work of this kind has not been continued in recent years, probably because other techniques, such as freeze drying and chromatography, have proved to be more effective.

Preparation of pure ice by zone refining prepurified water has been reviewed (127).

The equilibrium phase diagram of H_2O/D_2O indicates that k_0 for D_2O in H_2O is slightly greater than unity. Starting with 43.2 mole% D_2O, after 413 zone passes through a helical charge, an ingot was obtained that contained 46.2 mole% D_2O at the head and 39.5 mole% at the tail (see Section 2.4, Chapter 5) (128).

Ammonium, alkali–metal, and alkaline–earth salts have been studied extensively as neat materials, as cryohydrates, and as eutectics (see Section 2, Chapter 7). The cryohydrate of calcium nitrate was zoned in horizontal glass tubes and 3-cm zones were passed through 12-cm ingots at 1.94×10^{-4} cm s^{-1} (0.7 cm h^{-1}).

1.3.2   Alkali–Metal and Alkaline–Earth Salts

Single-pass zoning has been applied to ammonium nitrate in horizontal boats, for determination of the effective distribution coefficients of Cs, Rb, and K. The same technique was also applied to sodium nitrate (129). More recently, both sodium nitrate and potassium nitrate have been zoned in 0.8-cm Pyrex tubes at 1.7×10^{-3} cm s^{-1} (6 cm h^{-1}), with periodic reversal of axial rotation to promote mixing in the molten zone (see Section 2.5, Chapter 5). The effective distribution coefficient for chromium in sodium nitrate, added as 10^3 ppm of sodium chromate, was found to be $10^{-3} - 10^{-2}$ (130), and to depend on rotation rate and reversal period (130). In a related study, ^{137}Cs and ^{89}Sr were distributed in ingots of sodium nitrate in a horizontal refiner, for measurement of k_{eff} (131). The distribution of copper ions in sodium sulfate has been reported (132).

Barium and strontium fluorides were purified and converted to single crystals by zone refining in graphite boats, in vacuum, inert atmospheres, or reactive gases (133). Strontium and barium were removed from calcium nitrate by zone refining (134).

The alkali halides have elicited efforts at purification and crystal growth for many years, because of their importance as optical materials. Early efforts of Gründig (135) were followed by an exhaustive study of k_{eff}'s in various halides (136) (see Chapter 10). Zoning has generally been carried out in evacuated quartz tubes under about 5×10^4 Pa (0.5 atm) of the corresponding halogen. In the absence of the free halogen, the salts were found to adhere strongly to the quartz surface, apparently because of an exchange of oxygen between oxide impurities in the melt and the silica surface. Moisture had to be excluded rigorously, because the halides can hydrolyze to hydroxides at their melting temperatures. Despite these problems, fused silica remains the preferred container material.

Zoning was carried out at 10^{-4} cm s^{-1} (0.37 cm h^{-1}), with a 2-cm zone in a 20-cm ingot. Monovalent and divalent impurities were added at 10^3 and 10^2 ppm, respectively. Dopants were added singly and multiply; no interactive effects were noted (136). More recently, similar experiments on KCl and KBr gave crystals with impurity concentrations of 10^{-6}–10^{-8} mole fraction (137). Distribution coefficients of Rb and Cs were determined in KCl by zone refining (138).

Sodium iodide has been zone refined in quartz tubes of 1–2 cm diameter and 25–30 cm length. Concentrations of rubidium and cesium were reduced by two orders of magnitude in about half the ingot, and that of potassium was reduced by one order of magnitude (139).

1.3.3 Silver Salts

Purification of silver halides by zone refining is troubled by three properties of the materials: large thermal expansion coefficients of the solids, considerable expansion upon melting, and adhesion to container walls. The mechanical expansion led to abandonment of attempts to purify and crystallize the halides in vertical tubes (140). Successful purification was achieved in quartz tubes of circular or square cross section in a nearly horizontal orientation. The apparatus was mounted with a slope of 4° to counteract matter transport resulting from the expansion that accompanies melting; that is to say, the end of the ingot at which the zones began was the higher. Charges of 200 g were processed in vacuum or inert atmosphere, at a zoning rate of 2.08×10^{-3} cm s^{-1} (7.5 cm h^{-1}). The adhesion of the chemicals to container walls was overcome by careful removal of contaminating silver or silver oxide and/or by maintaining an appropriate halogen atmosphere over the charge. Since these halides are light sensitive, they had to be shielded from irradiation, including that originating in the zone heaters.

Table 8.3. Distribution Coefficients of Various Impurities in AgCl and AgBr

Halide	Impurity	Atmosphere	Distribution Coefficient, k_{eff}
AgCl	Cu	Cl$_2$ or vacuum	0.4–0.6
AgCl	Pb	Cl$_2$ or vacuum	0.4
AgCl	Ni	Cl$_2$ or vacuum	1.4
AgCl	Fe	Cl$_2$	~0.7
AgCl	Fe, Mn, Cd	Vacuum	>1
AgCl	Sn, Al, Sr	Vacuum	<1
AgBr	Cu	Vacuum	<1
AgBr	Fe, Ni, Mn	Vacuum	>1

Chemically purified AgCl was doped with copper, nickel, lead, and iron, to study the segregation of these solutes. Copper and lead were strongly rejected, with $k_{eff} < 1$. Nickel, on the other hand, showed $k_{eff} > 1$. The distribution of iron depended on the atmosphere in which the zoning was carried out: in vacuum, k_{eff} is >1, while in a chlorine atmosphere, k_{eff} is <1 (140). The distribution coefficients are assembled in Table 8.3.

Lichkov et al. (78) purified silver nitrate preliminarily, by recrystallization from 30% nitric acid and passage through silver oxide. After this the levels of copper, iron, chromium, manganese, lead, and bismuth were $10^{-4}–10^{-5}\%$. The nitrate was then zone refined in evacuated plastic containers at 8.3×10^{-3} cm s^{-1} (3 cm h^{-1}). After 40 passes impurities were not detectable by emission spectrography but could be assayed by mass spectrometry and resistance-ratio measurement. The nitrate was converted to chloride which was in turn reduced to silver sponge, using hydrogen at 673–723 K. The product was converted to single crystals which had a resistivity ratio of about 2×10^4 and impurity content $<10^{-5}\%$ (see Section 1.1).

1.3.4   Cuprous Oxide

Cuprous oxide is of interest as a p-type semiconductor; its electrical and optical properties are sensitive to chemical and physical defects. The high reactivity of the substance with container materials has made it difficult to purify and crystallize. Recent work has used the float-zone method, but because the polycrystalline starting material has high resistivity, it is not directly amenable to radiofrequency induction heating. Instead a copper-wire susceptor was wrapped around the feed rod, which was 0.8 cm in diameter and 15 cm long. Upon application of radiofrequency power, the copper melted and heated the cuprous oxide until its conductivity became high enough to couple to the radiofrequency field. After establishment of a stable molten zone near the bottom of the rod, in a nitrogen atmosphere, one end of the rod was rotated at

0.083–0.167 rev s^{-1} and the heating coil was raised at 10^{-3} cm s^{-1} (141). In other work on Cu_2O, float zoning was carried out with an arc-image furnace. Optical microscopy and X-ray topography showed that best results were obtained by growing along the [011] axis (142).

Cu_2O was also float-zoned by the plasma-arc method in a low-pressure oxygen atmosphere at 27 Pa (0.2 torr) at 3.4×10^{-4} cm s^{-1} (1.2 cm h^{-1}). Trace impurities were measured by atomic absorption spectroscopy along the length of the ingot. Chromium, magnesium, silver, and iron were present at levels of <2.5, <0.5, <1.5, and <2.5 ppm, respectively, over the entire ingot. The unrefined cuprous oxide contained 12, 9, 59, and <2.5 ppm, respectively. In experiments in which iron was added at the start of the ingot, it was found that the center contained <2.5 ppm and the top 5300 ppm after starting with 11,000 ppm. Evidently, Fe is transported by zoning and is lost from the ingot by evaporation (141).

1.3.5 Organometallics

Süe and Nouaille first suggested that inorganic and organometallic compounds might be purified by zone refining, not only because of interest in the compounds themselves, but to get pure metals. Two advantages accrue to the scheme:

- The compounds, having lower melting points than the metals, are easier to melt and contain.
- Pairs of metals that are not readily segregated may be converted to segregable derivatives.

In an early application of the idea, small amounts of metal ions, in the form of chelates, were zone refined in ingots of the corresponding unchelated ligand. Bis(benzoylacetonato)Cu (II) (mp 433 K, dec.) added at 0.52 wt% to benzoylacetone (mp 330 K), gave $k > 1$. With bis(benzoylacetonato)Fe (III) (mp 463 K, dec.) at 283 ppm in the same host, $k < 1$. Tris(8-hydroxyquinolinato) Fe (III) at 470 ppm in 8-quinolol, gave $k \ll 1$ (143).

Imamura (108) converted impure metals to low-melting derivatives with acetylacetone or dipivaloylmethane and zoned the resulting compounds. Zinc acetylacetonate (mp 397–403 K) was subjected to 10 downward zone passes in a glass tube, at 1.1×10^{-3} cm s^{-1} (4 cm h^{-1}) under inert gas. The concentrations in ppm of several impurities, measured by emission spectrography on portions of a 20-cm ingot, are collected in Table 8.4.

Copper was purified by 20 zone refining passes over an ingot of its dipivaloylmethane derivative (mp 471 K), at 1.4×10^{-3} cm s^{-1} (5 cm h^{-1}). Again, substantial reduction of impurity was achieved, as indicated by the impurity concentrations in a 70-cm ingot, tabulated in Table 8.4.

The most extensive study of purification of metals as their chelate derivatives has been carried out by Ueno and his co-workers (144–148). They rejected 8-quinolol as a ligand because of the thermal instability of many metal quin-

Table 8.4. Impurity Concentrations in Zone-Refined Metal Complexes, in ppm

	Zinc Acetylacetonate					Copper Dipivaloylmethanate				
	Pb	Fe	Cd	Sn	Cu	As	Sb	Bi	Pb	Fe
Top of ingot	0.3	0.3	0.1	—	0.2	—	—	—	—	—
Center of ingot	2.0	1.0	0.5	—	1.0	10	20	2	3	1
Starting material	30	20	20	10	20	30	100	50	50	100

olinolates. Pentane-2,4-dionates (acetylacetonates, XAA) are more stable, but only the chromium (III), beryllium (II), and aluminum (III) derivatives appear to be able to survive the prolonged heating encountered in zone refining. $Cr(AA)_3$ was zoned and the following ions were effectively segregated: Al^{3+}, Fe^{3+}, Cu^{2+}, Mn^{3+}, Ni^{2+}, Co^{3+}, and Rh^{3+}. Of these only the rhodium chelate showed $k > 1$ (147). Fluorinated pentane-2,4-dionates, while lower-melting than the unsubstituted compounds, still failed to survive zoning. Some alkyl-substituted pentane-2,4-diones, on the other hand, gave stable chelates.

Fifty-five products were prepared, from Cu(II), Be(II), Pd(II), Co(II), Co(III), Ni(II), Al(III), Fe(III), Mn(III), and Cr(III) with ligands

where R is ethyl, *n*-propyl, *iso*-propyl, *iso*-butyl, *tert*-butyl, and *n*-valeryl. It is important that the ligands are symmetrical, so that the chelates will have unique coordination structure. Many of the products were stable at temperatures slightly higher than the melt endotherm, at least in the absence of air. A study of the phase relationships among pairs of chelates showed that solid solutions result with common derivatives (i.e., of the same ligand) of central ions differing by less than 0.06 nm in radius, if the two substances have identical coordination structures.

Deliberately contaminated samples of the stable chelates were zone refined in a conventional zoner (five heaters traveling downward at $5.6 \times 10^{-4}\,\text{cm s}^{-1}$ ($2\,\text{cm h}^{-1}$)) and in a high-speed apparatus (four heaters traveling upward at 3.3×10^{-3} to $2.3 \times 10^{-2}\,\text{cm s}^{-1}$ (12 to $84\,\text{cm h}^{-1}$)), with the sample tube subjected to rotation with frequent reversal. Portions of the zoned ingots were decomposed with mixed acids and analyzed by atomic absorption spectroscopy.

Both $Fe(AA)_3$ and $Al(AA)_3$ were removed from the head of an ingot of $Be(AA)_2$ by zoning. When $Al(AA)_3$ was precipitated from a solution containing 3600 ppm of Fe(III), Cu(II), Zn(II), Ni(II), Co(II), and Mn(II), only Fe, Cu, and

Zn were detectable (at 3100, 70, and 50 ppm respectively); these ions were all removed effectively from the head of the ingot by zoning. The other ions (Ni, Co, Mn) were not detectable even in the tail of the zoned ingot, indicating that prior removal during precipitation of the major compound was indeed efficient.

Commercially available zirconium salts are usually contaminated with 1–2% hafnium. The chelate of Zr(IV) with heptane-3,5-dione was effectively freed of all contaminants except for Hf(IV) which was, however, concentrated to some extent. The measured k_{eff} (0.65) indicated that the Hf(IV) concentration in the purest fifth of an ingot, at ultimate distribution, should be $(1.7 \times 10^{-3} C_0)$. The material that was used had $C_0 = 1.5$ wt%, so that the purest fifth of the ingot should have contained only 0.11% of the original contamination. In fact, this segment contained 0.36%, leading to the conclusion that zone melting gives no better yield of purified product than other methods, although it is faster (145).

1.3.6 Miscellaneous

Acid tungstates have been zone refined for the removal of isomorphous molybdates. In $Na_2W_2O_7$ and $K_2W_4O_{13}$, k_{eff} decreased with increasing Mo concentration. Zone refining was also effective in removing W from $Na_2Mo_2O_7$ and $K_2Mo_4O_8$.

Lanthanum hexaboride elicits lively interest because of its usefulness as a cathode material for electron guns. Hot-pressed LaB_6 rods were prepared in graphite dies, and zone refined by arc float zoning. After three passes, the ingots were single crystals with circular cross section; these contained less than 30 ppm total impurity exclusive of N, O, H, and C, as measured by spark-source mass spectrometry. Vacuum fusion analysis showed no detectable N or H, and 20 ppm of O. Combustion analysis showed 55 ppm of C. Gravimetric analysis for B and La gave a ratio of 5.98 ± 0.02 for the starting material; the ratio fell to about 5.72 after the first zone pass and then remained constant during two additional passes. These data point to a preferential loss of B from the molten zone. The measured melting point dropped from 2988 to 2857 K on going from one to three passes (the reported melting point for stoichiometric LaB_6 is 2988 K). X-ray and metallographic searches for a second phase (possibly LaB_4) in these crystals gave negative results, leading to the conclusion that the crystals were of a single nonstoichiometric phase (149).

1.4 Organic Compounds

1.4.1 Aliphatic Hydrocarbons

Saturated and unsaturated hydrocarbons have been investigated for single crystal growth and as test substances for low-temperature zone refining.

Kieffer used circulating liquid propane to maintain cold zones at 93 K in an apparatus for purification of low-melting dienes (150, 151). Eight substances were processed in this apparatus; the resulting purifications are outlined in Table 8.5.

A different approach was used for purification of some other aliphatics. The

Table 8.5. Purification of Dienes by Zone Refining (150, 151)

Product	Melting Point, K	Number of Zone Passes	Purity of Starting Material, %	Purity of Different Fractions[b]				
				1	2	3	4	5
2,3-Dimethyl-1.3-butadiene	197	7	94.85	93.2	93.2	93.2	95.0	96.0
		21	94.85	94.2	94.2	94.2	97.0	98.5
		49	94.85	96.3	96.3	96.3	97.4	98.9
		84	94.85	96.3	96.3	96.3	97.4	99.6
		7	99.43	97.0	98.8	99.7	99.7	99.82
		21	99.43	96.8	98.6	99.7	99.88	99.80
		35	99.43	96.6	98.6	99.8	99.90	99.95
		49	99.43	96.6	98.6	99.84	99.92	99.98
		98	99.43	96.5	98.4	99.92	99.95	99.98
1,3-Pentadiene	185.7	9	68	62.2	65.3	67.2	68	70
		35	92.8	85.5	89.5	93.3	93.5	94.2
		7	98.7	96.7	98.7	99.7	99.8	99.8
		7	98.7	94.5	98.5	99.85	99.91	99.90
		7	98.7	95.4	98.6	99.87	99.21	99.89
		7	98.7	95.6	98.5	99.89	99.93	99.91
2,4-Dimethyl-1,3-pentadiene		7	97.9	99.1		97.9	97.7	97.7
		14	97.9	94.4	99.3	97.8	97.8	97.8
		35	97.9	99.6	99.2	97.8	97.8	97.8
2,4-Hexadiene	192.4	7	98.7	98.4		98.8	98.9	98.9
		14	98.7	98.2		98.5	99.1	99.1
		35	98.7	98.0		99.1	99.5	99.5
2,4-Heptadiene		7	98.3	97		97.6	98.6	98.6
		35	98.3	96.6	96.6	98.4	99.4	99.3
3-Methyl-1,3-hexadiene		35	98.2			No purification		
4-Ethyl-1,3-hexadiene		14	98.0			No purification		
4-Methyl-1,3-heptadiene		14	98.6			No purification		

[a]Zoning speed 6.9×10^{-4} cm s^{-1} (2.4 cm h^{-1}), $L/l \simeq 20$.

376

sample was charged into the annular outer chamber of a double-walled quartz cylinder; heaters were moved within the inner cylinder, while the exterior was cooled by immersion in a liquid bath cooled by liquid nitrogen. After 12 passes through an ingot of cyclohexane of 98.93% initial purity, a product 99.91% was obtained. Higher purity (99.99%) has been attained by zone refining in conventional cylindrical containers (152) and by directional crystallization (153). A similar procedure was used with n-hexane and n-octane, with good results (154).

Gas chromatographic analysis of commercially available $C_{19}H_{40}$, $C_{20}H_{42}$, and $C_{22}H_{46}$ showed the presence in each of all hydrocarbons from C_5 to C_{26}. The nearest-neighbor higher and lower homologs were most abundant. Recrystallization from hydrocarbon solvents did not provide pure material. Distillation at low pressure and high reflux ratio resulted in decomposition, with increasing amounts of low-molecular-weight impurities. Zone refining removed low-molecular-weight impurities efficiently, but close homologs were leveled at zoning rates above 2.78×10^{-4} cm s^{-1} (1 cm h^{-1}). Effective distribution coefficients were measured at various rates by directional crystallization of hydrocarbon samples doped with radiotracer impurities (see Chapter 10), and the results indicated that homolog pairs are intrinsically separable. Extensive zoning (250 passes) at slow speed (4.2×10^{-5} cm s^{-1}, 0.15 cm h^{-1}) gave very pure product in the upper one-third of each ingot (155).

1.4.2 Fatty Acids and Their Derivatives

Palmitic acid has been zoned in a horizontal rotation-convection refiner at 5.78×10^{-5} cm s^{-1} (5 cm d^{-1}), and impurity level was found to be less than 0.1% by gas chromatography. The purified acid was converted to the anhydride by dicyclohexyl carbodiimide. An acyl urea by-product was removed by zone refining, using a procedure similar to that used with the acid. The anhydride was used in the preparation of dipalmitoylphosphatidylcholine, for study of the transition from gel to liquid crystal (156).

Fatty acids of 99% purity have been used as starting materials for further purification by zone refining. Purities >99.95% were attained by zoning 100-g batches in tubes of 1.0-cm diameter, in a horizontal refiner, at a zoning rate of 8 cm d^{-1} (157). Myristic, lauric, palmitic, and stearic acids were purified by zone refining until gas chromatography of their methyl esters showed no impurity in the purest fractions. Stearic acid was used as a test compound in a study of the effect of magnetic stirring during zoning. Stirring made it possible to zone effectively at speeds up to 3 cm h^{-1} (158; see Section 2.5, Chapter 5). Application of zone refining to fatty acids has been reviewed (159).

Cholesteryl stearate was more effectively purified by zone refining than by crystallization from solution. Experiments pointed to the presence of an impurity which raised the solid/smectic transition temperature, and which was only slowly removed by zoning (160).

Trimethylacetic (pivalic) acid has been purified in a multizone refiner which provided mixing of the molten zones. Product of 6N purity was used for crystal growth by the Bridgman method (Section 4.1, Chapter 4), and effects of impurities on crystal perfection were studied (161).

Analysis of directionally crystallized methyl methacrylate (mp 230.7 K) gave a value of 0.4 for the average distribution coefficient of the impurities. This suggested that substantial purification would result from a small number of zone passes. Zoning was carried out in a metal trough immersed in liquid nitrogen; the molten zone was generated by direct immersion of a heating coil which was moved through the charge at 1.7×10^{-2} cm s^{-1} (60 cm h^{-1}). Five passes produced a 10% yield of product containing 0.14% impurity from starting material containing 0.8%; after 10 passes the impurity content was 0.05%. In these experiments purification did not reach that expected on the basis of the measured k_{eff}. This is not surprising in view of the very high zoning rate used. In fact, the observed reductions in impurity concentration correspond to a k_{eff} of approximately 0.7 to 0.8 (162).

1.4.3   Aromatic Hydrocarbons

Polycyclic aromatic hydrocarbons such as naphthalene, phenanthrene, and anthracene have been studied extensively with respect to their impurity concentration, largely because of interest in their spectroscopic properties. Naphthalene of coal-tar origin is commercially available in a variety of purities. Gas chromatographic analysis of several commercial grades, using flame-ionization detection, revealed the presence of benzo[b]thiophene (22 to 6300 ppm) and 2-methylnaphthalene (< 10 to 870 ppm). The limit of detection of benzo[b]thiophene was 20 ppm by flame-ionization detection and 1 ppm by flame-photometric detection. For 2-methylnaphthalene the limit of detection was 10 ppm by flame-ionization detection. No other impurities were detected in the starting material. In a concentrate obtained by zone refining, taking 10% of the zoned ingot, eight additional peaks were detected and identified by gas chromatography/mass spectroscopy (163) (see Table 8.6).

A related study of other commercial naphthalenes showed the presence of several other impurities (164). Measurements of luminescence spectra at 4.2 K indicated that the concentrations of methylnaphthalene and benzothiophene were approximately 0.1 ppm in two-thirds and one-third of the ingot, respectively (166). The distribution coefficient of 2-methylnaphthalene was found to be 0.12 for zoning at 3×10^{-5} cm s^{-1} (0.11 cm h^{-1}) and 0.33 for zoning at 1.2×10^{-4} cm s^{-1} (0.44 cm h^{-1}) with $C_0 = 0.12\%$ (164).

Extensive zone refining of commercial naphthalene was monitored by measurement of the decay time of the delayed fluorescence in the sealed zoning tube. The decay time increased with increasing number of zone passes, to a limit of 105 ms, corresponding to a triplet lifetime of 210 ms (165).

The system naphthalene/2-naphthol has been used as a model for zone refining of solid solutions in which the solute has $k > 1$. In one study, the effective distribution of 2-naphthol was studied as a function of zoning rate and stirring conditions (rate of rotation), to investigate the validity of a modified BPS equation which included a term specifying the fraction of interface melt

Table 8.6. Trace Impurities in Naphthalene

	Concentration, ppm	
Impurity	Maruyama et al.[a]	Grazhulene et al.[b]
Indane	<0.4–1.5	2–200
Indane	<0.5–33	3–1400
1,2-Dihydronaphthalene		10–620
1,4-Dihydronaphthalene		12–50
Tetralin	<0.4–2.3	—
1-Methylindene	<0.4–5	—
1-Methylnaphthalene	<1.0–3	11–1200
2-Methylbenzo[b]thiophene	<1.5–2.5	—
Quinoline	<1.5–10	—
Dimethylnaphthalene	<2–12	100
Biphenyl	<2–12	—
Diphenyl ether	—	800
Indole	—	1600

[a]GC conditions: 0.3 cm × 2 m glass column, 8% PEG 20 M silanized Chromosorb-W (80–100 mesh) at 433 K, using N_2 carrier gas.
[b]GC conditions: 0.2 cm × 1.5 m stainless-steel column, 10% PEG 40 M on Chromaton N (0.2–0.25 mm), with programmed temperature.

occluded in the advancing crystal front (see Equation 3.25). Measured effective distribution coefficients varied from 1.25 at 10 cm h^{-1} (without rotation) to 1.51 at 2 cm h^{-1} (with rotation at 200 rev min^{-1}). The equilibrium distribution coefficient for 2 wt% of 2-naphthol in naphthalene was estimated, from published phase diagrams, to be 5.5 (166). From the discrepancy between this value and the k's estimated from the zoning experiments, the fraction of occluded interface melt was calculated to be 0.87, a high value indeed. Earlier work with this system showed that separation efficiency was enhanced by including an axially disposed heat conductor, centered in the zone-refining container (167).

By contrast, 1-naphthol forms a simple eutectic system with naphthalene, with $k = 0.23$ (168).

A similar study of phenanthrene (169) indicated that commercial samples contained total impurity content in the range 2–8%. Preliminary examination included elemental analysis, which showed that both nitrogen and sulfur were present. Ultraviolet absorption spectra revealed the presence of anthracene. Gas chromatography detected 12 other impurities ranging in concentration from 0.01 to 4%. Two additional impurities were detected only after preconcentration by zone refining. The concentration profiles of various impurities gave clues to their chemical nature and ultimate segregability even before they were identified. Most of the impurities were strongly concentrated in the lower third of the ingot; hence they must differ considerably from phenanthrene in shape and/or size and have distribution coefficients much less than unity. Figure 8.1 shows profiles for several of the impurities. Curves 1 and 2 refer to impurities identified as biphenyl and dibenzothiophene, respectively, and are characteristic of strongly segregated materials. Curve 3 refers to anthracene, which has a distribution coefficient greater than unity and is consequently enriched at the head of the ingot. A third category of impurity is represented by curve 4, which refers to fluorene, a material that is only slightly segregated by zone refining. Information on chemical identity of the impurities, derived from chromatography and zone refining concentration profiles, was used to select other methods of purification appropriate to the chemical nature of the individual impurities. For example, oxidation, dimerization, and adduct formation with maleic anhydride were considered for removal of anthracene. Reaction with maleic anhydride was found to be most efficient, producing final concentrations not detectable by UV spectrophotometry (<0.01 ppm). Anthracene and nitrogen and sulfur-containing compounds may be removed from phenanthrene by fusion with alkali metals. Although fusion with alkali metal has the disadvantage of introducing new impurities (hydrogenated species) it was found that these could be removed by distillation, zone refining, or chromatography. The overall result of the work led to the adoption of the following sequence:

- Maleic anhydride treatment
- Fusion with sodium at 483 K
- Chromatography and vacuum distillation
- Treatment with Raney nickel (to reduce further the dibenzothiophene content, if necessary)

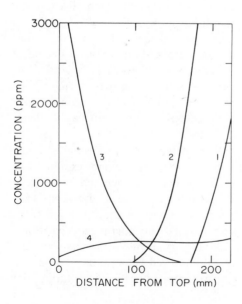

Fig. 8.1 Distribution of impurities in phenanthrene after zone refining. 1—biphenyl, 2—dibenzothiophene, 3—anthracenee 4—fluorene.

Of eight major impurities in the starting material, four were found to have distribution coefficients less than 0.6; the remaining four were identified as anthracene ($k = 1.6$), fluorene ($k = 0.95$), dibenzothiophene ($k = 0.8$), and dihydrophenanthrene ($k = 0.8$). After chemical removal of those impurities not effectively segregated by zone refining, an impurity content of < 1 ppm was attained by zone refining part of the purified ingot. Prolonged heating of phenanthrene at temperatures 20–30 K above the melting point caused formation of an impurity tentatively identified as dihydrophenanthrene.

It has been pointed out that spectrofluorimetric measurement of very low concentrations of anthracene in phenanthrene is made difficult by interference by phenanthrenequinone and fluoranthene. The problem is overcome by zone refining, since anthracene has $k > 1$ while the other two solutes have $k \ll 1$; hence, zoning separates the two categories and provides a simple binary mixture at the head of the ingot. Sensitivity was enhanced by analyzing concentrated solutions in xylene; spectra were measured at 483 nm, with excitation at 378 nm in a solution containing 50 ng anthracene mL^{-1}. The phenanthrene concentration was 10^6 times higher. Anthracene was determined in the range 0.05–10 ppm (170).

The purity of anthracene probably has been studied more intensively than that of any other organic material, and was the object of an early application of zone refining to a chemical problem (171). In the impurity-enriched ends of zoned ingots, 12 impurities were detected and identified. Most of these were not detectable in the starting material. Distribution coefficients were measured for the principal impurities. These values are useful in predicting the concentration profiles in doped crystals prepared by the Bridgman technique. More important from the standpoint of chemical purity, knowledge of the distribution coefficients makes it possible to estimate attainable impurity levels using published concentration profiles for relevant geometry and number of passes. Table 8.7, for example, shows the concentrations of five impurities in the upper half of an anthracene ingot after 10 passes. The inadequacy of zone refining for removal of carbazole is evident, while the other impurities are reduced to extremely low levels. Other studies have confirmed the need to use ancillary purification

Table 8.7. Impurity Concentrations Attainable by Zone Melting Anthracene, mole%

Impurity	k_{eff}	C_0	$C(0.5, 10)^a$
Tetracene	0.10	5×10^{-3}	9×10^{-9}
Anthraquinone	0.02	2×10^{-1}	1×10^{-10}
Fluorene	0.2	5×10^{-2}	2×10^{-5}
Phenanthrene	0.1	5×10^{-2}	9×10^{-8}
Carbazole	0.9	1×10^{-1}	8×10^{-2}

^{a}Concentration in the upper half of an ingot after 10 zone passes.

methods in addition to zone refining to attain the intrinsic spectroscopic properties of anthracene (172, 173).

Zone refining has been used to preconcentrate the ubiquitous impurity anthraquinone, prior to polarographic analysis. The preconcentration made it possible to achieve a tenfold reduction in the determinable level, to $3.5 \times 10^{-4}\%$ (174).

Pyrene, derived from coal tar, has been zone refined without chemical prepurification. Analysis of the resulting ingots for nitrogen and sulfur shows that nitrogen concentration is higher at both head and tail than in the center, while sulfur concentration increases gradually from head to tail. These simple results lead directly to the conclusion that at least two nitrogen-containing impurities are present, and at least one of these has $k > 1$. The sulfur compound evidently has k close to unity. These distributions imply that the nitrogen- and sulfur-containing contaminants are structurally similar to pyrene and must be heterocyclic analogs. In fact, careful analysis shows that small amounts of two isomeric azapyrenes are present and that 4,5-phenanthrylene sulfide is usually present in amounts up to 2% (175).

Several aromatic hydrocarbons from coal tar (naphthalene, anthracene, durene, biphenyl, acenaphthene and hexamethylbenzene) were zone refined before gas chromatographic analysis. The limits of detection were reduced by one to two orders of magnitude in each case (176).

1.4.4   Heterocyclic Aromatics

Aromatic hydrocarbons are extensively studied because of their presence as environmental contaminants and for their spectroscopic interest. Materials of coal-tar origin are usually contaminated with heterocyclic materials of similar molecular size. These in turn have been objects of detailed study.

Carbazole has been purified by a scheme involving extraction with maleic anhydride, which made it possible to remove anthracene by a later extraction of its adduct with alkali. Subsequent zone refining, after crystallization from 1,2-dichlorobenzene, resulted in a product of 5N purity (177).

1.4.5   Zone Refining of Radiochemicals

Zone refining has been applied to the separation of radioactive contaminants from organic compounds. Such separations are often carried out chromatographically and can yield product of high chemical purity. Nevertheless, high *radiochemical* purity may not be reached because tagged compounds similar to the product may not be separated. Separation of tracer-level acetamide from propionamide was studied as a model for this situation. Mixtures containing only 0.02% acetamide were subjected to various programs of zone refining, and it was found that the concentration of the tagged compound could easily be reduced by a factor of 10^4. Had the major compound been radioactive propionamide, it could easily have been freed from radioacetamide (178).

Another application of zone refining to radioactive materials can be envisioned. If a small amount of a radioactive product A^* is formed by a nuclear reaction in a large mass of another compound B, its presence can be verified by adding the $A*B$ mixture to a large quantity of nonradioactive A. The original $A*B$ mixture is now present as an impurity in carrier material A, which is freed from chemical impurities by zone refining. Hence $A*$ can be obtained in high purity, diluted in carrier A, and free from B and other possible contaminants. The uniform radioactivity along most of the ingot of A is a good indication that the radioactive species is in fact $A*$ (178).

1.4.6 Pharmaceuticals

A number of papers have been published (179, 180, 181) describing attempts to purify materials of pharmaceutical interest. For the most part, these have not gone beyond orienting studies of relatively simple molecules. In part this fact results from the widespread thermal instability of pharmacologically active materials (182). These negative results do not by any means close the book on application of zone refining to such materials. In the first place, many pharmaceuticals are sold as salts of amines, because the amines themselves have undesirable physical properties, such as low melting point. This implies that the precursor amine may in fact be amenable to exhaustive purification by zone refining, after which a pure salt may be prepared. Even if the ultimate and penultimate products are unstable, it may nevertheless be advantageous to zone refine an intermediate, to ease final purification. Ultimately, added-component techniques (Chapter 7) may be applicable even to products that melt with decomposition.

2 GROWTH OF SINGLE CRYSTALS BY ZONE MELTING

Growth of single crystals by zone melting offers three advantages:

- Simplicity
- Possibility of uniform doping
- Reduced contamination of the melt by the container

Four techniques have been used:

- Horizontal boat
- Enclosed vertical container
- Vertical floating zone
- Temperature-gradient zone melting

Some applications of these methods are mentioned in Section 1 and will not be discussed here. It should be noted, however, that attempts to combine purification and crystal growth in a single zone-melting operation are not likely to be successful, because the transport rates commonly used to promote solute segregation are too high for growth of single crystals.

2.1 Horizontal Boat Method

The earliest applications of the zone-melting technique were to growth of single crystals (183, 184, 185, 186). The method is identical to that described as zone leveling (see Section 1.3, Chapter 5, Figure 5.46) except that an oriented seed is used at the start of growth; most recent applications of this method have been to the growth of metal and metal–alloy crystals.

Pfann and Olsen (187) grew germanium crystals from zone-refined starting material in a fused-silica boat, the interior of which was coated with soot to prevent adhesion of the crystal to the boat.

Single crystals of bismuth/antimony alloy were grown in a wide range of compositions by the horizontal single-pass zone-leveling method. The starting charge consisted of two portions: a small (35-g) "zone bar," and a succeeding "charge bar" (150 g), contained in a quartz boat of semicircular cross section (2 cm $\times$ 20 cm) sealed under 5×10^4 Pa of hydrogen. To begin, the zone bar was completely melted by an external heater. Even without seeding, a large fraction of the resulting ingot was a single crystal. This result is doubtless related to the slow growth rate used, which ranged down to about 10^{-6} cm s^{-1}. The growth gradient used in this work was about 40 K cm^{-1}.

Tarjan (188) has described the use of the boat method for growth of low-melting metals (Sn, Cd, Zn, In). Polycrystalline charges were melted in contact with oriented seeds and growth was carried out at 10^{-3} cm s^{-1} in flowing argon.

Hurle has studied the impurity striations that appear in indium antimonide and other materials solidified in this way (189). The striations did not appear in crystals grown in a horizontal magnetic field applied perpendicular to the growth direction, when the field strength was above a critical value. Direct measurement of temperature in the molten zone revealed oscillations which diminished in amplitude with increasing field, until the critical field, H_c, was reached. At H_c, the temperature varied sinusoidally with time over a 2 K range. Above H_c, the amplitude dropped to <0.25 K. These field-effect studies were carried out without attempting to grow single crystals. InSb has been grown in a variety of low-index orientations, but grows most readily on $\langle 111 \rangle$ in graphite or graphitized silica boats that are not wetted by the InSb melt. Single crystals can be grown even without seeds. Because of the large difference in the densities of solid and melt, the boat must be tilted at about 13° from the horizontal to attain crystals of uniform cross section.

The growth of gallium arsenide crystals in horizontal boats poses particular problems because of the high vapor pressure of the system. The problem is overcome by keeping the entire container hotter than the condensation temperature of the volatile components. In turn, this requires operation under high pressure, and several methods have been used to do this. Silica vessels survive at pressures of 1.5×10^7 Pa at 1273 K; silica tubes enclosed in graphite containers filled with graphite powder have been used to about 2100 K (190, 191).

Anthracene has been purified and converted to a single crystal in a closed, horizontal apparatus in which the solid was not allowed to cool much below the solidification temperature. New zones entered the ingot at its closed end: supercooling of the entering liquid zone was prevented by heat conduction along a central platinum wire (Figure 8.2). Tube breakage did not take place because the solid zones were sufficiently plastic to allow displacement of the entire ingot. Intermittent rotation of the tube provided good mixing and crystals of uniform cross section were obtained (192).

Schleifman and Kirgintsev (193) investigated the relationships between temperature gradient at the interface and interface shape in the horizontal boat method. Heat-flow calculations showed that the gradient was smaller when the interface was inside the heater rather than adjacent to it or outside it; experiments indicated that single-crystal growth was more certain with smaller gradients and that interface shape was less critical. Figure 8.3 shows temperature distribution and gradient for three interface positions. With a heater of 3-cm length, single crystals grew reliably when the interface was 0.8 cm inside the heater (Figure 8.3c) but not when it was 0.4 cm inside the heater (Figure 8.3b) or outside it (Figure 8.3a).

Another theoretical analysis of temperature distribution in crystal growth by zone refining in horizontal boats used the Laplace equation in cylindrical form

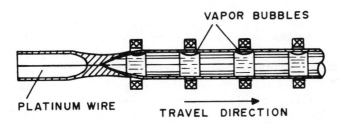

Fig. 8.2 Single-crystal growth by zone melting with minimum undercooling of solid.

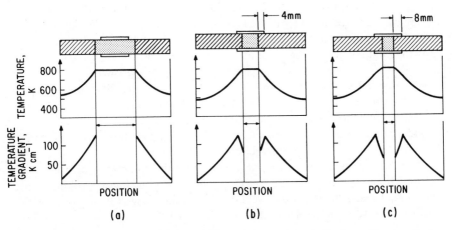

Fig. 8.3 Effect of interface position on single-crystal growth by the horizontal-boat method.

(194). It was concluded that single-crystal growth is favored when the solid at the interface is planar or convex, and as close as possible to the edge of the zone heater. To obtain the desired interface shape, the freezing temperature must be greater than $[T_c + (T_h - T_c)/2]$, where T_c is the overall ingot temperature and T_h is the temperature of the zone heater (194).

Isotactic polypropylene has been zone melted in high gradients (300 K cm^{-1}), at low rates (3 μm min^{-1}). The process is referred to here as zone *melting*, since no refining (solute redistribution) was noted. Oriented α-type spherulites were obtained only by nucleation. β-type nuclei formed only rarely, but were easily initiated by transformation along the α-solid/liquid front. Growth kinetics were investigated for both phases (195, 196).

2.2 Enclosed Vertical Method

Alkali halide crystals, 2 cm in diameter × 30 cm long, have been grown in vertical, sealed quartz tubes under pressures of about 5×10^2 Pa of the corresponding hydrogen halide. The molten zone was provided by an induction-heated graphite ring. The optimum zone velocity was 1.7×10^{-3} cm s^{-1} (6 cm h^{-1}). At higher rates, purification was less effective; lower rates allowed impurity diffusion from the container into the melt (197). In related work, oxygen-containing impurities were sparged from KCl by passage of carbon tetrachloride vapor in argon carrier gas before the start of zoning. The purified charge, after directional solidification in its silica container, was subjected to about 20 passes at 1.7×10^{-3} cm s^{-1} through a single resistance heater. A final zone pass at 3×10^{-5} cm s^{-1} gave a crystal that was free of inclusions. It should be noted that the silica tube in which the zoning was carried out was previously heated at 1273 K and treated at this temperature with CCl$_4$ vapor in argon carrier. Decomposition of the halide deposited a transparent layer of carbon on the wall. The purified melt did not wet this surface.

Vertical zone melting in a closed tube has been applied to the preparation of anthracene crystals. After downward zoning at 8×10^{-3} to 3.3×10^{-2} cm s^{-1} in a conventional procedure in a Pyrex tube of 1-cm diameter, the direction of zoning was reversed just before the lowermost heater reached the bottom of the tube. A single upward pass was executed at 1.4×10^{-4} cm s^{-1}. During upward growth, solidification takes place at the lower interface of each zone, which is normally flat and free of bubbles. Anthracene crystals grown by this method were comparable in quality to those produced by the Bridgman method. Moreover, the cleavage of these crystals was horizontal. This desirable orientation is rarely achieved in growth of anthracene from the melt (198).

Homogeneous single crystals of peritectic compounds have been grown by a hybrid zone leveling/Bridgman technique (199).

Another technique applicable to materials with high vapor pressure is the introduction of a molten zone at the closed bottom of the ingot container, where the solid charge acts as a vapor seal during a single upward zone pass. Bi$_2$Te$_3$ and CdTe have been grown in this way (200, 201).

2.3 Vertical Floating Zone

Establishment of molten zones in free-standing ingots is described in Section 2.4, Chapter 5. This "floating-zone" method has been widely used for preparation of single crystals of metals and metal oxides. It offers an advantage in that growth is not constrained by a container, and mechanical stress is reduced or eliminated. Further, good mixing in the molten zone may be achieved by rotating one or both solid ends of the ingot.

Molybdenum and other high-melting metals have been crystallized in this way. Transit of a molybdenum ingot (0.45-cm diameter) through a stationary electron-beam zone at 5×10^{-3} to 5×10^{-2} cm s^{-1}, without stirring, gave single crystals. When the starting charge contained radioactive tungsten, it was possible to study the homogeneity of the resulting crystals by autoradiography of polished sections (202).

High-melting oxide crystals such as sapphire and garnets were prepared by yet another floating-zone method, in which a perforated iridium strip extended through the ingot. Heat applied to the external portions of the strip caused formation of a thin molten zone in contact with the portion of the strip within the ingot. The melt passed through the perforations in the strip. Single crystals were grown from a sintered feed rod above the strip onto a seed rod below it. The grown crystals were studied by etching and microscopy (203).

The immersed-strip method has been improved by using crystals rather than sintered powder as feed. Greater cross-sectional uniformity was achieved by increasing the local current density through careful placement of the perforations in the strip. The modified method has provided calcite crystals up to 2.5 cm in diameter and 3 cm long (204). Barium titanate (mp 1885 K) has been crystallized by the immersed-strip method, using rutile as the flux. Background temperature in the molten zone was about 1575 K and the temperature of the immersed strip was only about 50 K higher. Single crystals 0.8 cm in diameter were grown in lengths up to 2 cm (205). The basis of this fluxed method of growing single crystals is discussed in Section 1, Chapter 7.

Single crystals of several high-melting oxides (YFe_2O_4, $YbFe_2O_4$, $Yb_2Fe_2O_7$, Mg_2TiO_4, and $MgTiO_3$) have been grown by a floating-zone method making use of an IR heater (205).

2.4 Temperature-Gradient Zone Melting

The unusual properties of temperature-gradient zone melting are discussed in Section 2.7.1, Chapter 5, as related to purification and growth of crystals.

3 STUDY OF THERMAL STABILITY

Impurities may be transported to one end or another of a zone-refined ingot, depending on whether k is greater than or less than unity. The question then arises whether some or all of the impurities derive from the zone-refining process itself. An answer may be derived easily by the following experiment:

A zone-refining program is carried out covering the entire length of the experimental ingot. The ingot is then subjected to another zone-refining program covering only a portion of the ingot. If the impurities observed from the first program were in fact generated by the zoning, then a new deposition of impurity will occur at the terminus of the redefined ingot in the second program. Conversely, if all of the impurity originally deposited came from the sample with which the experiment was started, no additional impurity will be observed.

This technique can also give valuable information on photostability in the presence of, or with the rigorous exclusion of, atmospheric gases. Thus, for example, Figure 8.4 shows an aromatic amine that was subjected to 50 zone passes in a nitrogen atmosphere in darkness. The segregated impurities are clearly visible in the bottom. The same tube was then subjected to additional zone passes over a portion of its length, with illumination at progressively shorter wavelengths. The first illuminations, at the longest wavelengths, produced no perceptible effect. When the ingot was illuminated with light of 365-nm wavelength during the zone refining, an additional impurity, evidently different

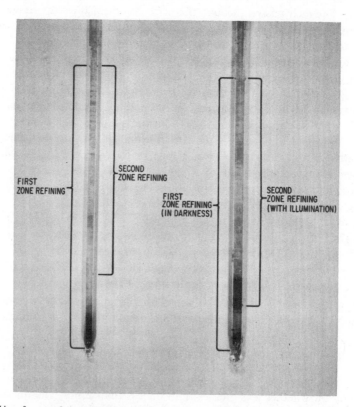

Fig. 8.4 Use of zone refining in the study of thermal stability. In the first case, a second zone-refining program covering only a portion of the ingot shows no additional impurity deposit. In the second case, repeated zone refining with illumination results in deposition of a segregated impurity.

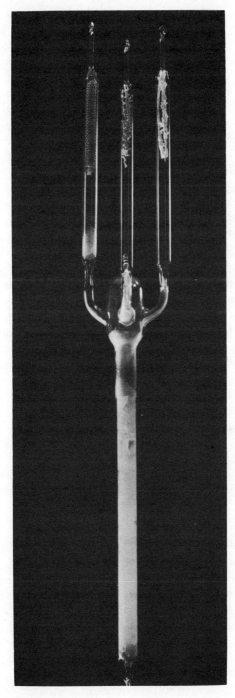

Fig. 8.5 Apparatus for testing the stability of a pure substance in contact with various materials of construction.

from that initially present, was deposited at the bottom of the zoned segment.

The experiment described above was carried out with illumination of both solid and liquid zones. In principle it is possible to determine the relative photostability of melt and crystal merely by illuminating only one or the other.

The procedure described above offers an advantage over direct illumination of a continuously molten sample in that the segregation produced by the zoning process offers enhanced sensitivity since the impurity is concentrated and becomes increasingly visible. Naturally, the absence of visible impurity is not conclusive; for greatest certainty several ingots might be illuminated at different wavelengths and each analyzed by chromatography and/or spectroscopy to detect the appearance of photochemical products.

A related procedure can be used to test the stability of pure substances in contact with various materials of construction, at varied temperatures. Figure 8.5 shows a zone-refined ingot in a glass vessel bearing three tubes, each containing a metal sample. The zone-refined chemical is melted and decanted into the three tubes. The content of each tube is heated for a fixed period and then zone refined. Any impurities formed by exposure of the metal to the melt are likely to be segregated into the capillary appendage. Attack on the metal can be detected by weighing it before and after exposure.

Decomposition during zone refining can result from the presence of a particular impurity, in the absence of which the parent compound can be zoned successfully. An example of the effect is seen in Figure 8.6, showing the black ingot resulting from zone refining commercial-grade 4-nitrodiphenylamine and the bright orange ingot given by once-sublimed solid.

4 DETERMINATION OF PHASE DIAGRAMS

Applications of zone-melting procedures to the study of phase relationships are discussed in Section 2.3, Chapter 2.

5 MISCELLANEOUS APPLICATIONS

5.1 Thermal Conductivity

It is often desirable to know the thermal conductivities of solids and liquids at their melting temperatures. The data can be obtained by extrapolating values measured at higher and lower temperatures, but this procedure is difficult and time consuming. A more direct method is based on measurement of the temperature gradients in liquid and solid (G and G_s) near the interface. From a knowledge of the latent heat of fusion at the melting temperature and the density of the liquid (ρ), the thermal conductivities of liquid and solid, K and K_s, can be determined from values of G and G_s measured at two rates of zoning or directional crystallization, from the relationship (207).

$$KG + \Delta H_f \rho V = K_s G_s \qquad (8.1)$$

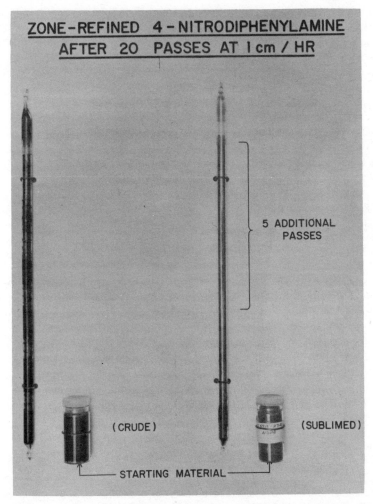

Fig. 8.6 The ingot on the left was prepared from commercial 4-nitrodiphenylamine; after 20 zone passes at $1\,cm\,h^{-1}$ it became steadily darker. The ingot on the right was prepared from sublimed product; only a small amount of visible impurity was deposited by the first zoning and no additional impurity was deposited in a second, partial zoning.

5.2 Diffusion Coefficients

TGZM (see Section 2.7.1, Chapter 5) offers the possibility of measuring liquid diffusion coefficients. In TGZM, the rate of advance of the zone is proportional to diffusivity. The flux F of the diffusing species is given by the product of the diffusivity and the concentration gradient; in turn the concentration gradient may be expressed as the product of the temperature gradient and the reciprocal of the slope of the liquidus:

$$D\frac{dc}{dx} = D\frac{dc}{dT}\frac{dT}{dx} = -F$$

The imposed temperature gradient and liquidus slope are readily measured; hence, it is only necessary to measure the rate of advance of the molten zone, which is proportional to F, to derive D. Because this method does not involve a sizable liquid zone, convection is negligible and does not interfere with measurement of D (208).

REFERENCES

1. N. McWhirter and R. McWhirter, *Guinness Book of World Records*, 13th ed., Bantam Books, New York, p. 170.

2. H. Schildknecht and U. Hopf, *Z. Anal. Chem.*, **193**, 401–411 (1963).

3. W. G. Pfann, *Zone Melting*, 2nd ed., Wiley, New York, 1966, Chapter 5.

4. V. N. Vigdorovich, *Purification of Metals and Semiconductors by Crystallization*, Freund Publishing House, Tel Aviv, 1971.

5. S. E. Maraev, E. I. Mudrova, and I. N. Varlamova, *Tr. Vses., Nauch.-Issled. Proekt. Inst. Alyum., Magnievoi Elektrodnoi Prom.*, **68**, 117–121 (1970); through CA**77**, 78120y.

6. W. D. Hannibal, G. Ibe, H. Pfundt, W. Reuter, and G. Winkhaus, *Metall.*, **27**(3), 203–211 (1973) (German).

7. T. Kino, N. Kamigaki, H. Yamasaki, J. Kawai, Y. Deguchi, and I. Nakamichi, *Trans. Jpn. Inst. Met.*, **17**(10), 645–648 (1976).

8. M. Isshiki, H. Nitta, Y. Noda, S. Itoh, K. Igaki, A. Mizohata, T. Mamuro, T. Tsujimoto, and S. Iwata, *Radioisotopes*, **28**(6), 349–354 (1979).

9. E. L. Anderson, *Chem. Ind.*, 131–136 (1975).

10. A. M. Nazar, M. Prates and T. W. Clyne, *J. Crystal Growth*, **55**, 317–324 (1981).

11. C. H. Li, *Phys. Stat. Sol.*, **15**, 3–56 (1956).

12. S. Joensson, A. Freund, and F. Aldinger, *Metall. (Berlin)*, **33**(12), 1257–1261 (1979).

13. V. V. Marychev, A. P. Kolesnev, V. I. Ponomarenko, and V. N. Vigdorovich, *Tsvet. Metal.*, **9**, 9–11 (1973).

14. V. N. Vigdorovich, A. E. Vol'pyan, and V. A. Ahavga, *Izv. Akad. Nauk SSSR, Metal.*, **2**, 106–108 (1972).

15. V. Pozsgai and Z. Horvath, *Banyasz. Kohasz. Lapok, Kohasz.*, **106**(7), 333–335 (1973); through CA**79**:148345t.

16. D. Durkovic, S. M. Petrovic, and R. M. Cosovic, *Glas. Hem. Drus., Beograd*, **38**(5–6), 391–396 (1973); through CA**81**:124598t.

17. V. N. Vigdorovich, *Purification of Metals and Semiconductors by Crystallization*, Freund Publishing House, Tel Aviv, 1971, p. 212.

18. G. A. Ivanov and A. S. Krylov, *Tsvetn. Met.*, **3**, 23–25 (1976); through CA**85** 66266c.

19. K. P. Tsomaya and F. N. Tavadze, *Bor. Poluch., Strukt. Svoistva, Mater. Mezhdunar. Simp. Boru, 4th (Pub. 1974)* **2**, 66–72 (1974); through CA**83**:106340q.

20. R. Kh. Karimov, Ya. M. Belder, and M. Kh. Khikmatillaev, *Geliotekhnika*, **6**, 49–51 (1974); through CA**82**:179165a.

21. G. V. Tsagareishvili, K. A. Oganezov, I. A. Bairamashvili, A. G. Khvedelidze, G. A. Mazmishvili, V. V. Chepelev, and M. L. Tabutsidze, *J. Less-Common Met.*, **67**(2), 419–424 (1979).

22. B. Schaub and C. Potard, *Proc. Int. Symp. Cadmium Telluride, Mater. Gamma-Ray Detectors*, 1971 (Pub. 1972), II-l-II-b.

23. G. Revel, J. L. Pastol, J. C. Rouchard, and A. Michel, *C.R. Acad. Sci., Ser. C*, **276**(2), 153–155 (1973).

24. D. Fort, B. J. Beaudry, D. W. Jones, and K. A. Gschneidner, Jr., Energy Res. Abstr., 4(16) (1979), Abstr. No. 43383.

25. K. Igaki, M. Isshiki, and K. Yakushiji, Trans. Jpn. Inst. Met., 20(11), 611–616 (1979).

26. V. G. Epifanov, A. N. Rakitskii, and V. P. Skvorchuk, Metallofizika, 37, 83–87 (1971); through CA77:91883n.

27. S. Hayashi, S. Ono, and H. Komatsu, Krist. Tech., 13(3), 263–267 (1978).

28. S. Hayashi, J. Echigoya, H. Hariu, T. Sato, Y. Nakamichi, and M. Yamamoto, J. Crystal Growth, 24–25, 422–425 (1974).

29. Yu. Kusaev, A. N. Zhukov, and R. K. Nicolaev, Izv. Akad. Nauk SSSR Met., 3, 33–36 (1978).

30. T. Inoue, J. Watanabe, T. Miura, and M. Yamamoto, Nippon Kinzoku Gakkaishi, 36(3), 256–262 (1972).

31. D. W. Jones, D. Fort, and D. A. Hukin, Rare Earths Mod. Sci. Technol., [Rare Earth Res. Conf.], 13, 309–314 (1978).

32. D. Fort, B. J. Beaudry, D. W. Jones, and K. A. Gschneidner, Jr., Rare Earth Mod. Sci. Tech., 2, 33–37 (1980).

33. V. N. Vigdorovich, G. E. Chuprikov, and K. I. Epifanova, Tsvet. Metal., 11, 49–50 (1973).

34. W. Wronski and L. Walis, Int. Symp. Autoradigr. [Proc.], 9, 76–81 (1978); through CA90:207816z.

35. M. Kusowski, M. Wajda, and Z. Horubata, Pol. Pat. 87,445, March 31, 1977; through CA90:125185d.

36. F. H. Horn, Jr., J. Appl. Phys., 32, 900–901 (1961).

37. A. N. Kirgintsev and Yu. A. Rybin, Izv. Sib. Otd. Akad. Nauk SSSR, Ser. Khim Nauk, 2, 38–44 (1971).

38. L. A. Prokhorov and E. Ya. Aladko, Izv. Sib. Otd. Akad. Nauk SSSR, Ser. Khim. Nauk, 6, 100–103 (1977); through CA88:113493y.

39. G. A. Goryushin, V. I. Dobrovol'skaya, D. I. Levinzon, V. N. Nefedov, Zh. P. Nikolashin, and D. G. Ratnikov, Tr. Vses. Nauch.-Issled. Inst. Tokov Vys. Chastoty, 11, 96–101 (1970); through CA76:64412n.

40. M. M. Nekrasov and G. V. Ryabchenko, Vop. Mikroelektron., 143–146 (1971); through CA76; 146650c.

41. J. Bogacki, Z. Horubala, and M. Kusowski, Pr. Nauk Inst. Chem. Nieorg. Metal. Pierwiastkow Rzadkich Politech. Wroclaw, 28, 255–268 (1976); through CA85:145228v.

42. Z. Horubala, M. Kusowski, L. Walis, and W. Wronski, Pr. Nauk Inst. Chem. Nieorg. Metal. Pierwiastkow Rzadkich Politech. Wroclaw, 28, 269–281 (1976); through CA85:145229w.

43. Z. Horúbala, M. Kusowski, and H. Mogielnicki, Pr. Nauk. Inst. Chem. Nieorg. Metal Pierwiastkow Rzadkich Politech. Wroclaw, 28, 249–254 (1976); through CA85:151880t.

44. L. Repiska and L. Kuchar, Sb. Ved. Pr. Vys. Sk. Banske Ostrave, Rada Hutn., 16(1), 25–44 (1970); through CA76:48550a.

45. C. W. Marschall, J. D. Myers, and G. W. P. Rengstorff, Metals Eng. Quart., 14(1), 19–24 (1974).

46. I. R. Baranov, L. P. Volkova, V. N. Gavrin, V. I. Glotov, D. S. Kamenetskaya, L. L. Koshkarov, I. B. Piletskaya, and V. I. Shiryaev, At. Energ., 47(3), 195–196 (1979); through CA92:80094w.

47. H. Kimura, S. Takagi, and H. Matsui, Japan. Kokai, 76, 73, 902, June 26, 1976.

48. S. G. Grayaznova and V. I. Bliznyuk, Deposited Doc. VINITI 1185-78 (1978); through CA91:126637k.

49. O. K. Chopra and P. Niessen, Acta Met., 21(10), 1451–1460 (1973).

50. J. C. Chaston, Int. Met. Rev., 148–152 (1976).

51. E. G. Kharitonova, V. A. Reznichenko, and B. A. Bochkov, *Protsessy Poluch. Refinirovaniya Tugoplavkikh Met.*, **204-209**, 253-260 (1975); through CA**84**:138886j.

52. S. P. Clough, S. J. Vonk, and D. F. Stein, *J. Less Common Met.*, **50**(1), 161-163 (1976).

53. V. A. Krakhmalev and G. A. Klein, *Rost Defekty Metal. Krist.*, 217-224 (1972); through CA**78**:58482f.

54. A. A. Yastrebkov and V. M. Lakeenkov, *Prib. Tekh. Eksp.*, (1), 235-236 (1973); through CA**78**:141161a.

55. P. Duzi, *Krist. Tech.*, **12**(1), 75-80 (1977).

56. V. G. Glebovskii, K. V. Kondakova, and Ch. V. Kopetskii, *Izv. Akad. Nauk SSSR, Metal.*, *1973* (5), 76-79.

57. J. C. Posey and H. A. Smith, *J. Am. Chem. Soc.*, **79**, 555-557 (1957).

58. E. M. Savitskii, G. S. Burkhanov, N. N. Raskatov, E. V. Ottenberg, G. D. Shnyrev, and N. N. Sergeev, *Strukt. Svistva Monokrist. Tugoplavkikh Metal.*, 5-10 (1973); through CA**80**:100856d.

59. S. Hayashi and H. Komatsu, *Cryst. Res. Technol.*, **14**(7), 761-764 (1979).

60. A. A. Samadi and M. Fedoroff, *Scr. Metall.*, **11**(6), 509-512 (1977).

61. M. Isshiki, K. Yakushiji, T. Kikuchi, M. Sato, E. Yanagisawa, K. Igaki, A. Mizohata, T. Mamuro, and T. Isujimoto, *Radioisotopes*, **30**(4), 211-216 (1981).

62. J. Barthel, K. H. Berthel, C. Fischer, G. Guenzler, J. Kunze, P. Mueh., H. Oppermann, R. Petri, G. Sobe, and G. Weise, East German Pat. 83,842, August 12, 1971.

63. R. E. Reed, *J. Vac. Sci. Technol.*, **9**(6), 1413-1418 (1972).

64. C. S. Pande, L. S. Lin, S. R. Butler, and Y. T. Chou, *J. Crystal Growth*, **19**(3), 209-210 (1973).

65. L. Kuchar, P. Duzi, and B. Wozniakova, *Neue Huette*, **21**(5), 297-300 (1976).

66. V. M. Azhazha, G. P. Kovtun, and V. L. Makarov, *Izv. Akad. Nauk SSSR, Metal.*, *1973* (6), 165-168; through CA**80**:64512c.

67. E. M. Savitskii, V. P. Polyakova, and N. B. Gorina, *Strukt. Svoistva Monokrist. Tugoplavkikh metal.*, 222-229 (1973); through CA**80**:113345m.

68. V. M. Azhazha, G. P. Kovtun, E. A. Elenskii, N. V. Volkenshtein, V. E. Startsev, and V. I. Cherepanov, *Fiz. Met. Metalloved.*, **41**(4), 888-890 (1976); through CA**85**:128905w.

69. J. Cremer and H. Cribbe, *Chemie-Ingenieur Technik*, **36**, 957-959 (1964).

70. J. S. Shah and D. M. Brookbanks, *Platinum Metals Rev.*, **16**(3), 94-100 (1972).

71. V. M. Azhazha, G. P. Kovtun, V. A. Elenskii, and V. Z. Kleiner, *Izv. Akad. Nauk SSSR, Met.*, (5), 41-43 (1975).

72. V. M. Azhazha, G. P. Kovtun, Yu. P. Bobrov, and V. A. Elenskii, *Poluch. Anal. Veschestv. Osoboi Chist.* [*Dokl. Vses. Konf. 1976*], **5**, 131-133 (1978); through CA**91**:110687r.

73. G. P. Kovtun, V. A. Elenskii, V. S. Belovol, and L. A. Tomskaya, *Izv. Akad. Nauk SSSR, Met.*, (1), 95-99 (1981); through CA**94**:212173z.

74. W. Keller and A. Muhlbauer, "Floating-Zone Silicon," in W. R. Wilcox, Ed., *Preparation and Properties of Solid State Materials*, Vol. 5, Dekker, New York, 1981.

75. V. N. Vigdorovich, *Purification of Metals and Semiconductors by Crystallization*, Freund Publishing House, Tel Aviv, 1971, Chapter 4.

76. D. Morvan, J. Amouroux, and G. Revel, *Rev. Phys. Appl.*, **15**(17), 1229-1238 (1980).

77. J. E. Kunzler and J. H. Wernick, *Trans. A.I.M.E.*, **212**, 856-860 (1958).

78. N. V. Lichkova, Yu, V. Korneenkov, and R. K. Nikolaev, *Izv. Akad. Nauk SSSR, Met.*, **1975**, 29-31.

79. A. V. Vakhobov, V. G. Khudaiberdiev, M. Z. Dusmatova, and T. A. Kogan, *Dokl. Akad. Nauk Tadzh. SSR*, **17**(10), 38-41 (1974); through CA**82**:114566n.

80. F. Feher and H. D. Lutz, *Z. Anorg. Allg. Chemie*, **334**, 235-241 (1965).

81. F. Feher, H. D. Lutz, and K. Obst, *Fresenius Z. Anal. Chem.*, **224**, 407–413 (1967).

82. H. Suzuki, K. Higashi, and Y. Miyake, *Bull. Chem. Soc. Japan*, **47**, 759–760 (1974).

83. H. Tsubakitani, K. Aono, and S. Dohi, *Boei Daigakko Rikogaku Kenkyu Hokoku*, **8**(1), 47–54 (1970); through CA**76**:132784n.

84. S. Dohi, T. Matsuyama, and H. Tsubakiani, *Proc. Jap. Congr. Mater. Res.*, **16**, 18–21 (1972); through CA**79**:119263j.

85. J. Barthel and M. Krumnacker, *Kristall Technik*, **11**, 955–968 (1976).

86. V. N. Vigdorovich, A. E. Vol'pyan, and V. V. Marychev, *Tsvetnye Metally*, **1978**, 52–54, No. 4.

87. J. Bogacki, Z. Horubala, and M. Kusowski, *Pr. Nauk Inst. Chem. Nieorg. Metal. Pierwiastkow Rzadkich Politech. Wroclaw*, **28**, 241–247 (1976); through CA**85**:153172n.

88. D. A. Hukin, R. C. C. Ward, D. K. Morris, and K. Davies, *Rare Earths Mod. Sci. Tech.*, **2**, 25–30 (1980).

89. Z. Wojtasek, R. Lehman, and J. Dubowy, *Rudy Met. Niezelaz.*, **20**(9), 442–445 (1975); through CA**84**:182886h.

90. R. Lehman and Z. Wojtaszek, *J. Less-Common Met.*, **65**(2), 271–277 (1979).

91. L. Repiska and L. Komorova, *Zb. Ved. Pr. Vys. Sk. Tech. Kosiciach*, 137–145 (1977).

92. Y. Hoshino and T. Utsonomiya, *Sep. Sci. Tech.*, **15**(8), 1521–1531 (1980).

93. I. I. Gorbacheva, A. N. Kirgintsev, and G. A. Kozhukhovskya, *Izv. Sib. Otd. Akad. Nauk SSSR, Ser. Khim. Nauk*, (3), 138–140 (1974); through CA**81**:109034n.

94. B. A. Solov'ev, I. M. Selivanov, and A. N. Kirgintsev, *Tsvet. Metal.*, (7), 82–83 (1973); through CA**80**:39550f.

95. A. N. Kirgintsev, I. I. Gorbacheva, and G. A. Kozhukhovskaya, *Izv. Sib. Otd. Akad. Nauk SSSR, Ser. Khim. Nauk*, (1), 137–140 (1976).

96. B. A. Solov'ev, *Nauchn. Tr.-Tsentr. Nauchno-Issled. Inst. Olovyannoi Prom-sti.*, **8**, 27–30 (1978); through CA**92**:26130s.

97. V. A. Fedotov, V. N. Lyubimov, and A. N. Kirgintsev, *Izv. Akad. Nauk SSSR Met.*, **2**, 61–63 (1980).

98. S. G. Gryaznova and L. A. Kondratenko, Deposited Doc., 1974, VINITI, 2157–2174; through CA**86**:182537m.

99. I. G. Yudelevich, L. A. Kondratenko, and Z. P. Kostrova, *Izv. Sib. Otd. Akad. Nauk SSSR, Ser. Khim. Nauk*, (6), 119–123 (1979).

100. Y. Hoshino and T. Utsonomiya, *Sep. Sci. Technol.*, **15**(8), 1521–1531 (1980).

101. N. V. Bereznikova and V. A. Reznichenko, *Protsessy Poluch. Rafinirovaniya Tugoplavkikh Met.*, **200–204**, 253–260 (1975); through CA**84**:168322q.

102. J. Mensik and L. Kuchar, *Sb. Ved. Pr. Vys. Sk. Banske Ostrave, Rada Hutn.*, **22**(1), 9–20 (1976); through CA**87**:187637.

103. J. J. Quenisset and R. Naslain, *J. Less-Common Met.*, **79**(2), 169–180 (1981).

104. L. P. Chupyatova, K. P. Morozova, E. P. Sidokhin, M. V. Pikunov, and V. V. Shishkov, *Rost Defekty Metal., Krist.*, 200–205 (1972); through CA**79**:46437g.

105. T. Gronek and A. Maciejewski, *Arch. Hutn.*, **24**(2), 193–202 (1979).

106. G. N. Grishkov, G. A. Klein, S. M. Mikhailov, and V. P. Sokolova, *Strukt. Svoistva Monokrist. Tugoplavkikh, Metal*, 31–36 (1973).

107. V. O. Eain and N. V. Belova, Deposited Doc., 1975, VINITI 936-75; through CA**86**:148913w.

108. T. Imamura, U.S. Pat. 3,301,660, January 31, 1967.

109. M. V. Pikunov, V. V. Shishkov, M. N. Rivkin, Yu. A. Kostyukhin, Yu. V. Batanov, S. V. Zhidovinova, A. Ya. Dubrovskii, and N. I. Strigina, *Nauch. Tr., Nauch.-Issled. Proekt. Inst. Redkometal Prom. No. 42*, 98–105 (1972); through CA**80**:87977x.

110. G. Barnes, *Met. Trans.* **4**(2), 549–551 (1973).

111. O. P. Kolchin, I. K. Berlin, and N. V. Presnetsova, *Tsvetn. Metal.*, **36**(9), 59–65 (1963).

112. J. Bressers, R. Creten, and G. Van Holsbeke, *J. Less-Common Met.*, **39**(1), 7–16 (1975).

113. I. I. Ivantsov, V. T. Mashkarova, V. S. Pavlov, and E. M. Saenko, *Izv. Akad. Nauk SSSR, Metal.*, (1), 114–118 (1972).

114. G. E. Chuprikov, A. I. Petukhova, and N. I. Moreva, *Redkozemel. Metal. Ikh Soedin., Mater. Vses. Simp.*, 91–95 (1968); through CA**78**:116072q.

115. V. D. Grigor'ev, V. N. Vigdorovich, and A. S. Yaroslavtsev, *Tsvet. Metal.*, (1), 21–22 (1973); through CA**79**:56319j.

116. V. D. Grigor'ev, V. N. Vigdorovich, and A. S. Yaroslavtsev, *Izv. Akad. Nauk SSSR, Metal.*, 64–66 (1973).

117. O. A. Troitskii, *Phys. Stat. Sol. (a)*, **35**, K151 (1976).

118. O. A. Troitskii, *Phys. Stat. Sol.*, **48**, 229–234 (1978).

119. K. B. Yurkevich, A. V. Vanyukov, N. N. Khavskii, V. K. Tolpyto, M. N. Dubrovin, A. I. Kuryatnikov, and V. A. Gromov, *Primen. UE'-trazvuka Met. Protsessakh*, 106–110 (1972); through CA**78**:6741z.

120. V. M. Amonenko, V. M. Azhazha, P. N. V'yugov, and L. N. Voronina, *Met. Metalloved. Chist. Metal.*, No. 9, 20–23 (1971).

121. B. G. Lazarev, V. I. Makarov, V. M. Azhazha, A. L. Donde, P. N. V'yugov, and A. A. San'kov, *Dohl. Akad. Nauk SSSR*, **201**(2), 321–323 (1971).

122. C. D. Thurmond, "Control of Composition in Semiconductors by Freezing Methods," in N. B. Hannay, Ed., *Semiconductors*, Reinhold, New York, 1959, pp. 145–191.

123. M. Oswaldowski, H. Szweycer, T. Berus, J. Goc, and M. Zimpel, *Vide, Couches Minces*, **201** (Suppl. Proc. Int. Vac. Cong., 8th V1), 207–210 (1980).

124. R. A. Cole and N. L. Skinner, U.S. Pat. 3,909,246, 30 September 1975.

125. Yu. L. Il'in and I. V. Isaenko, *Tezisy Dokl. Vses Soveshch. Rostu Krist.*, *5th*, **2**, 158–159 (1977); through CA**93**:86810w.

126. H. Schildknecht, *Zone Melting*, Academic Press, New York, 1966, pp. 194–198.

127. M. Seki and K. Kobayashi, *Kotai Butsuri*, **15**, 756–760 (1980); through CA**95**:681456u.

128. H. A. Smith and C. O. Thomas, *J. Phys. Chem.*, **63**, 445–447 (1959).

129. P. Süe, J. Pauly, and A. Nouaille, *Bull. Soc. Chim. France*, **5**, 593–602 (1958).

130. T. Utsonomiya, T. Yasukawa, and Y. Hoshino, *Report of the Res. Lab. of Eng.*, No. 1, 119–126 (1976); through CA**86**:173468n.

131. A. N. Kirgintsev and L. I. Isaenko, *Deposited Publ.* 6416–6473 (1973); through CA**85**:12442p.

132. E. Kirkova and P. Iliev, *God. Sofii. Univ., Khim. Fak.*, **66**, 479–486 (1971–1972); through CA**84**:114325z.

133. B. V. Sinitsyn, V. A. Kas'yanov, and T. V. Uvarova, *Nauchn. Tr., Gos. Nauchno-Issled. Proektn. Inst. Redkomet. Prom-sti*, **45**, 101–103 (1972); through CA**85**:179379t.

134. M. V. Mokhosoev, T. T. Got'manova, and I. F. Kokot, *Russ. J. Inorg. Chem.*, **9**, 1360–1364 (1964).

135. H. Gründig, *F. Physik*, **158**, 577–594 (1960).

136. M. Ikeya and N. Itoh, *Jap. J. Appl. Phys.*, **7**, 837–845 (1968).

137. R. I. Gindina, A. Maaroos, L. Ploom, and N. A. Yanson, *Tr. Inst. Fiz. Akad. Nauk Est. SSSR*, **49**, 45–89 (1979).

138. H. Nagasawa, *Science*, **152**, 767–769 (1966).

139. A. B. Blank, L. I. Afanasiadi, and E. S. Zolovitskaya, *Probl. Chist. Soversh. Ionnykh Krist., Mater. Pribalt. Semin. Ionnym. Krist.*, **1969**, 16–19; through CA**78**:21023k (1973).

140. F. Moser, D. C. Burnham, and H. H. Tippins, *J. Appl. Phys.*, **32**, 48–54 (1961).

141. A. S. Kakar, "Zone Refining and Single-Crystal Growth of Cuprous Oxide," Ph.D. Thesis, Wayne State University, Detroit, MI. *Diss. Abstr. Int. B. 1979*, **39**(10), 4908–4909.

142. J. L. Loison, M. Robino, and C. Schwab, *J. Crystal Growth*, **50**(4), 816–822 (1980).

143. K. Ueno, H. Kaneko, and Y. Watanabe, *Microchem. J.*, **10**, 244–249 (1966).

144. K. Ueno, H. Kobayashi, and I. Yoshida, *Memoirs of the Faculty of Engineering, Kyushu Univ.*, **38**(1), 83–114 (1978).

145. I. Yoshida, H. Kobayashi, and K. Ueno, *Talanta*, **24**, 61–63 (1977).

146. I. Yoshida, H. Kobayashi, and K. Ueno, *Bull. Chem. Soc. Japan*, **49**, 1874–1878 (1976).

147. H. Kaneko, H. Kobayashi, and K. Ueno, *Talanta*, **14**, 1403–1409 (1967).

148. I. Yoshida, H. Kobayashi, and K. Ueno, *Talanta*, **24**, 58–60 (1977).

149. M. A. Noack and J. D. Verhoeven, *J. Crystal Growth*, **49**, 595–599 (1980).

150. R. Kieffer, *Bull. Soc. Chim. France*, **8**, 3029–3030 (1967).

151. R. Kieffer, J. C. Maire, A. Deluzarche, and A. Maillard, *Bull. Soc. Chim. France*, **11**, 3271–3273 (1965).

152. H. Hawthorne and J. N. Sherwood, *Trans. Faraday Soc.*, **66**(7), 1783–1791 (1970).

153. G. M. Hood and J. N. Sherwood, *Brit. J. Appl. Phys.*, **14**, 215–217 (1963).

154. G. M. Dugacheva and A. G. Anikin, *Russ. J. Phys. Chem.*, **38**(1), C36–C39 (1958).

155. R. S. Narang and J. N. Sherwood, *J. Crystal Growth*, **49**, 357–362 (1980).

156. N. Albon and J. M. Sturtevant, *Proc. Nat'l. Acad. Sci. USA*, **75**, 2258–2266 (1978).

157. S. P. Kochar, J. M. V. Blanshard, and W. Derbyshire, *Lebensm.-Wiss. Technol.*, **9**(5), 284–288 (1976).

158. N. V. Avramenko, G. M. Dugacheva, and A. G. Anikin, *Russ. J. Phys. Chem.*, **42**(5), 673 (1968).

159. R. Kurkela, *Ann. Acad. Sci. Fennicae Ser. A II* (126), 74 pp., (1964); through CA**61**:9688c.

160. A. W. Neumann and L. J. Klementowski, *J. Thermal Anal.*, **6**, 67–77 (1974).

161. M. Brissaud, C. Dolin, J. LeGuigou, B. S. McArdle, and J. N. Sherwood, *J. Crystal Growth*, **38**(1), 134–138 (1977).

162. A. G. Anikin, G. M. Dugacheva, V. M. Presnyakova, and S. P. Bykova, *Russ. J. Phys. Chem.*, **36**(9), 1115–1116 (1962).

163. M. Maruyama and M. Kakemoto, *J. Chem. Soc. Japan* (*Chem. Ind. Chem.*), **9**, 1430–1435 (1976).

164. S. S. Grazhulene and G. F. Telegin, *J. Appl. Chem. USSR*, **50**, 840–843 (1977).

165. K. W. Benz, *Z. Naturforsch.*, **24a**, 298 (1969).

166. T. Iwano and T. Yokota, *J. Chem. Soc. Japan* (*Chem. Ind. Chem.*), **1**, 21–26 (1974).

167. W. R. Wilcox, U.S. Pat. 3,189, 419, 15 June 1965.

168. N. B. Singh, P. Rastogi, and N. B. Singh, *Krist. Tech.*, **13**(10), 1169–1174 (1978).

169. B. J. McArdle, J. N. Sherwood, and A. C. Damask, *J. Crystal Growth*, **22**, 193–200 (1974).

170. M. Furusawa and M. Tachibana, *Bull. Chem. Soc. Japan*, **54**, 2968–2971 (1981).

171. G. J. Sloan, *Mol. Cryst.*, **1**, 161–194 (1966).

172. A. Sasaki, Y. Iwai, K. Abe, T. Akutagawa, and S. Hayakawa, *Japan. J. Appl. Phys.*, **14**(9), 1387–1388 (1975).

173. A. Matsui and Y. Ishii, *Japan. J. Appl. Phys.*, **6**(2), 127–134 (1967).

174. S. S. Grazhulene and S. A. Poluyanova, *J. Anal. Chem.* (*USSR*), **34**(5), 786–788 (1979).

175. G. J. Sloan, unpublished results.

176. V. N. Kovrena, G. Z. Kruglick, and V. A. Utkin, *Akad. Nauk SSSR, Sib. Otd. Izv. Ser. Khim. Nauk*, (12), 115–119 (1974).

177. K. Kihara, Y. Ishida, Y. Suzuki, and T. Takeuchi, *J. Chem. Soc. Japan, Ind. Chem. Sect.*, **73**, 2630–2633 (1970).

178. B. Diehn, F. S. Rowland, and A. P. Wolf, *Anal. Chem.*, **40**, 60–64 (1968).

179. R. Friedenberg and P. J. Jannke, *J. Pharm. Sci.*, 657–658 (1965).

180. W. G. Walter, "Purification of Some Organic Medicinal Compounds by Zone Melting," Ph.D. Thesis, University of Connecticut, Storrs, CT, 1962.

181. R. Friedenberg, "Ultra Purity & Ultrapurification of Pharmaceuticals by Zone Melting," Ph.D. Thesis, *Diss. Abstr.*, **25**(4), 2246, 1964.

182. P. Jannke, "Ultrapurity in Pharmaceuticals," in M. Zief and W. R. Wilcox, Eds., *Fractional Solidification*, Dekker, New York, 1967, Chapter 19.

183. J. D. Bernal and W. A. Wooster, *Ann. Repts. Prog. Chem.*, **28**, 262–321 (1931).

184. J. D. Bernal and D. Crowfoot, *Faraday Soc. Trans.*, **29**, 1031–1049 (1933).

185. D. L. Mackay, *Trends Biochem. Sci.*, **4**(2), N33 (1979).

186. P. Kapitza, *Proc. Roy. Soc.*, **119A**, 358–386 (1928).

187. W. G. Pfann and K. M. Olsen, *Phys. Rev.*, **89**, 322 (1953).

188. I. Tarjan and M. Matrai, *Laboratory Manual on Crystal Growth*, Akademiai Kiado, Budapest, 1972 (UNESCO), p. 187.

187. H. S. Peiser, Ed., *Crystal Growth*, Proc. Inter. Conf. on Crystal Growth, Boston, 20–24 June 1966, Pergamon, 1967.

190. W. D. Lawson, S. Neilsen, E. H. Putley, and A. S. Young, *J. Phys. Chem. Solids*, **9**, 325–329 (1959).

191. A. G. Fisher, *J. Electrochem. Soc.*, **106**, 838–839 (1959).

192. D. Fischer, *Mat. Res. Bull.*, **8**, 385–392 (1973).

193. B. A. Shleifman and A. N. Kirgintsev, *Deposited Doc.*, **1974**, VINITI 588–674.

194. J. S. Cook, D. R. Mason, and P. H. Smith, *Electrochem. Technol.*, **1** (9–10), 300–303 (1963).

195. A. J. Lovinger, J. D. Chua, and C. C. Gryte, *J. Phys. E. Sci. Instr.*, **9**, 927–928 (1976).

196. A. J. Lovinger, J. O. Chua, and C. C. Gryte, *J. Polym. Sci., Polym. Phys. Ed.*, **15**, 641–656 (1977).

197. F. Rosenberger, "Preparation of Ultrapure Alkali Halide Single Crystals," in *Crystal Growth*, H. S. Peiser, Ed., Pergamon, Oxford, 1967, Chapter B20.

198. C.-H. Ting, *J. Chinese Chem. Soc.*, **18**, 5–7 (1971).

199. D. R. Mason and J. S. Cook, *J. Appl. Phys.*, **32**, 475–477 (1961).

200. F. K. Heumann, *J. Electrochem. Soc.*, **109**, 345–346 (1962).

201. M. R. Lorenz and R. E. Halsted, *J. Electrochem. Soc.*, **110**, 343–344 (1963).

202. J. Barthel and R. Scharfenberg, "On Crystal Growth of High-Melting Metals by Electron Beam Zone Melting", in *Crystal Growth*, H. S. Peiser, Ed., Pergamon, Oxford, 1967, Chapter B19.

203. Ibid. Chapter B14.

204. C. E. Turner, N. H. Mason, and A. W. Morris, *J. Crystal Growth*, **56**, 137–140 (1982).

205. National Institute for Research in Inorganic Materials, *Muki Zaiken Nyusu*, **40**, 1–3 (1976).

206. J. C. Brice and P. A. C. Whiffin, *Solid State Electron.*, **7**, 183–187 (1964).

207. M. G. Milvidskii and V. V. Eremeev, *Soviet Phys.–Solid State*, **6**, 1549–1552 (1965).

208. J. H. Wernick, *J. Chem. Phys.*, **25**, 47 (1956).

209. Y. S. Dement'ev, V. A. Fedorov, N. I. Bletskan, Y. A. Okunev, and V. N. Severtsev, *Izv. Akad. Nauk SSR, Neorg. Mater.*, **16**(7), 1164–1167 (1980).

210. A. Nguyen Van Mau, M. Averous, and G. Bougnot, *Mat. Res. Bull.*, **7**, 857–864 (1972).

MISCELLANEOUS MELT-CRYSTALLIZATION METHODS

Fractional crystallizations can be roughly divided into two classes:

- Processes using crystal suspensions
- Processes producing coherent crystal layers

The first class includes the Phillips pulsed-column method, the Schildknecht column crystallizer, and the Brodie purifier. The second class embraces zone melting, drum crystallizers, sweat pans (Proabd procedure), ripple-film crystallizers, and centrifugal crystallizers. Most of these methods are used primarily at industrial scale and detailed description of their operation is outside the scope of this book. Nevertheless we think it worthwhile to describe their basic principles because in some cases it may be possible to reverse the usual sequence and adapt industrial practice to laboratory scale.

1   DRUM CRYSTALLIZERS

Rotating drums are commonly used in industry to convert liquid products to flake form. A melt or solution is applied to the cylindrical surface of a cooled rotating drum and the solid film that forms is removed by a scraper. The intent of this process is to provide a solid product in a form suitable for packaging, transport by conveyor, or ultimate use. Heat transport in this process has been subjected to extensive mathematical analysis (1), toward the goal of predicting the rate of growth of the solid layer. If a melt is crystallized on the surface of a drum rotating at a suitably low rate, with adequate mixing of the melt, impurity may be rejected from the growing solid. Most drum crystallizers for laboratory study of the process use horizontal cylinders, internally cooled by gas or liquid at controlled temperature (see Figure 9.1). The drum is immersed to a variable extent in a tray containing melt at constant temperature. The drum is rotated at a constant rate, which can be varied from run to run. Generally an agitator of some kind is included near the drum to promote mixing at the crystallization front. After formation of a solid layer, the product is collected by scraping or by melting with a radiant heater disposed parallel to the drum axis, near its surface.

The major variables in drum crystallization are drum temperature, melt temperature, drum rotation speed, agitation of the melt, and degree of immersion of the drum in the melt. In addition, the concentration of impurity in the melt can influence the purification through changes in the morphology of the growing layer.

1.1   Experimental Studies

Early studies showed that separation can indeed be effected by drum crystallization. It was also found that two ancillary measures enhanced the separation: partial remelting ("sweating") brought about by radiant heating, and extraction by a pressure plate or a second drum, both improved product purity (2, 3).

Chaty (4) crystallized naphthalene/benzoic acid mixtures ranging from 80 to

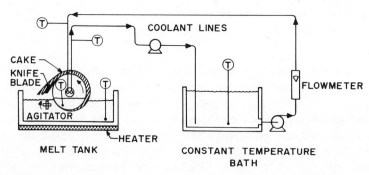

Fig. 9.1 Schematic view of a laboratory assembly for drum crystallization. T indicates thermocouple.

95% naphthalene. At constant drum speed of 6×10^{-3} rev s^{-1} and melt temperature of 353.9 K, separation efficiency E, the ratio of impurity rejected from the melt to the initial concentration, was found to increase linearly with agitation rate. For a given agitation rate, E increased with increasing coolant temperature. At constant agitation rate and constant coolant temperature, E went through a maximum when melt temperature was 12 K higher than coolant temperature. A similar effect was subsequently observed in the naphthalene/biphenyl system (5). The simple eutectic system, p-xylene/m-xylene, likewise showed an increase in E with increasing agitation and decreasing rate of rotation (4). A solid-solution system, naphthalene/2-naphthol, showed a much lower efficiency, consistent with the phase diagram. Table 9.1 gives purification efficiency for the three systems studied. Repeated drum crystallization would be required to achieve high product purity (4).

Lower drum speed yields higher efficiency, since the resulting slower growth approaches equilibrium freezing conditions. Moreover, low rotation speed allows more complete drainage. Removal of adhering liquid can be further promoted by using a portion of the purified product to wash the solid layer on the drum surface (6, 7). It is clear that the morphology of the crystalline layer growing on the drum can have a profound effect on the entrapment of impure interface liquid. In turn, the morphology is influenced by the efficiency of mass transport at the interface, that is, by agitation of the melt. A condition for obtaining a "closed" crystal layer is that the concentration enrichment at the phase boundary must remain below a critical value (6).

Gel'perin et al. have described a multistage crystallizer which incorporates five rotating drums with the solid scraped from each drum serving as feed for the next. The temperatures of the melt- and drum coolants were independently variable in each stage. The frame serving as a support for the array was vibrated to promote mixing. Mixtures of naphthalene with benzene, toluene, biphenyl, 1-methylnaphthalene, and benzoic acid were purified, as well as mixtures of the isomeric methylnaphthalenes. In addition, the system fluorene/2-methylnaphthalene was treated. Separation efficiency went through a maximum as melt temperature was increased (8).

Experiments in crystallization of organic and inorganic melts on a rotating

Table 9.1. Efficiency of Purification by Drum Crystallization

System (Major/Minor Component)	Purity of Feed, %	Product Purity, %	E, %
Naphthalene/benzoic acid	95	99.8	96
	80	91.3	57
p-Xylene/m-xylene	89.5	98.1	82
	71.8	93	75
Naphthalene/2-naphthol	56	66.1	23

drum have been reported, along with a detailed theoretical analysis of heat flow from the melt to the cooled drum wall. The experimental drum had a 50-cm od and was 49 cm long. The setup allowed variation of degree of immersion (0–160°) and rotation speed (0–0.117 rev s^{-1}), and accurate control of temperatures of drum and melt. In experiments with paraffin, ionol, and stearic acid, it was found that best agreement between experimental and calculated layer thicknesses was found for minimum time of growth. In experiments with $LiBr_2 \cdot 2H_2O$, however, thirty-sevenfold variation of duration of growth changed the thickness by only a factor of 1.4. The discrepancy was explained by considering the overall growth in two stages. In the first, crystallization is rapid, because of high ΔT between melt and drum surface. Evolution of heat of fusion on the drum's outer wall causes a large difference between the heat flux entering the wall and that abstracted by the coolant. After thermal equilibrium is established, a second (slower) stage of crystallization ensues. When both stages were considered in the calculation, good agreement was found between experiment and calculation (9).

2   SWEAT PANS

A batch crystallization may be carried out on a cooled surface until all of the charge has solidified, or an enriched impurity fraction may be drained before solidification is complete. In either case, the cooling cycle is followed by circulation of hot fluid through the heat-exchange elements, to effect partial remelting of the solid. The liquid formed during meltback is always less pure than the crystal layer, and "sweats" through the solid layer, hence the name sweat pan. The sweatpan apparatus usually consists of a rectangular tank containing a number of heat-exchange elements; these may be plates or finned tubes, through which heat-exchange fluid is circulated. The apparatus may be built in any size, from laboratory units of 5-L capacity to production units containing tens of cubic meters. The impure liquid is drained at the bottom of the vessel. Care must be taken to assure that the drainage space remains unobstructed by solid. This may be achieved by filling the bottom of the vessel with an inert liquid that is immiscible with the charge or by keeping the bottom of the tank somewhat warmer than the space containing the heat exchangers (10).

In the absence of stirring within the tank, the stagnant interface layer within which matter transport is diffusion controlled, can become large. To counter this effect, the cooling can be programmed to decrease the crystallization rate. However, the increased cycle time (2 days to 1 week) is an economic disadvantage. It is therefore desirable to find optimum crystallization and meltback rates and to carry out the overall process in several discrete stages. Erdmann (11) has analyzed the critical factors in deciding crystallization rates and the extent to which crystallization and meltback should proceed. Qualitatively it may be seen that removal of a large fraction of the charge as meltback will lead to a small yield of very pure product. Conversely removal of only a small fraction will lead to a higher yield of less-pure product.

The attractiveness of the sweat pan method lies in its simplicity. In operation the tank is charged with the mixture to be separated, in molten form. Coolant is passed through the heat-exchange elements and directional crystallization takes place. A commercial application of this technology to purification of naphthalene and p-dichlorobenzene has been developed by Société Proabd (12).

Recent work on devices of this kind has been directed toward calculating optimum crystallization times and temperature programs for multistage units (13). For example, it was calculated that a naphthalene feed containing 10% of 2-methylnaphthalene would give an 83% yield of product containing only $10^{-2}\%$ impurity and a waste fraction containing 60% impurity, from a three-stage installation.

3 COLUMN CRYSTALLIZATION

Columns have long been used to effect countercurrent mass transfer between two liquid phases or between a liquid and a gas. Analogous solid/liquid processes in columns, while less well developed, have been explored for purification in both batch and continuous modes. Many column designs have been described, but all have the following general features (see Figure 9.2):

1. A freezing section in which primary crystallization takes place; crystals generated here are then mechanically transported through a purification section.

2. A purification section, usually adiabatic, in which the crystals are in differential countercurrent contact with melt generated in a melting section.

3. A melting section in which crystals from 2 are melted; part of the melt is withdrawn as product, and part is returned to 2. The three sections and their relationships are shown schematically in Figure 9.2.

If the crystals are substantially denser than the melt, they may be allowed to settle under gravitational force. Most column crystallizers, however, use a mechanical transport device in the purification section. The mechanical device that moves the crystals may be an auger or a helix rotating in an annular space, about a stationary central column.

The separation attainable in a column crystallizer depends on the phase relationships of the system. With a simple eutectic system, the initial crystallization should produce pure crystals of the major component, in a melt that becomes progressively richer in the minor component, until the eutectic is reached. In practice, impurity is retained within crystal defects and the crystals retain impure melt on their external surfaces. Adhering melt may be removed by countercurrent contact with pure melt. Internally retained impurities are unlikely to diffuse to the crystal surface in reasonable times; remelting and recrystallization are required for their removal.

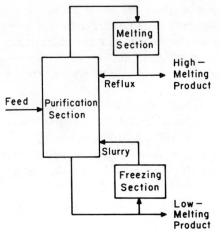

Fig. 9.2 Schematic representation of the components of a column crystallizer.

In systems that form solid solutions, it is not possible to obtain a pure product in a single crystallization and several stages of partial melting and refreezing will be required to attain high purity. Such multistage crystallizations can be carried out in a column if a suitable temperature gradient is provided along the length of the column. Crystals are transported to regions of progressively higher temperature, where they melt.

Column crystallization offers a number of practical advantages beyond the intrinsic benefits of energy saving and a high degree of efficacy. Columns can be built to handle charges ranging from a few milligrams to hundreds of kilograms. Further, batch column crystallizers can be converted to continuous operation more readily than can conventional zone refiners. It is only necessary to provide an inlet at some central point, and product and waste ports at the extremities of the column.

3.1 Columns with Screw Transport

Frevel described a column crystallizer containing a pair of mating, inter-meshed feed screws whose outer surfaces extend nearly to the inner wall of the column, to break adhering crystals from the wall during rotation. The screws are of opposite pitch, so that while counter-rotating they both transport solid upwardly. The screws, while rotating on their axes, are also rotated as a pair about the central axis of the column, to traverse the entire inner wall of the column. Crystals carried over the top of the column are collected in a gutter, where they are melted; part of the melt is collected as product and the remainder is returned to the column as reflux, to wash the ascending crystals (14). The Frevel machine was used to purify benzene, cyclohexane, xylene, and other organic compounds as well as inorganic salts.

A modified transport design has been patented, which can be built in a wide range of sizes, up to 0.5-m diameter (15). The basic member of the device (Figure 9.3) is a rotatable shaft 2, in a stationary container 1. Blades 3 are welded to 2 in

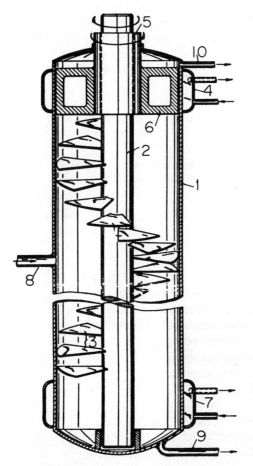

Fig. 9.3 Internal view of a Frevel column crystallizer. The components are identified in the text.

a spiral flight, with the vertical projection of each blade overlapping those of the blades above and below. The blades may be flat, but if their opening angle (i.e., the angle between the blade surface and the vertical axis) is large, it is preferable to use blades in the form of helical segments. The periphery of each blade may be a circular arc, or for more uniform approach to the inner wall of column 1, an elliptical arc or polygon. A cooler 4 jackets the upper part of 1; the inner wall is scraped free of crystals by paddle 6, attached to sleeve 5, which rotates independently of shaft 2. Heater 7 surrounds the lower part of the column. Feed is introduced centrally through tube 8 and product and waste are collected through tubes 9 and 10, respectively. In variations of the basic design, vertical deflectors may be attached to the blades to impede centrifugation of the crystals to the wall of column 1; other deflectors may be attached vertically to the inner wall of 1 to interrupt the downward transport of crystals. The latter arrangement increases the velocity of the crystal slurry relative to shaft 2 and blades 3, in the lower section of the column.

In operation, the blades impart an impulse to the crystals in the desired sense, that is, from the cooled end to the heated end of the device. As in other column crystallizers, liquid is transported countercurrently. The strong current of liquid impedes agglomeration of crystals. No examples of application of this device are provided.

A screw conveyor column was used for continuous fractionation of mixtures of benzene/heptane and binary and ternary mixtures of xylenes (16). The degree of separation was studied as a function of starting composition, composition of the mother liquor, and column diameter.

A similar device has been described in which fins mounted on a centrally rotating shaft convey crystals upwardly (17). Purification of p-dichlorobenzene, p-xylene, and naphthalene was effected by feeding impure material to the bottom of the column and collecting product overhead.

Later studies of column crystallizers using Archimedean screw transporters, in which the helix is affixed to a central shaft, showed lower separation efficiency that did devices based on helical coils (18). Nevertheless study of Archimedean screws has continued, in part because of the difficulty of forming and using very large, freely suspended helical coils. The separation performance of various rotor designs has been tested against a standard criterion, namely the throughput attainable in producing a 95% yield of benzene with a minimum crystallization point of 278.66 K, starting with a feed having a crystallization point of 278.2 K. The performance depended on helical pitch and on the cross section of

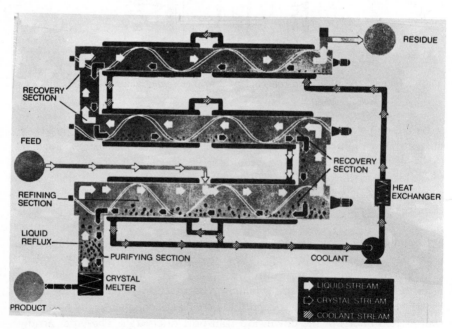

Fig. 9.4 Flow diagram of the Brodie purifier.

the flight, and attainable throughput varied with rotation speed. However, the speed range over which performance was optimum was higher than with helical coils. It seems clear that the Archimedean screw is not intrinsically deficient in separating power, and that suitably designed screws can perform as well as helical coils. On the basis of this work, a column of 23-cm diameter has been constructed (19).

Horizontal helix-swept columns are the basis of the commercial crystallizer known as the Brodie Purifier. In Figure 9.4, feed is introduced at the center of the lowest horizontal column, within which crystallization is induced by a cooling jacket. The crystals settle and are conveyed toward the refining section on the left by a slowly rotating helical ribbon. The melt flows countercurrently to the right. In the recovery section, additional crystals are formed and transported toward the refining section. Crystals entering the purifying section from the refining section fall under gravity into a heated sump. A portion of the descending crystal mass is collected while the remainder is melted and rises through the purifying section to provide liquid reflux. Commercial plants have been constructed for purification of naphthalene and p-dichlorobenzene.

3.2   Helical Coil Devices

In a series of papers between 1959 and 1967, Schildknecht and co-workers described construction and use of column crystallizers in which a helical coil with lenticular cross section rotates in an annular sample space. A schematic cross section of such a column is shown in Figure 9.5. The helical coil transports crystals from the freezing section at the top through the adiabatic purification section to the melting section at the bottom. The crystals are melted and displacement of melt by the arrival of further solid causes a countercurrent stream of reflux liquid to rise. The column may be operated with continuous feed through the central feed port, or it may be operated at total reflux as a batch unit.

As laboratory devices, columns of the Schildknecht design offer a number of advantages over earlier designs (19):

1. Better mixing is attained, resulting in more efficient interphase mass transfer.
2. The helix, freely suspended between inner and outer cylinders, can make very close contact with both confining surfaces. In this way, interphase contact is further enhanced.
3. The width of the free helix is only a small fraction of the outer diameter. Typically, the ratio of flight width to od is 0.3, and the ratio of flight width to pitch is 0.75. In experiments with flights of different width, it was found that results deteriorate above certain flight width/diameter ratios.

Schildknecht et al. described two microcolumns for purification of samples as small as a few tens of milligrams. One of these used a helix of 0.15-cm od, 5 cm

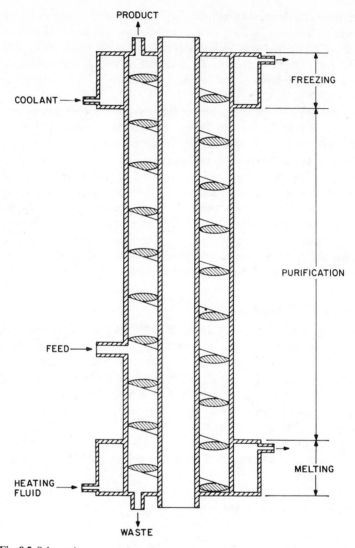

Fig. 9.5 Schematic cross section of a Schildknecht helical coil column crystallizer.

long, rotating around a stationary stainless-steel wire (20). The other, still smaller, used a 0.1-cm × 2-cm od coil, with no center tube or rod (21). In both cases, coil rotation speeds ranged from 8.3 to 33.3 rev s^{-1}. In a micro *tour de force*, one of these units separated azobenzene from stilbene, to steady-state distribution, in 20 s.

Macrocolumns were built in diameters to 2.5 cm and lengths to 80 cm. The largest of these contained about 200 cm^3 of material. The transport helix was driven at 0.5–2.5 rev s^{-1}. A number of these columns were provided with central feed ports, allowing continuous operation with throughputs of 0.8 to 200 g h^{-1}

(22, 23). The conditions used by Schildknecht to purify 20 chemicals have been summarized (19).

A column of the Schildknecht design was used in a study of purification of aromatic amines (24). The helical coil, 60 cm long, had inner and outer diameters of 1.5 cm and 3.5 cm, respectively. It was driven at 1.2 rev s^{-1}, while feed was admitted at 24 mL h^{-1}. Aniline, toluidines, and dimethylanilines were purified in the temperature range from 243 to 288 K.

Bolsaitis (25) described a modified column crystallizer in which it was possible to vary the fraction of the helix length occupied by the stationary central tube which comprised the enriching section. Further, the central section was fitted with external heating coils and not operated adiabatically. The column stripping efficiency was found to decrease both as temperature was lowered and as spiral rotation rate was increased.

Henry et al. (26) described a Schildknecht-type column crystallizer in which the annular sample space of 60-cm length was defined by an outer tube of 2.8-cm id and a stainless-steel inner tube of 1.27 cm od. The pitch of the helix was 1 cm. The freezing, purification, and melting sections were 8, 50, and 5 cm long. The freezing and melting sections were located at the bottom and top of the column, respectively. The purification section was provided with valves on 6-cm centers to facilitate removal of samples during continuous operation. Feed was introduced and waste removed by proportioning pumps. Product was collected by means of an overflow tube near the top of the column. The helix was rotated at 1 rev s^{-1}. These workers imposed an axial oscillation on the helix rotation with a frequency and amplitude of 5 s^{-1} and 0.1 cm, respectively. Benzene containing 1500 or 28,000 ppm cyclohexane was purified under varying operating conditions. The results obtained are shown in Table 9.2.

By far the most extensive study of helical-coil column crystallization is that reported by Betts and Girling (19), who addressed two goals:

- To examine the applicability of this method to systems of industrial importance, at industrial scale
- To optimize system design by identifying critical parameters

A glass column of 2.5-cm diameter and 72-cm length was built and used with steel inner tubes of 1.18-, 1.5-, and 1.85-cm od. A 5-cm column, similarly

Table 9.2. Separation of Benzene/Cyclohexane Mixtures by Column Crystallization (26)

C_0, ppm	28,000			1,500		
Feed rate, g s^{-1}	0.070	0.067	0.088	0.088	0.078	0.058
Product rate, g s^{-1}	0.011	0.017	0.038	0.030	0.020	0.005
Waste composition, ppm	33,900	37,000	37,000	2,090	2,540	2,140
Product composition, ppm	40	60	370	16	4	1

proportioned, was also built. Helical coils of various cross sections and pitches were evaluated. A second 2.5-cm column was built with a glass outer tube, sealed by O-rings to a bronze end fixture bearing a bronze center tube. The apparatus allowed easy disassembly.

The various devices were applied to eutectic and solid-solution systems. The degree of separation attained with the latter allowed calculation of the number of theoretical or equilibrium stages, from a knowledge of the appropriate phase diagram. For the 2.5-cm column at total reflux, the HETS (height equivalent to a theoretical stage) was found to be 22 cm, from runs with a naphthalene/thianaphthene test mixture. With the 5-cm column, HETS ranged from 31 to 67 cm, and for the 10-cm column, from 53 to 91 cm. Although the measurements of HETS were subject to substantial scatter, effects of helix pitch were discernible for the 5-cm column; 1.9-cm pitch gave lower HETS than 2.5-cm pitch. Helix geometry appears to vary in importance, depending on the nature of the material being processed. For materials in which transport is difficult, ratios of flight width to helix diameter greater than 0.2 were unsatisfactory. Likewise, ratios of flight width to pitch greater than 0.4 were unsatisfactory. Of course these ratios may not be reduced to arbitrarily small values, for mechanical reasons.

Other experiments showed that helical coils with square section were less efficient than those with round section, other things being equal. In the 2.5-cm column, helix rotation speed in the range 1.67–3.34 rev s^{-1} had little effect on degree of separation. In the 5-cm column, however, separation efficiency varied considerably with helix rotation rate, showing a pronounced maximum at about 1 rev s^{-1}.

Crystal flux, and hence reflux ratio, were estimated from both the heat supplied to the melting section and the heat abstracted by coolant. With feed containing 1% impurity, product of 99.9% purity was obtained from the 2.5-cm column when about 15% of the total crystal flow was returned as reflux. When reflux was increased to 40%, product of 5N purity was obtained. With the 5-cm column, throughputs up to 5 L h^{-1} were attained, and it was speculated that columns could be built in much larger sizes for industrial operation.

Helical-coil columns were built in 7.5-, 10-, and 14.3-cm diameters, in lengths up to 1.14 m. However, as column size was increased, it became progressively more difficult to fabricate and operate coils with the necessary small clearance between the helix and the column wall. In this context, it should be mentioned that simulation studies indicated that the helical coil method is applicable only to systems giving rise to crystals larger than the clearance between rotor and walls. Separation efficiency cannot be expected to improve continuously with increasing crystal size, however. Theoretical and experimental studies of the benzene/thiophene system showed that solid-state diffusion can limit purification. The recrystallization of small crystals into large ones reduces mass-transfer area and, hence, the efficiency of purification (27). Subsequent studies on large-diameter columns incorporating Archimedean screws showed that separation efficiencies could be obtained which were comparable with those afforded by helical coils.

Matz (28) has described experiments with a column crystallizer with a helical transporter of the type reported earlier by Leister (29). This unit differs from column crystallizers of the Schildknecht type in that the central shaft upon which the helix is fitted, can be rotated independently of the helix. A further distinction is that in the Leister–Matz apparatus, a temperature gradient was imposed upon the column by three separate thermostat jackets, while Schildknecht columns are operated adiabatically. The column used in this work was 5 cm in diameter and 1 m long with an annular sample chamber of about 1-L volume. The device was used for purification of two binary systems, one of which (cyclohexane/1,2-dibromoethane) is a simple eutectic and the other of which (1,2-dibromoethane/1,2-dichloroethane) forms solid solutions.

At an arbitrary feed rate of 2 L h^{-1} the column performance was equivalent to two theoretical stages with the solid-solution system. The feed, containing 91% of the desired component, was enriched to a maximum of 99.4%. With the simple-eutectic system, essentially pure cyclohexane was produced at rates up to about 0.4 T m^{-2} h^{-1}.

Theoretical analyses of mass-transfer processes usually refer to an efficiency, defined in terms of a number of theoretical equilibrium stages. In the case of crystallization of a melt that forms solid solutions, the number of stages required to effect a desired purification may be obtained from the solid/liquid phase diagram. For a eutectic-forming system, a single-stage crystallization should, in theory, give a pure product. However, eutectogenic systems always yield crystals that contain or retain some of the minor components, and theoretical analyses must reflect this fact. Both kinds of system have been treated, using one or both of two approaches:

- Differential countercurrent contacting, which posits two phases whose compositions change continuously in the direction of flow
- Countercurrent stagewise contacting

The first of these approaches was used by Powers et al. (30) for both eutectic and solid-solution systems. Eutectic systems were treated with the assumptions that no crystals of eutectic composition are formed in the freezing section. Flow rates and the height of a transfer unit, H, are independent of position in the purification section of the column. The column is operated at total reflux and steady state. The following relation between concentration and column position was derived.

$$\frac{C_l(Z)}{C_l(0)} = \exp -\left(\frac{Z}{H}\right) \tag{9.1}$$

where $C_l(Z)$ is the solute concentration in the free liquid at height Z in the column and $C_l(0)$ is the concentration at the start of the purification section. H depends on liquid and crystal flow rates, impurity diffusivity in the melt, solid/melt interfacial area, melt density, column geometry, and volume fraction of liquid. The exponential distribution implicit in Eq. 9.1 was observed in the column crystallization of azobenzene (30). This treatment has been extended to relate the degree of purification obtained in the column to product flow rates in a centrally fed column crystallizer in continuous operation. The equations derived were not verified experimentally. Further extension of this theory to solid-solution-forming systems gave the equation

$$\frac{Z}{H} = \int_{C_l(0)}^{C_l(Z)} \frac{dC_l}{C_s - C_l} \tag{9.2}$$

where C_s and C_l are the equilibrium solute concentrations in solid and liquid, respectively. This equation was applied to several solid-solution systems using a 2.5-cm diameter $\times$ 72-cm-long column and values of H in the range 5–50 cm were obtained (19).

The alternative theoretical approach based on countercurrent stagewise contacting, usually applied to distillation columns, was applied by Anikin (31) to column crystallizers operating at total reflux and steady state. No experimental verification of this theory has been attempted.

Using another approach to the characterization of efficiency in column crystallization of eutectic-forming systems, Gel'perin et al. defined a "degree of extraction" $\Psi = C_p/C_{p,\text{theor.}}$. C_p is the concentration of the desired substance in the product stream and $C_{p,\text{theor.}}$ is the theoretical (maximum) concentration, given by the equilibrium phase diagram for a given starting concentration. The following equation was derived

$$\Psi = 1.08 \frac{W^{0.6}}{n} \left(\frac{\Delta H_f}{RT_f}\right)^{-0.55} (1 - C_0) \frac{v}{v_{\text{col}}} \tag{9.3}$$

where

W = ratio of amount of melt to amount of solid in the column
$n = (C_{p,\text{theor}} - C_0)/(C_{p,\text{theor}} - C_e)$
C_0 = mole fraction of the desired product in the original mixture
C_e = mole fraction of the desired product in the eutectic
v = feed rate of the original mixture
v_{col} = effective volume of the column

and other terms have their usual meanings. This equation agreed well with experimental data. The average relative deviation of calculated values of Ψ from the values measured in more than 320 experiments was $\pm 4.8\%$.

3.3 Miscellaneous Column-Crystallization Techniques

A number of other techniques have been devised making use of varied methods to promote mass transfer between solid and liquid phases. Among the physical forces used to promote matter transport are periodic pressure pulses via a piston, agitation by freely rotating spheres resting on perforated plates distributed along the length of the column, and mixing with an insoluble liquid coolant to generate a crystalline slurry.

3.3.1   Pressure Pulse System

Work in the Phillips Petroleum Co. laboratories on crystallization extended through the 1950s beginning with the patents of Arnold (32, 33). These efforts culminated in a crystallizer design which can be scaled from a small pilot plant to that of the largest industrial process. All of the units are characterized by a horizontal scraped-surface crystallizer separate from the purification column. The crystal slurry enters the vertical purification column, which may contain a reciprocating piston with a porous or solid face. In the former case low-melting mother liquor passes through the plate during the compression stroke; crystals in the slurry are driven toward the melting section. In the latter case (solid face) the mother liquor is driven through a perforated or porous section of the column wall just below the piston (see Figure 9.6a). In a more recent version of this technique no internal piston is used within the column. Instead, a hydraulic pressure pulse is applied periodically in the liquid circuit outside the column (see Figure 9.6b). Columns of this type have been broadly applied to concentration of beer, flavor constituents, and edible products such as orange juice (34). The most important application is to the production of 99 + % p-xylene from an isomer mixture containing as little as 22% para (34).

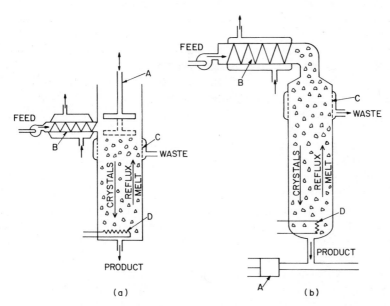

Fig. 9.6 Pressure pulse systems. (a) In this case pressure is applied directly to the crystal slurry by piston A. (b) Pressure pulses are applied periodically to the liquid circuits outside the column by hydraulic piston A.

Moyers and Olson (35) have described a center-fed column crystallizer with a reciprocating porous piston in which crystals are formed within the column. This column embodied a novel design feature in that the rotating scraper, which generates feed slurry at the central feed point, is attached to the reciprocating porous piston. Both elements are connected to a common shaft, rotating and reciprocating as a unit. Column performance was studied using an acetamide/water mixture. Product of $>99.95\%$ purity was consistently obtained from feed ranging from 90 to 97% acetamide. The quality of the product was insensitive to feed rate, product rate, and feed composition, indicating a high degree of separation power. The apparatus is simpler than helical-coil columns, yet appears to offer equivalent potential for scaleup.

A mathematical model was formulated to characterize the temperature distribution in the column and the related concentration profile. A nonadiabatic, plug-flow, axial-dispersion model was selected in preference to an extraction model of the type used in earlier studies. Models of the latter type require knowledge of the apportionment of mass among crystals, adhering liquid, and free liquid. The former type of model requires only one system parameter, the liquid axial-dispersion coefficient, which is experimentally measurable. The measured temperature profiles were well matched by curves computed with axial-dispersion coefficients of 0.2–$0.3 \, cm^2 \, s^{-1}$.

3.3.2   The TNO Crystallizer

In the column crystallizer developed by Arkenbout and Smit, (36), the purification section contains a number of perforated disks, each of which is covered with a layer of inert spheres (Figure 9.7). The spheres are agitated to cause crystal attrition. In laboratory versions the agitation is caused by vibrating the entire column; in larger-scale units, the sieve disks are all mounted on a single shaft which transmits the vibration.

The combination of grinding and sieving promotes local interphase mass transport and enhances axial countercurrent transport of crystals and melt. In other respects, the process resembles earlier column crystallizers. A slurry is first formed in the coaxial scraped crystallizer, A, of Figure 9.7. The crystals descend through purification section B until they reach heater C. A portion of the crystal mass is melted and collected as product while the remainder rises and washes the descending crystals.

While crystal grinding causes an increase in surface area, the smallest, least perfect crystals tend to melt as they descend; larger, more nearly perfect crystals grow in a region of purer melt. This melt/regrowth sequence provides the effect of multistage crystallization, with an efficiency of 5 plates m^{-1}, or 0.5 theoretical plate per disk.

Columns of this design have been built with 8-cm diameter in lengths of 0.5–1.5 m. Product flows of 1–$10 \, kg \, dm^2 \, h^{-1}$ have been achieved. In a series of crystallizations of the solid-solution system benzene/thiophene, sieves were used with 0.06-cm square openings. Each 8-cm sieve disk carried 30 metal spheres of 1.2-cm diameter. The column was vibrated at 50 Hz with an amplitude of

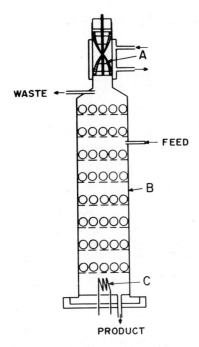

WASTE

FEED

A

B

C

PRODUCT

Fig 9.7 The TNO crystallizer uses vibrating trays bearing agitation spheres.

0.03 cm. Under conditions of total reflux, with a crystal flux of $0.6\,kg\,h^{-1}$, product/waste concentrations ranged from 0.002%/0.94% to 0.045%/23.6% when column conditions and thiophene concentration were varied. Since the distribution coefficient for this system was found to be 0.4, the separation results correspond to 5.0 to 8.4 theoretical plates, and efficiencies of 0.4 to 0.7 theoretical plates per disk (36, 37, 38, 39). The column has also been used to separate the systems H_2O/salt, naphthalene/benzoic acid, and naphthalene/2-naphthol (40). An improved version has been described (41).

3.3.3 Newton Chambers Process

In the Newton Chambers process the molten feed is brought into contact with an immiscible liquid coolant to induce rapid crystallization. After the desired fraction of the feed has crystallized, the resulting slurry is separated into a crystal mass and a liquid mixture of coolant and mother liquor, either by decantation or centrifugation. The crystals are washed with a portion of coolant at a temperature slightly higher than their melting point. This causes partial melting, which removes adhering mother liquor. The mother liquor may be further purified in a second crystallization stage, using a coolant at lower temperature than before. The crystals from this second stage can be melted and recycled into the feed for the first stage.

The method has been applied to purification of benzene, using brine as coolant (4). The presence of the added liquid gives a mobile slurry, even after a large fraction of the feed has been crystallized. Direct contact of the feed with the

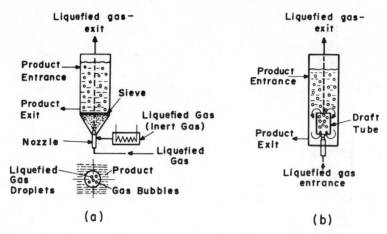

Fig. 9.8 Crystallization induced by evaporative cooling. (*a*) Crystallization is brought about by injection of a liquefied gas into the column charged with melt. (*b*) draft tube crystallizer in which crystallization is controlled by admission of a liquefied gas into the draft tube.

coolant leads to efficient heat transfer and consequent reduction in refrigeration costs.

Liquefied gases have been introduced into melts to bring about crystallization, but the spontaneous evaporation of the liquefied gas leads to strong supercooling near the inlet nozzle. Two devices have been described to retard and control this effect. In one of these the inlet nozzle is placed at the bottom of a conical inlet chamber positioned below the crystallization column and separated from it by a sieve. A portion of the liquefied gas is preevaporated in a heat exchanger and forms a mist stream which is fed into the melt (see Figure 9.8*a*). Since the droplets are no longer in direct contact with the melt, the heat abstraction is slower, as shown in the inset in Figure 9.8*a*. The other device makes use of a draft tube within the body of the crystallizer. The liquefied gas is sprayed into the draft tube (Figure 9.8*b*) and the resulting density gradient mixes the melt and limits excessive supercooling (43).

4 PARTIAL MELTING

Most single-stage laboratory crystallizations are carried out directionally in order to achieve a monolithic solid from which impure melt is maximally excluded. It is also possible to solidify a mass of liquid to a rather large extent (more than 40% solid), giving a dendritic solid, and then remove the entrapped liquid. The simplest approach is to allow the melt to exude from the dendritic solid under gravitational force. The effectiveness may be enhanced by remelting part of the solid and/or by centrifuging the semisolid product.

The simplest of devices can be used. For example Gel'perin et al. (44) used a cold, finned cylinder with a central core 1 cm in diameter by 7 cm tall, bearing six vertical fins which were 0.8 cm wide by 0.3 cm thick, to freeze a portion of an unstirred 70-g melt contained in a vessel of 4-cm diameter. The system was allowed to stand for 15 min after immersion of the cold body; the remaining melt was drained and analyzed. The adhering solid was then reheated until an additional portion of melt formed; it too was collected and analyzed. Only the last quarter to fifth of the solid to melt was purer than the starting liquid. This distribution corresponds to an effective distribution coefficient between 0.8 and 0.9. Hence, it is clear that the effectiveness of such a single-stage partial melting of solid formed rapidly from an unstirred melt is minimal.

A continuous process has been described for purification of aluminum by centrifuging a partially solidified melt (45). The advantage of solidifying extensively and then remelting may be seen from Figure 3.4, which shows that solute concentration does not rise sharply in a stirred melt until about 80% of the charge has solidified. Yet another approach uses isothermal compression of the semisolid to exclude remaining liquid.

Mastrangelo and Aston (46) compared the effectiveness of partial melting of hydrocarbon mixtures, with and without compression. For the system 2,2,4-trimethylpentane containing 5% n-heptane, the system using compression gave a threefold increase in yield of product of $>99.98\%$ purity, with a sixfold reduction of required time.

In recent work much higher pressures were used in a fairly simple apparatus for purification of tin/lead alloys. A brass cylinder with an exterior heating jacket is provided with a central stainless-steel well to contain the material to be purified. The bottom of the well is closed by a porous alumina filter with 0.2–0.3-mm pores. A close-fitting plunger closes the well; it is driven into the well at controlled, constant speeds of 10^{-1}, 10^{-2}, and 10^{-3} cm s^{-1} to generate pressures up to 21 MPa (2.9×10^3 lb in^{-2}). The sample to be purified is machined to fit closely in the well. Test meltings of various tin/lead alloys provided data for evaluation of a parameter called the *refining ratio*, defined as C_c/C_0 where C_c is the solute concentration in the final cake and C_0 is the initial solute concentration. Experiments are carried out in two ways:

1. The sample is heated in the well in an inert atmosphere to a preselected temperature to produce a semisolid structure. The plunger is advanced at a constant rate, and the resultant load is recorded to generate a stress-strain curve, to a maximum pressure of 21 MPa. Liquid expressed from the solid, after passage through the porous ceramic filter, is quenched by a flow of helium (47). For compression to 21 MPa, the refining ratio ranged from about 0.1 for small fractional solidification (i.e., $<10\%$ of starting charge solidified) to about 0.2 for 60% solidification. Compression to 10.9 MPa produced less favorable results. The refining ratio was 0.4 for 40% solidification and about 0.65 for 60% solidification.

2. The sample is heated at constant rate under zero load, to a temperature just below the eutectic temperature. Near the eutectic temperature a constant load is applied; constant pressure is maintained by adjusting plunger speed. Heating is continued until a desired terminal temperature is

reached, and expressed liquid passes through the filter and is quenched (48). A series of experiments were carried out on Sn–15% Pb alloy heated from T_E (456 K) to temperatures as high as 483 K, under constant pressure of 21 MPa. The refining ratio ranged from about 0.1 for 60% solidification to about 0.3 for 80% solidification.

In both types of experiment, the compressed cake was characterized by a "wetness factor," which serves as a measure of retained liquid phase. After compression to 21 MPa, wetness factors of 2 to 3% are evident. Metallographic studies of the compressed samples showed that both methods gave fine-grained, single-phase solids.

5   SEPARATION OF ENANTIOMER MIXTURES

Crystalline racemates can occur as racemic compounds or as conglomerates. Enantiomers that form racemic compounds can only be separated through diastereomeric interactions, as in formation of diastereomers with optically active resolving agents or chromatography on an optically active adsorbent. Enantiomeric conglomerates, on the other hand, can be separated either by diastereomeric interaction or by crystallization without any optically active agents. Most such crystallizations have been carried out from solutions, but it is also possible to separate conglomerates by crystallization from the melt (49).

A conglomerate is a mechanical mixture of crystalline enantiomers, fully equivalent to a mixture of two compounds A and B, forming a simple eutectic (see Figure 9.9). Conglomerate phase diagrams show the special feature of symmetry: the pure enantiomers have identical melting temperatures and enthalpies of fusion. The liquidus lines $T_D E$ and $T_L E$ may be extended beyond E into the metastable region at lower temperatures. If the supercooling is not excessive, a conglomerate may be cooled through E to a point such as M, without spontaneous nucleation. Introduction of a seed of L enantiomer will cause additional crystals of L to form, and the remaining liquid will become richer in D. A mole of racemate will give (MN/NU) mole of pure crystalline L and MU/NU mole of liquid of composition n. After removal of L, the nonracemic liquid may be seeded with D enantiomer, causing additional crystals of D to form. The process of seeding alternately with L and D can be continued

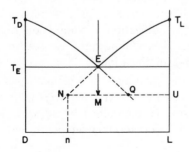

Fig. 9.9 Phase diagram for separation of enantiomer mixtures; see text for symbols.

until the melt is consumed. The yield in each cycle rises with increasing supercooling, but so does the risk of spontaneous nucleation of the nonseeded isomer. The goal of experimentation is to find optimum conditions of cooling and stirring, to allow a maximum yield of resolved components in minimum time.

Lactams **1, 2** and **3** form low-melting conglomerates which can be supercooled readily. They were resolved semiautomatically in the device shown in Figure 9.10. Chambers 1 and 2 are charged with conglomerate, which is melted, stirred, and cooled to a temperature below T_E. The two chambers are seeded with L and D, respectively; after crystallization is complete, the liquids are filtered into vessels 3 and 4. The liquid in 3, containing excess D, is transferred to 2, while that in 4 is simultaneously transferred to 1. The D-enriched melt in 2 is seeded with D and the L-enriched melt in 1 is seeded with L (49), and the process is continued.

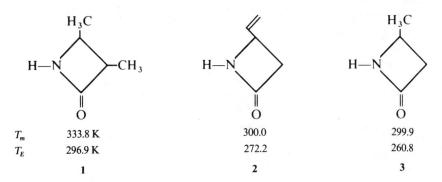

T_m	333.8 K	300.0	299.9
T_E	296.9 K	272.2	260.8
	1	**2**	**3**

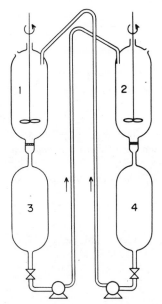

Fig. 9.10 Apparatus for separation of low-melting conglomerates by crystallization.

REFERENCES

1. B. Covelli, J. Wochele, and F. Widmer, *Chimia*, **30** (6), 318–324 (1976).
2. T. Baron, *Continuous Production of Paraffin Wax*, B. S. Thesis, University of Illinois, Urbana, IL (1943).
3. B. L. Graham, U.S. Patent 2,651,922, 15 September 1953.
4. J. C. Chaty, "Rotary-Drum Techniques," in M. Zief and W. R. Wilcox, Eds., *Fractional Solidification*, Dekker, New York, 1967, p. 409.
5. N. I. Gel'perin, G. A. Nosov, and A. V. Makotkin, *Khim. Khim. Tekhnol., Tr. Yubileinoi Konf.*, 414–415, (1970) (Pub. 1972).
6. K. Wintermantel and W. Kast, *Chem.-Ing. Technik*, **45**, 728–731 (1973).
7. K. Wintermantel and W. Kast, *Chem.-Ing. Technik*, **45**, 284–289 (1973).
8. N. I. Gel'perin, G. A. Nosov, and A. V. Makotkin, *Moscow Inst. Tonk. Khim. Tekh. Trudy*, **2**, 167 (1972).
9. V. G. Ponomarenko, G. F. Potebnya, V. I. Bei, and K. P. Trachenko, *J. Appl. Chem. USSR*, **52**(9), 1871–1875 (1979).
10. W. S. Calcott, "Sweating," in J. H. Perry, Ed., *Chem. Engr. Handbook*, 3rd ed., 1950, p. 1075.
11. H.-H. Erdmann, *Chem.-Ing. Technik*, **48**, 793 (1976).
12. J. G. D. Molinari, "The PROABD Refiner," in M. Zief and W. R. Wilcox, Eds., *Fractional Solidification*, Dekker, New York, 1967, p. 393.
13. E. J. Eisenbraun, C. J. Moyer, and P. Vuppalapaty, Chem. Ind., 229–230 (1978).
14. L. K. Frevel et al., U.S. Patent 2,659,761, November 17, 1953.
15. Sulzer Frères, French Patent 2,103,803, (1972).
16. M. Sh. Dzhashiashvili, N. I. Gel'perin, and N. N. Zelenentskii, *Tr. Gruz. Inst. Subtrop. Khoz.*, **142**, 643–652 (1969).
17. Kureha Kagaku Kogyo KK, British Patent 1,349,958, 10 April 1974.
18. H. Schildknecht and H. Vetter, *Angew. Chem.*, **73**, 612–615 (1961).
19. W. D. Betts and G. W. Girling, "Continuous Column Crystallization," in E. S. Perry and C. J. Van Oss, Eds., *Progress in Separation and Purification*, Vol. 4, Wiley-Interscience, New York, 1971.
20. H. Schildknecht, *Chimia*, **17**, 145–157 (1963).
21. K. Maas and H. Schildknecht, *Z. Anal. Chem.*, **236**, 451–461 (1968).
22. H. Schildknecht and K. Maas, *Wärme*, **69**, 121–127 (1963).
23. H. Schildknecht and J. E. Powers, *Chemiker-Ztg./Chem. App.*, **90**, 135–141 (1966).
24. B. Pouyet, "Purification of Aromatic Amines," in M. Zief, Ed., *Purification of Inorganic and Organic Materials*, Dekker, New York, 1969, p. 121.
25. P. Bolsaitis, *Chem. Eng. Science*, **24**, 1813–1825 (1969).
26. J. D. Henry, Jr., M. D. Danyi, and J. E. Powers, "Reduction of Cyclohexane Content of Benzene in a Column Crystallizer Under Steady Flow Conditions," in M. Zief, Ed., *Purification of Inorganic and Organic Materials*, Dekker, New York, 1969, p. 107.
27. G. G. Devyatykh, V. M. Vorotyntsev, Yu. E. Elliev, and E. M. Shcheplyagin, *Teor. Osn. Khim. Teknol.*, **10**(2), 302–304 (1976).
28. G. Matz, *Verfahrenstechnik*, **5**, 507–512 (1971).
29. K. Leister, W. German Patent 1,200,788, 16 September 1965.
30. R. Albertins, W. C. Gates, and J. E. Powers, "Column Crystallization," in M. Zief and W. R. Wilcox, Eds., *Fractional Solidification*, Dekker, New York, 1967, p. 343.
31. A. G. Anikin, *Dokl. Akad. Nauk SSSR*, **151**(5), 1139–1142 (1963).
32. P. M. Arnold, U.S. Patent 2,540,997 (1951).

33. P. M. Arnold, U.S. Patent 2,540,083 (1951).

34. D. L. Mackay, "Phillips Fractional-Solidification Process," in M. Zief and W. R. Wilcox, Eds., *Fractional Solidification*, Dekker, New York, 1967, p. 427.

35. C. G. Moyers, Jr. and J. H. Olson, *AICHE J.*, **20**, 1118–1124 (1974).

36. G. J. Arkenbout, *Chem. Tech.*, 596–599 (1976).

37. G. J. Arkenbout, A. Van Kuijk, and W. M. Smit, *Chem. Ind.*, 139–142 (1973).

38. G. J. Arkenbout, A. Van Kuijk, and W. M. Smit, *Dechema Monogr.*, **73**, 277–291 (1974).

39. G. J. Arkenbout, A. Van Kuijk, and W. M. Smit, *Sep. Sci.*, **3**, 501–517 (1968).

40. G. J. Arkenbout, A. Van Kuijk, and W. M. Smit, *TNO Nieuws*, **27**, 767–774 (1972).

41. G. J. Arkenbout, A. Van Kuijk, J. Van der Meer, and L. H. J. M. Schneiders, Eur. Pat. Appl. 34,852, 11/16/83. (Related U.S. Patent 4,400,189.)

42. J. G. D. Molinari, "Newton Chambers Process," in M. Zief and W. R. Wilcox, Eds., *Fractional Solidification*, Dekker, New York, 1967, p. 401.

43. C. Casper, *Chem.-Ing. Tech.*, **53**(3), 210–211 (1981).

44. N. I. Gel'perin, G. A. Nosov, and V. V. Fillipov, Khimicheskaya promyshlennost' 923–925, No. 12 (Chemical Industry, USSR) (1973).

45. J. DeNo Martin, Spanish Patent No. 388,488 (1973).

46. S. V. R. Mastrangelo and J. G. Aston, *Anal. Chem.*, **26**, 764–766 (1954).

47. A. L. Lux and M. C. Flemings, *Met. Trans. B*, **10B**, 71–78 (1979).

48. A. L. Lux and M. C. Flemings, *Met. Trans. B*, **10B**, 79–84 (1979).

49. A. Collet, M.-J. Brienne, and J. Jacques, *Chem. Rev.*, **80**(3), 215–230 (1980).

50. H. Jensen, German Patent 1,807,495, 1970.

Chapter **X**

DETERMINATION OF THE DISTRIBUTION COEFFICIENT

To estimate in advance the probable efficacy of a zone-refining or directional-crystallization experiment, it is necessary to know the effective distribution coefficients of the relevant impurities. In principle, it is possible to get this information from phase diagrams. In practice, however, accurate phase diagrams are rarely available. This is especially true with respect to the "ends" of the phase diagram, that is, the regions relating to nearly pure host containing a small amount of impurity. Other methods are available for determining *k* directly or indirectly, and experimental knowledge of *k* sheds light on phase relationships more often than the reverse. That is, one learns about the phase diagram from measured values of *k* more often than one learns *k* from phase diagrams. A number of approaches to the determination of *k* are outlined here.

1   DETERMINING *k* FROM THE PHASE DIAGRAM

If an equilibrium phase diagram is available, k_0 is obtained from the ratio C_s/C_l, at a given temperature. From k_0 and the growth conditions, k_{eff} can be estimated using the Burton, Prim, and Slichter relationship (1). In this way,

422

Table 10.1. Distribution Coefficients from Phase Diagrams, Compared with Experimental Values (from ref. 2)

System	From Phase Diagram $k_0 \times 10^5$	Measured $k_0 \times 10^5$	A	B
Ge/Sb	400	300	11,050	6.75
Ge/Cu	0.98	1.5	9,200	2.59
Si/Cu	32	45	7,120	0.69

distribution coefficients k of antimony and copper in germanium and of copper in silicon have been estimated from published phase diagrams, for a number of temperatures close to the melting point of the major component. Plots of log k_0 against $1/T$ are linear and are described by the equation

$$\log k_0 = B - \frac{A}{T} \tag{10.1}$$

where A and B are constants. Extrapolation of the linear plots to the melting point of the major component gives a limiting value of k_0 for a very dilute solution (2). These limiting values are shown in Table 10.1 together with values measured experimentally by directional crystallization, and the constants of Equation 10.1. The behavior observed in these systems is atypically ideal; tin

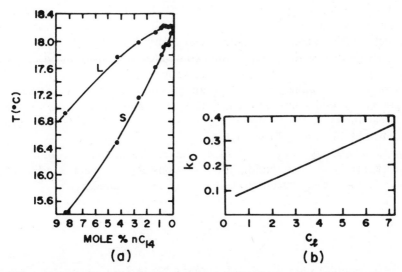

Fig. 10.1 (a) Partial phase diagram of n-hexadecane/n-tetradecane, derived from cooling curves measured by the thin-film method. (b) Distribution coefficients of tetradecane in hexadecane, derived from isothermal tie lines.

does not conform to this simple relationship in either germanium or silicon.

Gouw and Jentoft (3) prepared an accurate phase diagram for a small range of compositions in the n-hexadecane/n-tetradecane system (91–100% n-$C_{16}H_{34}$) from cooling curves measured by the thin-film method (see Section 2.2.4, Chapter 2). Distribution coefficients obtained from isothermal tie lines were found to vary by a factor of about four in the concentration interval studied (see Figure 10.1). A plot of the measured k_0's against $1/T$, however, is only linear down to 291 K, corresponding to 4% solute. In fact, it has been asserted that the simple $\ln k$ against $1/T$ relationship can only be expected to apply to systems for which $k_0 < 10^{-2}$ (4).

Phase diagrams were determined for limited concentration ranges of several binary systems of aromatic hydrocarbons and heterocyclics. An empirical relationship was derived, relating k_0 to concentration: $k_0 = a \pm bC_s$, where a and b are constants, and the $+$ sign applies to k_0 less than unity. The measured values were as follows: naphthalene/benzothiophene (0.01–8.26 wt%, $a = 0.10$, $b = 0.0338$), phenanthrene/fluorene (0.05–14.2, 0.66, 0.0224), phenanthrene-/carbazole (0.07–7.1, 4.2, −0.249), anthracene/carbazole (0.05–2.8, 0.58, 0.0261) (5).

2 DETERMINING k FROM THERMODYNAMIC RELATIONSHIPS

If a complete phase diagram is not available in the concentration range of interest, then k_0 may be estimated from thermodynamic considerations. In our earlier discussion of the thermodynamics of ideal solid solutions and liquid solutions (Section 1.2.2, Chapter 2) it was pointed out that in dilute solutions, the logarithm of the distribution coefficient is a linear function of the reciprocal of absolute temperature (Eqs. 2.10 and 2.11). For a nearly pure major component containing a small amount of impurity, the distribution coefficient of the higher-melting component approaches unity while that of the impurity is less than unity. For the lower-melting component, the distribution coefficient approaches unity while that of its impurity is greater than unity. These statements recapitulate the information contained graphically in the phase diagram for a solid-solution system; in turn the phase diagram gives the relative molar amounts of the components in equilibrium at any temperature, as described by Equations 2.10 and 2.11.

If a dilute liquid solution of B in A behaves ideally, and on cooling gives a similarly ideal solid solution, the concentrations of component B in liquid and solid solutions at equilibrium at temperature T are related by

$$\ln \frac{X_{B,s}}{X_{B,l}} = \frac{\Delta H_B}{R}\left(\frac{1}{T} - \frac{1}{T_B}\right) \tag{10.2}$$

where $x_{B,s}$ and $X_{B,l}$ are the mole fractions of B in solid and liquid solutions, respectively, and T_B is the freezing temperature of pure B.

Similarly, for dilute solutions of A in B,

$$\ln \frac{X_{A,s}}{X_{A,l}} = \frac{\Delta H_A}{R} \left(\frac{1}{T} - \frac{1}{T_A} \right) \tag{10.3}$$

Thus only the heats of fusion and freezing temperatures of the pure components are needed for a calculation of the relative concentrations in solid and liquid phases at temperatures near the freezing points of the pure components. The relative concentration X_s/X_l is, of course, the distribution coefficient of the minor component, expressed in mole fraction. Only a few systems conform to this simplified law.

In a somewhat more realistic, but still restricted case, the liquid solution is assumed to be ideal and in equilibrium with a regular solid solution, giving (6)

$$\ln \frac{X_{B,s}}{X_{B,l}} = \frac{\Delta H_B}{R} \left(\frac{1}{T} - \frac{1}{T_B} \right) - \frac{\Delta H_m}{RT} \tag{10.4}$$

where ΔH_m is the heat of mixing of the components in the solid state. More complicated equations have been derived for still less ideal systems (6). A similar thermodynamic approach was used by Thurmond and Struthers (2) to calculate distribution coefficients in germanium for solutes with limiting distribution coefficients (near zero solute concentration) less than 0.1. These authors provided a general formulation of k in terms of thermodynamic quantities:

$$\ln k = \frac{\Delta H_f - \Delta H_2}{RT} + \frac{\sigma - \Delta S_f}{R} + \ln \gamma \tag{10.5}$$

where ΔH_f and ΔS_f are the enthalpy and entropy of fusion of the solute at its freezing point, ΔH_2 is the differential heat of solution of the solute, σ is a vibrational-entropy term, γ is the activity coefficient of the solute in the liquid solution, and T is the temperature at which the crystal is grown. Weiser (7) later showed that σ is the logarithm of the ratio of the Debye temperature of the solvent to that of the solute, and that its contribution is negligible. Since

$$\Delta H_f - T \Delta S_f = \Delta G_f$$

Equation 10.5 becomes

$$k = \gamma \exp\left(\frac{\Delta G_f - \Delta H_2}{RT} \right) \tag{10.6}$$

Equation 10.6 has been used with calculated differential heats of solution to provide theoretical values of k which agree well with experimental results (8).

Major effort is now being directed toward calculation of phase diagrams by computers, with basic thermodynamic data as input (9, 10). Complete programs have been published (11, 12) and a new journal is devoted entirely to this topic (*Calphad: Computer Coupling of Phase Diagrams and Thermochemistry*, Pergamon Press Ltd., Headington Hill Hall, Oxford OX30BW, UK).

3 DETERMINING *k* BY CYLINDRICAL/AXIAL DIRECTIONAL CRYSTALLIZATION

It is possible to determine k_{eff} directly by experiment. Essentially, one need only solidify a quantity of a mixture and determine the solute concentration in the resulting ingot as a function of fraction solidified. A logarithmic plot of the measured concentrations against g (the fraction solidified) gives a line whose slope and intercept are both measures of k_{eff} in accordance with Eq. 3.2. If the experiment is carried out under conditions of mixing and heat flow similar to those to be used in zone melting, then the resulting value of k_{eff} will be more appropriate than a value derived from the phase diagram. Moreover, if the experiment is carried out at two or more growth rates, one can derive k_0 by graphical means and, hence, establish the limiting (most favorable) distribution attainable.

In practice, the situation is not quite so straightforward. Several experimental problems must be recognized and overcome in order to derive reliable values of k_{eff} from directional crystallization or zone-melting experiments:

- The directional crystallization equation was derived under the assumption that k does not vary with C_0. This is rarely the case and much study has been devoted to experimental and theoretical methods of dealing with varying k.

- For solid solutions in which k approaches unity, it is difficult to get adequate precision in the determination of concentration. Relatively small uncertainty in the derived k can lead to large errors in estimating purification attainable by multipass zone refining.

- Radial segregation can cause enhancement or depletion of surface concentration, with the result that the net axial distribution does not represent the result of directional crystallization. One cause of radial inhomogeneity is faceted growth, which can give $k > 1$ in one orientation and $k < 1$ in another (see Section 1.4, Chapter 3).

- Anomalous values of k can result from directional crystallization if there is solute segregation at grain boundaries.

- Most directional crystallization experiments are carried out with ascending solid/liquid interface, while most zone-refining experiments are carried out with descending interface. If thermally driven convection is the main source of mixing in either or both experiments, quite different k_{eff}'s may prevail in the two cases. The result, then, is that k_{eff} derived from directional crystallization may not be useful in predicting the effectiveness of zone refining.

The manner in which these difficulties have been treated is shown in the following case histories of determination of k_{eff} from directional crystallization.

The method proposed by Pfann (13) was applied by Matsui (14) to anthracene containing tetracene. Figure 10.2 shows a plot of C against $(1 - g)$ for a $4.8 \times 10^{-5} M$ solution, from which k_{eff} may be estimated at about 0.1. Figure 10.3 shows values derived from similar experiments at several speeds. The values of k_{eff} derived from the intercept vary little while those derived from the slope increase markedly with decreasing speed. The latter effect is opposite to what is expected and may be an analytical artifact deriving from slow formation of an impurity in the melt during prolonged crystallization. The same methodology was applied by Sloan (15) to several impurities in anthracene; for

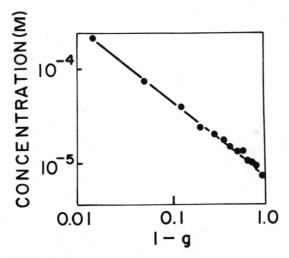

Fig. 10.2 Distribution of tetracene in anthracene, with $C_0 = 4.8 \times 10^{-5} M$.

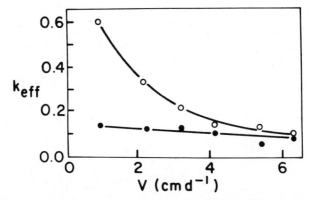

Fig. 10.3 Distribution coefficient of tetracene in anthracene, as a function of growth rate ○ from slope, ● from intercept.

example, tetracene was dissolved in anthracene at a concentration of 0.10 wt% in silicone-treated Pyrex tubes (see Section 2.8.1, Chapter 5). Solidification was carried out at various rates from 2.8×10^{-5} to 1.7×10^{-4} cm s^{-1} (0.1–0.6 cm h^{-1}) both with and without stirring. The grown ingot was removed from its Pyrex tube; after removal of the ingot surface, it was sectioned and analyzed. Tetracene was determined spectrophotometrically in dioxane solution, from the intensity of absorption at 473 nm. Figure 10.4 shows a plot of C/C_0 against $-\log(1 - g)$ for an unstirred melt (a); k derived from the intercept was 0.10, while k derived from the slope was 0.23. The discrepancy was attributed to variation of k with concentration. The stirred melt gave more nearly linear plots of C/C_0 against $-\log(1 - g)$ even at higher rates (b). Figure 10.5 shows a plot of

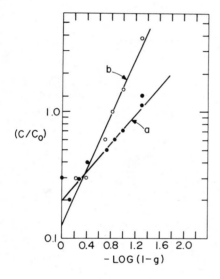

Fig. 10.4 Effect of stirring on distribution tetracene in anthracene: (a) unstirred; (b) stirred.

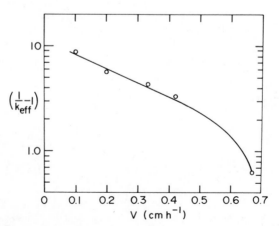

Fig. 10.5 Burton–Prim–Slichter plot of distribution of tetracene in anthracene.

$1/k_{\text{eff}} - 1$ against V for the unstirred melt. Extrapolation to $C = 0$ gives a value of about 0.08. This is in fair agreement with the value (0.01) determined by Matsui from zone-melting experiments at a lower concentration (0.01 wt%).

The effect of growth-temperature gradient upon impurity distribution in a solid-solution system was studied by Krivandina and Kostyleva (16). Crystals of bibenzyl, heavily doped (2%) with tolane, were prepared by a modified Bridgman–Stockbarger method in which a stationary crystal-growth tube was traversed by a furnace moving at $5.6 \times 10^{-6}\text{cm s}^{-1}$ (0.02 cm h^{-1}) with gradients of 2.5, 5.0, 10, and 30 K cm^{-1}. Cylindrical crystals, 1 cm in diameter and 9–11 cm long, were sectioned and weighed. Samples were dissolved from the surface of each section for spectrophotometric analysis. In crystals grown at the highest gradients (5–30 K cm^{-1}), impurity concentration increased monotonically along the length of the crystal, while the concentration profile of the crystal grown at 2.5 K cm^{-1} showed a roughly linear segment in the concentration profile. The first three curves agreed quite well with distributions calculated for convective mixing, with $k_{\text{eff}} = 0.44, 0.22$, and 0.28, while the fourth curve did not conform strictly to either the diffusive or convective case.

The distribution coefficient was found to decrease with increasing gradient between 2.5 and 10 K cm^{-1}; this was said to imply a transition from diffusive to convective transport of impurity with increasing gradient. It should be noted, however, that the experimental data used in this work refer to the surfaces of the grown crystals, and no attempt was made to study radial distribution. In the light of later studies (17) of just this point, it is known that the surfaces of crystals grown by the Bridgman–Stockbarger method are quite unrepresentative of the bulk composition; hence, the conclusions drawn from these experiments are open to question. The bibenzyl/tolane system has also been studied by multipass zone melting (see Section 4.2).

Feederle (18) analyzed another mechanism to explain the differing k's from slope and intercept of directional-crystallization plots. The treatment assumes that solute can be incorporated in a growing crystal by two mechanisms. In the central core region, solute is incorporated as solid solution in the host or by entrapment at defects; its concentration will be $k_{\text{eff}}C_l$. At the periphery of the crystal, interface liquid is drawn into a gap formed between the contracting crystal and the wall of the container; this liquid has concentration C_l. The two processes were treated in a material balance which led to the following directional crystallization equation

$$C_s = C_0 k_{\text{eff}}(1 - g)^{(k_{\text{eff}} + M - 1)} \tag{10.7}$$

where M is the factor relating interface solute concentration to bulk solute concentration. For $k_{\text{eff}} < 1$, M is greater than unity and vice versa.

Distribution coefficients close to unity can be determined by directional crystallization if a sufficiently precise analytical method is available for characterization of the grown ingot. Methods based on radiochemical analysis generally meet this criterion. To facilitate a study of distribution of eicosane ($C_{20}H_{42}$) in closely related linear hydrocarbons, the solute was prepared with a

Table 10.2. Distribution Coefficients of Eicosane in Homologous Alkanes (19)

Rate of Crystallization $cm^{-1} \times 10^4$	Host		
	Octadecane	Nonadecane	Docosane
27.8	1	1	1
5.97	0.91	0.95	0.90
4.17	0.86	0.90	0.85
3.83	0.82	0.86	0.82
1.44	0.72	0.80	0.74
k_0	0.59	0.70	0.66

tritium label by catalytic reduction of 1-eicosene. Dilute solutions (10^{-2}%) of the tritiated solute were prepared in octadecane, nonadecane, and docosane, as ingots of 1-cm diameter, 10 cm long. These were solidified in a shallow gradient (5 K cm^{-1}) at various rates, with reverse-rotation stirring. The sectioned ingots were analyzed and the data were fitted to Eq. 3.2. The resulting values of k_{eff} are shown in Table 10.2, along with extrapolated values of k_0 (19).

Scott et al. (20) attempted, with some success, to relate the distribution coefficient to the ease with which a dopant may be incorporated into a lattice. As an independent measure of mutual "fit" between host and dopant, they derived a quantity χ which was defined as the ratio of nonoverlapping area to overlapping area, when models of host and dopant were superimposed. Figure 10.6 shows a

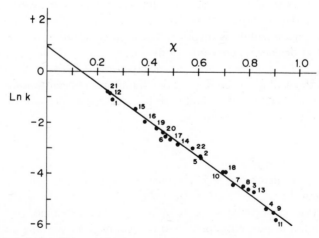

Fig. 10.6 Distribution coefficients of binary organic systems, as a function of host–guest geometric "fit" (see text).

plot of χ against k; the linear plot was related to an equation from which the bulk modulus of the host was calculated. The values of k were determined by analyzing the first-grown portions of crystals grown by the Bridgman–Stockbarger method (see Section 4.1, Chapter 4). The ratio of the measured concentration to that of the original melt gave k_{eff}; the individual values are included in Table 10.3.

Table 10.3. Relationship Between Molecular-Fit Parameter χ and Distribution Coefficient k of Dopants in Cyclododecane

Name of Compound	χ	k
Acenaphthene	0.26	0.34
Acenaphthylene	0.61	0.037
Acridine	0.80	0.0099
2-Aminoanthracene	0.87	0.0045
2-Aminofluorene	0.61	0.036
Aminopyrene	0.47	0.081
Anthracene	0.74	0.012
2-Bromonaphthalene	0.78	0.011
9-Bromophenanthrene	0.90	0.004
Carbazole	0.70	0.0198
Chrysene	0.92	0.0030
o-Dinitrobenzene	0.25	0.43
9,10-Dinitrophenanthrene	0.82	0.009
Fluorene	0.52	0.060
Indole	0.35	0.23
Naphthalene	0.39	0.138
Perylene	0.49	0.071
Phenanthrene	0.71	0.020
Picolinic acid	0.44	0.11
Pyrene	0.46	0.091
Resorcinol	0.24	0.46
Triphenylene	0.58	0.050

Table 10.4. Equilibrium Distribution Coefficients of Sodium Chloride in Water

NaCl, wt%	$k_0 \times 10^5$
2.5×10^{-3}	0.5
5.0×10^{-2}	4.0
0.12	6.0
0.38	10.0

Zharinov et al. (21) used vertical directional crystallization to measure k_{eff} for the segregation of sodium chloride from water. Solutions containing several concentrations of salt were solidified, with stirring, at several rates. From the rate dependence of k_{eff}, k_0 was determined for each concentration, according to the BPS equation. The values of k_0 are given in Table 10.4.

These values imply that there is a small, finite solubility of NaCl in ice. This probably reflects inclusion of solute at crystal defects rather than true solid solution. At each concentration, k_{eff} increased markedly with rate of crystallization (Figure 10.7); from the data of this figure it can be seen that for a given rate of crystallization, k_{eff} increases with concentration.

Kirgintsev and Isaenko (22) measured distribution by horizontal directional crystallization for ternary systems in which a host crystal was contaminated with a major and a minor impurity. For example, $NaNO_3$ was doped with $Sr(NO_3)_2$ and $CsNO_3$, as the two impurities and NaCl was doped with $SrCl_2$ as major impurity and CsCl as minor. The distribution coefficients of both major and minor impurities were found to be independent of one another under equilibrium conditions. Distribution coefficients have been measured for some anions in $NaNO_3$; for NaCl in the concentration range 1–4.7%, $k_0 = 0.05 \pm 0.01$ (23).

Gnilov and Nashelskii (24) have proposed a simplified method for determining k_{eff} on the basis of only two samples from a directionally crystallized ingot. At positions g_1 and g_2, the ratio of corresponding impurity concentrations C_1 and C_2 is given by

$$\frac{C_2}{C_1} = \frac{(1 - g_2)^{k-1}}{(1 - g_1)^{k-1}} \qquad (10.8)$$

where g and k have the usual meanings.

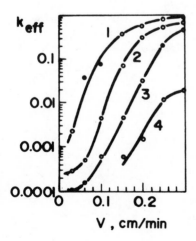

Fig. 10.7 Distribution of NaCl in ice, for various concentrations, decreasing in order 1 through 4.

If one of the sampled positions is the origin, then $g_1 = 0$ and the ratio yields

$$k = 1 + \frac{\log(C_2/C_1)}{\log(1 - g_2)} \qquad (10.9)$$

It is noteworthy that melt concentration does not appear in Equation 10.9, so that k may be established merely by determining the relative concentrations at two points in the solid. Hence, any analytical technique is usable, even if no calibration is available, so long as the response is linear with concentration. A nomogram was prepared for approximate determination of k from one value of g (abscissa) and one value of relative concentration (ordinate) in the ranges $0.1 < k < 0.9$ and $k > 1$ (Figure 10.8).

Metallurgical applications of this method abound. Magnesium alloys containing copper and other impurities were studied by Yue and Clark (25). Figure 10.9 shows a plot for magnesium/0.1 wt% copper, solidified at several rates.

Sen and Wilcox (26) carried out directional crystallizations in which k_{eff} varied with concentration. The grown crystals were sectioned longitudinally, polished, etched, and then analyzed by the electron-microprobe method at 0.1-cm intervals. This analytical method made it possible to study both radial and axial distribution, since the sampled area is far smaller than the radius of the crystal. In some cases, plots of relative concentration against fraction unsolidified were linear, but the values of k derived from the intercept were quite different from those derived from the slope; in fact the latter were sometimes negative. Both second-order and sixth-order polynomials were used to represent the composition data. While the latter gave better fits to the concentration profiles, the resulting polynomials gave rise to oscillations in the numerical

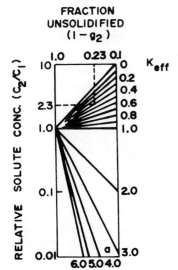

Fig. 10.8 Nomogram allowing approximate determination of k from one value of g and one value of relative concentration.

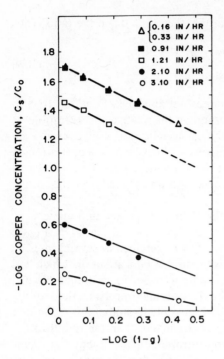

evaluation of k. Hence second-order (quadratic) approximations were used, with Eqs. 3.7 and 3.8. These equations give a family of k-against-g curves with $k_{initial}$ as parameter. The decision as to which curve is "correct" is made on the basis of which accords best with the physical requirement that k must tend to unity as g approaches unity. In Figure 10.10 it may be seen that curve 2 best fits this requirement.

Lyubalin (27) carried out a careful study of the distribution of antimony between a germanium melt and the individual faces of a single crystal during growth by the Czochralski method. Polycrystalline Ge of resistivity >40 ohm-cm was melted under argon in quantities of 600 g with a very small amount of added Sb. Growth was carried out along predetermined directions at 2.5×10^{-3} cm s^{-1} (9 cm h^{-1}) to yield crystals 2.0 cm in diameter × 10 cm long. The crystal was rotated at 0.5 rev s^{-1} and the crucible at 1 rev min^{-1}. The crystal was sectioned and its Sb content was determined by measuring electrical resistivity, making use of published data on the relationship between charge-carrier mobility and electrical resistivity. Because k is very small (5×10^{-3} to 10^{-2}), it was assumed that the amount of dopant in the melt did not change during growth of a crystal. Six crystals were grown in each of the orientations studied, and k_{eff} was found to vary by about 5% in each orientation; hence, experimental error is much smaller than the variation observed among the several growth directions.

A rough correlation was noted between $k_{eff}(hkl)$ and the supercooling

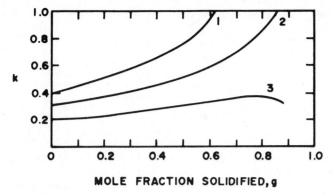

Fig. 10.10 Second-order quadratic approximations of Eqs. 3.7 and 3.8, showing variation of *k* with *g* and k_{initial} as parameter.

$\Delta T(hkl)$. That the correlation was not complete may have resulted from errors in the direct measurement of supercooling. More probably, the discrepancy results from the fact that the experimentally measured supercooling is an overall value, while the rate of impurity incorporation depends on the component of supercooling along the microscopic growth direction, which may not be perpendicular to the macroscopic solid/liquid interface.

Kralina and Sazonova (28) looked into the effect of growth direction on impurity distribution in Czochralski-grown nickel crystals. It was found that distribution of carbon close to the crystallization surface led to cellular-dendritic morphology and that its k_{eff} varied by a factor of about two, between (100) and (110). A related study on orientation dependence of k_{eff} in the bibenzyl/tolane system was carried out by Krivandina, see Section 5 (29).

Kosyakov and Kirgintsev (30) have evaluated the error in determination of *k* occasioned by curvature of the crystallization front; they concluded that the error is <0.5% in commonly used methods of directional crystallization.

The distribution coefficients of several ions have been measured by directional crystallization of sodium nitrate and potassium chloride (see Table 10.5) (31). More recently, cesium iodide crystals have been grown with a variety of anion dopants, for determination of the distribution coefficients of the latter. From the absorption coefficients of the dopants and Beer's law, it was possible to derive *k* from the axial variation of the absorption, for BO_2^-, CNO^-, CNS^-, CO_3^{2-}, IO_3^-, NO_2^-, NO_3^-, OH^-, and SO_4^{2-} (32).

Hurle et al. have considered the effects of temperature fluctuations during directional crystallization upon the effective distribution coefficient (33). Their analysis evaluates k_{eff} in terms of local concentration in the interface solid, averaged over one period of the fluctuation. In turn, the fractional variation of k_{eff} during a cycle is expressed in terms of the frequency of the fluctuation, ω, and the amplitudes of interface displacement and melt concentration, ϕ_ω, and C_ω, respectively. A marginal state is defined, such that the overall growth rate $V = \omega|\phi_\omega|$, and the fluctuation just fails to cause meltback. In this regime, a

Table 10.5. Distribution Coefficients of Trace Ions in $NaNO_3$ and KCl (30)

Host	Trace Element	Ionic Radius	Distribution Coefficient
$NaNO_3$		0.98	
	Ag^a	1.13	7.5×10^{-1}
	K	1.33	2.5×10^{-1}
	Rb	1.49	2.5×10^{-2}
	Cs	1.65	4.8×10^{-4}
KCl		1.33	
	Rb	1.49	7.9×10^{-1}
	Cs	1.65	3.4×10^{-1}

^{a}This value was determined by single-pass zone melting; see Section 4.1 Chapter 10.

fluctuation of any amplitude or frequency, in a system with $k_0 \gg 1$, will result in a measured value of k_{eff} equal to only half the correct value. Likewise, extrapolation of such experimental results according to the BPS theory will give a value of k_0 equal to half the true value. If the true value of k_{eff} is less than 2, the value measured by directional crystallization in the presence of temperature fluctuation may be less than unity, corresponding to a reversal of the sense of solute transport. More generally, for any value of k, temperature fluctuations in the melt reduce the value of $|1 - k_{eff}|$, that is, they reduce the effective solute segregation. There is experimental evidence that temperature fluctuation leads to growth striations and that melting back commonly occurs during the crystal rotation cycle in Czochralski growth.

4   DETERMINING k FROM ZONE MELTING

As we have seen, it is possible to measure k_{eff} through directional crystallization experiments. This method is inaccurate for k_{eff}'s that approach unity. Moreover, the k_{eff}'s determined in this way are sometimes not useful for calculating solute distributions after zone refining, because the physicochemical processes governing the two methods may differ markedly. Consequently, a family of methods has emerged for determining k_{eff} directly from zone melting. These are described here approximately in order of their frequency of use:

1. From single-pass distributions, using a homogeneous ingot or the starting-charge-only method (see Section 1.1.2, Chapter 5)
2. By comparison of multipass distributions with calculated curves (see Section 1.2, Chapter 5) and Appendix I
3. From the ultimate distribution (see Section 1.2, Chapter 5)

4. From distributions after n and $n + 1$ passes

5. From zone leveling

4.1 Determining *k* from Single-Pass Distributions

4.1.1 Homogeneous Ingot

Assuming a homogeneously doped ingot of uniform cross section traversed by a molten zone of constant length, solute distribution is described by Equation 5.5. This expression may be rewritten as follows:

$$X = f(k_{\text{eff}}) = \frac{1}{1 - k_{\text{eff}}}\left(1 - \frac{C_s}{C_0}\right) \tag{10.10}$$

and

$$Y = \phi(k_{\text{eff}}) = \exp\left(\frac{x}{l}k_{\text{eff}}\right) \tag{10.11}$$

To evaluate k_{eff} it is only necessary to measure relative concentration (C_s/C_0) at a single position (x/l). Function Y is then plotted against k_{eff} for the selected value of (x/l), as in Figure 10.11 (34). Function X is represented by a hyperbola,

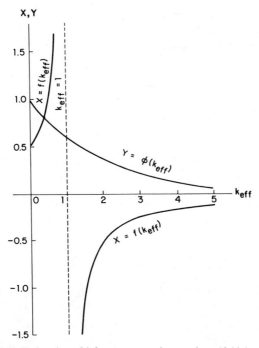

Fig. 10.11 Estimation of k from parametric equations 10.11 (see text).

one branch of which refers to $k_{eff} > 1$ and another to $k_{eff} < 1$. The intersection of curves X and Y gives k_{eff}. The hyperbola is centered at 1, 0 and asymptotically approaches the abscissa axis and the line $k = 1$. In the example shown in Figure 10.11, $k_{eff} \simeq 0.33$. Of course, k_{eff} may be estimated by comparing a measured solute profile with profiles calculated from Eq. 5.5.

Distributions of monovalent and divalent cations in alkali halides have been studied by single-pass zoning of uniformly doped ingots (35). Divalent ions were added at about 10^2 ppm and monovalent ions at about 10^3 ppm. Zone length was about 1/10 ingot length and the zoning rate was 10^{-4} cm s^{-1} (0.37 cm h^{-1}). The results are given in Table 10.6. Similar experiments with monovalent impurities are summarized in Table 10.7, along with values calculated from thermodynamic data, through Eq. 10.5.

The systems naphthalene/catechol and naphthalene/1-chloro-4-nitrobenzene were subjected to single-pass zoning over large concentration ranges, and it was

Table 10.6. Distribution Coefficients of Divalent Impurities in Alkali Halides (35)

	Solvent			
Solute	NaCl	KCl	NaBr	KBr
Ba	—	0.20	—	—
Ca	0.95	0.40	0.95	0.35
Cd	0.10	2×10^{-3}	—	—
Mn	0.30	0.04	0.10	10^{-2}
Ni	0.13	4×10^{-3}	2×10^{-3}	—
Pb	0.25	0.20	—	—
Sr	0.40	0.27	—	—
Zn	$<10^{-3}$	—	—	—

Table 10.7. Experimental and Theoretical Distribution Coefficients for Monovalent Impurities in Alkali Halides (35)

	NaCl		KCl		NaBr		KBr	
Solute	k_{exp}	k_{theor}	k_{exp}	k_{theor}	k_{exp}	k_{theor}	k_{exp}	k_{theor}
Ag	0.25	0.40	—	—	0.09	0.43	0.09	—
Cs	0.25	—	—	—	—	—	—	—
K	0.008	0.05	—	—	0.03	0.069	—	—
Li	0.20	0.15	—	—	0.07	0.26	0.004	0.0028
Na	—	—	0.11	0.15	—	—	0.20	0.19
Rb	0.002	0.0017	0.6	0.55	0.001	0.0034	0.4	0.48

found that the k's were 0.48 and 0.42, respectively. The concentration profiles were described by the empirical equation

$$C = C_0 a \exp \frac{bx}{l} \tag{10.12}$$

where constants a and b were 0.48 and 0.045 for catechol and 0.42 and 0.034 for 1-chloro-4-nitrobenzene (36).

4.1.2 Starting-Charge-Only Method

If solute is added only to the first zone length of an ingot, the distribution achieved by passing a molten zone through the entire ingot is described by

$$C_s = k_{\text{eff}} C_0^{-k_{\text{eff}} x/l} \tag{10.13}$$

where

l = the zone length
x = the distance along the ingot, measured from the start
C_s = the solute concentration in the ingot at position x
C_0 = the concentration in the starting zone $(0 < x < l)$

k_{eff} is determined by plotting $\ln C$ against x/l (position in the ingot). In measuring solute concentration along the ingot, it is necessary to recall that surface and bulk concentrations may differ substantially.

Süe et al. (37) described early applications of this method, and showed that it is possible to determine k_{eff} for more than one solute after a single one-pass experiment. A mixture of 30 mg of ^{184}CsNO$_3$ and 30 mg of ^{86}RbNO$_3$ was deposited atop an ingot comprising 30 g of KNO$_3$ and a single molten zone was passed through the ingot. Because the energies of the β particles emitted by the two isotopes are different, it was possible to determine their concentrations independently. A plot of these concentrations as a function of x gave straight lines whose slopes and intercepts were measures of k_{eff}. The ratio $k_{\text{RbNO3}}/k_{\text{CsNO3}}$ was found to be 2.5.

Sorensen (38) also used this method and plotted integral impurity $q(x)$ (i.e., total impurity contained in the ingot, from point x to the bottom) against position in the ingot, to derive k according to

$$q(0) - q(x) = \int_0^x C(x)\, dx = -q(0)e^{-kx/l} + q(0) \tag{10.14}$$

Effective distribution coefficients were obtained for 4-nitrophenol in phenol by casting an ingot of phenol in a tube (1.2-cm id × 60 cm), overlaying it with 1 mL of phenol containing a known amount of 4-nitrophenol, and then passing one molten zone downward through the ingot. The resulting ingots were sectioned and the segments were analyzed spectrophotometrically in dilute aqueous alkali at 405 nm. Figure 10.12 shows plots of concentration against position (in terms of zone lengths) for zoning at three speeds, using partially purified phenol of mp 313.4 K. The plots are linear and show the expected increase of k_{eff} toward unity with increasing speed. The standard deviation of the measured values was said to be 10–20%. Sorensen reported related experiments (38) in which the phenol host was prepurified by multipass zone refining and the resulting purified ingot, of mp 313.9 K was doped with 4-nitrophenol as described above. The measured k_{eff}'s showed that segregation was more effective than from the less-pure phenol.

Nojima (39) used the single-pass, starting-charge-only method to measure k_{eff} for the system stilbene/bibenzyl and found values ranging from about 1.3 to 2.5, depending on starting concentration and zoning speed. From the measured data, a value of 5.08 was derived for k_0, the equilibrium distribution coefficient. The corresponding values obtained from the phase diagram and from thermo-dynamic theory were 5.23 and 5.08, respectively. It should be noted that k_{eff} was derived from both slope and intercept of the experimental plots with excellent agreement. Van Mau et al. (40, 41) have used the single-pass zoning method to study the distribution of both selenium and tellurium in gallium antimonide. Seeds of GaSb were cut along various crystallographic directions from mono-crystalline ingots of 2–3 cm diameter, which had been grown by the Czochralski method. The seeds were used to grow variously oriented single crystals of 0.7-cm diameter. These were doped by passage of a single zone which contained the selected contaminant (either Se or Te), through the crystal. Thin disks were cut from the doped crystals and the dopant concentration was obtained from measurement of electrical resistivity under the assumption that the number of charge carriers is proportional to the dopant concentration. This electrical

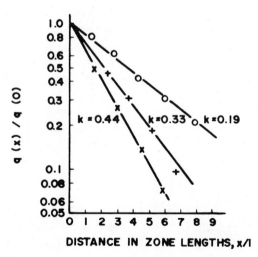

Fig. 10.12 Estimation of k_{eff} by the integral-impurity method (see text).

method was found to be applicable only to dopant concentrations that gave more than 10^{18} carriers cm^{-3}. At 1.67×10^{-3} cm s^{-1}, k_{eff} for Te was found to be 0.40, independently of orientation. From a BPS plot ($[1/k_{eff} - 1]$ against V) k_0 was found to be 0.30.

In the case of Se, both n-type and p-type host were studied and a substantial difference in k_{eff} was noted (0.49 and 0.39, respectively, at 1.67×10^{-3} cm s^{-1}); k_0 was found to be 0.013 independently of orientation.

Li derived a relationship describing the single-pass profile when k is not constant. Application of this method to the system Ge/Sb showed that k differed by only 0.7% from the value calculated by use of a constant $- k$ equation (42). Nonconstant k can, of course, be a result of equilibrium phase relationships. In addition, however, any experimental aberrations can perturb the zone-refining process. Wilcox (43) has demonstrated the appearance of bubbles, voids, cracks, and insoluble particles by microscopic examination of a solidifying interface. Irregular growth rates resulting from nonuniform heat flow can drastically reduce impurity segregation.

4.2   Determining k from Multipass Distributions

Fairly accurate estimates of k_{eff} can be made by sectioning and analyzing ingots after small numbers of zone passes, and then comparing the observed concentration profiles with those calculated from Equations 5.17 and 5.18 (and plotted in Appendix I). This method indicates whether k_{eff} changes with concentration, since each ingot, starting at a given concentration, spans a differing overall concentration range. Analysis of segments from ingots with differing starting concentrations, but taken from the same position in each ingot after equal number of passes, gives a number which characterizes the change of k_{eff} with concentration. The following relationship has been proposed (5)

$$A = \frac{(C_n/C_0) - (C_n'/C_0')}{C_n'/C_0'} \qquad (10.15)$$

where C_0 and C_0' are the initial concentrations of the two ingots and C_n and C_n' are the concentrations after n passes at a given position (x cm from the start) in both ingots. If k_{eff} does not vary with concentration, $A = 0$. Typical values for some organic binary systems are given in Table 10.8. The concentration dependence of k_0 was found to conform to the equation $k_0 = a \pm 10^{-2}bC_s$; values of a and b are given in Table 10.8; see also Section 1, Chapter 5. Multipass zone melting has been widely used to determine k_{eff}'s for metallic systems and a number of values are cited in Section 1, Chapter 8.

Iwano and Yokuta (44) studied the distribution of 2-naphthol (3%) in naphthalene as a function of zoning rate and rotation rate. At zoning rates between 1.1×10^{-3} cm s^{-1} (4 cm h^{-1}) and 2.8×10^{-3} cm s^{-1} (10 cm h^{-1}), k_{eff} converged to values close to 1.45 for rotation speeds ranging from 0 to 3.3 rev s^{-1}. The extrapolated BPS plot (see Section 1.2.3, Chapter 3) of the data gave

Table 10.8. Variation of A During Zone Refining for Various Systems (C, mass %) (5)

x, cm	n	A, %	x, cm	n	A, %
Naphthalene-Benzothiophene			Phenanthrene-Carbazole		

x, cm	n	A, %	x, cm	n	A, %
Naphthalene-Benzothiophene $(C_0 = 0.51; C_0' = 1.34)$ $a = 0.1, 10^2 b = 3.38$			Phenanthrene-Carbazole $(C_0 = 1.12; C_0' = 1.78)$ $a = 4.20, 10^2 b = -24.9$		
4.0	6	23.3	2.0	12	3.5
7.5		15.3	10.0		10.9
14.0		11.5	16.0		21.4
4.0	12	41.4	. 2.0	30	5.2
7.5		26.3	10.0		17.1
14.0		15.1	16.0		43.2
Phenanthrene-Fluorene $(C_0 = 1.96; C_0' = 3.81)$ $a = 0.66, 10^2 b = 2.24$			Anthracene-Carbazole $(C_0 = 0.85; C_0' = 1.35)$ $a = 0.58, 10^2 b = 2.61$		
2.0	12	13.7	2.0	10	18.8
6.0		11.9	5.0		16.2
16.0		8.8	10.0		2.7
2.0	30	23.3	2.0	18	39.8
6.0		18.7	5.0		26.6
16.0		9.8	10.0		14.8

$k_0 = 1.57$. This value differs markedly from that obtained from the phase diagram, namely $k = 5.5$. The discrepancy results from the fact that under the experimental conditions, most of the solid forming at the interface arose from entrapped melt and consequently had the same composition as the melt. The same technique was applied to the system phenol/3% m-cresol, with a resultant value of $k_0 = 0.40$.

The system bibenzyl/diphenylacetylene(tolane) was studied by multipass zoning and the phase diagram was found to have a rather flat minimum-melting region centered around 26.4% diphenylacetylene. Hence, the distribution coefficient for diphenylacetylene is expected to be slightly greater than and slightly less than unity in hyper- and hypoeutectic regions, respectively. To cope with this difficult situation, hyper- and hypoeutectic ingots were zoned to ultimate distribution. From a determination of the crossover point (i.e., the position at which the concentration equals that of the initial mixture), it was possible to obtain the constant B of Eq. 5.8 and from this, k_{eff}. The values thus established ranged from 0.61 to 0.99 for the hypoeutectic mixture and from 1.10 to 1.01 for the hypereutectic mixture (45).

The distribution of tantalum in tungsten has been studied by electron-beam zone melting. Activation analysis was used to measure the concentration profile of ^{182}Ta after solidification at 5×10^{-3} cm s^{-1}. The value of k_{eff}, 0.830, agreed well with the value measured from the phase diagram, 0.809 (46).

Tracer studies have been carried out on ^{203}Hg and ^{35}Se in tellurium (47). Zoning at 5.6×10^{-4}, 3.1×10^{-4}, and 4.4×10^{-4} cm s^{-1} (2.0, 1.1, and 1.6 cm h^{-1}) gave values of $k_0 = 0.11$ for the former and 0.34 for the latter.

The distribution of dopants in semiconductors has been analyzed theoretically with regard to behavior in intrinsic and extrinsic regions. In the transition from intrinsic to extrinsic conditions, k_{eff} becomes concentration dependent and a simple linear relationship between donor-ion concentration and position in the ingot was predicted (48). This theoretical treatment was applied to earlier experiments on tellurium-doped gallium antimonide and the expected linear relationship was found in the extrinsic region.

As a very rough approximation, solute concentration at the head of an ingot after n zone passes may be taken to be

$$C = k_{\text{eff}}^n C_0 \qquad (10.16)$$

4.3   Determining k from the Ultimate Distribution

A distribution coefficient can be obtained from the solute profile that exists at ultimate distribution. For this, it is necessary to evaluate the constants A and B of Equations 5.7 and 5.9. Constant A is merely $C_\infty[0]$; that is, the concentration at the head of the ingot at ultimate distribution. B is the slope of a plot of

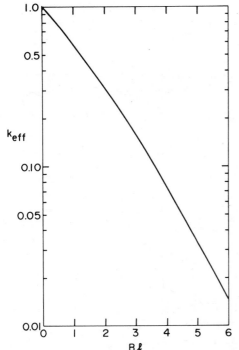

Fig. 10.13 Evaluation of approximate value of k from ultimate distribution.

Equation 5.8, in coordinates $\ln C_\infty$ and x.

An approximate value of k_{eff} can be derived from a plot of Equation 5.8, that is, $\log k_{\text{eff}}$ against Bl, as shown in Figure 10.13. Segments of the curve can be approximated by straight lines and for $0.01 < k_{\text{eff}} < 0.2$,

$$\log k_{\text{eff}} = -0.3Bl \tag{10.17}$$

while for $k_{\text{eff}} > 0.2$,

$$\log k_{\text{eff}} = -0.25Bl \tag{10.18}$$

4.4 Determining k from Distributions After n and n + 1 Passes

It has been demonstrated that the concentrations at some point in an ingot, after n and $(n + 1)$ passes, can be approximated by

$$C_n(x + l) - C_{n+1}(x) \simeq [C_{n+1}(x + l) - C_{n+1}(x)] + [C_n(x) - C_{n-1}(x)] \tag{10.19}$$

where l is the zone length. The distribution coefficient can be related to these measurable quantities, as follows:

$$k_{\text{eff}} = l \frac{\partial C_{n+1}(x)}{\partial x} \frac{1}{C_n(x + l) - C_{n+1}(l)} \tag{10.20}$$

The zone length l is known and the derivative

$$\frac{\partial C_{n+1}(x)}{\partial x}$$

may be approximated by the slope of the tangent to the solute profile after $(n + 1)$ passes, at point x (Figure 10.14) (49). The denominator may be measured

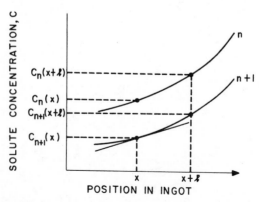

Fig. 10.14 Determination of k from solute distribution after n passes and $(n + 1)$ passes.

from the figure. Assuming that the two concentration profiles are congruent, the denominator may be approximated by

$$C_n(x + l) - C_{n+1}(x) \simeq [C_{n+1}(x + l) - C_{n+1}(x)] - [C_{n+1}(x) - C_n(x)] \quad (10.21)$$

The two quantities on the right can be evaluated graphically.

4.5   Determining k from Zone Leveling

Equation 5.24 can be solved for k_{eff}, giving

$$k_{eff} = \frac{C_f}{(L/l)(C_0 - C_f) + C_f} \quad (10.22)$$

where L and l are ingot and zone lengths and C_0 and C_f are initial and final concentrations in the ingot. The ratio C_f/C_0 approaches unity as k approaches unity; hence this method is less useful for k_{eff} close to unity.

5   DETERMINING k FROM KYROPOULOS AND CZOCHRALSKI CRYSTAL GROWTH

Krivandina (29) has investigated the distribution coefficients of the bibenzyl/diphenylacetylene system as a function of crystallographic orientation and of supercooling. To do this, a small seed of bibenzyl was introduced into the binary melt at small, known supercoolings in the range 0.1–0.5 K. Crystals were grown to 0.2–0.4 g and individual crystal faces were analyzed, with the results shown in Table 10.9. The measured data show that there is little variation of k with supercooling and diphenylacetylene concentration up to 12%, but that a small, real orientation dependence exists. The average value of k_{eff} for all

Table 10.9. Distribution Coefficients of Diphenylacetylene in Bibenzyl (50)

Diphenylacetylene content of melt, wt%	Supercooling, K	Effective Distribution Coefficient for Faces[a]		
		(201)	(001)	(111)
2.0	0.1	0.45	0.42	0.39
	0.2	0.43	0.39	0.37
3.52	0.3	0.48	0.47	0.42
	0.5	0.45	0.43	0.38
Average k_{eff}		0.45	0.43	0.39

[a]The uncertainty was estimated to be ± 0.01.

orientations measured in this way was about 0.42, which is close to the value obtained from the phase diagram, viz. 0.43 ± 0.02. Thus the measured values of k_{eff} are close to k_0, the equilibrium value. The anisotropy of growth rate was larger; for the (111), (201), and (001) faces the rates were related as 3.3:1.9:1.

Casalboni et al. have determined the distribution coefficient of thallium in NaBr crystals grown by the Kyropoulos method. They were able to determine the value of k_{eff} as 0.0164 and additionally reported that the impurity concentration in the crystal decreased with time because evaporative loss was larger than the increase of impurity concentration resulting from impurity distribution (51).

REFERENCES

1.   J. A. Burton, R. C. Prim, and W. P. Slichter, *J. Chem. Phys.*, **21**, 1987–1996 (1953).

2.   C. D. Thurmond and J. D. Struthers, *J. Phys. Chem.*, **57**, 831–835 (1953).

3.   T. H. Gouw and R. E. Jentoft, *Anal. Chim. Acta*, **39**, 383–391 (1967).

4.   C. D. Thurmond, "Control of Composition in Semiconductors by Freezing Methods," in N. B. Hannay, ed., *Semiconductors*, Reinhold, New York, 1959, pp. 145–191.

5.   A. L. Sobolevskii and V. V. Rode, *Russ. J. Phys. Chem.*, **54**(2), 262–263 (1980).

6.   J. Hildebrand and R. L. Scott, *The Solubility of Non-electrolytes*, Reinhold, New York, 1950.

7.   K. Weiser, *J. Phys. Chem. Solids*, **7**, 118–125 (1958).

8.   T. B. Douglas, *J. Chem. Phys.*, **45**, 4571–4585 (1966).

9.   R. Gallagher, H. D. Nuessler, and P. J. Spencer, *Physica B + C*, **103** (1), 8–20 (1981).

10.   M. Hillert, *Physica B + C*, **103**(1), 31–40 (1981).

11.   A. S. Andrew and J. Linde, *Computers & Geosciences*, **6**, 227–236 (1980).

12.   M. M. Kimberley, *Computers & Geosciences*, **6**, 237–266 (1980).

13.   W. G. Pfann, *Zone Melting*, 2nd ed., Wiley, New York, 1966, p. 9.

14.   A. Matsui, *J. Phys. Soc. Japan*, **19**, 578 (1964).

15.   G. J. Sloan, *Mol. Cryst.*, **1**, 161–194 (1966).

16.   E. A. Krivandina and E. E. Kostyleva, *Sov. Phys. Crystallogr.*, **18**(4), 560–561 (1974).

17.   A. R. McGhie and G. J. Sloan, *J. Crystal Growth*, **32**, 60–67 (1976).

18.   H. Feederle, Diplomarbeit, University of Stuttgart, 1964, cited in N. Karl, "High Purity Organic Molecular Crystals," in H. C. Freyhardt, Ed., *Crystals, Growth, Properties, Applications*, Vol. 4, Springer Verlag, New York, 1980, p. 34.

19.   R. S. Narang and J. N. Sherwood, *J. Crystal Growth*, **49**, 357–362 (1980).

20.   R. A. M. Scott, S. R. Laridjani, and D. J. Morantz, *J. Crystal Growth*, **22**, 53–57 (1974).

21.   V. I. Zharinov, E. E. Konovalov, Sh. I. Paezulaev, and T. V. Kayurova, *Russ. J. Phys. Chem.*, **48**(4), 572–574 (1974).

22.   A. N. Kirgintsev and L. I. Isaenko, *Izv. Sib. Otd. Akad. Nauk SSSR, Ser. Khim. Nauk*, **1977** 148–149.

23.   V. A. Isaenko, N. D. Mironova, and A. N. Kirgintsev, *Izv. Sib. Otd. Akad. Nauk SSSR Ser. Khim. Nauk*, **1979** 74–76.

24.   S. V. Gnilov and A. Ya. Nashelskii, *Industrial Laboratory*, **44**, 241–243 (1978).

25.   A. S. Yue and J. B. Clark, *Trans. AIME*, **212**, 881–884 (1958).

26.   S. Sen and W. R. Wilcox, *Mat. Res. Bull.*, **13**, 293–302 (1978).

27.   M. D. Lyubalin, *Inorganic Materials (USSR)*, **13**, 938–941 (1977).

28. A. A. Kralina and V. A. Sazonova, *Sov. Phys. Crystallogr.* **22**, 824–830 (1977).

29. E. A. Krivandina, *Sov. Phys. Crystallogr.*, **23(2)**, 205–208 (1978).

30. V. I. Kosyakov and A. N. Kirgintsev, *Izv. Sib. Otd. Akad. Nauk SSSR Ser. Khim. Nauk*, **1971** 107–110.

31. H. Nagasawa, *Science*, **152**, 767–769 (1966).

32. W. Bollman, *Cryst. Res. Technol.*, **16(9)**, 1051–1054 (1981).

33. D. T. J. Hurle and E. Jakeman, *J. Crystal Growth*, **5**, 227–232 (1969).

34. V. N. Vigdorovich, *Purification of Metals and Semiconductors by Crystallization*, Freund Publishing House, Ltd., Tel Aviv, 1971.

35. M. Ikeya, N. Itoh and T. Suita, *Jap. J. Appl. Phys.*, **7**, 837–845 (1968).

36. R. P. Rastogi, N. B. Singh, and N. B. Singh, *Indian. J. Chem. Sect. A*, **15A(9)**, 819–820 (1977).

37. P. Süe, J. Pauly, and A. Nouaille, *Bull. Soc. Chim. France*, **5**, 593–602 (1958).

38. P. Sorensen, *Chem. Ind.*, 1593–1595 (1959).

39. H. Nojima, *Bull. Chem. Soc. Japan*, **51(9)**, 2513–2517 (1978).

40. A. Van Mau, M. Averous, and G. Bougnot, *Mat. Res. Bull.*, **7**, 857–864 (1972).

41. A. C. G. Van Genderen, C. G. De Kruif, and H. A. J. Oonk, *Z. Phys. Chem. Neue Folge*, **107**, 167–173 (1977).

42. C. H. Li, *J. Appl. Phys.*, **38**, 3793–3795 (1967).

43. W. R. Wilcox, *Sep. Science*, **4(2)**, 95–109 (1969).

44. T. Iwano and T. Yokuta, *J. Chem. Soc. Japan* (*Chem. Ind. Chem.*), **1**, 21–26 (1974).

45. H. Nojima and S. Akehi, *Bull. Chem. Soc. Japan*, **53**, 2067–2073 (1980).

46. K. A. Jackson and C. E. Miller, "Periodic Regrowth During Crystal Growth," in R. L. Parker, A. A. Chernov, G. W. Cullen, and J. B. Mullin, Eds., *Crystal Growth 1977*, Proceedings of the 5th International Conference on Crystal Growth, Cambridge, MA, July 1977, North Holland, Amsterdam, 1977.

47. Sh. Movlanov and A. A. Kuliev, *Izv. Akad. Nauk Azerb. SSSR, Ser. Fiz. Mat. i Tekhn. Nauk*, **3**, 55–62 (1961).

48. J. Bloem and L. J. Giling, *Mat. Res. Bull.*, **9**, 265–272 (1974).

49. E. D. Tolmie and D. A. Robbins, *J. Inst. Metals*, **85**, 171–176 (1957).

50. E. A. Krivandina, *Kristallografiya*, **23(2)**, 372–376 (1978).

51. M. Casalboni, U. M. Grassano, A. Scacco, and A. Tanga, *J. Crystal Growth*, **41**, 175–176 (1977).

MEASUREMENT OF PURITY

Zone melting and related crystallization processes are usually carried out for the purpose of purification. It is natural then that criteria of purity are sought, to measure the degree of success achieved in the purification step. In many cases the purification is undertaken for the removal of specific, known contaminants. In such cases analytical assays for the contaminants may well be known beforehand. If the pure material is desired for some physical measurement which is impurity sensitive, that measurement may itself be the best criterion of purity.

More often, little or nothing is known about the amount or identity of impurity present. While it would be pleasant to have at hand a universal measure for absolute purity of materials, no such panacea is known. The experimenter is thus constrained to use chemical intuition in deciding whether to measure purity or instead to measure the levels of individual impurities and obtain purity by difference. We will review briefly some of the many methods available for detection and quantitation of impurities, and then discuss thermal methods for measuring total impurity.

1 ELEMENTAL ANALYSIS

The precision of elemental analysis is not normally high enough to allow detection of molecules containing the same atoms as the host. For example, 1% of palmitic acid in stearic acid reduces the carbon content by only 0.01%, far less than a detectable difference. However, molecules containing foreign atoms can be detected readily and sensitively. The detection of foreign atoms can be a valuable clue in identifying molecular impurities (see Section 1.4.3, Chapter 8).

Emission spectra from arcs, sparks, and flames provide a rapid screen for many elements, using small samples. Many elements can be determined quantitatively at low levels. Atomic absorption, physically the reverse of emission spectroscopy, is suited to the detection of many metals in a wide variety of matrices. Chemical interferences (from anions) and matrix effects are not usually troublesome, and the method offers part-per-million sensitivity or better. Unfortunately, atomic absorption is not applicable to nonmetallic elements (N, O, halogens), but sensitive coulometric methods are available for N, S, and P. Extensive literature is available on trace analysis (1, 2).

2 FUNCTIONAL GROUP ANALYSIS

Scores of ingenious methods are available for chemical detection of functional groups (3). Some of these can be carried out on very small samples with the simplest of apparatus (4). Infrared spectrophotometry is applicable to identification of contaminants containing characteristic functional groups. In solid samples, however, short path lengths are usually mandated by strong bands of the major component, which often interfere. Hence, impurity detection by this method has been restricted to abundances greater than 1%. The situation is somewhat better for liquids and gases. Improved sensitivity for all categories of materials is provided by Fourier-transform IR (FTIR) techniques (5). Especially when used with a computer, these techniques make it possible to measure "difference" spectra of much smaller amounts of impurity.

3 GAS CHROMATOGRAPHY AND GAS CHROMATOGRAPHY WITH SPECIFIC DETECTION

"Gas chromatography" denotes an extended family of separation techniques in which a gas transports the components of a mixture through a separating medium. The latter may be an adsorptive solid (gas–solid chromatography, GSC) or a nonvolatile liquid (gas–liquid chromatography, GLC). The nonvolatile liquid may be spread on the surface of an inert, particulate support (packed-column GLC) or on the inner wall of a long, slender tube (capillary GLC).

An enormous array of separating phases is available, but they fall into a small number of families and nearly all separation problems can be handled with relatively few stationary phases (see Section 2.6.5, Chapter 3). Other things being equal, each compound emerges from the column after a characteristic retention time. The separating media depend on different properties for their effectiveness: for example, some separate on the basis of boiling point and some on the basis of polarity. Consequently, the retention time of an impurity, relative to that of the major component, can provide important insights into the properties and structure of the impurity. Gas chromatography has been applied to materials which are gases, liquids, and solids at room temperature. The sample must be stable at inlet and column temperatures. Retention-time data do not provide unambiguous identification of an impurity. Nevertheless, if an unknown has the same retention time as a known substance on two columns with differing mechanisms of separation, it may be concluded that the unknown is chemically identical to the known.

Early gas chromatographs used thermal conductivity detectors, which respond to the effect of an eluting peak on the thermal conductivity of the carrier gas. These detectors offer a large dynamic range (up to 10^6) and they are reasonably nonselective in their response to differing substances. Several other detectors are now available, offering higher sensitivity and/or selective response to certain elements; these are listed in Table 11.1.

The ultimate detector for the gas chromatograph is a mass spectrometer. The latter compensates for the nonspecific nature of the former. That is, the chromatographic separation does not provide identification, except as noted above, but the spectrometer does so. Modern combinations of the two instruments make it possible to acquire mass spectra (to about 500 amu) at a rate of one spectrum per second. The mass spectrometric peaks and fragmentation patterns generally provide exact structures of impurities. Even if an impurity is only partially resolved chromatographically, it may be possible to obtain its mass spectrum by subtracting the mass spectrum of the major

Table 11.1. Commonly Used Detectors for Gas Chromatography

Detector	Minimum Detectable Level $(MDL)^a$
Thermal conductivity (TCD)	5×10^{-10} g cm^{-3}
Flame ionization (FID)	10^{-12} g (carbon) s^{-1}
Electron capture (ECD)	10^{-16} moles cm^{-3}
Flame photometric (FPD)	10^{-10} g (sulfur) s^{-1}
	2×10^{-12} g (phosphorus) s^{-1}
Alkali flame (AFID)	5×10^{-14} g (nitrogen) s^{-1}
	5×10^{-15} g (phosphorus) s^{-1}
Photoionization (PID)	5×10^{-11} g cm^{-3}
GC/MS with selected-ion monitoring (SIM)	10^{-13} g s^{-1}
GC/MS with total-ion monitoring	10^{-9} g s^{-1}

^{a}MDL for TCD, ECD, and PID is defined in terms of concentration in the detector, while the others are defined in terms of mass flow rate.

component of the peak from that of the "shoulder" containing the impurity. Needless to say, powerful data systems are required to carry out such manipulations (6, 7). Trace analysis by mass spectrometry and gas chromatography has been reviewed (8, 9).

It is often impossible to distinguish among isomers by mass spectrometry; this shortcoming has been addressed by recent combination of gas chromatography with FTIR detection (see Section 2, above) (10–14).

4 LIQUID CHROMATOGRAPHY

Liquid chromatography, which is normally carried out at or near room temperature, lifts the thermal-stability and volatility requirements imposed by gas chromatography. In their stead, a requirement of solubility is imposed. Like gas chromatography, liquid chromatography embraces a diversity of experimental practices. Mixtures may be separated by differing adsorbability on a stationary solid phase. Alternatively, separation may be by partition between mobile and stationary liquid phases. Ion-exchange and size-exclusion phases are effective in some cases.

Classical liquid chromatography is carried out at or near atmospheric pressure on relatively coarse solid particles. In this regime, analytical or preparative separations require hours or days. Starting in the mid-1960's, a revolution has taken place in liquid-chromatographic practice. Separating phases have been introduced in ever-smaller particles sizes to provide enhanced resolution. To move analyte solutions through such columns, high pressure (around 2×10^7 Pa) is needed. At these pressures, analyses are usually complete in a few minutes or tens of minutes. Sensitive detectors are available, based on a variety of chemical and physical principles. The most widely used of these respond to changes in optical absorption or refractive index in the liquid leaving the column. Others measure fluorescence, electrochemical properties, IR spectra, or heat of adsorption. Liquid chromatographs have also been coupled to mass spectrometers and Fourier-transform IR spectrometers (13). Trace analysis by liquid chromatography has been reviewed (15).

5 ELECTRICAL MEASUREMENTS

The electrical resistivity of metals can be thought to result from three factors: thermal lattice vibrations, lattice imperfections (such as dislocations, vacancies, and grain boundaries), and impurity effects. The first of these is temperature dependent and approaches zero at 0 K; the last two are essentially independent of temperature. Since the resistivity resulting from lattice imperfections is usually much smaller than that from impurities, a low-temperature measurement shows impurity effects nearly uniquely. While the method is fairly general, some impurities have much greater influence than others. Moreover, only atoms present in solid solution are effective scattering sites, and even these are active only at concentrations below about 200 ppm. With these cautions in mind, resistivity measurements are useful probes of overall purity. To avoid the difficulties of making absolute measurements (which require accurate knowledge of sample dimensions), one normally measures only the *ratio* of resistance at a high temperature (usually 298 K) to that at a low temperature (usually 4.2 K). For impure metals the ratio is < 10; for pure metals, it is $10^2 - 10^5$.

Electrical properties of certain organic crystals can be equally informative. Kepler introduced an optical-pulse technique in which a short flash from a lamp or laser generates charge carriers that move through the crystal under an applied electric field (16). Impurities in the crystal act as trapping sites for the carriers and reduce their lifetimes. The method is sensitive to parts-per-million of impurity, or less. If charge carriers are generated in a crystal at low temperature, they may fill stable trap sites. Heating the crystal results in "detrapping" and the crystal becomes conductive. The resulting "thermally stimulated current" (TSC) is a sensitive test of purity (17).

6 OPTICAL SPECTROSCOPY

Absorption spectra resulting from electronic transitions can reveal the presence of as little as 10 ppm of impurity, in favorable systems. To reach this sensitivity requires that there be no overlap between the bands of the major and minor components. Fluorescence and phosphorescence spectra can detect about 100 times less impurity. The enhanced sensitivity results from the fact that the excitation responsible for emission has a much longer lifetime than the excited state in absorption. During this lifetime, the excitation wanders through the crystalline host until it is trapped, traversing regions of micrometer dimension. The lifetime of the excitation is greater in pure crystals, free of trapping sites. In delayed fluorescence, emission takes place only when two excitations meet and combine; the sensitivity of this process is yet another two orders of magnitude greater than that of prompt fluorescence (18).

7 THERMAL METHODS

7.1 Classical Methods of Freezing-Point Determination

Organic chemists have traditionally used sharpness of melting point as a qualitative indication of the purity of a sample. The earliest attempt to base a quantitative assay of purity on freezing behavior was made by White (19). White's freezing-point method was used by Schwab and Wichers to measure impurity content in benzoic acid with a precision of 10^{-3} mole% (20). Since these early efforts, numerous groups have used both freezing and melting in measuring purity. Both methods are based on the following equation, which relates the equilibrium temperature T_f, in a partially molten binary system, to the mole fraction of the solute, X_2, the heat of fusion of the major component, ΔH_f, and fraction melted, f:

$$T_f = T_{0,1} - \frac{RT_{0,1}^2 X_2}{\Delta H_f} \frac{1}{f} = \frac{A}{f} X_2 \tag{11.1}$$

where $T_{0,1}$ is the freezing point of the pure major component and A is the cryoscopic constant. A graph of $1/f$ against T_f gives a straight line whose intercept on the temperature axis gives the freezing temperature of the pure solvent and whose slope gives the mole fraction of the solute. Equation 11.1 was derived under the following assumptions:

- Solute concentration is small.
- No solid solution is formed.
- The solute is completely miscible in the major component in the liquid state.
- f is close to unity.

In principle, Equation 11.1 applies equally to melting and freezing, and work on both approaches has continued to recent times. A variety of devices and methods has been used with various sample sizes and solute concentrations.

A more general relationship, applicable to systems forming solid solutions, was derived graphically by Smit (21, 22) for a binary system of known phase diagram, and thermodynamically by Mastrangelo and Dornte (23).

$$\frac{1 - k_0}{(1 - f)k_0 + f} X_2 = \frac{-\Delta H_f}{R T_{0,1}^2}(T_{0,1} - T_f) \tag{11.2}$$

Rigorous application of Smit's graphical method requires that the identity of the solute and its phase diagram with the host be known. In this special case, addition of a small known amount of the impurity followed by remeasurement of the melting curve, allows calculation of the original impurity content (22).

Both Equations 11.1 and 11.2 are based on the assumption of complete compositional equilibrium of solid and liquid phases. Because diffusion in the solid is so much slower than in the liquid, this condition is rarely fulfilled. Van Wijk and Smit (21) derived an equation relating temperature and fraction solidified under conditions encountered earlier, in directional crystallization (see Section 1.2.1, Chapter 3). In this regime, a liquid of uniform composition is in equilibrium with only the surface layer of solid, a condition referred to as *partial equilibrium*.

$$(1 - k_0)f^{k_0 - 1} X_2 = \frac{-\Delta H_f}{R T_{0,1}^2}(T_{0,1} - T_f) \tag{11.3}$$

The melting of a gradually frozen sample, which will always have a concentration gradient, is likewise described by Equation 11.3.

Both Equations 11.2 and 11.3 have been used in an experimental study of the benzene/thiophene system, which forms solid solutions. Plots of measured values of equilibrium temperature against fraction melted were compared with theoretical curves calculated from Equation 11.3, for a mixture containing 0.094 mole% thiophene. The calculated curves fit the experimental data well, with

$k_0 = 0.05$. On the other hand, a solution containing 1.15 mole% thiophene was better described by Equation 11.2, with $k_0 = 0.3$. While it is possible that k_0 changes substantially with concentration, it is also possible that one or another of these experiments was not carried out at equilibrium (24).

Saylor and Ross have analyzed the effects that decrease the accuracy of cryometric measurements based on freezing and melting curves (25). The problems associated with freezing are:

1. Because supercooling is unavoidable, the rise in temperature after the onset of crystallization is not instantaneous and temperature equilibrium is reached slowly. Since the heat of fusion released after supercooling is proportional to the amount of solid formed and this amount is in turn approximately proportional to the square of the supercooling, recovery can never be complete.

2. For most substances, total sample volume decreases during freezing, reducing the area radiating heat to the surroundings and causing temperatures recorded in the later stages of freezing to be erroneously high.

3. If solid forms on the container wall, it acts as insulation between the external heat sink and the internally evolved latent heat, again causing the temperatures recorded late in the run to be erroneously high.

4. In unstirred systems the distribution of solid may change the conduction pattern during freezing. In stirred systems, the energy imparted to the system by stirring increases with increasing degree of solidification.

Melting curves are affected by other problems. Because the liquid that is forming is less dense than the starting solid, it rises. The solid, which comprises a heat sink, remains at the bottom of the container. In the liquid remote from this heat sink, temperatures will be higher than the equilibrium value for the whole sample. The liquid is not homogeneous, especially in the early stages of melting, when stirring is impossible. If the liquid in contact with the pure solid remains purer than the bulk liquid, the local solid/liquid equilibrium temperature will be high. This effect would increase the steepness of the time/temperature curve and decrease the estimate of impurity in the sample. Very slow melting minimizes this error, but introduces other problems.

These problems were largely overcome by use of a cell containing a group of horizontal, thermally conductive vanes. The outer edges of the vanes were curved upward to form miniature circular pans. Thus the vanes served both as heat conductors and as dividers, distributing the sample in thin layers so that no part of the sample could be more than tenths of a millimeter from a conductive surface. The sample volume, about 1.5 mL, was sufficient to fill all of the pans, which were gold-soldered to a central platinum tube. The assembly of pans was sealed into a thin-walled soft-glass cell with a sidearm through which the cell was attached to a transfer manifold for filling. The charged cell was sealed off and suspended by the thermocouple leads from above and by a glass fiber from below, centered within an evacuable jacket.

The content of the sealed cell was frozen; both jackets were evacuated and the container was immersed in a constant-temperature bath. The heating rate was adjusted by varying the pressure in the vacuum jackets, and the time/temperature curve was recorded automatically. The analysis of the experimental curves was based on certain assumptions regarding an ideal melting curve; the time/temperature curve under conditions of compositional, thermal, and thermodynamic equilibrium should be hyperbolic. One asymptote of the hyperbola will be the melting temperature of the pure substance, $T_{0,1}$, and the other will represent the heating rate of the original (solid) sample and cell. The intersection of the hyperbola with the liquid-warming curve gives T_f, the freezing point of the sample, and the asymptote of the hyperbola gives $T_{0,1}$. The theoretical hyperbola was manipulated optically to fit portions of the measured melting curve. In contemporary work, it would be easier to match the two curves by numerical manipulation with a computer. In any case, the temperature difference $(T_{0,1} - T_f)$ is proportional to the mole fraction of impurity in the sample. The method and apparatus were applied to a benzene sample of known purity (99.797 mole%); in runs with very slow melting, agreement was within 10^{-2} mole%.

Recent work on cryometric measurement of purity has been directed toward the use of small samples. Chavret and Merlin (26) have described a relatively simple method for measuring melting curves of 1-gram samples in thin-walled glass tubes of 0.7-cm diameter, sealed under vacuum. Temperature was monitored by a thermocouple in a central well. Making the usual assumption that the temperature difference between the sample and its surroundings was much larger than the change in sample temperature during fusion, then the rate of fusion can be considered constant (25). Hence f, the fraction melted, becomes proportional to time. Samples of m-xylene (mp 225.29 K), benzene (mp 278.69 K), and biphenyl (mp 343.6 K) were evaluated in pure form and after addition of n-heptane (to m-xylene and benzene) and anthraquinone (to biphenyl). In the range 0.1–2.5 mole%, the measured values agreed with the known impurity contents to better than 10%.

One experimental approach to the use of small samples and the attainment of solid/liquid equilibrium involves the use of a sample in the form of a thin film (22). This work is discussed in Section 2.2.4, Chapter 2. This thin-layer method has been embodied in two devices, suited for measurement of impurity concentrations of ± 0.1 and ± 0.01 mole%, respectively (27).

Very small freezing-point depressions have been measured with a quartz-crystal thermometer (28). Such thermometers show high reproducibility and sensitivity, and offer temperature resolution of 10^{-4} K. The temperature of initial crystallization of a melt can be measured very accurately and reproducibly if conditions of supercooling and nucleation are carefully controlled. Hence, the method is suitable for comparison of various batches of a material or for monitoring the progress of a purification scheme. In one application of this method, benzene (fp 278.69 K) was stirred in a bath at 276.30 ± 0.01 K and seeded when the sample temperature reached 277.60 K. Commercial product of fp 278.60 K was easily followed by quartz-crystal thermometry to a final freezing temperature of 278.69 K.

A device has been described recently in which melting or freezing of a 1–2 g sample can be determined with an accuracy of ± 2 mK (29).

Kroeger et al. (30) described an elaborate freezing-curve calorimeter in which a 13-gram sample was frozen during a 10–12 h period in a gold crucible while temperature was monitored by a platinum resistance thermometer with a sensitivity of better than 5×10^{-4} K. The rate of cooling was controlled by a heat shield and a vacuum jacket. The design did not lend itself to mechanical stirring, but numerous radial gold fins were incorporated to promote thermal equilibrium in the liquid and furnish a large area on which solid could form.

To record a freezing curve, the crucible and its shield were heated above the melting temperature of the sample. The crucible temperature was stabilized just above the melting point and shield temperature was reduced to establish a fixed ΔT (typically 5 K). Power to the crucible was interrupted; crucible temperature was monitored while shield current was manually adjusted to maintain constant shield temperature. Supercooling was always observed before nucleation; slight superheating was observed for 15–30 min after nucleation. This superheating was attributed to the rapid build-up of impurity caused by sudden freezing after supercooling, followed by slow diffusion of the impurity into the bulk melt. The equilibrium freezing temperature T_0 was obtained by extrapolating the flat portion of the curve back to the cooling slope of the liquid. If the freezing temperature of a totally pure sample is defined as T'_0, the freezing point depression $\Delta T_0 = (T'_0 - T_0)$ will be positive and can be used as a measure of purity. Further, the freezing-temperature depressions after varying solidification times (measured as a fraction of the time required for total solidification) were calculated and found to agree well with measured values. The measured depression was 5×10^{-3} K, corresponding to 8×10^{-3} mole%. However, this value reflected the results of contamination by attack on the crucible, and it was estimated that the true depression might have been an order of magnitude lower.

The methodology of this kind of calorimetry represents heroic effort: loading a new sample required that the crucible's glass support-tube be broken; the crucible wiring, thermocouple, and heater had to be removed and a new closure had to be electron-beam welded in place. In spite of these measures, sample/container interaction took place (see above) and the final result remained open to question.

7.2 Differential Scanning Calorimetry (DSC)

In recent years, sensitive automatic calorimeters have come into use for measurement of melting curves. These instruments have been interfaced to time-sharing or dedicated computers. Special programs detect peak onset and return to baseline and record the differential melting curve of a small sample (1–3 mg), prepare a linearized plot of 1/f against temperature, and calculate mole fraction of impurity (31). This dynamic DSC method requires a cell calibration coefficient, obtained from a pure standard material; indium is widely used for this purpose. The calibration run is used to calculate the heat of fusion and the leading-edge slope of the indium curve, which is used to determine the instantaneous sample temperature during the later run with an unknown mixture. The melting endotherm of the sample (Figure 11.1) is integrated to give total peak area ($ACBA$), corresponding to fusion of the entire sample. The region between f = 0.1 and f = 0.5 is divided into a number (typically 20) of segments and the partial area of each segment is measured, for example, $ADEA$ in Figure 11.1. Each partial area is divided by the total area to give f, the corresponding fraction melted. The associated sample temperature, T_f, is obtained by applying line DF, the slope of the indium calibration curve. For area $AEDA$, the temperature given by this slope is F. A plot of these fractional areas should give a straight line with intercept $T_{0,1}$ and slope $-RT_{0,1}^2 X_2/H_f$. In practice, however, the plot is usually concave upward (see Figure 11.2). Several experimental circumstances have been found to contribute to this deviation. These include: large sample size, high heating rate, and solid-solution formation. Even in the absence of these factors, inadequate instrumental sensitivity may still lead to "undetected melting" before the observed endotherm (32). Attempts have been made to measure a "primary liquid fraction," already present at the temperature taken for the start of melting (33). This endotherm may be associated with eutectic melting that always precedes the major endotherm in non-solid-solution systems. It has been argued that failure to include eutectic melting has been responsible for the need to include arbitrary corrections of the melt baseline and heat of fusion. The measured curve is usually corrected empirically by addition of an amount of heat to each partial area sufficient to linearize the 1/f against T_f plot (see Figure 11.2). In addition to mole fraction

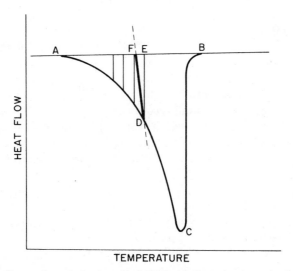

Fig. 11.1 Geometric analysis of a typical DSC endotherm for determination of purity.

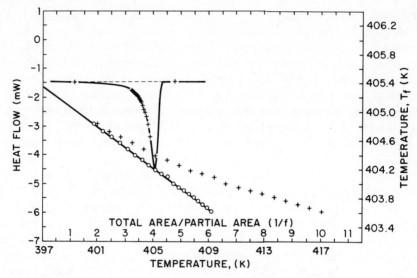

Fig. 11.2 Experimentally determined DSC melting endotherm (left ordinate) with partial area plot vs melting temperature (right ordinate). See text for details.

impurity, the calorimetric measurement provides ΔH_f, $T_{0,1}$, and ΔT_f (melting-point depression of the impure sample).

It is generally asserted that the DSC method is applicable to non-solid-solution samples containing up to about 3% impurity (32); Plato has provided a survey of the application of this method to 64 organic compounds (34) meeting these criteria. With a known solid-solution system (benzene/thiophene), dynamic DSC gave consistently low results for impurity content. Known thiophene concentrations of 0.53, 1.15, 1.90, and 2.42 mole% gave measured values of 0.36, 0.42, 0.61, and 0.82 mole%, respectively (32).

A fundamentally different method has been proposed for the treatment of data from dynamic DSC measurements (35). The new method requires measurement of first, second, and third time derivatives of the fusion endotherm. After identification of the temperature and fraction melted at the point of maximum second derivative (i.e., maximum acceleration of melting), a tangent is drawn to the plot of $1/f$ against T, to yield a slope from which X_2 is calculated from Equation 11.1. The intercept of the tangent with the temperature axis gives $T_{0,1}$. This method is said to be applicable to the entire concentration range from $< 10^{-4}$ mole% to 50 mole% (35).

Many papers have been published on the use of DSC to measure the impurity content of mixtures prepared from a pure host substance and a single eutectic-forming dopant. It is usually argued that agreement between the amount of impurity added and the amount measured by DSC is an adequate validation of the method. Implicit in this argument are the assumptions that in real, multiply contaminated samples, the impurities do not interact, and their effects on fusion of the host are additive. A recent study has extended this method to a ternary system in which two contaminants (benzamide and p-aminobenzoic acid) were added in varying ratios to a pure host (phenacetin). It is known that each of the additives forms a simple eutectic system with the host. Mixtures were prepared by direct weighing into aluminum pans, to a total weight of 3 mg. A weighing error of 1 μg causes an error of ± 0.05 mole% in measured purity. Analyses of mixtures containing 0.5–1.5 mole% total impurity are given in Table 11.2, along with the correction required for linearization of the plots of T against $1/f$, and $T_{0,1}$, the extrapolated melting point of the pure host (36).

Staub and Perron have introduced a modified DSC method which is said to be applicable to higher impurity concentrations and to obviate the need for empirical linearization of the $1/f$ against T_f plot (37). The modification consists in applying the heat of fusion in stepwise fashion. After each temperature increase, one waits until equilibrium has been reestablished (i.e., until the heat-flux trace has returned to baseline). This procedure is thus similar to classical, static calorimetry.

In the step-heating method, the peak areas before eutectic melting and after final melting are due only to the differing heat capacities of the sample and reference, since no latent heat is consumed. These peaks provide a "background area" which must be subtracted from areas measured in the fusion interval. The sum of the corrected partial heats of fusion is equal to the total heat of fusion. By adding partial heats of fusion up to a temperature T_f, it is possible to determine $1/f$. Figure 11.3 shows a step-heated fusion. Peak 1 is a pre-eutectic background peak. Peak 2 reflects the melting of the eutectic; peaks 3, 4, and 5 show the beginning of posteutectic melting, measured with 10 K temperature increments. Because the peaks become progressively larger with uniform-temperature steps, the later phase of the fusion was carried out with 0.5 K steps (peaks 6–13). Peak 14 is a postmelting background peak. With the stepwise method, impurity levels up to 10% are measurable with 5% relative accuracy (37).

Table 11.2. DSC Purity Determination of Doped Phenacetin (36)

Dopant 1 Benzamide, mole%	Dopant 2 p-aminobenzoic acid, mole%	Linearization correction, %	T_0, K	ΔT, K	Impurity, mole%
0.25	0.25	14.6	407.4	0.22	0.48_2
0.50	0.50	13.0	407.4	0.43	1.00_7
0.75	0.75	13.2	407.5	0.67	1.47_7
0.67	0.33	5.3	407.5	0.44	1.02_4
0.33	0.67	9.9	407.4	0.47	1.09_7
0.67	0.33	11.4	407.5	0.46	$1.03_4{}^a$
0.33	0.67	30.4	407.6	0.49	$1.09_0{}^a$

^{a}10-mg sample.

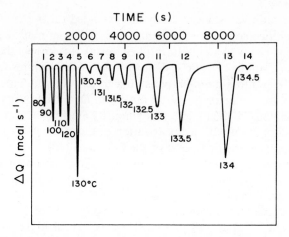

Fig. 11.3 Step-heating method for determination of purity by DSC.

Ramsland has extended the stepwise method to allow for solid-solution systems. An equation was developed that did not require the assumption that the solution is dilute:

$$T_f = \frac{T_{0,1}}{1 + \dfrac{RT_{0,1}}{\Delta H_f} \ln\left(\dfrac{k_0 X_{0,l} - k_0 f + f}{X_{0,l} - 1 + k_0 + f - k_0 f}\right)} \tag{11.4}$$

where $X_{0,l}$ is the mole fraction of the major component in the original sample. Heat capacities of solid and liquid, and instrumental blanks were considered in evaluating accurate values of $1/f$ at various temperatures. Iterative, multiple-regression analyses of these data give accurate purity values at doped impurity levels as high as 10% with a maximum relative error of 10% (38).

8 MISCELLANEOUS

A wide range of properties can be correlated with purity. For example, the rate of crystallization of amorphous films has been found to be sensitively and reproducibly dependent on purity. Iwashima et al. (39) related the impurity concentration in tetrabenzo(a,cd,j,lm)perylene to the time required for amorphous films to crystallize at room temperature:

Time to Crystallize, h	Impurity Concentration, mole/mole
∞	10^{-1}
24	10^{-4}
24	10^{-5}
9	10^{-6}
0	10^{-8}

Crystallization occurred sooner at elevated temperature, but the time was still purity dependent.

Mechanical properties can also be influenced by impurities. Subtle effects can arise from changes in spatial distribution of impurities. For example, segregation at grain boundaries can cause gross differences from homogeneous solid solution.

The use of distribution coefficients to calculate impurity levels below analytical detectability is discussed in Chapter 8.

REFERENCES

1. G. H. Morrison, Ed., *Trace Analysis*, Interscience, New York 1965.

2. M. Pinta, *Detection and Determination of Trace Elements*, Ann Arbor-Humphrey Science Publishers, Ann Arbor, 1970.

3. S. Siggia, *Quantitative Organic Analysis via Functional Group Analysis*, 4th ed. Wiley-Interscience, New York 1979.

4. F. Feigl and V. Anger, *Spot Tests in Organic Analysis*, 7th ed., Elsevier, Amsterdam, 1966.

5. P. R. Griffiths, *Chemical Infrared Fourier Transform Spectroscopy*, Wiley, New York, 1975

6. N. Karl, *J. Crystal Growth*, **51**, 509–517 (1981).

7. M. C. Ten Noever de Brauw, *J. Chromat.*, **165**, 207–233 (1979).

8. A. J. Ahearn, Ed., *Trace Analysis by Mass Spectroscopy*, Academic Press, New York, 1972.

9. H. Hachenberg, *Industrial Gas Chromatographic Trace Analysis*, Heyden & Son, Ltd., London, 1973.

10. P. R. Griffiths, *NATO Adv. Study Inst. Ser., Ser. C*, **57** (*Anal. Appl. FT-IR Mo. Biol. Syst.*), 149–155 (1980).

11. D. Kuehl, G. J. Kemeny, and P. R. Griffiths, *Appl. Spectrosc.*, **34**(2), 222–224 (1980).

12. P. R. Griffiths, Report, EPA/600/4-79/064, Order No. PB80-122575 (1979).

13. D. T. Kuehl and P. R. Griffiths, *Anal. Chem.*, **52**(9), 1394–1399 (1980).

14. M. M. Gomes-Taylor and P. R. Griffiths, *Anal. Chem.*, **50**(3), 422–425 (1978).

15. J. F. Lawrence, *Organic Trace Analysis by Liquid Chromatography*, Academic Press, New York, 1981.

16. R. G. Kepler, *Phys. Rev.*, **119**, 1226–1229 (1960).

17. N. Karl, "High Purity Organic Molecular Crystals," in H. C. Freyhardt, Ed., *Crystals, Growth, Properties, Applications*, Vol. 4, Springer Verlag, New York, 1980, p.82.

18. Ibid., p. 78.

19. W. P. White, *J. Phys. Chem.*, **24**, 393–416 (1920).

20. F. W. Schwab and E. Wichers, U.S. Dept. of Commerce, Ntl. Bureau of Stds. Research Paper RP1647; Part of *Journal of Research of the Ntl. Bureau of Stds.*, **34** (1945).

21. W. M. Smit, *Z. Electrochem.*, **66**(10), 779–787 (1962).

22. W. M. Smit, *Recueil*, **75**, 1309–1320 (1956).

23. S. V. Mastrangelo and R. W. Dornte, *J. Am. Chem. Soc.*, **77**, 6200–6201 (1955).

24. A. Bylicki and Z. Bugajewski, in I. Buzas, Ed., *Thermal Analysis*, Vol. 2, Heyden, London, 1975, pp. 497–504.

25. C. P. Saylor and G. S. Ross, *J. Res. N.B.S.*, **68C**(1) 35–39 (1964).

26. M. Chavret and J. C. Merlin, *Analusis*, **8**(5), 191–195 (1980).

27. R. Stober and W. Haberditzl, *Z. Physik. Chem.* (*Leipzig*), **238**, 60–68 (1968).

28. F. C. Youngken, *Anal. Chem.*, **44**, 559–565 (1972).

29. L. Crovini, P. Marcarino, and G. Milazzo, *Anal. Chem.*, **53**, 681–686 (1981).

30. F. R. Kroeger, C. A. Swenson, W. C. Hoyle, and H. Diehl, *Talanta*, **22**, 641–647 (1975).

31. S. A. Moros and D. Stewart, *Thermochim. Acta*, **14**, 13–24 (1976).

32. E. F. Palermo and J. Chiu, *Thermochim. Acta*, **14**, 1–12 (1976).

33. A. Bylicki and Z. Bugajewski, *Thermochim. Acta*, **34**, 35–41 (1979).

34. C. Plato, *Anal. Chem.*, **44**(8), 1531–1534 (1972).

35. G. M. Gustin, *Thermochim. Acta*, **39**, 81–93 (1980).

36. J. P. Elder, *Thermochim. Acta*, **34**, 11–17 (1979).

37. H. Staub and W. Perron, *Anal. Chem.*, **46**, 128–130 (1974).

38. A. C. Ramsland, *Anal. Chem.*, **52**(9), 1474–1479 (1980).

39. S. Iwashima, H. Honda, J. Aoki, and H. Inokuchi, *Mol. Cryst. Liq. Cryst.*, **59**, 207–218 (1980).

Appendix I

CALCULATED ZONE-MELTING CONCENTRATION PROFILES

Concentration profiles showing relative solute concentration against position are presented in this appendix for various numbers of passes, distribution coefficients, and ratios of ingot length to zone length, L/l. The curves were computed by the method of Burris, Stockman, and Dillon (1), dividing the ingot into 500 cells for the computation (see Section 1, Chapter 5). Note that many of the computed curves show a discontinuity at the start of the last zone length of the ingot. This arises because the profile in the last zone length is calculated from the equation for directional crystallization (see Section 1, Chapter 3) rather than the zone-refining equation. We are indebted to Mr. S. W. Rementer for writing the computer program used in generating the curves.

The first group of curves (Figs. I.1 to I.5) shows single-pass distributions for values of k between 0.1 and 0.9, with L/l equal to 3, 5, 10, 20, and 50, respectively, up to $C_0 = 1$. Note that in these profiles, the solute segregation decreases markedly with decreasing zone length: a small zone can only hold a small amount of solute before the solute concentration reaches the saturation value, C_0/k_0.

The second group of curves (Figs. I.6 to I.9) shows single-pass distributions for values of k between 1.1 and 2.0, with L/l equal to 3, 5, 10, and 20, respectively.

The next section (Figs. I.10 to I.18) shows logarithmic multipass profiles for values of k between 0.1 and 0.9, with L/l equal to 3 in each case. In these and ensuing multipass figures, a curve is shown for ultimate distribution (i.e., infinite number of passes). Only four curves are visible in Figures I.9 through I.16 and five in Figures I.17 through I.19, because the distributions for higher numbers of passes coincide with ultimate distribution. Succeeding sections (Figs. I.19 to I.27, I.28 to I.37, I.38 to I.48, and I.49 to I.57) show multipass profiles for the same values of k, but with L/l having values of 5, 10, 20, and 50, respectively.

The next sections (Figs. I.58 to I.63, I.64 to I.71, I.72 to I.78, and I.79 to I.85) show multipass profiles for values of k between 1.1 and 2.0 with L/l equal to 3, 5, 10 and 20, respectively. Departure from smooth profiles in certain of these figures is an artifact of the computation and has no physical meaning.

Figures K.86 to I.90 show "highest attainable purity" curves for various values of k (less than unity) and L/l. These curves give the relative concentration at the head of an ingot as a function of number of zone passes. The values of C/C_0 range from unity to that prevailing at ultimate distribution. With increasing zone length, the curve for a given k "bottoms out" after progressively larger numbers of passes, but at lower values of C/C_0. For example, when $L/l = 3$ and $k = 0.3$, there is practically no change in concentration at the head of the ingot after 10 passes; the terminal concentration is 8×10^{-3}. For $L/l = 50$, the relative concentration does not reach its ultimate value even after 200 passes, but after 50 passes it is already down to about 10^{-13}.

Figures I.91 to I.94 show "maximum enrichment" curves for various values of k (greater than unity) and L/l. Again, large zones attain ultimate distribution sooner than small ones, but at smaller enrichments.

If yield is of primary concern, rather than ultimate purity, reference should be made to the "crossover" curves of Figs. I.95 and I.96. These show the position in an ingot at which (C/C_0) reaches unity.

1.   L. Burris, Jr., C. H. Stockman, and I. G. Dillon, *Trans. AIME*, **203**, 1017 (1955).

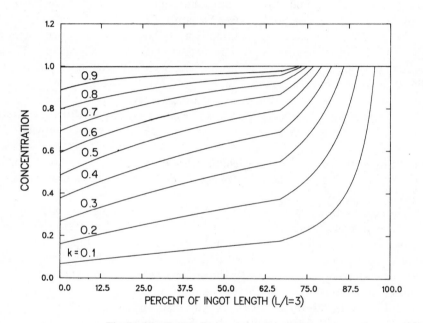

Fig. I.1 Distribution for $n = 1$, $k = 0.1$ to 0.9.

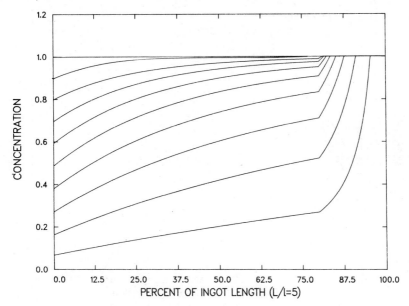

Fig. I.2 Distribution for $n = 1$, $k = 0.1$ to 0.9.

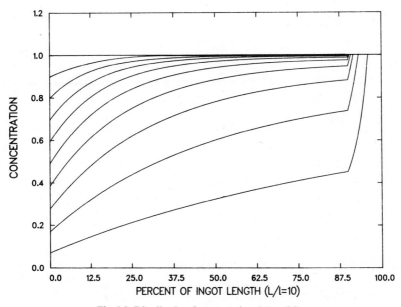

Fig. I.3 Distribution for $n = 1$, $k = 0.1$ to 0.9.

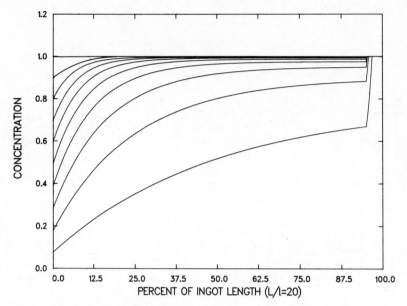

Fig. I.4 Distribution for $n = 1$, $k = 0.1$ to 0.9.

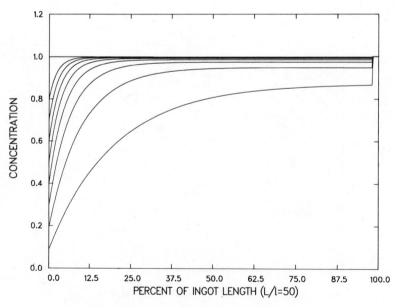

Fig. I.5 Distribution for $n = 1$, $k = 0.1$ to 0.9.

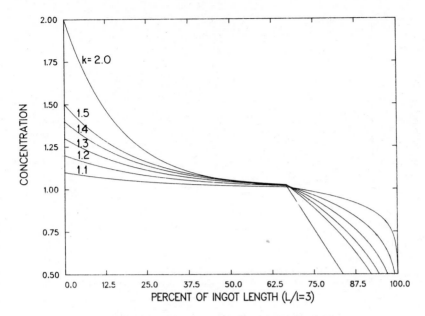

Fig. I.6 Distribution for $n = 1$, $k = 1.1$ to 1.5 and 2.0.

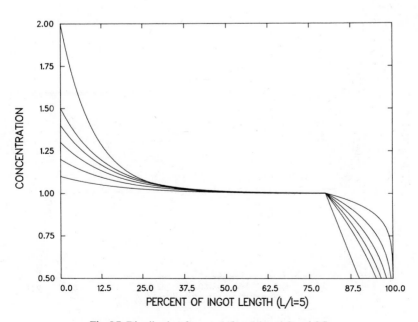

Fig. I.7 Distribution for $n = 1$, $k = 1.1$ to 1.5 and 2.0.

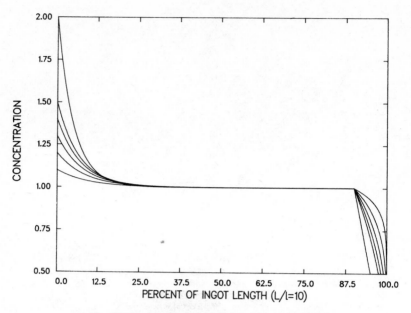

Fig. I.8 Distribution for $n = 1$, $k = 1.1$ to 1.5 and 2.0.

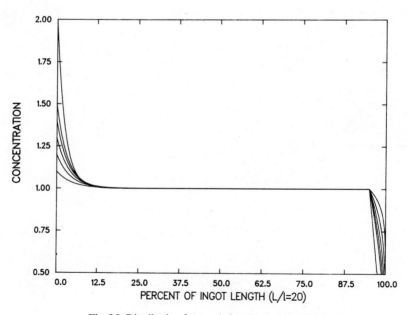

Fig. I.9 Distribution for $n = 1$, $k = 1.1$ to 1.5 and 2.0.

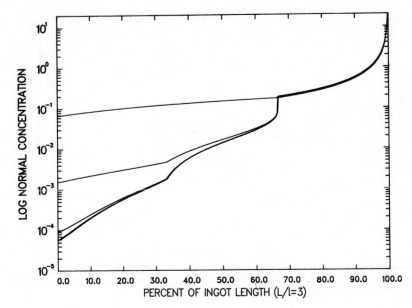

Fig. I.10 $k = 0.10$, $n = 1, 3, 5, 10, 20$ passes.

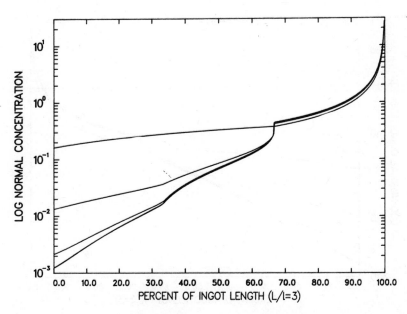

Fig. I.11 $k = 0.20$, $n = 1, 3, 5, 10, 20$ passes.

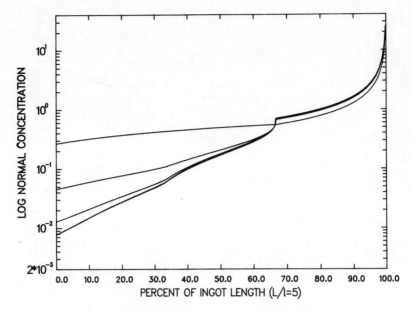

Fig. I.12 $k = 0.30$, $n = 1, 3, 5, 10, 20$ passes.

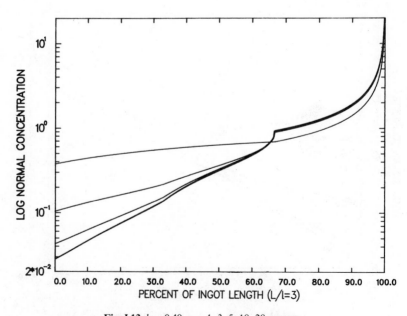

Fig. I.13 $k = 0.40$, $n = 1, 3, 5, 10, 20$ passes.

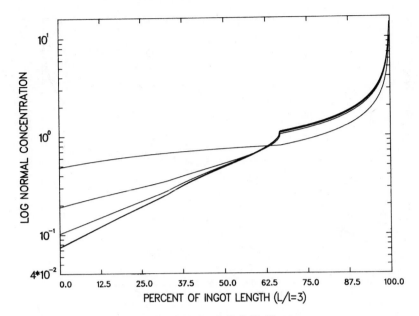

Fig. I.14 $k = 0.50$, $n = 1, 3, 5, 10, 20$ passes.

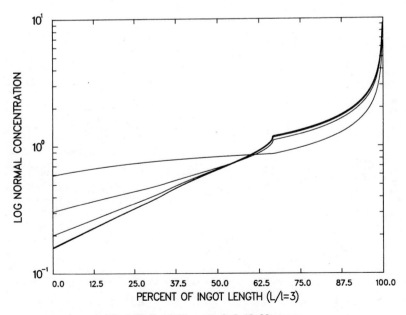

Fig. I.15 $k = 0.60$, $n = 1, 3, 5, 10, 20$ passes.

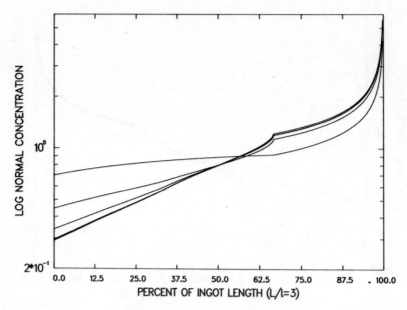

Fig. I.16 $k = 0.70$, $n = 1, 3, 5, 10, 20$ passes.

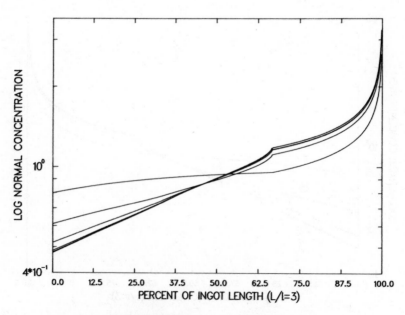

Fig. I.17 $k = 0.80$, $n = 1, 3, 5, 10, 20$ passes.

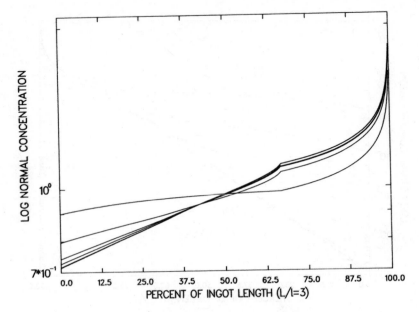

Fig. I.18 $k = 0.90$, $n = 1, 3, 5, 10, 20$ passes.

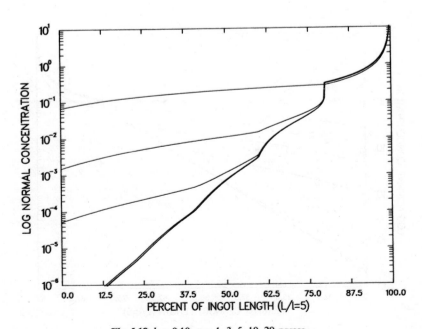

Fig. I.19 $k = 0.10$, $n = 1, 3, 5, 10, 20$ passes.

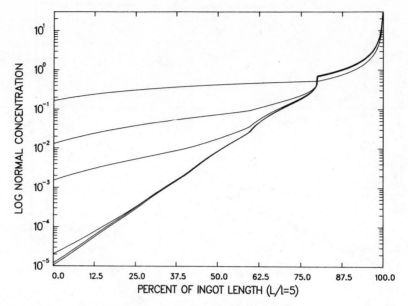

Fig. I.20 $k = 0.20$, $n = 1, 3, 5, 10, 20$ passes.

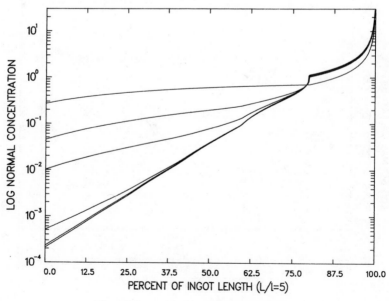

Fig. I.21 $k = 0.30$, $n = 1, 3, 5, 10, 20$ passes.

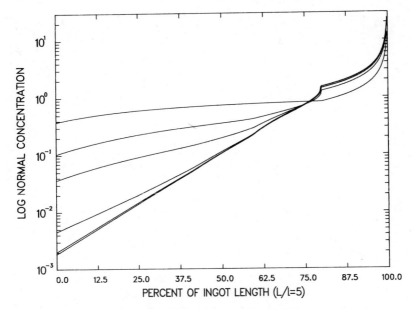

Fig. I.22 $k = 0.40$, $n = 1, 3, 5, 10, 20$ passes.

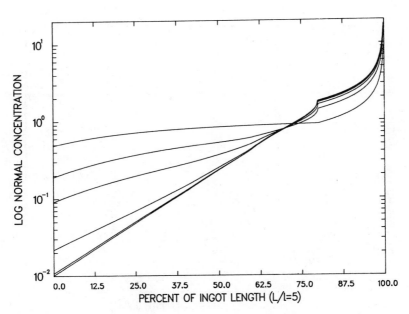

Fig. I.23 $k = 0.50$, $n = 1, 3, 5, 10, 20$ passes.

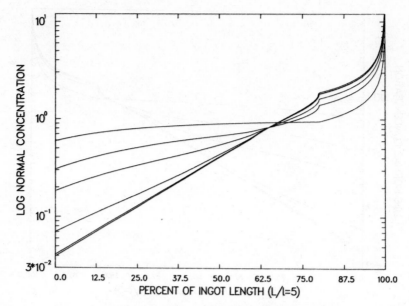

Fig. I.24 $k = 0.60$, $n = 1, 3, 5, 10, 20$ passes.

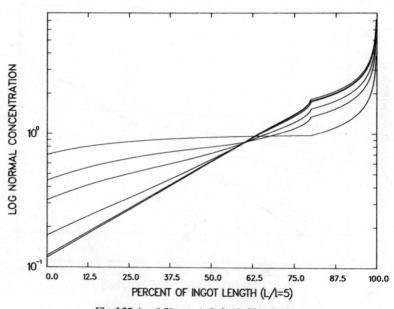

Fig. I.25 $k = 0.70$, $n = 1, 3, 5, 10, 20$ passes.

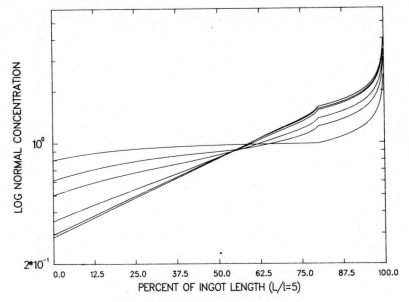

Fig. I.26 $k = 0.80$, $n = 1, 3, 5, 10, 20$ passes.

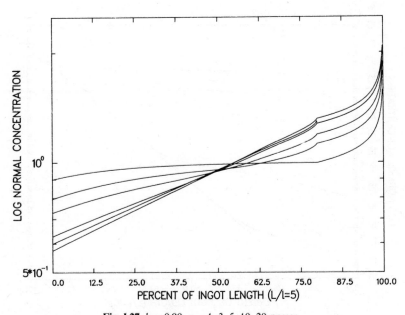

Fig. I.27 $k = 0.90$, $n = 1, 3, 5, 10, 20$ passes.

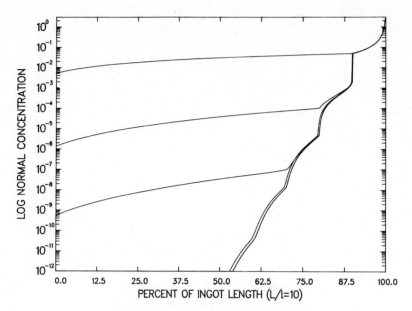

Fig. I.28 $k = 0.01$, $n = 1, 3, 5, 10, 15, 20$ passes.

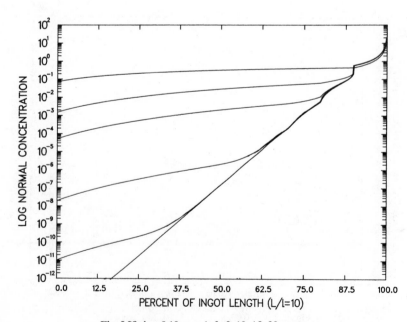

Fig. I.29 $k = 0.10$, $n = 1, 3, 5, 10, 15, 20$ passes.

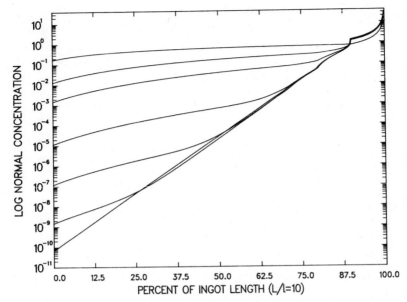

Fig. I.30 $k = 0.20$, $n = 1, 3, 5, 10, 15, 20$ passes.

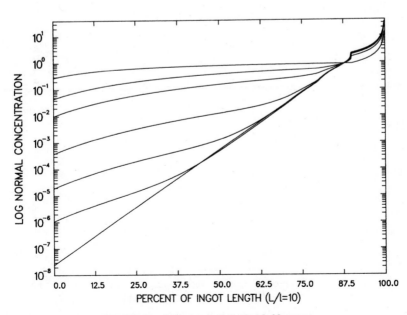

Fig. I.31 $k = 0.30$, $n = 1, 3, 5, 10, 15, 20$ passes.

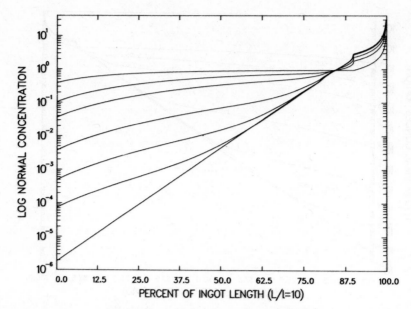

Fig. I.32 $k = 0.40$, $n = 1, 3, 5, 10, 15, 20$ passes.

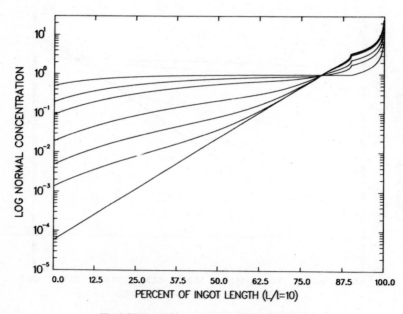

Fig. I.33 $k = 0.50$, $n = 1, 3, 5, 10, 15, 20$ passes.

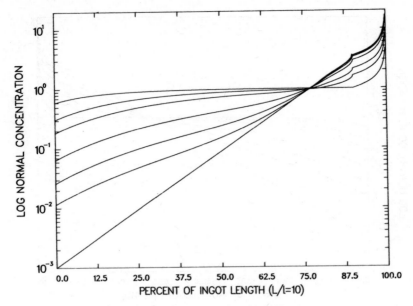

Fig. I.34 $k = 0.60$, $n = 1, 3, 5, 10, 15, 20$ passes.

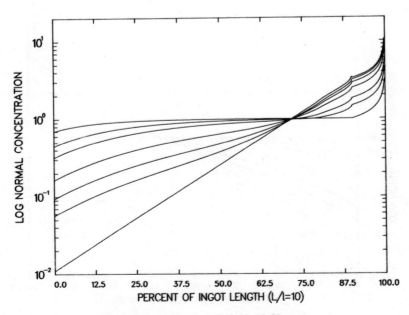

Fig. I.35 $k = 0.70$, $n = 1, 3, 5, 10, 15, 20$ passes.

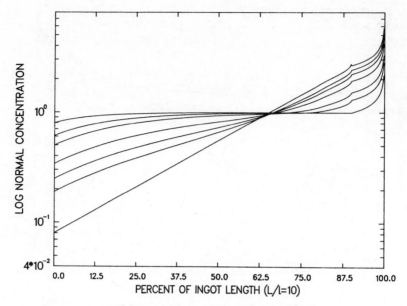

Fig. I.36 $k = 0.80$, $n = 1, 3, 5, 10, 15, 20$ passes.

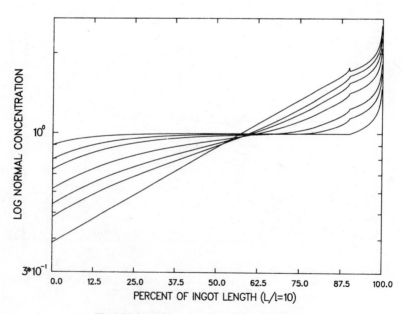

Fig. I.37 $k = 0.90$, $n = 1, 3, 5, 10, 15, 20$ passes.

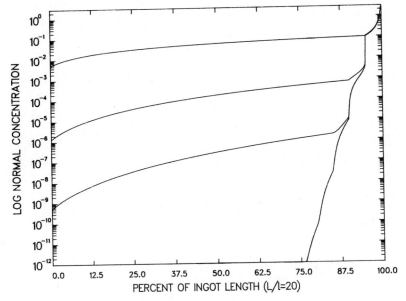

Fig. I.38 $k = 0.01$, $n = 1, 3, 5, 10, 15, 20$ passes.

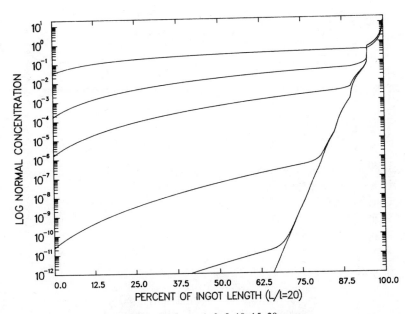

Fig. I.39 $k = 0.05$, $n = 1, 3, 5, 10, 15, 20$ passes.

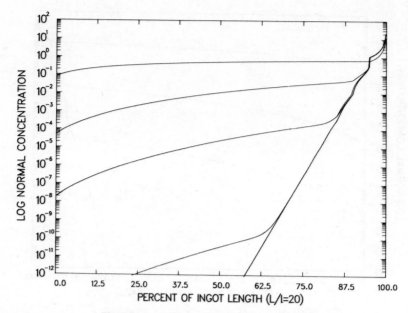

Fig. I.40 $k = 0.10$, $n = 1, 5, 10, 20, 50, 100$ passes.

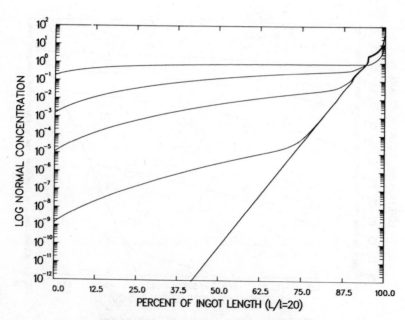

Fig. I.41 $k = 0.20$, $n = 1, 5, 10, 20, 50, 100$ passes.

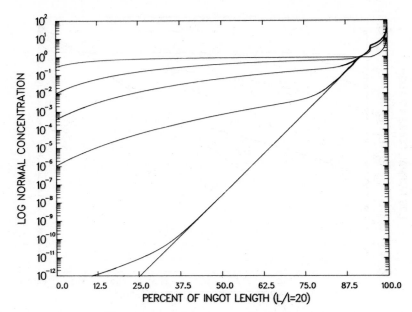

Fig. I.42 $k = 0.30$, $n = 1, 5, 10, 20, 50, 100$ passes.

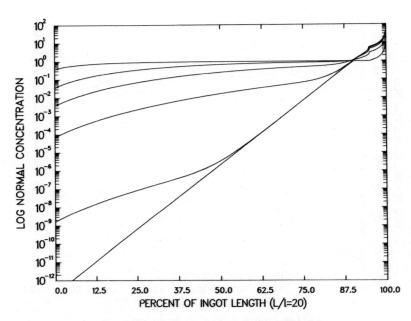

Fig. I.43 $k = 0.40$, $n = 1, 5, 10, 20, 50, 100$ passes.

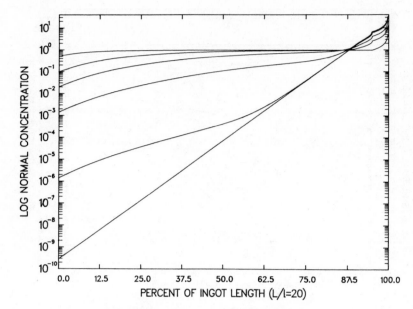

Fig. I.44 $k = 0.50$, $n = 1, 5, 10, 20, 50, 100$ passes.

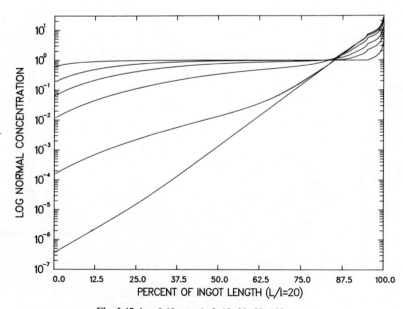

Fig. I.45 $k = 0.60$, $n = 1, 5, 10, 20, 50, 100$ passes.

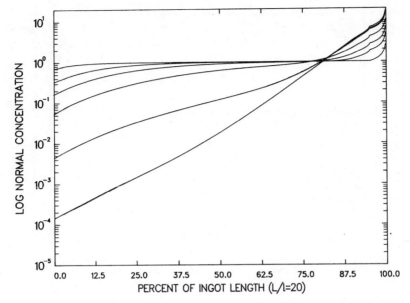

Fig. I.46 $k = 0.70$, $n = 1, 5, 10, 20, 50, 100$ passes.

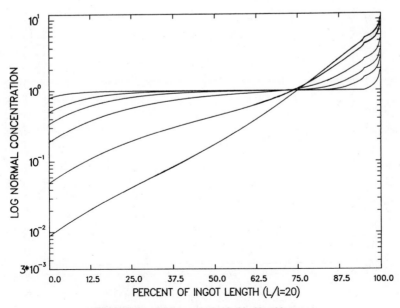

Fig. I.47 $k = 0.80$, $n = 1, 5, 10, 20, 50, 100$ passes.

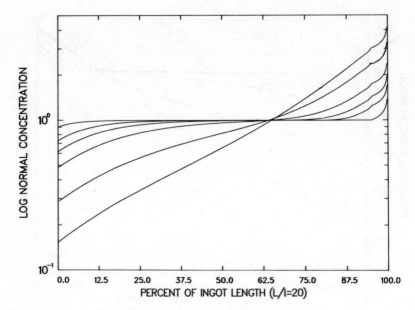

Fig. I.48 $k = 0.90$, $n = 1, 5, 10, 20, 50, 100$ passes.

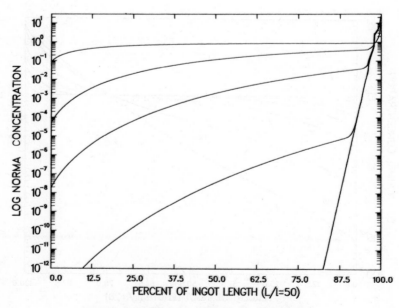

Fig. I.49 $k = 0.10$, $n = 1, 5, 10, 20, 50, 100, 200$ passes.

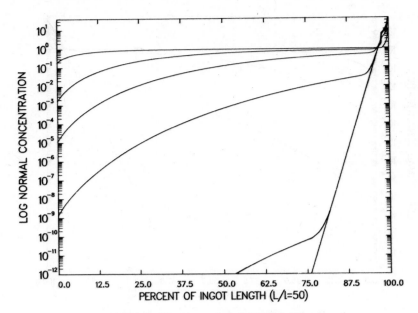

Fig. I.50 $k = 0.20$, $n = 1, 5, 10, 20, 50, 100, 200$ passes.

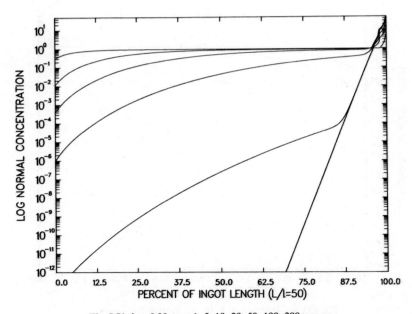

Fig. I.51 $k = 0.30$, $n = 1, 5, 10, 20, 50, 100, 200$ passes.

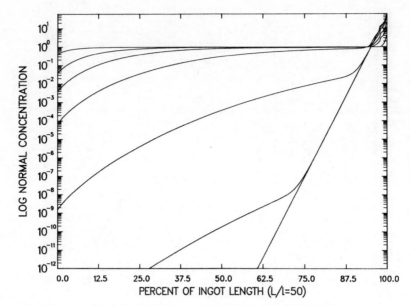

Fig. I.52 $k = 0.40$, $n = 1, 5, 10, 20, 50, 100, 200$ passes.

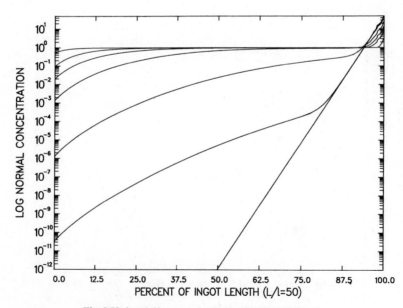

Fig. I.53 $k = 0.50$, $n = 1, 5, 10, 20, 50, 100, 200$ passes.

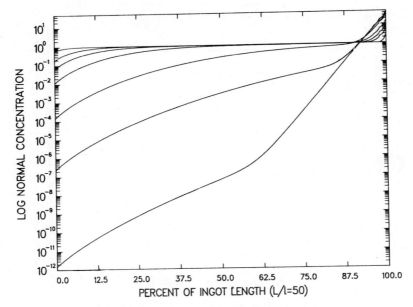

Fig. I.54 $k = 0.60$, $n = 1, 5, 10, 20, 50, 100, 200$ passes.

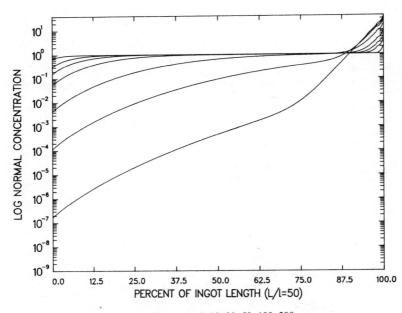

Fig. I.55 $k = 0.70$, $n = 1, 5, 10, 20, 50, 100, 200$ passes.

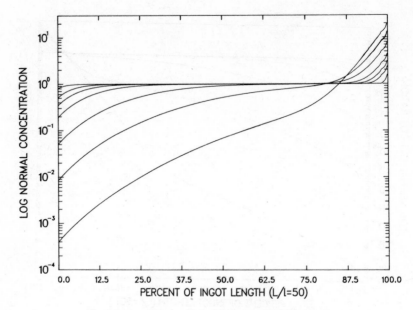

Fig. I.56 $k = 0.80$, $n = 1, 5, 10, 20, 50, 100, 200$ passes.

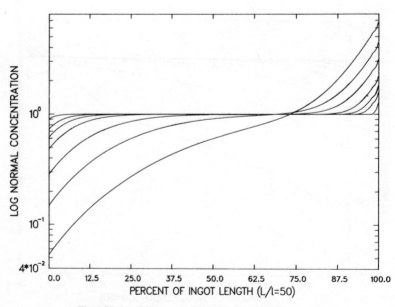

Fig. I.57 $k = 0.90$, $n = 1, 5, 10, 20, 50, 100, 200$ passes.

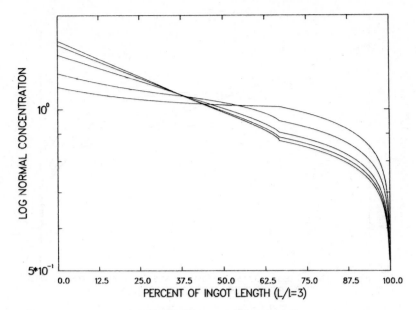

Fig. I.58 $k = 1.1$, $n = 1, 2, 4, 8$ passes.

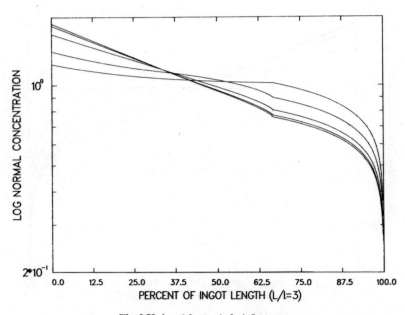

Fig. I.59 $k = 1.2$, $n = 1, 2, 4, 8$ passes.

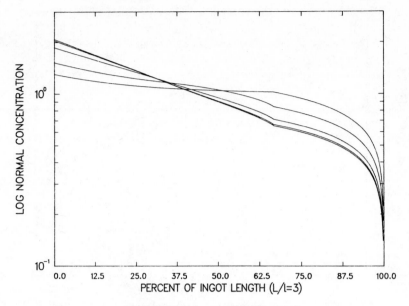

Fig. I.60 $k = 1.3$, $n = 1, 2, 4, 8$ passes.

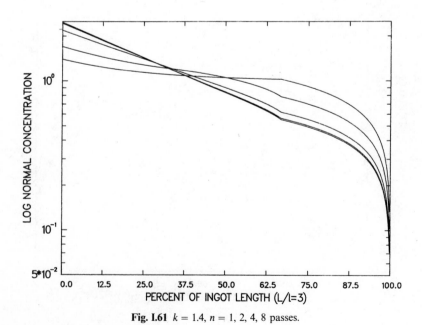

Fig. I.61 $k = 1.4$, $n = 1, 2, 4, 8$ passes.

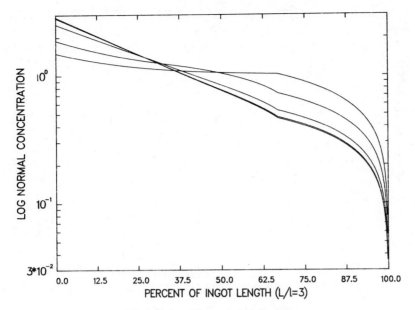

Fig. I.62 $k = 1.5$, $n = 1, 2, 4, 8$ passes.

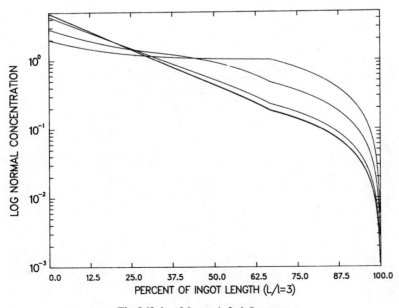

Fig. I.63 $k = 2.0$, $n = 1, 2, 4, 8$ passes.

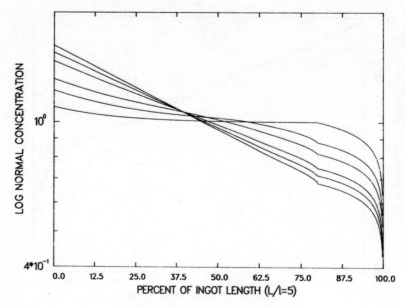

Fig. I.64 $k = 1.1$, $n = 1, 3, 5, 10, 20$ passes.

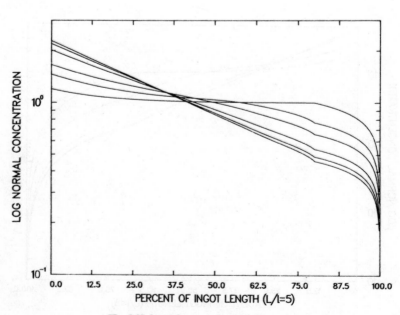

Fig. I.65 $k = 1.2$, $n = 1, 3, 5, 10, 20$ passes.

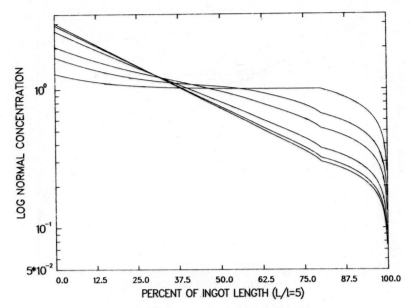

Fig. I.66 $k = 1.3$, $n = 1, 3, 5, 10, 20$ passes.

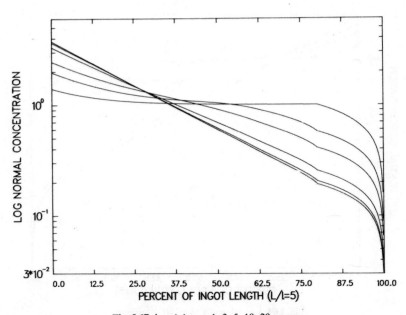

Fig. I.67 $k = 1.4$, $n = 1, 3, 5, 10, 20$ passes.

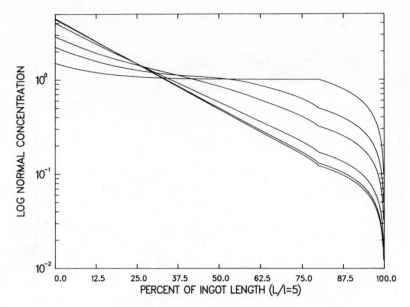

Fig. I.68 $k = 1.5$, $n = 1, 3, 5, 10, 20$ passes.

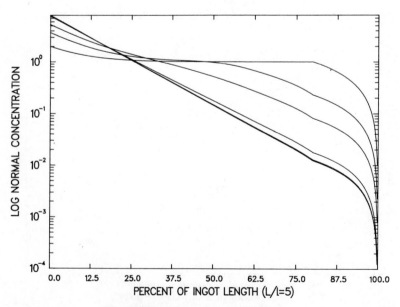

Fig. I.69 $k = 2.0$, $n = 1, 3, 5, 10, 20$ passes.

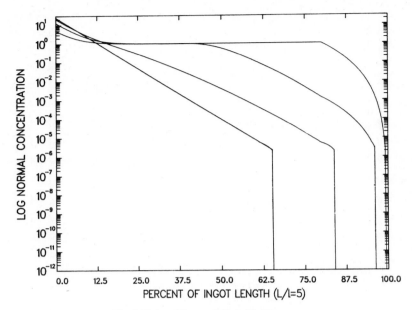

Fig. I.70 $k = 5.0$, $n = 1, 3, 5, 10, 20$ passes.

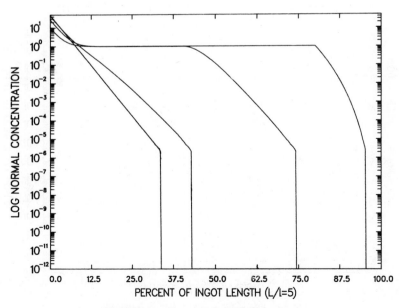

Fig. I.71 $k = 10$, $n = 1, 3, 5, 10, 20$ passes.

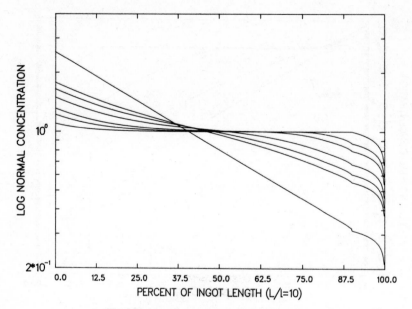

Fig. I.72 $k = 1.1$, $n = 1, 3, 5, 10, 15, 20$ passes.

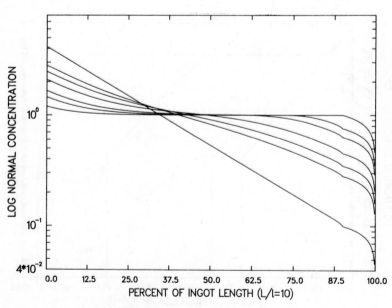

Fig. I.73 $k = 1.2$, $n = 1, 3, 5, 10, 15, 20$ passes.

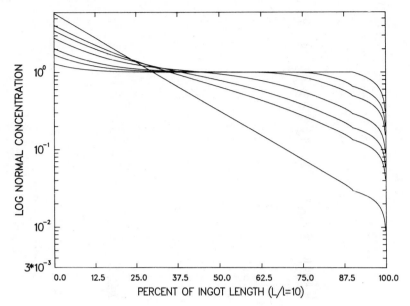

Fig. I.74 $k = 1.3$, $n = 1, 3, 5, 10, 15, 20$ passes.

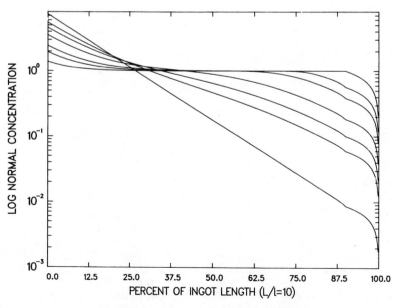

Fig. I.75 $k = 1.4$, $n = 1, 3, 5, 10, 15, 20$ passes.

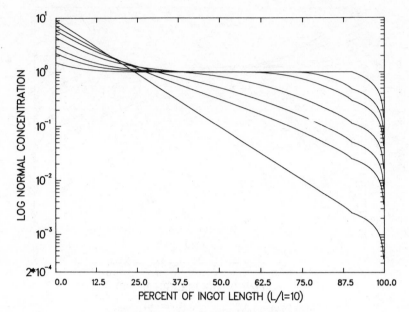

Fig. I.76 $k = 1.5$, $n = 1, 3, 5, 10, 15, 20$ passes.

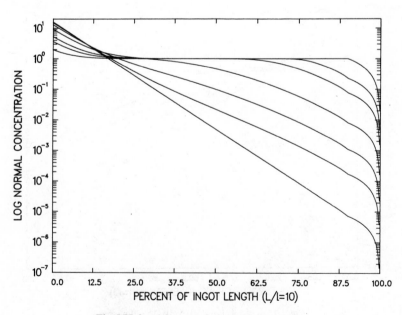

Fig. I.77 $k = 2.0$, $n = 1, 3, 5, 10, 15, 20$ passes.

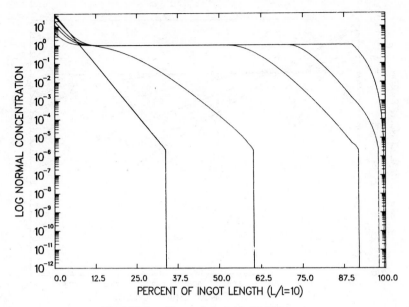

Fig. I.78 $k = 5.0$, $n = 1, 3, 5, 10, 15, 20$ passes.

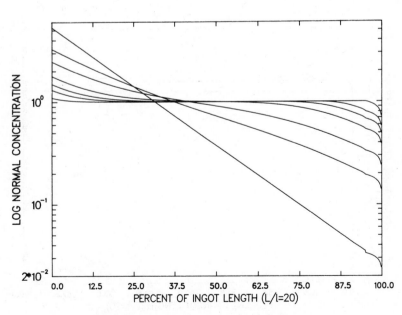

Fig. I.79 $k = 1.1$, $n = 1, 5, 10, 20, 50, 100$ passes.

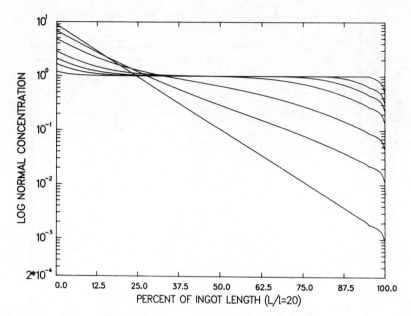

Fig. I.80 $k = 1.2$, $n = 1, 5, 10, 20, 50, 100$ passes.

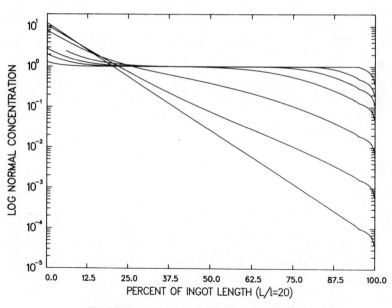

Fig. I.81 $k = 1.3$, $n = 1, 5, 10, 20, 50, 100$ passes.

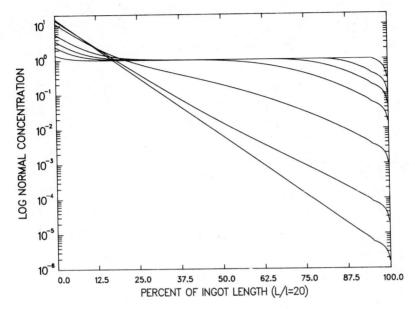

Fig. I.82 $k = 1.4$, $n = 1, 5, 10, 20, 50, 100$ passes.

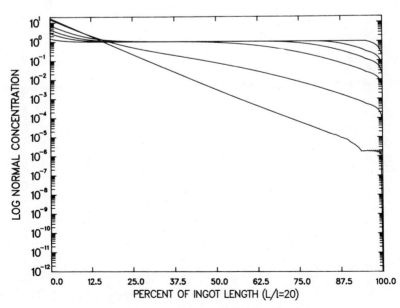

Fig. I.83 $k = 1.5$, $n = 1, 5, 10, 20, 50, 100$ passes.

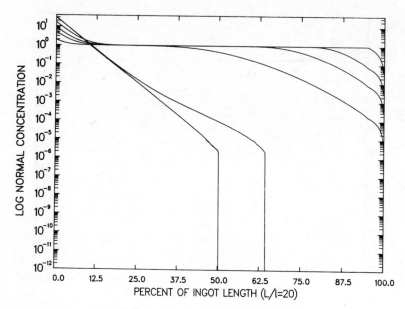

Fig. I.84 $k = 2.0$, $n = 1, 5, 10, 20, 50, 100$ passes.

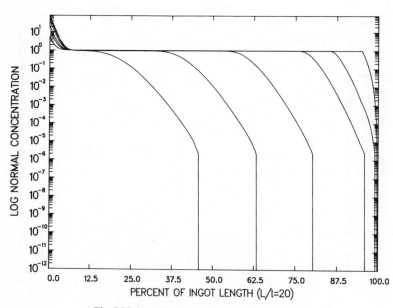

Fig. I.85 $k = 5.0$, $n = 1, 3, 5, 10, 15, 20$ passes.

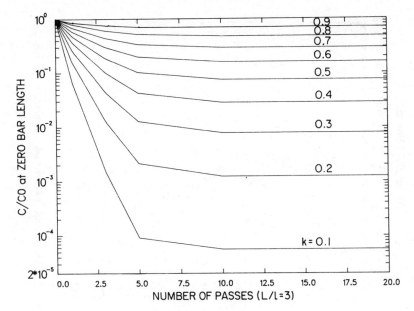

Fig. I.86 $k = 0.1$ to 0.9 by 0.1.

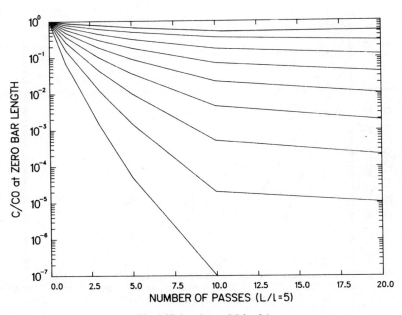

Fig. I.87 $k = 0.1$ to 0.9 by 0.1.

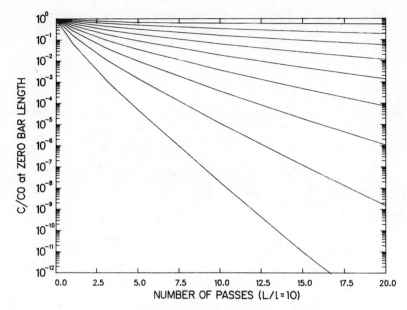

Fig. I.88 $k = 0.1$ to 0.9 by 0.1.

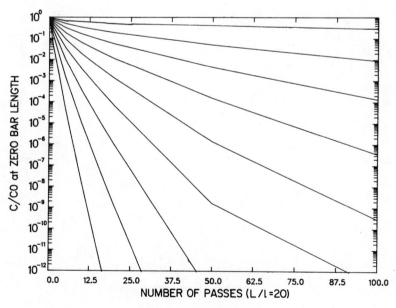

Fig. I.89 $k = 0.1$ to 0.9 by 0.1.

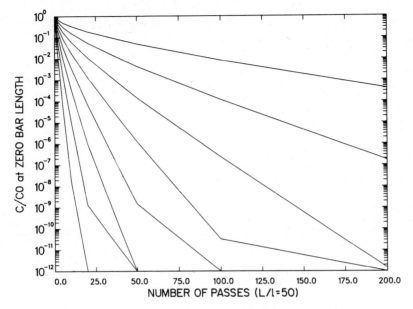

Fig. I.90 $k = 0.1$ to 0.9 by 0.1.

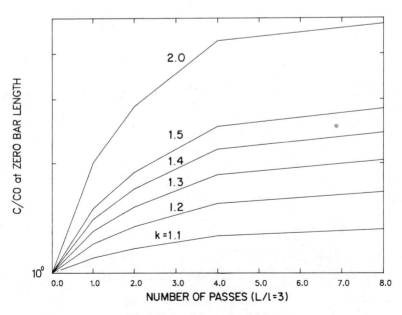

Fig. I.91 $k = 1.1$ to 1.5 and 2.0.

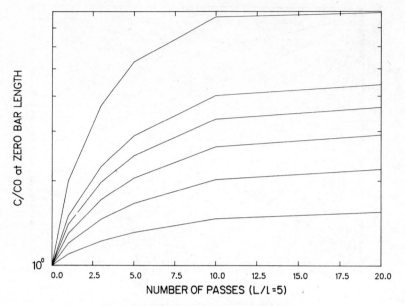

Fig. I.92 $k = 1.1$ to 1.5 and 2.0.

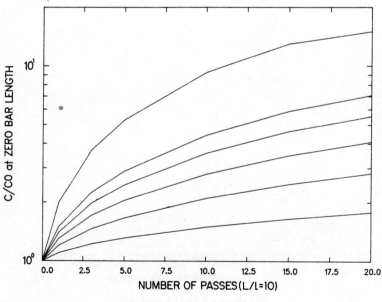

Fig. I.93 $k = 1.1$ to 1.5 and 2.0.

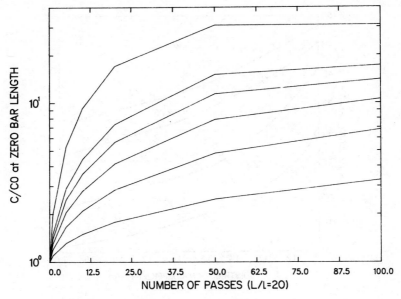

Fig. I.94 $k = 1.1$ to 1.5 and 2.0

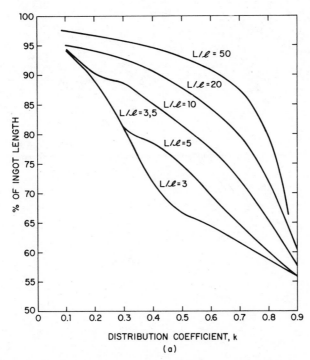

Fig. 1.95 Position in the ingot at which $C = C_0$ as a function of k ($k < 1$), with zone length (L/l) as parameter.

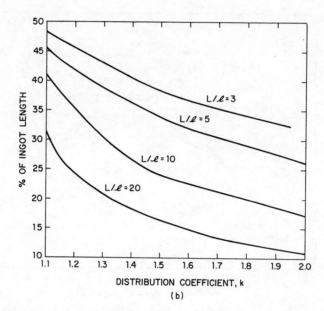

Fig. I.96 Position in the ingot at which $C = C_0$, as a function of k ($k > 1$), with zone length (L/l) as parameter.

APPARATUS AND SUPPLIES

We list here selected sources of items useful in crystal growth and purification. More extensive information is available in a variety of buyers' guides including:

1. *Solid State Processing and Production Buyers Guide and Directory.* Published by SYMCON, 14 Vanderventer Ave., Port Washington, NY 11050.
2. *Analytical Chemistry Lab Guide.* Published each August by American Chemical Society, 1155 16th St. N.W., Washington, DC 20036.
3. *Industrial Research and Development Telephone Directory.* Published by Industrial Research and Development, 1301 South Grove Ave., Barrington, IL 60010.
4. *Guide to Scientific Instruments.* Published by American Association for the Advancement of Science, 1515 Massachusetts Ave., N.W., Washington, D.C. 20005.
5. *American Laboratory, Laboratory Buyers' Guide Edition.* Published by International Scientific Communications, Inc., 808 Kings Highway, Fairfield, CT 06430.

Reference materials and information about them may be obtained from:

1. Office of Standard Reference Materials
   Room B311, Chemistry Building
   National Bureau of Standards
   Washington, D.C. 20234
   Telephone (301) 921-2045
2. Office of Reference Materials
   National Physical Laboratory
   Teddington, Middlesex TW110LW
   United Kingdom
3. BDH Chemicals, Ltd.
   Poole, Dorset BH 12 4 NN
   United Kingdom

4. The International Standards Organization (ISO) publishes *The Compendium of Certification Documents*, which describes standard reference materials available from various national standards organizations. This volume is available from ISO: Caisse Postal 56, CH-1211 Geneva 20 Switzerland; it is distributed in the United States by American National Standards Institute, 1430 Broadway, New York, NY 10018.

Baths, Constant-Temperature, Circulators

B. Braun Instruments
805 Grandview Drive
S. San Francisco, CA 94080
(415)589-9217

Brinkmann Instruments
Cantiague Road
Westbury, NY 11590
(516)334-7500

Forma Scientific
P.O. Box 649
Marietta, OH 45750
(614)373-4763

Haake, Inc.
244 Saddle River Rd.
Saddle Brook, NJ 07662
(201)843-2320

Neslab Instruments, Inc.
871 Islington St.
Portsmount, NH 03801
(603)436-9444

Techne, Inc.
3700 Brunswick Pike
Princeton, NJ 08540
(609)452-9275

Cements, Casting Materials, Insulating Pastes, and Foams

Aremco Products, Inc.
P.O. Box 429
Ossining, NY 10562
(914)768-0685

Dow Corning Corp.
Midland, MI 48640
(517)496-4000

Emerson and Cuming, Inc.
Canton, MA 02021
(617)828-3300

General Electric Co.
Silicone Products Div.
Waterford, NY 12188
(518)237-3330

Sauereisen Cements Co.
Pittsburgh, PA 15238
(412)781-2323

Crystal Growth Assemblies

Cambridge Instrument Co., Inc.
40 Robert Pitt Dr.
Monsey, NY 10952
(914)356-3331

Crystal Specialties, Inc.
15519 E. Arrow Highway
Irwindale, CA 91706

Crystalox
1 Limborough Road
Wantage
Oxfordshire OX12 9AJ
England

Engelhard Industries
70 Wood Ave., S.
Iselin, NJ 08830
(201)589-5000

Leybold Heraeus, Vacuum Systems, Inc.
200 Seco Rd.
Monroeville, PA 15146
(412)856-9030

Arthur D. Little, Inc.
20 Acorn Park
Cambridge, MA 02140
(617)864-5770

The Mellen Company, Inc.
RFD No. 5
Penacook, NH 03303
(603)648-2121

Nippon Electric Co., Ltd.
NEC Building
5-33-1 Shiba, Minako-Ku
Tokyo, Japan 108

Varian Associates, Inc.
121 Hartwell Ave.
Lexington, MA 02173
(617)861-7200

Furnaces

CM Manufacturing Co., Inc.
103 Dewey St.
Bloomfield, NJ 07003
(201)338-6500

Lindberg
304 Hart St.
Watertown, WI 53094
(414)261-7000

Thermolyne Corp.
2555 Kerper Blvd.
Dubuque, IA 52001
(319)556-2241

Heaters, Resistance

ARI Industries, Inc.
9000 King Street
Franklin Park, IL 60068
(312)671-0511

Glas-Col Apparatus Co.
711 Hulman St.
Terre Haute, IN 47802
(812)235-6167

Hotwatt, Inc.
128 Maple St.
Danvers, MA 01923
(617)777-0070

ITT Vulcan Electric
Kezar Falls, ME 04047
(207)625-3231

Thermcraft, Inc.
P.O. Box N-74, 800 Chatham Road
Winston-Salem, NC 27105
(919)772-3111

Watlow Electric Manufacturing Co.
12001 Lackland Road
St. Louis, MO 63141
(314)878-4600

Heaters, Induction

Axel Electronics
134-20 Jamaica Avenue
Jamaica, NY 11418
(212)291-3900

Lepel High Frequency Laboratories, Inc.
59-21 Queens Midtown Expressway
Maspeth, NY 11378
(212)426-4580

Insulating Materials

The Carborundum Co.
Refractories Division
P.O. Box 337
Niagara Falls, NY 14302
(716)278-2000

Johns-Manville
P.O. Box 5108
Denver, CO 80217
(303)394-2086

United States Gypsum
101 S. Wacker Dr.
Chicago, IL 60606
(312)321-4345

Zircar Products, Inc.
110 North Main St.
Florida, NY 10921
(914)651-4481

Insulating Tapes

3M Company
3M Center
St. Paul, MN 55101
(612)733-1110

The Connecticut Hard Rubber Co.
407 East St.
New Haven, CT 06509
(203)777-3631

Nomex, Aramid Fiber and Paper
E. I. du Pont de Nemours & Co., Inc.
1007 Market St.
Wilmington, DE 19898
(302)999-2791

Mechanical Components, Drives, Clutches, etc.

Metals Research, Ltd.
Melbourn
Royston, Herts SG86EJ
United Kingdom

Micro Slides, Inc.
629 Main Street
Westbury, NY 11590
(516)334-4180

Stock Drive Products
55 South Denton Ave
New Hyde Park, NY 11040
(516)328-0200

Thomson Industries
Manhasset, NY 11030
(516)883-8000

Velmex, Inc.
P.O. Box 38
East Bloomfield, NY 14443
(716)657-6151

Warner Electric Brake and Clutch Co.
Beloit, WI 53511
(815)389-3771

Motors, Instrument

Hansen Manufacturing Co., Inc.
Princeton, IN 47570
(812)385-3415

Hurst Manufacturing Corp.
Princeton, IN 47670
(812)385-2564

Mallory Timers Company
3029 E. Washington St.
Indianapolis, IN 46206
(317)636-5353

Motors, All Types

Bodine Electric Co.
2500 W. Bradley Place
Chicago, IL 60618
(312)478-3515

Electrocraft Corp.
1600 Second St. South
Hopkins, MN 55343
(612)935-8226

PMI Motors
5 Aerial Way
Syosset, NY 11791
(516)938-8000

Refrigerators and Other Cooling Devices

Gilson Medical Electronics, Inc.
P.O. Box 27
Middleton, WI 53562
(608)836-1551

Revco Refrigeration Products Div.
Rheem Manufacturing Co.
1100 Memorial Drive
West Columbia, SC 29169
(803)796-1700

Tenney Engineering, Inc.
1090 Springfield Rd.
Union, NJ 07016
(201)686-7870
(212)962-0332

Refrigerators, External Immersion Probe

Forma Scientific
Box 649
Marietta, OH 45750
(614)373-4763

FTS Systems, Inc.
P.O. Box 158, Route 209
Stone Ridge, NY 12484
(914)687-7664

Neslab Instruments, Inc.
25 Nimble Hill Rd.
Newington, NH 03801
(603)436-9444

Refrigeration Modules, Thermoelectric

Cambridge Thermionic Corp.
445 Concord Ave.
Cambridge, MA 02138
(617)491-5400

Melcor
990 Spruce St.
Trenton, NY 08648
(609)393-4178

So-Low Environmental Equipment Co.
10310 Spartan Drive
Cincinnati, OH 45215
(513)772-9410

Thermoelectrics Unlimited, Inc.
1202 Harrison Ave.
Wilmington, DE 19809
(302)798-5360

Temperature Controllers, Programmers, Thermocouples

Athena Controls, Inc.
20 Clipper Road
W. Conshohocken, PA 19428
(215)828-2490

Doric Scientific
3883 Ruffin Road
San Diego, CA 92123
(714)565-4415

Eurotherm Corporation
Isaac Newton Center
Reston, VA 22090
(703)471-4870

Fenwal, Inc.
400 Main Street
Ashland, MA 01721
(617)881-2000

Iveron Pacific Corp.
1152 Morena Blvd.
San Diego, CA 92110
(714)275-1500

Neslab Instruments, Inc.
871 Islington St.
Portsmouth, NH 03801
(603)436-9444

Omega Engineering, Inc.
One Omega Drive
Box 4047
Stamford, CT 06907
(203)322-1666

RFL Industries, Inc.
Boonton, NJ 07005
(201)334-3100

Research, Inc.
Process Systems Div.
Box 24064
Minneapolis, MN 55424
(612)941-3300

Science Products Corp.
280 Route 46
Dover, NJ 07801
(201)366-0827

Tetrahedron
5060A Convoy St.
San Diego, CA 92111
(714)277-2820

Theall Engineering Co., Inc.
P.O. Box 336
Oxford, PA 19363
(215)932-3488

Valco Instruments Co.
P.O. Box 19032
Houston, TX 77000
(713)688-9315

Valley Forge Instrument Co., Inc.
Research Products Div.
55 Buckwalter Rd.
Phoenixville, PA 19460
(215)933-1806

Thermal Analysis and Thermal Microscopy Equipment

AO/Reichert
American Optical
Box 123
Buffalo, NY 14240
(716)891-3000

E. I. du Pont de Nemours & Co., Inc.
Concord Plaza, Quillen Building
Wilmington, DE 19898
(302)772-5495

E. Leitz, Inc.
Industrial Division
Link Drive
Rockleigh, NJ 07647
(201)767-1100

Linkam Scientific Instruments
37 Pine Ridge
Carshalton Beeches
Surrey SM5 4QQ
England
01-647-7143

Maple Instruments
Philipsweg 1
6227 AJ Maastricht
The Netherlands
043-613900

Mettler Instrument Corp.
Box 71
Hightstown, NJ 08520
(609)448-3000

Netzsch Inc.
Thermal Analysis Dept.
Pickering Creek Industrial Park
Lionville, PA 19353
(215)363-8010

Omnitherm Corp.
48 E. University Ave.
Arlington Heights, IL 60004
(312)870-7007

Perkin-Elmer Corp.
Main Avenue, M5-12
Norwalk, CT 06856
(203)762-1000

Stanton Redcroft
Coppermill Lane
London SW 17 OBN
United Kingdom

Zone Refiners, Chemical

Astro Industries, Inc.
606 Olive St.
Santa Barbara, CA 93101
(805)963-3461

Atomergic Chemetals Co.
Div. of Gallard Schlesinger Chemical Mfg. Corp.
564 Mineola Ave.
Carle Place, L.I, NY 11514
(516)333-5600

Crystal Specialties, Inc.
15519 E. Arrow Hwy.
Irwindale, CA 91706

Crystalox
1 Limborough Road
Wantage
Oxfordshire OX12 9AJ
England

C. Desaga GmbH
Postfach 101969
Maasstrasse 26-28
D-6900 Heidelberg 1
German Federal Republic

Ecco High Frequency Corp.
7024 Kennedy Blvd.
North Bergen, NJ 07047
(201)869-7112

Enraf Nonius, Inc., distributed by
Kratos Scientific Instruments
24 Booker St.
Westwood, NJ 07675
(201)664-5702

Materials Research Corp.
Rt. 303 and Glenshaw St.
Orangeburg, NY 10962
(914)359-4200

SOURCES OF INFORMATION ON CRYSTAL GROWTH

1. Proceedings of the 8th International Conference on Crystal Growth, York, U.K., edited by B. Cockayne, J. H. C. Hogg, B. Lunn, and P. J. Wright, North-Holland Publishing Co., Amsterdam, 1986.

2. *Crystal Growth—Theory and Techniques*, Plenum, New York.

3. *Crystals in Modern Laser Technology* (in German), P. A. Arsenjew, S. Bagdasarow, K. Bienert, E. F. Kustow, and A. W. Potjomkin, Akademische Verlagsgesellschaft, Leipzig, 1980.

4. *Crystal Research and Technology*, Verlag Chemie International, Inc., Deerfield Beach, FL.

5. *Growth of Crystals* (Translations of Russian texts published by USSR Academy of Sciences), Plenum, New York.

6. *Journal of Crystal Growth*, North-Holland, Amsterdam.

7. *Journal of Electronic Materials*, Plenum, New York.

8. *Laser Crystals—Their Physics and Properties*, A. A. Kaminskii, Springer-Verlag, New York, 1981.

9. *Molecular Crystals and Liquid Crystals*, Gordon and Breach, New York.

10. *North Holland Series in Crystal Growth*, North Holland, Amsterdam and New York.

11. *Preparation and Properties of Solid State Materials*, W. R. Wilcox, Ed., Dekker, N.Y.

12. *Progress in Crystal Growth and Characterization*, B. Pamplin, Ed., Pergamon, Oxford.

13. Research Materials Information Center, Oak Ridge National Laboratory, P.O. Box X, Oak Ridge, TN 37831.

14. Shaped Crystal Growth—A Selected Bibliography, D. O. Bergin, *J. Crystal Growth*, **50**(1), 381–396 (1980); a review with 299 refs.

15. Solid State Physics Literature Guides

    a. Vol. 2, *Semiconductors, Preparation, Crystal Growth and Selected Properties*, T. F. Connolly, Ed., Plenum, New York, 1972.

    b. Vol. 10, *Crystal Growth Bibliography*, compiled by A. M. Keesee, T. F. Connolly, and G. C. Battle, Jr., Plenum, New York, 1979.

16. Spectroscopy of Molecular Crystals—A Bibliography for 1982, E. F. Sheka, V. S. Makarova, and T. A. Krivenko, *Mol. Cryst. Liq. Cryst.*, **114**, 305–391 (1984), and earlier bibliographies cited therein.

17. Liquid Crystals—A Bibliography for 1982, W. Fehrenbach, A. Stieb, and G. Meier, *Mol. Cryst. Liq. Cryst.*, **130**, 25–141 (1985).

18. *Springer Series in Solid-State Sciences*, Vol. 5, *Fundamentals of Crystal Growth I: Macroscopic Equilibrium and Transport Concepts*, F. Rosenberger, Springer-Verlag, New York, 1979.

INDEX

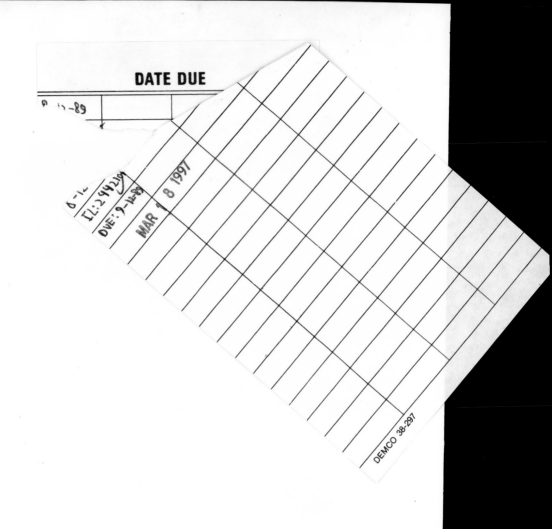